International Conference on Screw Machines (ICSM 2022)

IOP Conference Series: Materials Science and Engineering
Volume 1267

Dortmund, Germany
7 – 8 September 2022

ISBN: 978-1-7138-6426-4
ISSN: 1757-8981

Printed from e-media with permission by:

Curran Associates, Inc.
57 Morehouse Lane
Red Hook, NY 12571

Some format issues inherent in the e-media version may also appear in this print version.

Institute of Physics
Dirac House, Temple Back
Bristol BS1 6BE UK

Phone: 44 1 17 929 7481
Fax: 44 1 17 920 0979

techtracking@iop.org

Additional copies of this publication are available from:

Curran Associates, Inc.
57 Morehouse Lane
Red Hook, NY 12571 USA
Phone: 845-758-0400
Fax: 845-758-2633
Email: curran@proceedings.com
Web: www.proceedings.com

TABLE OF CONTENTS

IOP Conf. Series: Materials Science and Engineering 1267 (2022) 011001 doi:10.1088/1757-899X/1267/1/011001

tu technische universität dortmund

ICSM 2022

International Conference on Screw Machines 2022

Dear friends of screw machines,

Welcome to the 11th International Conference on Screw Machines (ICSM 2022) at TU Dortmund University. Whether screw spindle or vacuum pump, screw compressor or expander, experts from industry and research institutions meet here to get to know about the latest developments and research findings. I am particularly pleased that, despite the pandemic, we will once again be able to meet in person at ICSM 2022, which - in addition to the technical information - will also enable us to build up and expand personal relationships.

I would especially like to welcome Dr.-Ing. F.-J. Peveling and Prof. J.-G. Persson as honored guests at this year's ICSM. Mr. Peveling has already been a speaker at the first "Schraubenmaschinen-Tagung" in 1984 and has been a lecturer at the TU Dortmund University for several years. Mr. Persson has been a speaker at the second "Schraubenmaschinen-Tagung" in 1987 and will again enrich this year's conference with his own paper.

Four years have flown by again! At the last conference in 2018, we were able to present the world's first tested dual-lead screw compressor as a highlight. From my point of view, a highlight of this year's conference is a transparent industrial screw compressor with oil injection, with which the distribution of the oil in the entire machine can be visualized for the first time via high-speed videos. Furthermore, I am convinced that the many other exciting contributions at this conference will also provide you with new insights into the state-of-art on screw machines. As usual, the papers of all past conferences are available for free download on the conference homepage (www.icsm.tu-dortmund.de). In addition, all papers of the current conference are published in the open access IOP (Institute of Physics) conference series.

Finally, I would like to thank all authors and attendees for their active contribution to the conference. I would also like to thank our sponsors, PTG Holroyd and Kapp Niles, whose financial support has helped to keep the conference fee constant. Furthermore, I would like to express my special thanks to the entire team of the Chair of Fluidics, especially Mrs. Ilona Kokott, Mr. Manuel Grieb and Mr. Kevin Weidler. It is only through their tireless engagement and personal efforts that the conference is made possible the way it is.

Prof. Andreas Brümmer
Conference Chair

IOP Conf. Series: Materials Science and Engineering 1267 (2022) 011001 doi:10.1088/1757-899X/1267/1/011001

tu technische universität dortmund

ICSM 2022

CONFEENCE COMMITEE

Andreas Brümmer (general chair), TU Dortmund University, DE

Thomas Dreifert, Leybold GmbH, DE

Hans-Ulrich Fleige, Aerzener Maschinenfabrik GmbH, DE

Eckhard A. Groll, Purdue University, US

Knut Kauder (retired), TU Dortmund University, DE

Ahmed Kovacevic, City, University of London, UK

Johann Lenz, KÖTTER Consulting Engineers, DE

Laurenz Rinder (retired), TU Wien, AT

Jack Sauls (retired), Trane, US

CONTACT

icsm.ft.mb@tu-dortmund.de

Chair of Fluidics
Faculty of Mechanical Engineering
TU Dortmund University
Leonhard-Euler-Str. 5
44227 Dortmund
Germany

Phone: +49 (0)231 755 5721
Fax: +49 (0)231 755 5722

IOP Conf. Series: Materials Science and Engineering 1267 (2022) 011002 doi:10.1088/1757-899X/1267/1/011002

Peer Review Statement

All papers published in this volume have been reviewed through processes administered by the Editors. Reviews were conducted by expert referees to the professional and scientific standards expected of a proceedings journal published by IOP Publishing Publishing.

- **Type of peer review:** Double Anonymous
- **Conference submission management system:** ConfTool
- **Number of submissions received:** 25
- **Number of submissions sent for review:** 24
- **Number of submissions accepted:** 23
- **Acceptance Rate (Submissions Accepted / Submissions Received × 100):** 92
- **Average number of reviews per paper:** 2
- **Total number of reviewers involved:** 24
- **Contact person for queries:**
  Name: Manuel Grieb
  Email: manuel.grieb@tu-dortmund.de
  Affiliation: TU Dortmund University, Chair of Fluidics

Design and analysis of conical rotary compressor

Y Lu, K Hoang, D Noake and N Low

Vert Technology, Edinburgh, UK

E-mail: yang.lu@vertrotors.com

Abstract. Conical Rotary Compressor (CRC) is a screw type Positive Displacement Machine (PDM) consisting of one inner rotor rotating inside an outer rotor. The conical internal meshing rotors with the feature of variable pitch and rotor profile have an inherent capability to operation at high pressure ratio due to high built-in volume ratio and low leakage. CRC has potential to run in various applications such as Heating, Ventilation, and Air Conditioning (HVAC) system, vacuum pumps, and air compressors. The performance of the machine is heavily affected by rotor profile and rotor geometry. In this paper, non-parallel axis conical rotor is firstly introduced according to current successful design and manufacture practice, then parallel axis conical rotors are design based on meshing theory. Then, geometrical parameters such as meshing line length, cross section area, chamber volume and relative velocity are analysed. Finally, the experiments were conducted under different pressure ratio up to 26 and the volume flow rate, specific power, volumetric efficiency, and isentropic efficiency are compared. As a result of this study, CRC could maintain high level of volumetric and isentropic efficiency under wide range of pressure ratio. Improvement of the compressor design can be made to limit increases in temperature, reduce relative velocity and increase the overall efficiency of the machine.

1. Introduction

While there have been numerous conventional types of Positive Displacement Machine (PDM), such as scroll compressor, screw compressor, on the market, there has been relatively few innovations over the last 40 years. The Conical Rotary Compressor (CRC) is a novel type of PDM with the feature of variable pitch and profile as shown in Figure 1. The CRC has two helical internal meshing rotors which spin in the same direction at different speeds according to gear ratio. When the air travels down the conical rotors, the volume continually reduces and so the pressure continually increases. The CRC has advantages of high built-in Volume Index (Vi), continuous compression, low vibration, and noise.

Figure 1. CRC in National Museum of Scotland

The internal gear rotors machine types can be divided into straight and helical rotors. Helical rotors can be further divided into cylindrical and conical forms. For the cylindrical form, the rotational axes of inner and outer rotor are parallel while conical form can have either parallel or non-parallel axes. The pitch can be constant and variable. The rotational axes could be orbiting or fixed in the space. There are many studies on the geometry analysis of internally geared machines. The theory of rotor profile design can be referred to Litvin's book [1]. The cycloidal pump with rotor profile free of singularities has been computerized design [2]. Zhang [3]proposed a general mathematical model using Planar double-enveloping (PDE) method to generate two types of double-enveloping cycloid which are inward curve of epitrochoid and its outer conjugate envelope, another type is outward curve of epitrochoid and its inner conjugate envelope. Read [4–6] has carried out extensive research on the geared screw machines. Rolling and sliding velocity of trochoidal gearing profile in the Gerotor at the contact point has been analyzed by Ivanovic [7,8]. The ellipse has been used in the internal gear pump and flow rate, irregularity, specific sliding and torque were calculated by automated design program [9]. However, there is no research investigation on the geometry of CRC, such as rotor profile, contact line, sealing line, cross section area and chamber volume which are highly important.

In this paper, these geometrical parameters are calculated and analyzed using an in house-developed software package called VertSim which is specialized for the rotor profile design, thermodynamic calculation and performance optimization of CRC.

2. Profile design algorithm

Profile generation is based on meshing condition. The coordinate systems are initially built for the coordinate transformation. The matrices for the rotor profile generation and meshing condition to calculate the meshing angle are provided after this process. Finally, the 3D parallel and non-parallel rotors are displayed.

2.1. Coordinate systems

Coordinate systems are illustrated by Figure 2. Global coordinate system S_1 is fixed and rigidly connected to stationary casing. Two movable outer and inner rotor coordinate systems S_{01} and S_{02} respectively connecting to inner rotor and outer rotor which are both rotating clockwise. The auxiliary coordinate system S_2 is employed to transform coordinate between inner rotor and outer rotor coordinate systems. The center distance is E. The angle between z_1 and z_2 is γ which defines the parallel and non-parallel inner and outer rotational axis. Point I is the point of tangency of centrodes of the inner and outer rotors. The relative motion of centrodes is pure rolling about I.

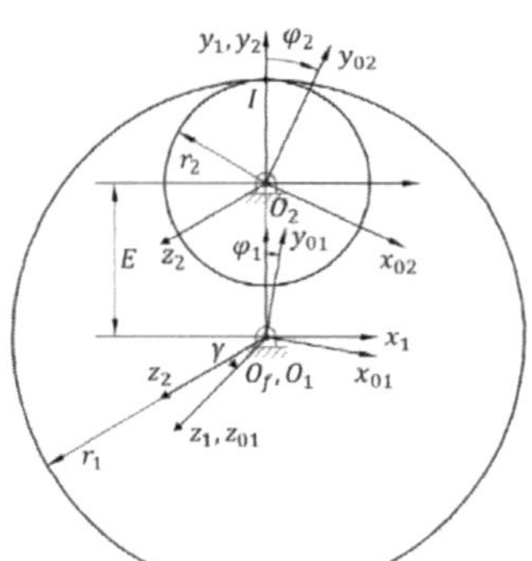

Figure 2. Coordinate system

Consider coordinate system $S_1(X_1, Y_1, Z_1)$ and $S_2(X_2, Y_2, Z_2)$ that are rigidly connected to outer rotor and inner rotor respectively and coordinate system $S_1(X_1, Y_1, Z_1)$ can also be used as the global coordinate system. And the local rotational coordinate system $R(X_{01}, Y_{01})$ rotates around Z_1 axis with outer rotor and the coordinate system $R(X_{02}, Y_{02})$ rotates around Z_2 axis with inner rotor. The rigid and rotational coordinate systems are used to define the cutting tool, inner rotor profile and outer rotor profile.

When the angle between inner and outer rotor axes satisfies $\gamma = 0$, the parallel and axes coordinate system can be obtained and the profiles of inner and outer rotors can be designed. For the non-parallel conical rotors, the center distance decreases along the z-axis while the center distance is same for the parallel conical rotors.

The rotation angles have the relationship as shown in equation 1.

$$m_{21} = \frac{N_1}{N_2} = \frac{N_1}{N_1 - 1} = \frac{r_1}{r_2} = \frac{\omega_2}{\omega_1} \tag{1}$$

Where m_{21} is the gear ratio and N_1 and N_2 are the tooth numbers of outer and inner rotors. The procedure starts with a generating surface $r_1(t, \varphi_1)$ for which a meshing, or generated surface is to be determined. A family of such generated surfaces is given in parametric form by: $r_2(t, \varphi_1, \varphi_2)$ where t is a profile parameter while φ_1 and φ_2 are motion parameters.

2.2. Coordinate transformation

With the coordinate system, the matrix of coordinate transformation provides the mathematical methods for generating profiles of the rack from the rotor profiles and conversely, generating the rotor profiles from the rack. With the coordinate systems showed in Figure 2, the coordinate transformation in transition from the main rotor profile coordinate S_1 to the gate rotor profile coordinate S_2 is based on the matrix equation:

$$\boldsymbol{r}_2 = M_{21}\boldsymbol{r}_1 = M_{2p}M_{pf}M_{f1}\boldsymbol{r}_1 \tag{2}$$

$$\boldsymbol{r}_1 = \begin{bmatrix} x_1 \\ y_1 \\ z_1 \\ 1 \end{bmatrix}, \boldsymbol{r}_2 = \begin{bmatrix} x_2 \\ y_2 \\ z_2 \\ 1 \end{bmatrix}, \boldsymbol{M}_{pf} = \begin{bmatrix} 1 & 0 & 0 & -E \\ 0 & cos(\gamma) & -sin(\gamma) & 0 \\ 0 & sin(\gamma) & cos(\gamma) & 0 \\ 0 & 0 & 0 & 1 \end{bmatrix} \tag{3}$$

$$\boldsymbol{M}_{2p} = \begin{bmatrix} cos(\varphi_2) & -sin(\varphi_2) & 0 & 0 \\ sin(\varphi_2) & cos(\varphi_2) & 0 & 0 \\ 0 & 0 & 1 & 0 \\ 0 & 0 & 0 & 1 \end{bmatrix}, \boldsymbol{M}_{f1} = \begin{bmatrix} cos(\varphi_1) & -sin(\varphi_1) & 0 & 0 \\ sin(\varphi_1) & cos(\varphi_1) & 0 & 0 \\ 0 & 0 & 1 & 0 \\ 0 & 0 & 0 & 1 \end{bmatrix} \tag{4}$$

Where $r_1(t, \theta)$ and $r_2(t, \varphi_1, \varphi_2)$ are the generating and generated surfaces respectively and t is a profile parameter while φ_1 and φ_2 are motion parameters of main rotor profiles and gate rotor profile respectively. M_{2p} and M_{f1} are rotational matrices corresponding to rotation with the Z_1 and Z_2 axes respectively and M_{pf} is a translational matrix corresponding to the transformation from coordinate system $S_1(X_1, Y_1, Z_1)$ to $S_2(X_2, Y_2, Z_2)$. Here these matrices are used to generate the rack from the given rotor profiles.

The conjugated inner rotor and outer rotor roll and slide over each other produces relative motion over the profile and transfer motion with the profile [1]. The tangency and unit normal at the contact point should satisfy the equation of meshing which is represented as equation 5.

$$f(\varphi_1, \varphi_2) = \left(\frac{\partial \boldsymbol{r}_2}{\partial \varphi_1} \times \boldsymbol{k}_2\right) \cdot \frac{\partial \boldsymbol{r}_2}{\partial \varphi_2} = 0 \tag{5}$$

Where $\boldsymbol{k}_2$ is the inner rotor unit vector of the z_2-axis.

2.3. Rotor profile generation

When the inner rotor and outer rotor rotational axis are parallel or non-parallel, the rotor profile should be designed in parallel or non-parallel coordinate system individually. The parallel coordinate system is utilized to show the rotor profile generation process. The cutting tool algorithm used to generate the inner and outer rotor profiles is explained in this section. A cutting tool is firstly created to generate the inner rotor profile, then outer rotor profile is generated from inner rotor profile by using enveloping method. The inner profile is generated through cutting tool which is red ellipse as shown in Figure 3. An initial black circle which covers the area where the cutting tool occupied is selected for the inner rotor. The cutting tool then removes area that it overlaps with base circle to produce the inner rotor. The cutting tool is rotated around the global origin and the base circle is rotated around the inner rotor coordinate origin based on gear ratio between the inner and outer rotors. The final inner rotor is shown in Figure 3 (c). The enveloping inner rotor is illustrated by Figure 4 (a).

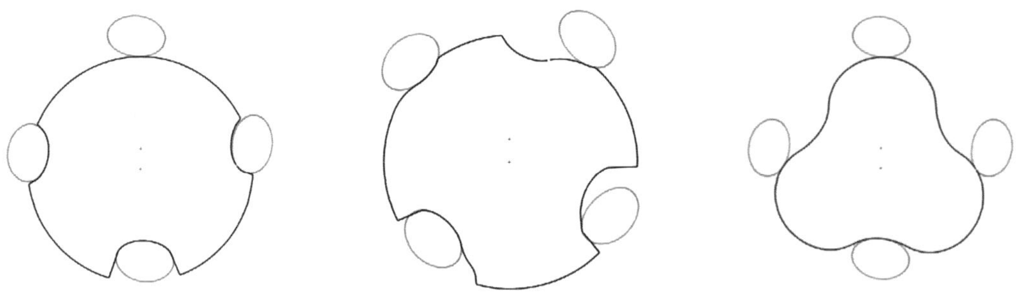

(a) Step 1 (b) Step 2 (c) Step 3

Figure 3. Cutting tools and base circle

Outer rotor profile is generated from inner rotor profile thorough enveloping method as shown in Figure 4 (b). Although in practice the rotation axis of each rotor is fixed, to envelope outer rotor, the algorithm considers the inner rotors orbiting relative to the stationary outer rotor. The inner profile is rotated around its own axis, then rotated around the global origin. Finally, the relative position of inner rotor and outer rotor are shown in Figure 4 (c).

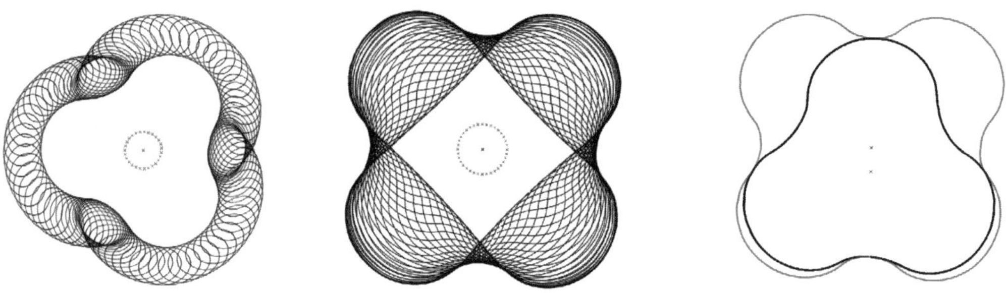

(a) Inner rotor (b) Outer rotor (c) Inner and Outer rotor

Figure 4. Rotor profile

2.4. Rotor generation

The rotor axes of inner rotor and outer rotor could be parallel or non-parallel. The 3D parallel axes rotors are more difficult to generate since rotor profiles are designed with same axis center in every cross sections. The angularities need to be checked in every cross section to avoid intersections between inner and outer rotors. The non-parallel rotors can be obtained by sweeping the initial designed rotor profiles along the helix line which define the scale rate and pitch rate. Figure 5 shows the 3D parallel and non-parallel inner and outer rotors. Figure 5 (a) shows non-parallel variable pitch rotors. It is clear there is an angle between the inner and outer rotor axes which intersects in remote distance. The center distance decreases with the rotor diameter decreasing along the rotor. Figure 5 (b) shows the 3D parallel constant pitch rotors, the centre distance keeps constant when the rotor diameter is decreased. It is worth to point that the diameter ratio of the parallel rotor could not be greater than 2 to avoid singularities. That means it is difficult to achieve higher Vi compared to non-parallel rotors however, this disadvantage could be compensated by utilising suction and discharge ports.

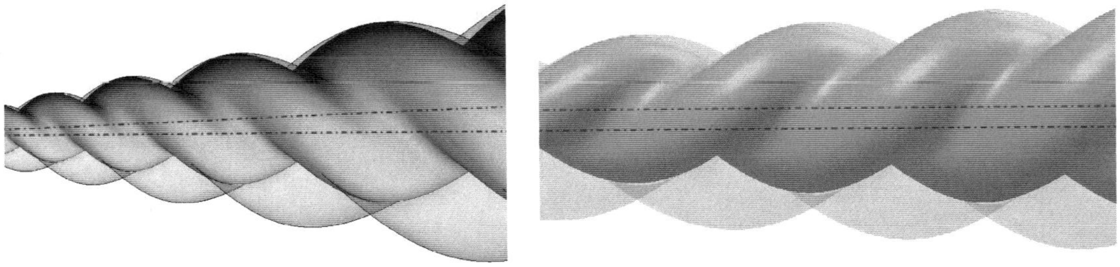

(a) Non-parallel variable pitch rotors (b) Parallel constant pitch rotors

Figure 5. 3D Inner and outer rotors

3. Geometry analysis

The rotor profile, contact line, sealing line, cross section area, chamber volume and relative velocity are calculated by using an in house-developed software package called VertSim. The rotor design parameters are listed in Table 1. Rotor length is 160mm. The lobe combination of inner and outer rotor is 3 by 2. Outer rotor outer diameter is 86.2 mm. Volume index is 4.6. q is diameter ratio of outer circle at suction and discharge side. The number of wrap of outer rotor is 1.52. The rotor geometry is variable pitch and rotor profile.

Table 1. CRC.60 geometrical parameters.

Parameters	Unit	Value	Parameters	Value
L	[mm]	160	Z_1/Z_2	3/2
D1	[mm]	86.2	Vi	4.6
Ec	[mm]	7.81	q	4.3
V_suc	[cm³]	30	Wrap1	1.52

3.1. Contact line

The contact points of inner and outer rotors are calculated based on the equation of meshing. The black line in Figure 6 is the contact line while the red and blue line are inner and outer rotor profile. The continuous contact line is the prerequisite and guarantee of lower leakage between the high-pressure chamber and the low -pressure chamber.

IOP Conf. Series: Materials Science and Engineering 1267 (2022) 012001 doi:10.1088/1757-899X/1267/1/012001

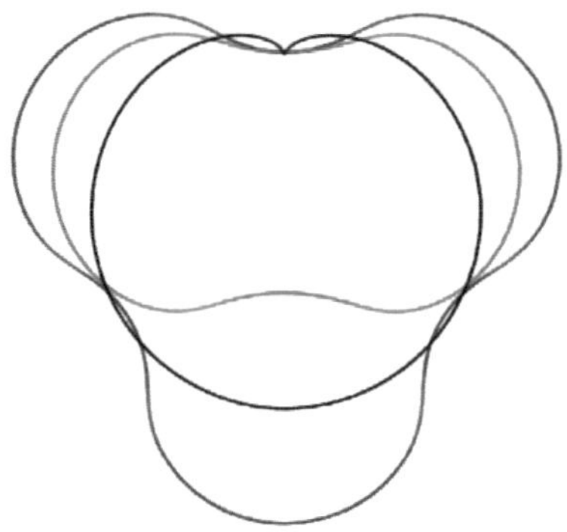

Figure 6. Rotor profile and contact line

3.2. Sealing line

The sealing line is produced by considering z coordinates to contact line. As shown in Figure 7, one chamber can share boundaries with up to 6 chambers because the wrap angle is greater than 360 degree. One chamber has leading sealing and trailing sealing line where leakage between the chambers will occur if there is a pressure difference. The details of the sealing line calculation can be referenced to the conference paper [10].

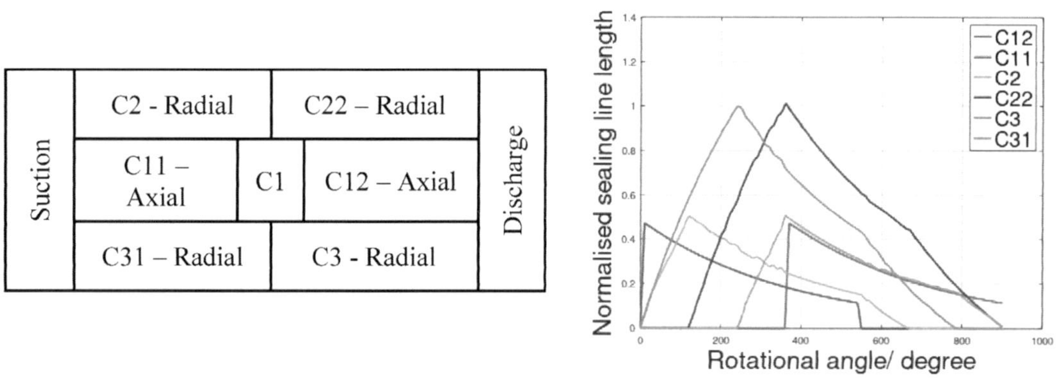

Figure 7. Sealing line

3.3. Cross-section area

The cross-section area of one chamber is calculated by integrating the area between the leading and trailing contact points. The normalised cross-section area of one chamber is shown in Figure 8. The cross-section area variation with rotational angle and Z-axis is not identical because of the conical geometry. Along Z-axis, the cross-section area in one closed chamber increases slowly to the maximum then decrease quicker to 0.

IOP Conf. Series: Materials Science and Engineering 1267 (2022) 012001 doi:10.1088/1757-899X/1267/1/012001

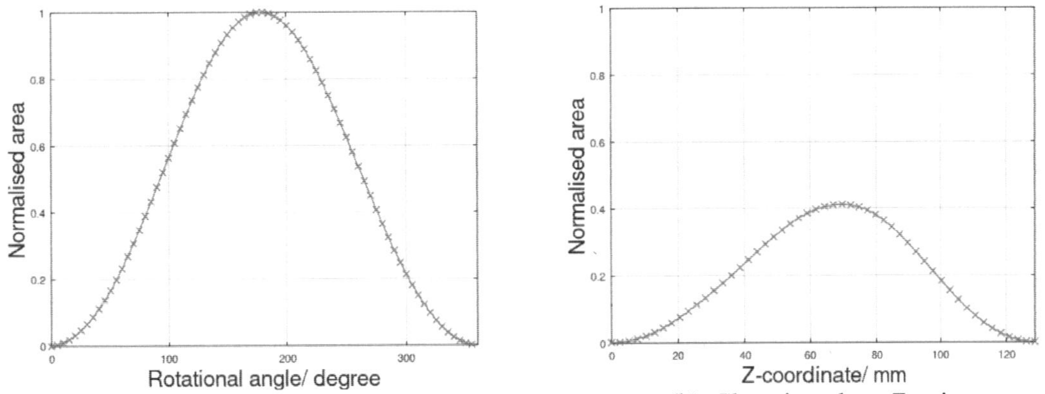

(a) Changing with rotation angle (b) Changing along Z-axis

Figure 8. Normalised cross section area of working chamber

3.4. Chamber volume

The chamber volume is calculated by integrating the cross-section area along Z-axis in one working chamber. The working process is defined in Table 2. The suction process and discharge process will take 360 rotational degree because there is no suction and discharge port plate. So, the volume index can be defined by calculation of the chamber volume at suction end position (360 degree) and discharge start position (wrap angle – 360 degree). The internal compression process depends on the wrap angle. The greater of the wrap angle, the longer of the internal compression process. The suction port and discharge port can also be designed for CRC to achieve higher volume index. It is highly important to precisely define suction, compression, and discharge process before thermal dynamic calculation.

Table 2. Working process of CRC machine

Process	Degree	Comments
Suction	[0, 360]	Chamber volume Vc increases from 0 to maximum then closed from suction at 360-degree.
Internal compression	[360, 547]	Chamber volume continues decrease leading to internal compression.
Discharge	[547, 907]	Chamber volume decreases to 0 after 360-degree.

Chamber volume is calculated from VertSim and illustrated in Figure 9. Rotation angle is based on outer rotor. According to the calculation, rotation angle ranges from 0 to 907 degree. Maximum chamber volume 50 cm^3 is obtained at 230 degree and 30 cm^3 chamber volume is obtained at a 360-degree angle when the working chamber closed form the suction chamber. The chamber volume is 7 cm^3 where discharge process starts. The volume index calculated in this case is 4.6. There is a potential to increase the Vi to 7.14 by positioning a suction port which can close the working chamber when the volume reaches the maximum. This also helps to decrease suction losses caused by back flow. There is no point to add a discharge port because the chamber volume is small enough and discharge losses will increase with the discharge port. The diagram of volume and rotational angle provides the reference for the suction and discharge ports design.

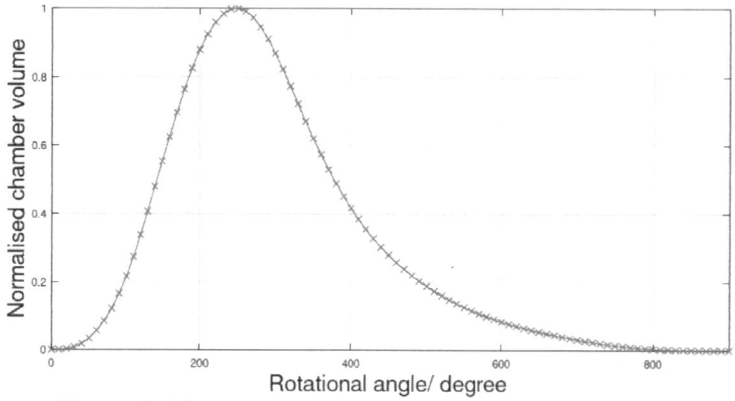

Figure 9. Chamber volume changing with rotational angle

3.5. Relative velocity

The relative velocity of inner and outer rotors is calculated at the contact points based on the following process [1]. Consider that two rotors are rotated about parallel axes with angular velocities ω_1 and ω_2 respectively. Rotor relative velocity is defined as a difference between rolling velocities at contact points. The expression for the sliding velocity is shown in Equation *6*. The subscript i indicates that the vector is represented in coordinate system S_i.

$$\boldsymbol{v}_{12} = -\boldsymbol{v}_{21} = \boldsymbol{v}_1 - \boldsymbol{v}_2 = [(\boldsymbol{\omega}_2 - \boldsymbol{\omega}_1) \times \boldsymbol{r}_i] + (\boldsymbol{E}_i \times \boldsymbol{\omega}_2) \tag{6}$$

Based on the constructed coordinate system, the vectors of expression 6 are represented in S_1 as following:

$$\boldsymbol{\omega}_1 = [0\ \ 0\ \ \omega_1]^T, \boldsymbol{\omega}_2 = [0\ \ 0\ \ \omega_2]^T$$

$$\boldsymbol{r}_1 = [x_1\ \ y_1\ \ z_1]^T, \boldsymbol{E}_1 = [0\ \ E\ \ 0]^T$$

Equation 6 with above expression yields

$$\boldsymbol{v}_{12} = [-y_1(\omega_1 - \omega_2) - E\omega_2]\boldsymbol{i} + [x_1(\omega_1 - \omega_2)]\boldsymbol{j}$$

where $\boldsymbol{i}$, $\boldsymbol{j}$ are the unit vectors of the S_1 coordinate system. The sliding velocity $\boldsymbol{v}_{12}$ can be expressed in coordinate system S_1 by the following expressions:

$$\boldsymbol{\omega}_1 = [0\ \ 0\ \ \omega_1]^T, \boldsymbol{\omega}_2 = [0\ \ 0\ \ \omega_2]^T$$

$$\boldsymbol{r}_1 = [x_1\ \ y_1\ \ z_1]^T, \boldsymbol{E}_{01} = [E\sin\varphi_1\ \ E\cos\varphi_1\ \ 0]^T$$

Equation 6 with above expression yields:

$$\boldsymbol{v}_{12} = [-y_1(\omega_1 - \omega_2) - \omega_2 E\sin\varphi_1]\boldsymbol{i} + [x_1(\omega_1 - \omega_2) + \omega_2 E\cos\varphi_1]\boldsymbol{j}$$

The sliding velocity calculated using the above equation for 3 by 2 rotor combination cases are shown in Figure 10. The pure rolling without sliding is zero when the contact points sit on the pitch circle. These results could be used to optimize the rotor profile design, such as centre distance, number of lobes and pitch circle, to reduce relative velocity in the future.

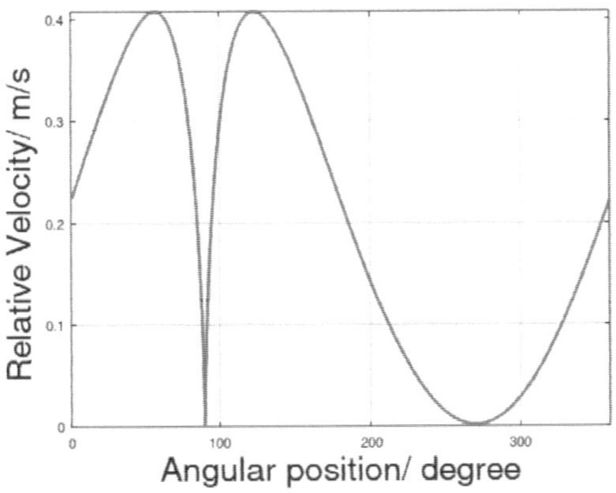

(a) Cartesian coordinate system

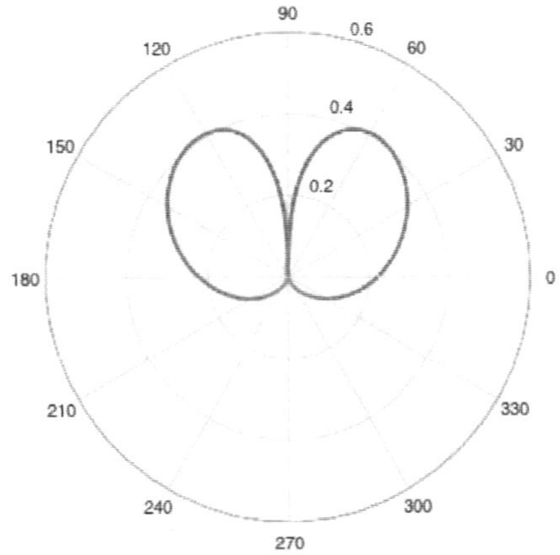
(b) Polar coordinate system

Figure 10. Relative velocity

4. Experiment results

The non-parallel and variable rotor profile and pitch is designed, manufactured, and tested in house. The base line model is CRC.60 with built-in Vi 4.6. The displacement is 60 cm^3/rev.

4.1. Test rig

The test rig is shown in Figure 11. Torque, air volume flow rate, inlet and outlet temperature and pressure of oil and air can be measured in this test rig. The rotational speed of inner rotor is 3000 rpm. Oil is directly injected to suction chamber with oil flow rate from 3-5 L/min depending on the pressure ratio.

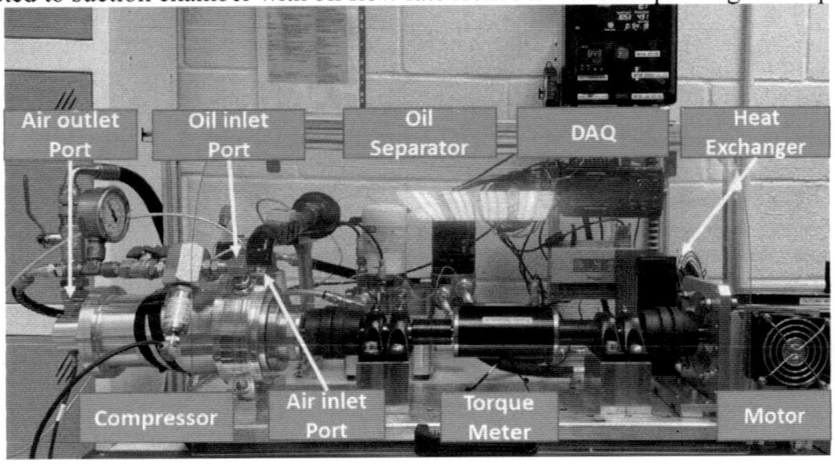

Figure 11. Chamber volume changing with rotational angle

4.2. Test results

The experimental results of the air inlet flow and specific power changing with outlet pressure of CRC.60 is shown in Figure 12. The air inlet flow is not affected by the outlet pressure before 20 barg while the specific power increases linearly with outlet pressure. The air inlet flow starts decreasing and specific power increasing quickly when the outlet pressure over 20 barg because under-compression and severe leakage. This test shows the great ability of CRC.60 working in high pressure ratio.

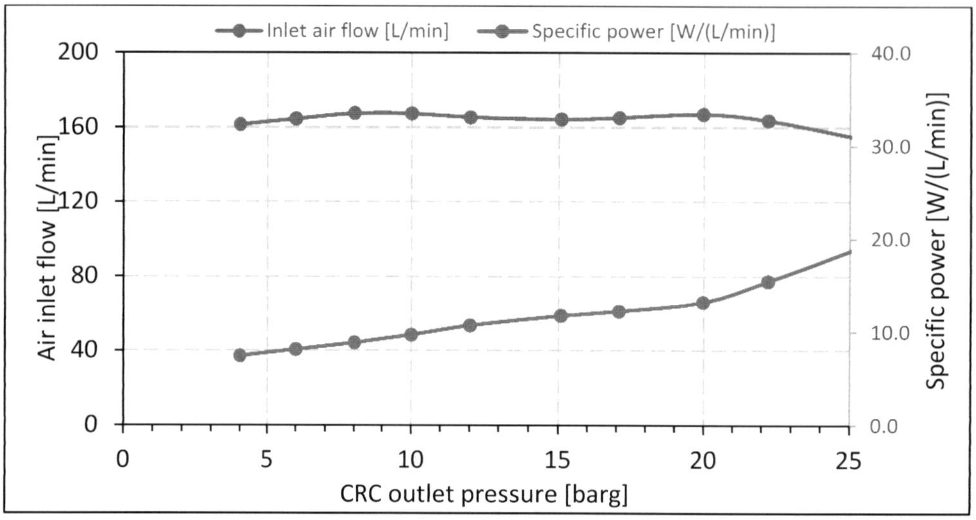

Figure 12. Experimental results

As shown in Table 3, shaft power keeps increasing with increasing of discharge pressure. The isentropic efficiency and volumetric efficiency are calculated and compared for different outlet pressure. Isentropic efficiency is 54% at 6 barg outlet and increases to as high as 62% with the increasing of discharge pressure and decrease to 48.1% at 25 barg. The lower isentropic efficiency at 6 barg and 25 barg may cause by over-compression and under-compression individually. Volumetric efficiency is quite steady between 90% to 92% with the increasing of discharge pressure and starts decreasing after 20 barg. The performance is quite steady between 10barg to 20 barg. High volumetric efficiency 86% can be achieved even at pressure ratio 26.

Table 3. Comparison of efficiency.

Outlet pressure [barg]	Inlet air flow [L/min]	Shaft power [W]	Isentropic efficiency [%]	Volumetric efficiency [%]
6	164.6	1339.1	54.0	90.5
10	167.5	1624.4	60.0	92.1
15	164.3	1930.1	61.0	90.8
20	167.2	2208.7	62.0	92.5
25	154.8	2931.0	48.1	86.0

5. Conclusions

Conical rotary compressor is a novel positive displacement machine which needs to explore to optimise the performance. In this paper, the rotor profile is generated based on the meshing theory. The parallel and non-parallel 3D rotors are demonstrated. Taking the non-parallel variable pitch and rotor profile as an example, the contact line, sealing line, cross-section area, chamber volume and relative velocity are calculated using the VertSim package. In the end, the experiments were conducted with the base line CRC.60. With the feature of variable pitch and rotor profile, CRC.60 works efficiently across a large range of pressure ratios.

Acknowledgements
I would like to thank reviewers for their comments to this paper. My thanks also to the Vert Technologies for permission to prepare and publish this paper. Finally, thanks to the Dortmund University for providing the conference for presentation of this work.

References
[1] Litvin F and Fuentes Alfonso 2004 *Gear Geometry and Applied Theory*
[2] Vecchiato D, Demenego A, Argyris J and Litvin F L 2000 *Geometry of a cycloidal pump*
[3] Zhang Y and Li X 2021 Geometry design and analysis of the double-enveloping cycloid drive *Journal of the Brazilian Society of Mechanical Sciences and Engineering* **43**
[4] Read M G, Smith I K and Stosic N 2017 Internally geared screw machines with ported end plates *IOP Conference Series: Materials Science and Engineering* vol 232 (Institute of Physics Publishing)
[5] Read M G, Stosic N and Smith I K 2020 The influence of rotor geometry on power transfer between rotors in gerotor-type screw compressors *Journal of Mechanical Design, Transactions of the ASME* **142**
[6] Read M G, Smith I K and Stosic N 2020 Influence of rotor geometry on tip leakage and port flow areas in gerotor-type twin screw compressors *Proceedings of the Institution of Mechanical Engineers, Part E: Journal of Process Mechanical Engineering*
[7] Ivanović Lozica and Josifović Danica 2006 *Specific Sliding of Trochoidal Gearing Profile in the Gerotor Pumps* vol 34
[8] Ivanović L, Devedžić G, Cuković S and Mirić N 2012 Modeling of the meshing of trochoidal profiles with clearances *Journal of Mechanical Design, Transactions of the ASME* **134**
[9] Bae J H and Kim C 2015 Design of rotor profile of internal gear pump for improving fuel efficiency *International Journal of Precision Engineering and Manufacturing* **16** 113–20
[10] Lu Y, Hoang K, Noake D and Low N 2021 Quasi 1D modelling of conical rotary compressors *25th International Compressor Engineering Conference at Purdue*

IOP Conf. Series: Materials Science and Engineering 1267 (2022) 012002 doi:10.1088/1757-899X/1267/1/012002

Systematic engineering design approach for improvement of oil-free twin-screw compressors

Thibaud Plantegenet[1], Thibault Tam[1], Ahmed Kovacevic[1], Sham Rane[1], Brijeshkumar Patel[1], Aleksander Krupa[1], Richard Leach[1], Donald Lawson[2], Manoj Heiyantuduwa[2], Billy Milligan[2], James Ure[2]

[1] City, University of London, Centre for Compressor Technology, EC1V 0HB, London, UK
[2] Howden Compressors, Old Govan Road, PA4 8XJ, Renfrew, UK

E-mail: thibaud.plantegenet@city.ac.uk

Abstract. Twin-screw compressors are widely used in industry, especially in compressed air, refrigeration, air-conditioning and process gas which consume a significant part of the world's energy. Nowadays, oil-injected compressors represent the majority of twin-screw compressors in the market due to their high efficiency and reliability. The oil-free compressor is potentially a better solution in the context of the net-zero CO2 target in 2050. However, due to its high thermal deformation and small clearances, this technology still suffers from reliability issues. To remedy this problem disruptive innovative solutions are required. In this purpose, the present study uses a systematic engineering design process to develop new concepts for the improvement of the oil-free twin-screw compressor. The paper is focused on the first two phases of the design process which are the definition of problem and the conceptual design. In the problem definition, main objectives are expressed and are divided into sub-objectives and weighed using an objective tree decomposition. Moreover, a thorough functional model of the oil-free compressor is detailed with a focus on the leakage paths and heat transfers. For the conceptual design, engineering characteristics extracted from the functional analysis have been assessed against the most important objectives using Quality Function Deployment matrices (QFD). Based on the developed problem definition, new concepts have been generated and three distinct concept categories have been further explored: Secondary flow; Surface features; Clearance control and monitoring. The evaluation, embodiment and detailed designs of the concepts, using experimental and numerical analyses will follow.

1. Introduction

The main purpose of this study is to present the use of a systematic engineering design method to improve the efficiency and reliability of oil-free twin-screw compressors and thus develop their market among oil-injected ones. This research is part of the project SECRET (Smart Efficient Compression: Reliability and Energy Targets) which is funded by Royal Academy of Engineering and Howden. To understand the relevance and importance of this research, it is therefore essential to recall the context of global warming and compressor market by citing some key figures.

The impact of human activities on climate change and global warming are now known by everyone and it is a major societal preoccupation. Following COP21 in 2015 in Paris and COP26 in 2021 in Glasgow, most countries are in agreement to try to limit the rise in global temperature to 1.5 °C [1]. As a figure of example, the UK developed a strategy of net zero Greenhouse Gas (GHG) emissions by 2050 [2]. Recently the Intergovernmental Panel on Climate Change (IPCC) published its sixth assessment report that shows, among other things, that the increase of GHG since 1750 are caused by human activities, which in turn have raised global temperatures between 1 to 2 °C during the last decade when compared to records of 1850-1900 [3]. Moreover, in 2019, GHG were mainly composed by 64% of CO_2 coming from fossil fuel, 18% of CH_4 (Methane) and 11% of CO_2 coming for land use [4].

Based on BP statistical and IEA (International Energy Agency) it is known that fossil fuel represented, in 2019, 84% of the primary energy consumption in the world, as well as, 60% of the generated electricity [5]. Industry, among all the other sectors, is responsible for more than 40% of the total electricity consumption [6].

Compressor systems are widely used in the world, in every sector (industry, domestic, commercial...) and for a wealth of applications. There are too many applications to list them all but they can be categorised in two main groups: process gas applications (refining, chemical, exploration...) and compressed air applications. Various sources quote that 10% of industrial electricity consumption in European countries is used by compressed air systems [7, 8]. Two reports ordered by the European Commission published in 2014 and 2017, shows that, in 2010 in the EU, 42% of energy consumption in compressed air systems (driven by electric motors) was consumed by oil-injected screw and vane compressors (7-15 bar (g)), 31% by low pressure compressors (all types, 0.05-3.5 bar (g)), 26% by oil-free compressors (all types, 7-15 bar (g)), and to a lesser extent by reciprocating compressors (1%) [9, 10].

Correlation of the figures previously cited gives that around 1.7% of generated electricity is directly consumed by oil-injected screw compressors, 1.3% by low pressure compressors and 1.1% by oil-free compressors[1]. These figures clearly demonstrated that oil-injected twin-screw compressor is the most common and most energy consuming compressor, which can be explained by the maturity of this technology that shows high efficiency and high reliability [11,12]. Figures for oil-free twin-screw compressors are not expressed directly and are included in low pressure compressors (with roots blower, turbo, scroll...) and oil-free compressors (piston, water-injected screw, scroll...).

Although based on the same principle of using helical screws to reduce the chamber volume and increase the pressure of gas, oil-free and oil-injected twin-screw compressors are very different [13]. Oil-injected compressors operate with one driven shaft and direct contact between screws. They use oil as a lubricant to limit wear, to seal the internal clearances, and to limit and carry away the heat generated by the compression process and minimise thermal deformation. Moreover, bearings are usually lubricated with the same oil supply which eliminates the need for mechanical seals between the gas chamber and the bearings. Oil-free compressors are synchronised using a timing gear and operational clearances prevent contact between screws. Clearances are relatively larger than the oil injected ones to accommodate thermal deformation due to compression. To avoid contamination of the process gas by the lubricated bearings, seals are mandatory.

For oil-free twin-screw compressors, it is estimated that a reduction of 20% of internal leakages will lead to a reduction of 2% of energy consumption. This, associated with the needs of reliability and reduction of use of fossil fuel will allow the oil-free screw compressor to expand its market and become the preferred solution in compressed air and process gas sectors.

The project SECRET is born in this context. This five years project awarded to City,

[1] This correlation is based on the assumption than the European Union compressor market is representative of the world market.

University of London aims to pave the way for the future of compression technology. Thus, it is divided into three phases: I. Understanding the physic of leakage flows and Conjugate Heat Transfer (CHT) in clearance gaps; II. Building and testing an efficient and reliable prototype machine; III. Developing a smart efficient commercial screw machine.

To achieve the first phase, an experiment was developed based on a roots blower specifically modified to integrate an optical access to the radial clearance gap and observe the leakage flow during normal operation. This apparatus allowed the measure of surface and leakage flow field temperatures as well as flow velocity, boundary layer and vortex formation by various techniques. Surface temperatures were measured by high speed infrared thermography [14]. Flow velocity, boundary layer and vortex formation by Particle Image Velocimetry (PIV) [15,16]. Leakage flow field temperatures by Planar Laser-Induced Fluorescence (PLIF) for which the feasibility was demonstrated [17]. These experiments are exploited to validate the development of CFD tools to improve the modelling of CHT in leakage flows [18] which can then be used to model more complex geometries.

The second phase of the project consists of building and performing extensive laboratory tests on an oil-free screw compressor prototype. New concepts developed using the systematic engineering design process and presented here will be tested into the prototype and evaluated based on the performances. In parallel, research on innovative techniques of monitoring and control of clearance gaps applied to positive displacement machines is conducted by a PhD student [19]. This phase of the project will lead to the selection of concepts which demonstrated the best improvements in the prototype. The final phase will be in closer collaboration with Howden Compressors to develop and fully test a new commercial oil-free twin-screw compressor built with the selected concepts, and the monitoring and control system.

Finding innovative disruptive technological solutions to reach the goals of improving efficiency and reliability of oil-free twin-screw compressors is very challenging, especially due to the competitive level of efficiency that corresponds to this high technology readiness level. Thus, to develop new concepts which can be implemented in an oil-free compressor, a systematic approach of engineering design process has been employed. The next sections of this paper will present the systematic engineering design method, and subsequently describe how this method is applied to the oil-free twin-screw compressor design. A discussion will conclude on the use of this method and highlight future developments.

2. Systematic engineering design method

This systematic approach is made to rigorously decompose and fully define a complex problem for which solutions will be generated, compared and refined based on quantitative criteria. Various models defining systematic design process have been proposed in the literature but, the process map defined by G. Pahl et al. [20] is considered as the benchmark for a modern systematic design process [21]. Overall, this process can be divided in four steps, which are: 1) Planning and clarifying the task (or problem definition); 2) Conceptual design; 3) Embodiment design; 4) Detail design. Along this design process decisions are made to progress from step to step and the design is refined using iteration between steps.

This process map is general and can be applied to a wide variety of projects/product designs. In this study, as the initial goal is to define new concepts/solutions for an existing product the focus will be on the first two steps. These two steps can also be seen in the form of the symmetrical problem-solution model presented in Figure 1b. and given by N. Cross in his book [22]. This second model is interesting because it shows the decomposition of a whole problem, which is usually hard to approach, into sub-problems that are easily defined. These sub-problems are used to generated sub-solutions, and sub-solutions are evaluated and combined to form the global solution.

A mix of these two models was adopted to develop the concepts in two distinctive phases: Problem definition and Conceptual design. Problem definition regroup: Identifying opportunities; Clarifying objectives; Establishing functions; Setting requirements and Determining characteristics. Conceptual design is divided into Generating alternatives, Evaluating alternatives, and Improving details. In the problem definition, the identification of opportunities is not addressed in this study as it is related to the wider project SECRET. Conceptual design will be succinctly described in the paper as new concepts developed during this phase cannot be disclosed before the end of the project.

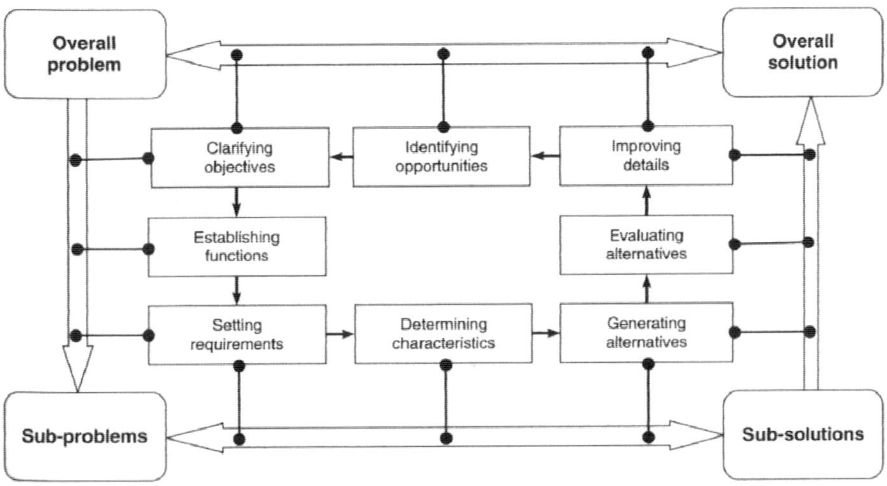

Figure 1. Symmetrical problem/solution model [22].

3. Problem definition applied to the oil-free twin-screw compressor

3.1. Objectives definition

The main objectives of the project SECRET are to improve efficiency and reliability of oil-free twin-screw compressors by keeping a relatively low cost. These objectives are global and hardly achievable directly. To make SMART (Specific, Measurable, Achievable, Realistic, Time-bound) objectives an objective tree was used as a decomposition tool.

Working groups have been formed in the team from City, University of London and Howden Compressors to build the objective tree for all three main objectives: Increase efficiency; Increase reliability; Minimise cost. Due to their large size, these trees are not directly presented in the paper, but results will be shown. In a tree decomposition, the objectives are hierarchical and are expressed in terms of Level, from Level 0 for main objectives to Level 5. Moreover, objectives on a level are rated on a scale from 0 to 1 (percentage scale) respectively to each other under the same objective. A second weight is established by multiplying all hierarchical predecessors with the objective. This second (product) weight is used for the global comparison of objectives on every level.

For example, the objective tree for "Increase efficiency" is presented in the Figure 2 with the relative and global weights for each sub-objective. Level 0 objective (Increase efficiency) is weighted with 0.3 and is divided into three Level 1 objectives: "Increase volumetric efficiency", "Increase adiabatic efficiency" and "Increase isothermal efficiency" with respective weights of 0.6, 0.2 and 0.2 and product weights of 0.18 (0.3*0.6), 0.06 (0.3*0.2) and 0.06 (0.3*0.2).

<table>
<tr><td>Level 0</td><td colspan="22">Increase efficiency</td></tr>
<tr><td>L0 weight</td><td colspan="22">0.3</td></tr>
<tr><td>Level 1</td><td colspan="10">Increase volumetric efficiency</td><td colspan="8">Increase adiabatic efficiency</td><td colspan="4">Increase isothermal efficiency</td></tr>
<tr><td>L1 weight</td><td colspan="10">0.600</td><td colspan="8">0.200</td><td colspan="4">0.200</td></tr>
<tr><td>Product</td><td colspan="10">0.180</td><td colspan="8">0.060</td><td colspan="4">0.060</td></tr>
<tr><td>Level 2</td><td colspan="6">Reduce leakage</td><td>Reduce thermal deformation</td><td colspan="3">Reduce pressure deformation</td><td colspan="4">Reduce power losses</td><td>Match built-in VI</td><td>Reduce suction pressure loss</td><td colspan="2">Reduce discharge pressure loss</td><td colspan="3">Improve cooling</td><td>Reduce frictional losses</td></tr>
<tr><td>L2 weight</td><td colspan="6">0.500</td><td>0.300</td><td colspan="3">0.200</td><td colspan="4">0.500</td><td>0.300</td><td>0.100</td><td colspan="2">0.100</td><td colspan="3">0.700</td><td>0.300</td></tr>
<tr><td>Product</td><td colspan="6">0.090</td><td>0.054</td><td colspan="3">0.036</td><td colspan="4">0.030</td><td>0.018</td><td>0.006</td><td colspan="2">0.006</td><td colspan="3">0.042</td><td>0.018</td></tr>
<tr><td>Level 3</td><td colspan="4">Reduce clearance</td><td>Reduce differential pressure across the gap</td><td>Improve sealing</td><td></td><td>Seals and bearings</td><td>Dynamic deflection from imbalance</td><td>Rotor deflection due to pressure</td><td>Reduce bearing losses</td><td>Reduce flow losses</td><td>Reduce sealing losses</td><td>Reduce drivetrain losses</td><td></td><td></td><td>Throttling loss</td><td>Discharge pressure pulsation</td><td>Cool gas</td><td>Cool rotor</td><td>Cool housing</td><td></td></tr>
<tr><td>L3 weight</td><td colspan="4">0.300</td><td>0.200</td><td>0.500</td><td></td><td>0.3</td><td>0.4</td><td>0.3</td><td>0.1</td><td>0.3</td><td>0.5</td><td>0.1</td><td></td><td></td><td>0.5</td><td>0.5</td><td>0.2</td><td>0.4</td><td>0.4</td><td></td></tr>
<tr><td>Product</td><td colspan="4">0.027</td><td>0.018</td><td>0.045</td><td>0.054</td><td>0.011</td><td>0.014</td><td>0.011</td><td>0.003</td><td>0.009</td><td>0.015</td><td>0.003</td><td>0.018</td><td>0.006</td><td>0.003</td><td>0.003</td><td>0.008</td><td>0.017</td><td>0.017</td><td>0.018</td></tr>
<tr><td>Level 4</td><td>Reduce axial clearance</td><td>Reduce interlobe clearance</td><td>Reduce tip clearance</td><td>Reduce blowhole</td></tr>
<tr><td>L4 weight</td><td>0.300</td><td>0.300</td><td>0.300</td><td>0.100</td></tr>
<tr><td>Product</td><td>0.0081</td><td>0.0081</td><td>0.0081</td><td>0.0027</td></tr>
</table>

Figure 2. Objective tree for "Increase efficiency" with relative weight.

For each level, objectives are ranked using the global weight. Rankings for Level 0 and Level 1 are presented in Figure 3 and Figure 4, respectively. On Level 0, "Increase reliability" is the most important objective as main issues with oil-free compressors are related to reliability. This is followed by efficiency and finally cost. Similarly on Level 1, objectives related to reliability are at the top ("Design for reliable operation", "Optimise reliability metrics", "Facilitate serviceability"). Overall, cost related objectives are less important especially at the early stage of the project, which also serves to initiate a study which is not restricted by cost.

Level 0	Weight
Increase reliability	0.5
Increase efficiency	0.3
Minimise cost	0.2

Figure 3. Ranking of the main (Level 0) objectives.

Level 1	Weight
Design for reliable operation	0.2
Increase volumetric efficiency	0.18
Optimise reliability metrics	0.125
Facilitate serviceability	0.125
Minimise operating costs	0.1
Increase adiabatic efficiency	0.06
Increase isothermal efficiency	0.06
Protect against external environmental conditions	0.05
Minimise capital (purchase) costs	0.04
Minimise maintenance cost	0.03
Minimise sales and admin costs	0.01
Minimise development costs	0.01
Minimise end of life costs	0.01

Figure 4. Ranking of the Level 1 objectives.

Level 3 objectives presented in Figure 5 are very specific and much more measurable and achievable than upper levels. In this list, objectives from Level 2 without descendants are also reported with the Level 3 objectives. This level regroups 78 objectives in total, top objectives are related to reliability like the reduction of thermal deformation, the safe start and stop operation and the safe operation in transient states. Efficiency related objectives are also on top, with the improvement of sealing for internal clearances, the reduction of power consumption or reduction of internal clearances. Level 1 objectives will be used in the definition and selection of functions. Level 3 objectives will be used for the requirements definition and the generation of concepts.

3.2. Functional model

For an oil-free twin-screw compressor, the main function is to compress the gas. The compressor can be defined as a black box with input flows and output flows transformed by this function as presented in the Figure 6. The inputs are the gas flow (at a certain pressure and temperature)

Level 3 with reported Level 2 alone	Weight
Reduce thermal deformation (Level 2)	0.054
Improve internal sealing	0.045
Reduce power consumption (Level 2)	0.04
Ensure safe start & stop operation	0.04
Safe transient behaviour (speed/load)	0.04
Monitor health in operation (Level 2)	0.03
Personnel damage	0.03
Reduce internal clearances	0.027
Manage thermal loads i.e. cooling requirements	0.025
Minimise cost of failure (Level 2)	0.024
Compressor design for field maintenance (Level 2)	0.02
Simplify / minimise compressor aux. systems (Level 2)	0.02
Conceive easy operation (Level 2)	0.02
Flexible to range of pressure ratio	0.02
Flexible to range of process gases	0.02
Minimise downtime (planned / unplanned)	0.02
Reduce pressure differential across the gap	0.018
Match built-in VI (Level 2)	0.018
Reduce frictional losses (Level 2)	0.018
Cool rotor	0.0168
Cool housing	0.0168
Resist to extreme conditions (Level 2)	0.0165
Adapt to site specific hazards (Level 2)	0.0165
Reduce sealing losses	0.015
High standard of control quality	0.015
Prevent mechanical failure	0.015

Figure 5. Ranking of the Level 3 objectives with reported level 2 objectives - First 26 objectives on a list of 78.

and the mechanical energy. This mechanical energy is used to transform the gas (to reach another pressure and temperature), which implies a transformation of this mechanical energy and generation of losses (gas leaks, heat). In addition, the compressor can receive and transform additional material flows (oil for bearing lubrication, inert gas for sealing, water for cooling...) and signal flows (monitoring and/or control of temperature, pressure, clearances...).

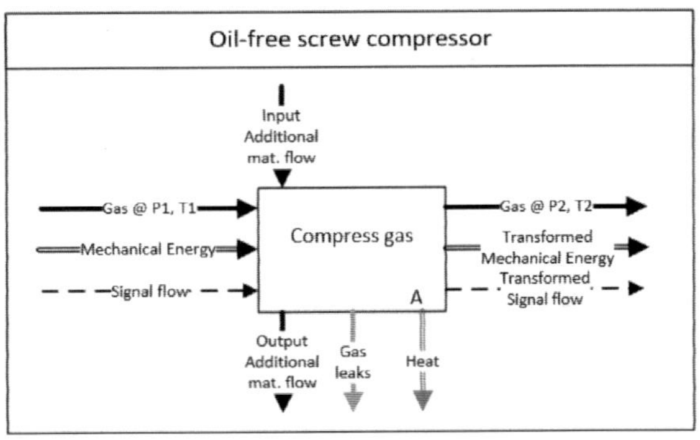

Figure 6. Simplified functional model of an oil-free twin-screw compressor - Main function.

The functional model of the compressor consists of a collection of functions that are completed to achieve the main function. However, as it is difficult to list every functions of a complex system such as the oil-free compressor, it will initially be divided into sub-domains, and functions will be defined for each sub-domain.

The functional model for the whole system (called "Global stream") is presented in the Figure 7. Following the mechanical energy flow (double black arrow on the left of the figure), the functional model starts with the motor shaft (sub-domain #13) which can be considered as an external or an internal part of the compressor. Then it goes to the gearbox (#14) which adapts the mechanical energy for the driving shaft (#15). The driving shaft transfers the mechanical energy to the synchronising gear (#16) and finally to the main (#10) and gate (#11) rotors. Both rotors are guided by bearings (#B) which transfer mechanical energy to the stator (#12). Moreover, bearing seals (#S5) which separate bearing lubricant from the environment and/or from processed gas, also transfer mechanical energy to the stator (#12).

In terms of material flow (top of the figure), which is mainly the processed gas (gas flow), this starts with the gas conditioning at suction (#1) (adjustment of temperature, composition, moisture...) and can be internal or external to the compressor. The gas flow enters the suction domain (#2) where the flow velocities are homogenised, then passes through the suction port (#3) which controls the intake volume and directs the flow into the central chamber (or main gas domain #4) where the compression is achieved by reduction of volume. Flow is released by the discharge port (#5) to the discharge domain (#6) and finally the gas can be conditioned at discharge (#7) (internally or externally).

Mechanical energy coming from the rotors (#10, #11) acts on the main gas domain where it is transferred into internal energy and heat of gas (increase of pressure and temperature), and into heat transferred to solids (#10, #11, #12).

In combination with the gas flow there is a leakage flow which can be defined as a loss of the gas flow and decomposed in several leakage paths. A distinction can be made between external leakage paths going to environment and internal leakage paths opposed to the gas flow and going from a high pressure sub-domain to a low pressure sub-domain. In this way, two external paths have been defined: the low pressure leakage at suction port (#3) and the high pressure leakage at discharge port (#5). These leakages are limited by the use of seals (#S1-S4) on both rotors at suction and discharge and also generate heat and mechanical losses.

Fleming and Tang [23] and more recently Buckney et al. [24] have defined internal leakage paths in twin-screw compressors. Based on that, six internal leakage paths have been identified between sub-domains of the compressor and associated with the existence of the following clearances:

1. Interlobe clearance between main and gate rotor.
2. Radial clearance between rotor tips (both rotors) and housing bores.
3. Blow hole area between rotors and cusps (high pressure side and low pressure side).
4. Axial clearance at discharge between rotors and discharge end plate at the intersection of the rotor shafts.
5. Axial clearance at suction between rotors and suction end plate.
6. Axial clearance at discharge between rotors and discharge end plate at the housing bore side.

The internal leakage paths associated with interlobe, radial and axial discharge clearances are represented in the Figure 8. The blow hole area is not represented in the Figure 8 and corresponds to the area between the two rotors and the housing cusp on the low pressure side (upper side of Figure 8) and on the high pressure side (lower side of Figure 8). The blow hole area is mainly dependant on the rotor profiles and will not be specifically addressed in this paper.

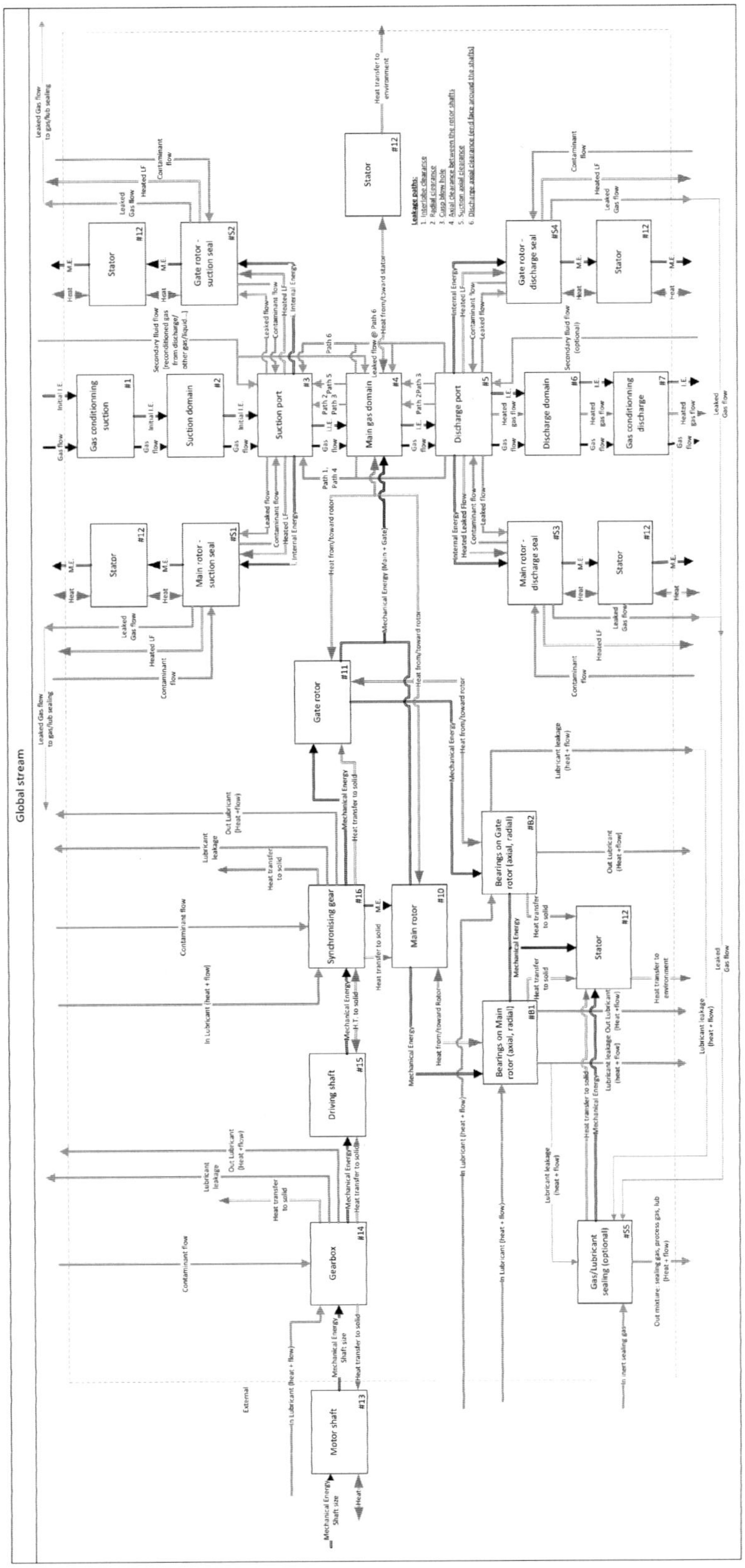

Figure 7. Functional model of the compressor.

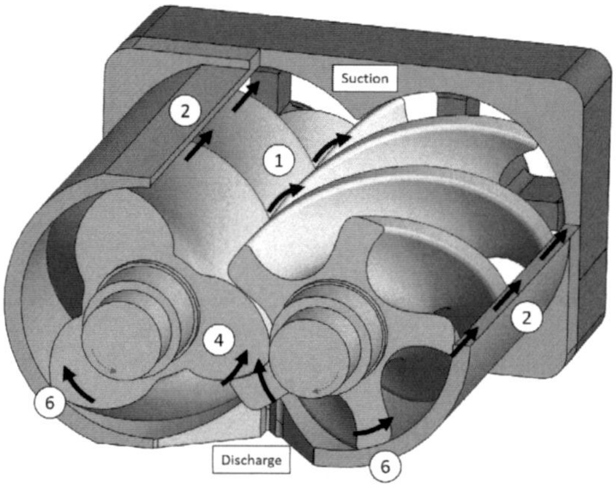

Figure 8. Definition of the internal leakage paths - 1. Interlobe; 2. Radial; 4. Axial discharge between shafts; 6. Axial discharge at the housing bore side. Leakage paths 3. Blow hole area and 5. Axial clearance at suction are not represented.

The domain of the compressor was divided in sub-domains and interactions between sub-domains (flow exchanges). For each sub-domain, functions are established. Overall, the compressor is defined with 22 sub-domains for around 100 sub-functions in total (some sub-functions are identical for several sub-domains). An initial list of 25 functions was drawn up by focusing the functional model to the domains which act on the gas flow. Furthermore, a Quality Function Deployment (QFD) method was used to rate the functions and extract those important functions for the objectives.

Quality Function Deployment 1 (QFD1): In this method, the relationship is evaluated between customer requirements and functional requirements. For this particular case, the customer requirements are the Level 1 objectives (presented in Figure 4) and the functional requirements are the 25 pre-selected functions. As presented in Figure 9, objectives are placed horizontally with their respective weights (defined in the objectives definition) and functions are positioned vertically, forming a matrix. The relationship is evaluated between each objective and function and is defined as: strong relationship ($\ominus = 9$), moderate relationship ($\bigcirc = 3$), weak relationship ($\blacktriangle = 1$), or no relationship (empty cell = 0).

All functions are rated based on the sum of products of relationship weight and objective weight. Consequently, functions related to higher importance objectives are likely to be the top rated functions. This QFD1 leads to the selection of functions which are regrouped as follow:

- Ensure functioning clearances.
- Leak through clearances (internal leakages).
- Leak through seals (external leakages).
- Seal the gas domain.
- Restrict radial movements.
- Restrict axial movements.
- Ensure an accurate positioning of rotors (machining, assembling).

As it can be seen, these functions are primarily related to clearances in terms of operation, leakages and sealing. Other functions are related to the main function of bearings and to the general manufacturing accuracy. "Leak through seals" and "Seal the gas domain" are similar but are expressed in opposition to each other. The first one is expressed more in terms of leakages and how a solution will act to reduce or minimise this leakage. The second one expresses the sealing aspect which is more related to 'how to limit exchange between domains, how to avoid contamination of the gas domain'.

Row #	Relative Weight	Weight / Importance	Demanded Quality (a.k.a. "Customer Requirements" or "Whats") \ Quality Characteristics (a.k.a. "Functional Requirements" or "Hows")	1: #1 #7 Condition the gas	2: #2 #6 Homogenise the flow	3: #3 Controle intake gas volume	4: #3 Direct the flow to the main domain	5: #3 #5 Leak through seals	6: #4 Reduce Chamber volume	7: #4 Transfer M.E. to I.E. in gas	8: #4 Transfer M.E. to heat in gas	9: #4 #5 Leak through path 1	10: #4 #5 Leak through path 2	11: #4 #5 Leak through path 3	12: #4 #5 Leak through path 4	13: #4 #5 Leak through path 6	14: #5 Release the compression gas	15: #5 Direct the flow to the discharge domain	16: #10 #11 #12 Support M.E.	17: #10 #11 #12 Support I.E. of gas	18: #10 #11 #12 Support Thermal energy	19: #10 #11 #12 #16 Ensure an accurate positioning	20: #10 #11 #12 #13 #14 #15 #16 Interface with other domains	21: #B Restrict radial movement	22: #B Restrict axial movement	23: #B Support Reverse load	24: #S Seal the gas domain	25: #3 #4 #5 #S Ensure functioning clearances
			Direction of Improvement: Minimize (▼), Maximize (▲), or Target (x)	x	x	x	x	▼	x	▲	▼	▼	▼	▼	▼	▼	▲	x	x	x	x	▲	x	▲	▲	▲	▲	▲
1	20.0	0.20	Design for reliable operation	▲				Θ	O	O	O	Θ	Θ	Θ	Θ	Θ			Θ	Θ	Θ	Θ		Θ	Θ	O		Θ
2	18.0	0.18	Increase volumetric efficiency	O	▲	Θ	O	Θ	▲			Θ	Θ	Θ	Θ	Θ						▲		O	O		Θ	Θ
3	12.5	0.125	Optimise reliability metrics	▲				▲	O	O	O								O	O	O	O	▲			▲	▲	O
4	12.5	0.125	Facilitate serviceability					▲															O	▲	▲	▲	O	▲
5	10.0	0.10	Minimise operating costs	▲				▲	▲	▲	▲	Θ	Θ	Θ	Θ	Θ						O					O	O
6	6.0	0.06	Increase adiabatic efficiency	O				O		Θ		O	O	O	O	O	O	▲						O	O		O	O
7	6.0	0.06	Increase isothermal efficiency	O	O		O	O			Θ	O	O	O	O	O	O	O						O	O		O	O
8	5.0	0.05	Protect against external environmental conditions					▲											▲		▲		▲				O	
9	4.0	0.04	Minimise capital (purchase) costs	▲																		O	▲			O	Θ	O
10	3.0	0.03	Minimise maintenance costs	O		▲		Θ			▲	▲	▲	▲	▲	▲			▲	▲	▲	Θ	O	Θ	Θ	▲	O	Θ
11	1.0	0.01	Minimise sales and admin costs	▲				O				▲	▲	▲	▲	▲							▲			▲	▲	▲
12	1.0	0.01	Minimise development costs	▲	▲	▲	▲	▲	▲								▲	▲				O	▲			▲		
13	1.0	0.01	Minimise end of life costs																				▲			▲		
			Difficulty (0=Easy to Accomplish, 10=Extremely Difficult)	3	4	3	3	8	1	1	8	7	8	6	9	7	3	3	4	4	8	7	2	5	5	3	8	7
			Weight / Importance	147.5	37.0	166.0	73.0	449.0	126.5	161.5	164.5	472.0	472.0	472.0	472.0	472.0	37.0	25.0	225.5	220.5	225.5	307.5	71.0	309.5	309.5	103.0	339.0	498.0
			Relative Weight	2.3	0.6	2.6	1.1	7.1	2.0	2.5	2.6	7.4	7.4	7.4	7.4	7.4	0.6	0.4	3.5	3.5	3.5	4.8	1.1	4.9	4.9	1.6	5.3	7.8

Figure 9. Quality function deployment 1 for selection of functions.

3.3. Setting requirements

The setting of requirements aims to detail the attributes of the desired product to satisfy the customer needs (objectives) and to define the limits of the range of acceptable concepts. In this study, an existing machine is used as a prototype in which the concepts will be implemented. This will help define attributes and limits.

The prototype machine is an oil-free twin-screw compressor with 3/5 "N" rotors and a 127 mm male rotor diameter designed for a delivered flow of 1000 m^3/h at a maximum discharge pressure of 2.5 bar (g) and a designed radial and interlobe clearance gaps of 180 μm. Rotors are guided by ball and roller bearings to support axial and radial loads, respectively. Sealing of the main gas domain is realised by the use of labyrinth seals at suction and double lip PTFE seals at discharge.

Based on the prototype, some limits are established: It is not considered to change the profile of the rotors and the blow hole area will not be of direct interest in this study. As the bearings and seals are very dependant of the application, they will not be modified in the prototype and will be further explored in the phase III of the project. Furthermore, the manufacturing capabilities will not be explored in terms of machining and assembly tolerances of rotors and casing.

Initial tests were realised for male rotor tip speeds from 50 to 120 m/s and pressure ratio from 1.5 to 2.5. In the worst case (low speed and high pressure ratio) the discharge temperature is maximum at 164 °C for a suction temperature of 30.7 °C. In this case, the specific power was measured at 3.1 $kW.m^{-3}.min$ with 61.9 % and 55.7 % of volumetric and adiabatic efficiencies, respectively. Husak et al. [25] performed a numerical study on this prototype and shows that for a discharge temperature of 180 °C, the radial clearance can decrease from 180 μm up to 100

μm on the male rotor and up to 140 μm on the female rotor. This study also shows that change of clearance is not uniform and is bigger at the position of opening of the discharge port.

These results justify the top rated Level 3 objectives: reduce thermal deformation, improve internal sealing, reduce power consumption, ensure safe start and stop operation, safe transient behaviour and monitor health in operation. Our goal is then to reduce the thermal deformation by reducing the discharge temperature. This will lead to the increase of specific power and thus an increase of volumetric efficiency (directly related to the reduction of internal leakages) and an increase of adiabatic efficiency (related to the thermodynamic behaviour of the machine). Moreover, this will ensure a better reliability (safety of transient operation) by limiting the clearance gaps variation and encouraging the reduction of the designed clearance gaps.

3.4. Determining characteristics

The characteristics, or more specifically the Engineering Characteristics (ECs) of a product are the physical parameters of functions that affect the requirements. In other words, the ECs defined the performance expected by the user of the parameters of a function. These ECs can be common for several functions but with different values. From the selected functions, 48 ECs were defined for the following categories of properties:

- Geometry: Clearance gap and length, surface roughness, machining accuracy...;
- Kinematic: Gradient and average flow velocity, surfaces velocity;
- Energy: Pressure, temperature, flow rate...;
- Forces: Mechanical and thermal deformation, rotor balancing...;
- Material: Viscosity and density of gas, Young modulus and thermal expansion coefficient of solids...

A second QFD was done to rate the ECs based on the relationship with the Level 3 objectives which reflects the setted requirements. This QFD is too extensive to be presented in the paper, but top rated ECs were find to be: Clearance variation, clearance gap, temperature difference of the gas between suction and discharge, heat flux in the gas flow, surfaces temperatures, bearing internal clearance, thermal deformation of rotors and casing, rotor balancing, Leakage mass flow rate, clearance lengths and shape, etc.

To summarise, the top rated ECs have a large influence on the most important customer requirements. Which means that innovative design should focused on the concepts that act on these ECs. The ECs will be important in addition of the customer requirements for the evaluation of concepts.

4. Conceptual design

The conceptual design starts with the generation of concepts followed by the evaluation and the development of concepts. This phase leads to a variety of concepts that appear to be promising and results in the design embodiment. Concept generation will be succinctly presented thereafter, but next steps will not, as outside of the scope of the current paper.

4.1. Generating alternatives

For the concepts generation, the morphological chart method was used. This chart consist of a matrix presenting several concepts for each function found with the QFD1. As presented in the setting requirements section, the functions related to bearings and seals, as well as manufacturing will not be covered. This leads to two main functions: "Ensure functioning clearances" and "Leak through clearances" which are applicable to the three most important clearances: 1. Interlobe clearance; 2. Radial clearances; and 4. & 6. Axial clearances at discharge.

A large number of concept can be generated with a morphological chart. To narrow our study and limit the number of concept variances, three streams for concepts were established based on the customer requirements and are succinctly presented below:

1. Concepts related to surfaces properties and shape, such as:
   - Sacrificial tip seal
   - Grooves/steps on tip
   - Roughness
   - Surface texturing
   - Surfacc coating
2. Concepts related to the use of a secondary flow in the clearances:
   - Re-injection of cold gas
   - Use of economiser port
   - Impingement through casing
   - Impingement through rotor
3. Concepts related to the monitoring and control of clearances:
   - Monitoring of operating conditions and control in case of unexpected behaviour
   - Active control of position of rotor by the use of hydrodynamic bearings
   - Active control of position of rotor by the use of electromagnetic bearings or actuators.

Three independent morphological charts were build for the three categories of concepts. Concepts coming from these charts are under evaluation using a decision matrix based on the fulfilment of the objectives.

5. Conclusion

This paper discusses the systematic engineering design method applied to the development of innovative concepts for the improvement of the oil-free twin-screw compressor. This method conducts to the expression of the specific objectives describing the requirements in a clear and quantifiable way. This was possible with the use of objective tree decomposition of the three main objectives: "Increase efficiency", "Increase reliability" and "Minimise cost". A functional model of the whole compressor detailing the sub-domains, sub-functions, material and energy flow interactions was built. This model was then focused on the functions on which improvements can be made, mainly related to the internal clearances i.e. ensure functioning clearances, leak through clearances, ensure accurate positioning, restrict movements. The setting of requirement further limit the functions of interest and engineering characteristics help to describe the important parameters. Finally the conceptual design depicted the first step of the development of solutions.

In a future work, the concepts will be evaluated and most promising concepts will be implemented and tested in the compressor prototype at City, University of London. The performances of the prototype with the new concepts will be thoroughly analysed. The best concept will be applied to the commercial prototype in the third phase of the project SECRET.

References

[1] COP26 2021 The Glasgow climate pact *UN Climate Change Conference* UK (https://ukcop26.org/wp-content/uploads/2021/11/COP26-Presidency-Outcomes-The-Climate-Pact.pdf)

[2] UK Government 2021 *Net Zero Strategy: Build Back Greener*, https://www.gov.uk/government/publications/net-zero-strategy, Accessed 3 August 2022

[3] IPCC 2021 *Working Group I Contribution to the IPCC 6th Assessment Report, Climate Change 2021: The Physical Science Basis* Summary for Policymakers (https://www.ipcc.ch/report/ar6/wg1/#SPM)

[4] IPCC 2022 *Working Group III Contribution to the IPCC 6th Assessment Report, Climate Change 2022: Mitigation of Climate Change* Summary for Policymakers (https://www.ipcc.ch/report/ar6/wg3/#SPM)

[5] BP 2021 *Statistical Review of World Energy* 70th edition
[6] IEA 2021 *Electricity Information: Overview* (https://www.iea.org/reports/electricity-information-overview)
[7] Saidur R, Rahim N A, Hasanuzzaman M, 2010 A review on compressed air energy use and energy saving. *J Renew Sustain Energy Rev* **14** 1135-1153
[8] Mousavi S Kara S Kornfeld B. 2014 Energy efficiency of compressed air systems *Procedia CIRP 15* 21st CIRP conference on life cycle engineering 313-318 (http://dx.doi.org/10.1016/j.procir.2014.06.026)
[9] DG Energy of the European Commission 2014 Ecodesign preparatory study on electric motor systems / compressors *ENER Lot 31 Final Report of Task 1-8* Brussels (http://www.eco-compressors.eu/index.html)
[10] DG Energy of the European Commission 2017 Ecodesign preparatory study on low pressure & oil-free compressor packages *ENER Lot 31 Final Report on low pressure and oil-free compressor packages of Task 1-7* Brussels (http://www.eco-compressors.eu/index.html)
[11] Stosic N, Smith I K, Kovacevic A 2005 Opportunities for innovation with screw compressors *Proceedings of the Institution of Mechanical Engineers, Part E: Journal of Process Mechanical Engineering* **217** (3) 157-170 doi:10.1177/095440890321700301
[12] Stosic N, Smith I K, Kovacevic A, Mujic E 2010 Three Decades of Modern Practice in Screw Compressors *International Compressor Engineering Conference* Paper 1942 Purdue (https://docs.lib.purdue.edu/icec/1942)
[13] Stosic N, Smith I K, Kovacevic A 2005 *Screw Compressors: Mathematical Modelling and Performance Calculation* (1st ed.) Springer
[14] Patel B, Rane S, Kovacevic A 2022 Infrared-thermography and numerical investigationof conjugate heat transfer in roots blower *International Compressor Engineering Conference* Purdue
[15] Singh G, Sun S, Kovacevic A, Li Q, Bruecker C 2019 Transient flow analysis in a Roots blower: Experimental and numerical investigations *Mechanical Systems and Signal Processing* **134** 106305
[16] Patel B, Kovacevic A, Krupa A 2021 On Measuring Velocity and Temperature in Leakage Flows of Oil Free Rotary Positive Displacement Machines *In: Karabegovi, I. (eds) New Technologies, Development and Application IV. NT 2021. Lecture Notes in Networks and Systems* **233** Springer, Cham.
[17] Patel B, Kovacevic A, Charogiannis A, Alam M N, Schtte M 2021 The use of laser-induced fluorescence to measure temperature in the leakage gaps of oil-free positive displacement rotary machines *Measurement* **185** 110057
[18] Rane S, Kovacevic A, Stosic N, Smith I 2021 Bi-directional system coupling for conjugate heat transfer and variable leakage gap CFD analysis of twin-screw compressors *IOP Conf. Ser.: Mater. Sci. Eng.* **1180** 012001
[19] Tam T, Kovacevic A, Milligan W, Sun T, Grattan K, Fabian M, Plantegenet T, Patel B, Bandyopadhyay S 2022 Experimental investigation of screw compressor clearance monitoring techniques *International Compressor Engineering Conference* Purdue
[20] Pahl G, Beitz W, Feldhusen J, Grote K-H 2007 *Engineering Design - A Systematic Approach* 3rd edition, Springer
[21] Haik Y, Shahin T M 2011 *Engineering Design Process* 2nd edition, Cengage Learning
[22] Cross N 2008 *Engineering Design Methods - Strategies for Product Design* 4th edition, Wiley
[23] Fleming J S, Tang Y 1995 The Analysis of Leakage in a Twin Screw Compressor and its Application to Performance Improvement *Proceedings of IMechE, Journal of Process Mechanical Engineering* **209** 125
[24] Buckney D, Kovacevic A, Stosic N 2011 Consideration of clearances in the design of screw compressor rotors. *7th International Conference on Compressors and their Systems* London UK
[25] Husak E, Kovacevic A, Karabegovic I 2019 Calculation of clearances in twin screw compressors *IOP Conf. Series: Materials Science and Engineering* **604** 012001

Acknowledgement
The authors acknowledge the Royal Academy of Engineering and Howden Compressors for their continuous support and engagement in this research funded by the RAEng grant RCSRF2021\11\43.

IOP Conf. Series: Materials Science and Engineering 1267 (2022) 012003 doi:10.1088/1757-899X/1267/1/012003

Laser-Optical Shear-Flow Analysis across the Annular Gap of a Simplified Displacement Compressor Model

Robin Leister[1], Andreas Brümmer[2], Jochen Kriegseis[1*]

[1] Institute of Fluids Mechanics (ISTM), Karlsruhe Institute of Technology (KIT), Kaiserstr. 10, 76131 Karlsruhe, Germany

[2] Chair of Fluidics (FT), TU Dortmund University (TUD), Leonhard-Euler-Str. 5, 44227 Dortmund, Germany

E-mail: kriegseis@kit.edu

Abstract. The present experimental feasibility study testifies the two flow measurement techniques Defocusing Particle Tracking Velocimetry (DPTV) and Interferometric Particle Imaging (IPI) for their applicability to measure the two-phase flow of thin (sub-millimeter) annular rotor-stator gaps such as occur across for the leakage flow e.g. in the housing gap of oil-injected rotary positive displacement compressors (RPDC). To provide unrestriced optical access to the annular gap and in turn eliminate secondary effects, a simplified displacement compressor model has been developed and fabricated from perspex. The proof-of-concept results of both experimental campaigns (DPTV & IPI) are discussed and avenues for future efforts towards a straight-forward and accurate applicability of either method are elaborated.

1. Introduction

The efficiency of oil-injected rotary positive displacement compressors (RPDC) is dominated by surge- and gap-flow phenomena between adjacent chambers. To improve both prediction and optimization strategies towards further RPDC advances, an in-depth knowledge of the involved flow patterns across e.g. the annular rotor-stator gap (rotor-housing gap) is of utmost importance. However, experimental data for RPDCs rely on integral measures as basis for modeling efforts and are at most supported by qualitative visualizations of the respective flow regimes. Valuable reports on such combined qualitative visualizations and integral torque measures have been provided e.g. by Xin *et al.* [1] and Vasuthevan & Brümmer [2], which as yet generally remain to be complemented with spatio-temporally resolved void-fraction and velocity information.

Typical flow-measurement techniques encounter limitations when applied to wide aspect-ratio gap flows, where a single microscopic characteristic length scale, i.e. the annular gap, is accompanied by macroscopic characteristic length scales along the lateral surface. The curved boundaries of the annular rotor-stator gap and limited optical access further render the given flow scenario particularly challenging for a quantitative analysis of the involved multiphase-flow phenomena as occur across such narrow gaps. Recently, a similarly challenging flow scenario inside the 540 µm wide rotor-stator gap of an open wet clutch has been successfully analyzed experimentally by Leister *et al.* [3] with phase-resolved defocusing particle tracking velocimetry (DPTV) [4, 5], where the recorded volumetric three-component velocity information uncovered vortical structures and related implications on the adverse drag-torque generation of the considered clutch.

As for RPDCs, additional complexity comes into play due to the two-phase character of bubble growth, bubble-bubble and bubble-wall interactions, all of which requires additional complementary measurement techniques to further account for the identification of the gas phase (onset) in the gap. For small gas particles, where shadowgraphy-based sizing of particles is limited (if possible at all), interferometric particle imaging (IPI) has proven applicability for image-based particle-size analysis [6], which is well established for spray characterization [7], but was also applied for bubble-sizing studies [8, 9].

The present study addresses the aforementioned lack of quantitative void-fraction and velocity data in the annular gap of RPDCs by means of a proof-of-concept measurement campaign. First, the DPTV approach – as developed by Leister *et al.* [3] for thin rotor-stator configurations – is adapted and tested for the investigation of curved, i.e. annular, gap flows. In a second step, this DPTV-setup is chosen to furthermore provide a first indicator of the general applicability of IPI measurements for bubble (-growth) investigations in thin annular rotor-stator gaps of RPDCs.

2. Experimental Procedure

First the test facility and experimental setup are introduced in Section 2.1, before the two considered measurement techniques, i.e. Defocusing Particle Tracking Velocimetry (DPTV) and Interferometric Particle Imaging (IPI) are briefly recapped in Section 2.2 and 2.3, respectively.

2.1. Experimental Setup

In order to provide a straight-forward applicable test scenario for the present study, a simplified RPDC model tabletop test facility has been built for the experiments, which is comprised of a rotating perspex cylinder (diameter 39 mm), a cylindrical perspex container (inner diameter 60 mm, wall thickness 8 mm) and a *Bosch PBD 40* tabletop drill press to operate the rotor at an adjustable angular velocity of $\Omega = 21\,\mathrm{s}^{-1}$. To avoid any tumbling issues during operation, the rotor was first assembled in the drill chuck, and then in-situ evened with metal-lathe tool bits and subsequently polished while spinning. The container was eccentrically adjusted on the base plate of the drill to provide a $h = 200\,\mu\mathrm{m}$ gap height at the smallest cross-section for the measurements. The experimental setup for the conducted DPTV and IPI measurements is shown in Figure 1, which further indicates the angular arrangement of the optical components.

A *PCO Pixelfly* camera (14 bit, maximum resolution 1392 × 1040 pixels) equipped with a *Nikon Nikkor* lens captured a 8.2 × 6.1 mm^2 field of view (FOV) of the smallest cross-section of the annular gap, where the optical axis was aligned with the radial coordinate of both rotor and stator cylinders. The FOV was illuminated with the raw beam of a *Quantel EverGreen* Nd:YAG ($\lambda = 532\,\mathrm{nm}$, 200 mJ) dual-cavity laser at an angle range of $\theta = 92° - 100°$ to the camera axis to ensure reasonable fringe intensities for the IPI results [8]. The (single-phase) DPTV experiments were conducted with a white mineral oil (density $\varrho_o = 850\,\mathrm{kg/m^3}$, dyn. viscosity $\mu_o = 0.0136\,\mathrm{kg/ms}$ at $40°C$, CAS No: 8042-47-5), which was seeded with fluorescent particles of mean diameter $d_\mathrm{p} = 9.84\,\mu\mathrm{m}$ (density $\varrho_p = 1510\,\mathrm{kg/m^3}$, particle response time $\tau_p = 0.6\,\mu\mathrm{s}$, emission wave length $\lambda_e = 584$ nm) to avoid reflections. To promote the formation of bubbles in the gap during the IPI tests, the oil was replaced with H_2CO_3-supplied water (density $\varrho_w = 993\,\mathrm{kg/m^3}$, dyn. viscosity $\mu_w = 6.5 \times 10^{-4}\,\mathrm{kg/ms}$ at $40°C$, refractive index $m = 1.33$) between the two experimental campaigns. The relevant experimental parameters are summarized in Table 1.

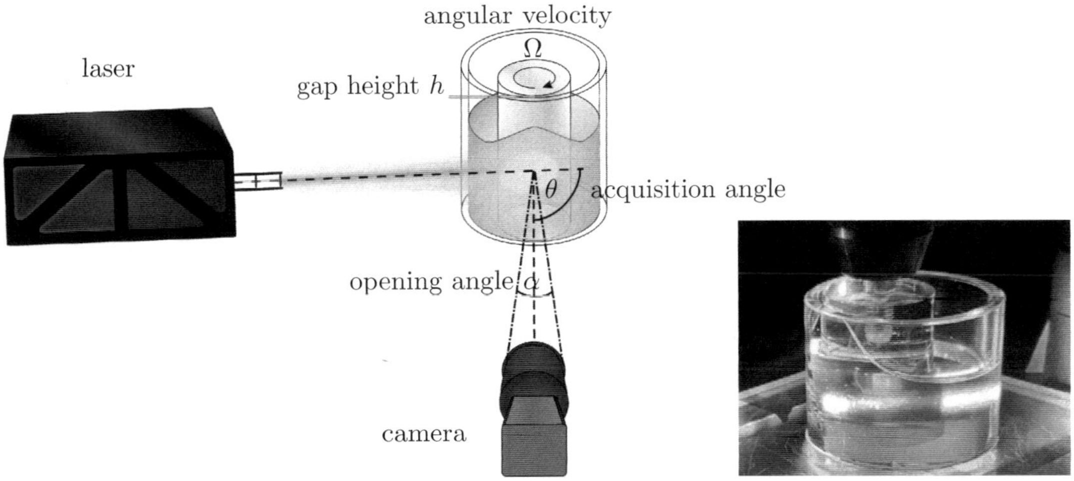

(a) Experimental scheme (b) close-up picture

Figure 1. Experimental setup for DPTV and IPI measurements; (a) Sketch of test setup, utilized equipment and relevant parameters; (b) picture during DPTV experimentation.

Table 1. List of experimental parameters and respectively chosen liquids for the DPTV and IPI measurements.

Exp.	Quantity	Value
DPTV & IPI	angular rotor velocity	200 rpm $\widehat{=}$ 21 s^{-1}
	inner gap radius R_1 (rotor)	19.5 mm
	outer gap radius R_2 (stator)	30 mm
	laser-pulse delay Δt	300 µs
	image magnification M	1.1
	Field of View	8.2×6.1 mm^2
	expected displacement ΔX at rotor	20.7 px
DPTV	particle image diameter at $z = 0$ (stator)	43.7 px
	particle image diameter at $z = h$ (rotor)	56.2 px
	radial gap height h	200 µm
	Nikon Nikkor lens	105 mm f/2.8
	$f_{\#}$-number	2.8
	aperture diameter D_a	37.5 mm
	white mineral oil: dyn. viscosity	$\mu_o = 0.0136$ kg/ms (at 40°C)
	density	$\varrho_o = 850$ kg/m^3
	CAS No.	8042-47-5
IPI	fringe-pattern diameter	≈ 42 px
	radial gap height h	400 µm
	Nikon Nikkor lens	200 mm f/4
	H_2CO_3-supplied water: dyn. viscosity	$\mu_w = 6.5 \times 10^{-4}$ kg/ms (at 40°C)
	density	$\varrho_w = 993$ kg/m^3
	refractive index	$m = 1.33$

2.2. DPTV Summary

DPTV is a single-camera measurement technique for volumetric velocity measurements, which uses purposely defocused particle images to gain additional depth-localization information out of one image pair acquisition [10]. First introduced by Willert & Gharib [11] in the approach gained popularity as macroscopic, but even more as microscopic measurement technique.

A theoretical mathematical description of the diameter estimation d_i of a particle image under absence of optical aberrations is a sum of three terms, that contribute significantly to the diameter change. This relationship has been summarized by Olsen & Adrian [12] and can be written as

$$d_{\mathrm{i}}{}^2 = \underbrace{M^2\, d_{\mathrm{p}}{}^2}_{\text{geometric image}} \underbrace{+ 5.95\,(M+1)^2\,\lambda^2 f_{\#}^2}_{\text{diffraction}} + \underbrace{\frac{M^2 z^{*2} D_{\mathrm{a}}{}^2}{(s_0 + z^*)^2}}_{\text{defocusing}}, \tag{1}$$

where M appears as magnification, d_{p} as physical particle diameter, λ as wave length of the emitted light and $f_{\#}$ presents the focal number of the lens. The third term is modeling the out-of-focus behavior and describes the diameter change of the particle on the image plane as function of the distance z^* between the particle and the focal plane, the distance s_0 between the lens and the focal plane, and the physical diameter of the used aperture D_{a}.

For practical defocusing measurements this equation can be simplified further, since $s_0 \gg z^*$ applies for any optical set-up of typical imaging applications, which reduces Equation (1) to the approximation

$$d_{\mathrm{i}}(z^*) \propto \left(\text{const.} + z^{*2}\right)^{1/2}. \tag{2}$$

This hyperbola can be approximated as linear interdependence when considering out-of-focus elements that have a sufficient distance to the focal plane. Equation (2) can thus be simplified to

$$d_{\mathrm{i}}(z^*) \propto z^*. \tag{3}$$

Consequently, a linear relationship between the imaged diameter and the z-axis position holds true. The actual proportionality coefficient between these two physical values can be estimated with the theoretical equation (1), but this formula is derived under the absence of optical aberrations. Accordingly, the direct use would lead to relatively high deviations, since a large array of imaging errors are present for every real camera set-up (like field curvature or spherical aberrations). To overcome this problem and link the diameter to a value in physical space the in-situ calibration approach of Fuchs *et al.* [4] is used, where the diameters are linked locally to the expected boundary conditions of the flow under consideration. Note that this approach in combination with the Hough-Transform [13] only compensates errors due to set-up misalignment or lens-inherent imperfections. Systematically induced astigmatic aberrations by e.g. the curvature of the test rig, however, requires additional more advanced correction methods, as outlined below.

2.3. IPI Summary

The idea and working principle of IPI goes back to the initial reports by e.g. König *et al.* [14] and Ragucci *et al.* [15], and takes advantage of the interference of two scattering orders if small particles, i.e. droplets in gas or bubbles in liquids, are illuminated with a single beam of coherent polarized light [6].

If imaged in the focal plane at a carefully chosen acquisition angle θ between illumination-beam and observation directions (cp. Figure 1(a)), the reflection glare point and the first refraction glare point (after crossing the particle) of the observed particle result in the focused image of two separated glare points of similar light intensities. Since the spatial distance of these two points is in the range of the particle diameter, the immediate quantification of the distance

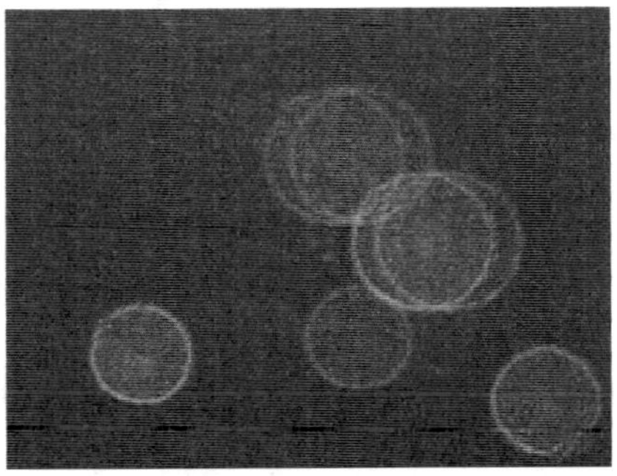

(a) color-coded raw image

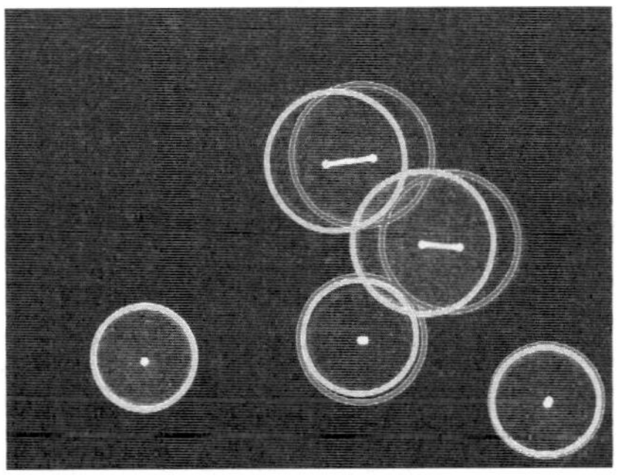

(b) processed image

Figure 2. DPTV results; (a) recorded elliptical particle image, (b) processed image with detected particles and calculated displacements ΔX (yellow).

suffers heavily from the limited spatial resolution of the imaging equipment at hand and the competing goal to observe sufficiently large wide aspect-ratio FOV at accordingly lower spatial resolution.

To overcome these conflicting requirements, the particles – and thus the corresponding pairs of two glare points – are recorded with sufficient displacement from the focal plane of the imaging device. If the distance to the focal plane is large in comparison to the glare-point spacing, both glare points collapse to a single quasi-circular defocused particle, which is further comprised of an interference fringe pattern. As already outlined in Section 2.2, the diameter of the resulting fringe patterns scales with the distance of the focal plane. The number of fringes N inside each defocused particle image, moreover, scales with the diameter of the respective particle, which can be determined for bubbles diameters d_b according to

$$d_b = \frac{2\lambda N}{\alpha}\left[m\cos\left(\frac{\theta}{2}\right) - \frac{m\sin\left(\frac{\theta}{2}\right)}{\sqrt{m^2 - 2m\cos\left(\frac{\theta}{2}\right) + 1}}\right]^{-1}. \tag{4}$$

In addition to the above-introduced quantities, α appears in Equation (4), which corresponds to the collecting angle of the chosen camera lens as indicated in Figure 1(a). It is furthermore important to note that Equation (4) is only valid for the presently considered bubbles in liquid, since internal refraction inside droplets and bubbles leads to different beam paths as elaborated thoroughly by Kawaguchi *et al.* [9].

3. Results

The circles of either experimental approach are detected by means of the Hough-Transform [13], which provides information on both diameter and center of the recorded circles. This procedure has been shown to have sub-pixel accuracy, such that no issues due to pixel locking occur due to the chosen processing strategy (cp. Leister & Kriegseis [16]).

Figure 2(a) shows a random double-frame image of the DPTV measurements, where both raw images are superimposed and color coded to indicate the particle displacement between the first (red) and second (green) frame (flow from right to left, cp. Figure 1(a)). The resulting circle estimations from the Hough-Transform are added to the double frame in Figure 2(b). Furthermore, the center locations of each circle are indicated with a yellow dot. Also, the

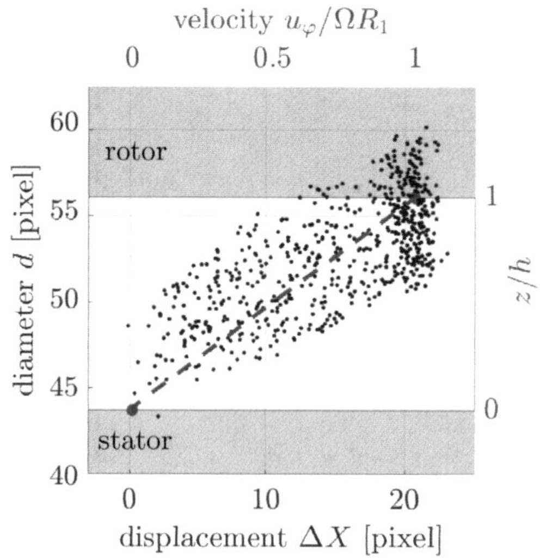

Figure 3. Resulting displacement distribution of the DPTV measurements and corresponding velocity profile across the gap; • raw data points, — — resulting velocity profile $u_\varphi(r)$, ■ transparent walls.

tracks of the particle displacements are plotted as yellow lines connecting the center locations. The images of Figure 2 already outline that the depth codification of the particle images in combination with the detected particle displacements reveal larger particle velocities in proximity of the rotor (large circles), whereas the near-constant locations of the smaller circle-pairs indicate the vicinity to the outer stator wall.

The results from the combined depth-codification and velocity estimation are shown in Figure 3. The scatter margin ($\approx 4\,\text{px}$) of the data points indicates the limited accuracy of the circle-detection based evaluation approach. Recall, that the curved stator wall introduces astigmatism to the optical system, which in turn leads to ellipse-shaped defocused particle images. Note that the high number of registered events for $u_\varphi/\Omega R_1 = 1$ furthermore stem from the number of particle, which were sticking to the rotor wall during experimentation. The vertically stretched shape of denser particle cloud around $u_\varphi/\Omega R_1 = 1$ further confirms the depth-codification estimation uncertainty to be the main contributor to the overall measurement uncertainty, since the small horizontal margin gives rise to an accurate velocity estimation of the sticking particles. The resulting fitted velocity profile, however, resembles a laminar (Poiseuille-) Couette flow, which is in accordance with Reynolds-equation considerations of the given flow scenario [17]. The pressure-driven Poiseuille flow component is to be classified as low in this case.

The feasibility tests of the IPI method for an applicability of bubble sizing and eventually tracking in thin curved annular rotor-stator gaps has been done with the identical DPTV setup, where the tracer-laden oil has been replaced by H_2CO_3-supplied water to promote the generation of bubbles in the shear flow. Some example IPI recordings are shown in Figure 4. Similar to the DPTV processing, the circular fringe pattern can be identified via the Hough-Transform [13]. Once identified, the fringe number can be quantified by means of gray-value evaluation across the diameter of the respective particle images, where a sine fit of the gray-value distributions also enables the quantification of fringe portions. The determined fringe numbers are provided above the respective fringe patterns of Figure 4, whereas the resulting bubble diameter estimations according to Equation (2) are provided below the respective patterns. Note that the most right fringe patterns belongs to the converging part of the annular gap at the far right of the observed FOV, where the gap height is significantly larger than the nominally investigated $h = 400\,\mu m$ gap height. Since the optical system and corresponding astigmatic effect on the optical transfer functions also influences the diameter determination process, the provided numbers below the respective fringe patterns, as yet, only serve as an order of magnitude estimation of the bubble-sizing effort.

4. Concluding Remarks

The presented proof-of-concept study reveals various insights regarding future efforts in the field of laser-optical flow diagnostics in RPDCs. Most importantly, the two tested DPTV and IPI appear promising experimental approaches for the investigation of the given flow scenario - even though it becomes obvious from today's perspective that either measurement technique needs to be substantially revised, improved and/or adapted for the given test scenario of wide aspect-ratio rotor-stator shear flow across curved annular sub-millimeter gaps.

The scatter margin of the Hough-Transform diameter-estimation indicates the impact of the astigmatic influence of the curved stator wall resulting particle images. However, even the present level of processing accuracy saliently demonstrates the applicability of DPTV for the investigation of velocity fields across thin annular rotor-stator gaps. Future efforts foresee particular extension of the standard Hough-Transform to the particular detection of ellipses, where the combination of the semi-major and semi-minor axes will be considered simultaneously. In addition, these efforts will also take complementary advantage of the recently developed data-processing knowledge of the astigmatism particle tracking velocimetry (APTV) [18, 19], which interprets the particle-image shape rather than its diameter for depth codification.

The presented IPI tests demonstrated the general persistence of the fringe pattern across the curved transparent wall. However, similar to the calibration of any advanced particle-image detection effort, also the IPI post processing requires a careful and in-depth revision and reconsideration of all involved processing steps before a quantitative evaluation of the recorded fringes becomes possible. For instance, the inclined illumination beam refracts towards the radial coordinate into the stator wall (and the fluid). As such, the bubbles do not encounter parallel light, which in turn requires further elaboration of the underlying geometrical optics for advance data processing beyond the present level of standard treatment as outlined in Equation (4). This influential factor is foreseen to be further affected by the laser-beam diameter and correspondingly varying refraction angle over the beam cross-section upon inclined impact onto

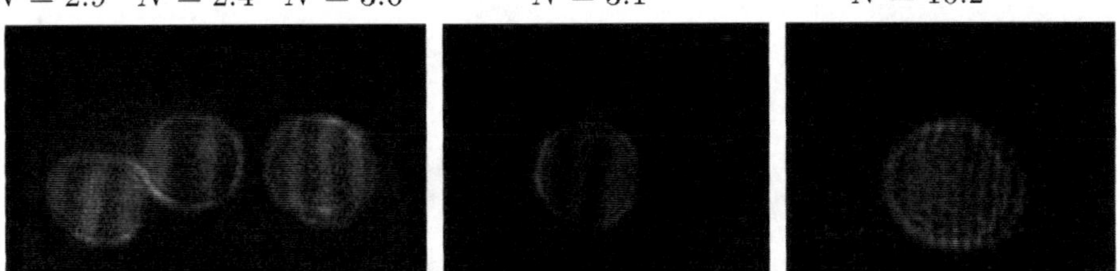

Figure 4. Raw data of IPI with corresponding fringe number N and approximated diameter d_b according to Equation (4).

a curve air-glass interface.

Once expanded towards a robust quantitative bubble-sizing approach for thin annular gaps, the IPI method seems also particularly candidating to provide void-fraction (-distribution) information, since location and diameter of the IPI-bubbles can be further converted into gas volumes in the known observed liquid volume. For statistically significant amounts of data such information is foreseen to lead to robust information of the void-fraction distribution.

As a final conclusion, despite the present premature quality of a feasibility study, the achieved results and uncovered insights of the investigation provide a clear outline of the IPI and DPTV to be an ideal combination to analyze complex wide aspect-ratio gap-flow scenarios. Future investigations will, therefore, address the outlined processing strategies so as to make the DPTV-IPI combination accurately applicable for both the presented annular gap flow (e.g. housing gap of RPDCs) but also for planar rotor-stator gaps (e.g. front gap of RPDCs).

References

[1] Xin D, Feng J, Jia X and Peng X 2010 An investigation into oil—gas two-phase leakage flow through micro gaps in oil-injected compressors *Proceedings of the Institution of Mechanical Engineers, Part C: Journal of Mechanical Engineering Science* **224** 925–933 URL `https://doi.org/10.1243/09544062JMES1704`

[2] Vasuthevan H and Brümmer A 2019 Generic experimental investigation of hydraulic losses within twin-screw machines *The 9th International Conference on Compressor and Refrigeration.* **Xi'an (China)** 10pp

[3] Leister R, Fuchs T, Mattern P and Kriegseis J 2021 Flow-structure identification in a radially grooved open wet clutch by means of defocusing particle tracking velocimetry *Experiments in Fluids* **62** 29 URL `https://doi.org/10.1007/s00348-020-03116-0`

[4] Fuchs T, Hain R and Kähler C J 2016 In situ calibrated defocusing ptv for wall-bounded measurement volumes *Measurement Science and Technology* **27** 084005 URL `https://doi.org/10.1088/0957-0233/27/8/084005`

[5] Fuchs T, Hain R and Kähler C J 2016 Uncertainty quantification of three-dimensional velocimetry techniques for small measurement depths *Experiments in Fluids* **57** 73 URL `https://doi.org/10.1007/s00348-016-2161-5`

[6] Tropea C 2011 Optical particle characterization in flows *Annual Review of Fluid Mechanics* **43** 399–426 URL `https://doi.org/10.1146/annurev-fluid-122109-160721`

[7] Fansler T D and Parrish S E 2014 Spray measurement technology: a review *Measurement Science and Technology* **26** 012002 URL `https://doi.org/10.1088/0957-0233/26/1/012002`

[8] Dehaeck S and van Beeck J 2008 Multifrequency interferometric particle imaging for gas bubble sizing *Experiments in Fluids* **45** 1432–1114 URL `https://doi.org/10.1007/s00348-008-0502-8`

[9] Kawaguchi T, Akasaka Y and Maeda M 2002 Size measurements of droplets and bubbles by advanced interferometric laser imaging technique *Measurement Science and Technology* **13** 308–316 URL `https://doi.org/10.1088/0957-0233/13/3/312`

[10] Raffel M, Willert C E, Scarano F, Kähler C J, Wereley S T and Kompenhans J 2018 *Particle Image Velocimetry* (Springer-Verlag GmbH) URL `https://doi.org/10.1007/978-3-319-68852-7`

[11] Willert C E and Gharib M 1992 Three-dimensional particle imaging with a single camera *Experiments in Fluids* **12** 353–358 URL `https://doi.org/10.1007/BF00193880`

[12] Olsen M G and Adrian R J 2000 Out-of-focus effects on particle image visibility and correlation in microscopic particle image velocimetry *Experiments in Fluids* **29** S166–S174 URL `https://doi.org/10.1007/s003480070018`

[13] Hough P V C 1959 Machine Analysis of Bubble Chamber Pictures *Proceedings, 2nd International Conference on High-Energy Accelerators and Instrumentation* **C590914** 554–558 URL `https://s3.cern.ch/inspire-prod-files-5/53d80b0393096ba4afe34f5b65152090`

[14] König G, Anders K and Frohn A 1986 A new light-scattering technique to measure the diameter of periodically generated moving droplets *Journal of Aerosol Science* **17** 157–167 URL `https://www.sciencedirect.com/science/article/pii/0021850286900637`

[15] Ragucci R, Cavaliere A and Massoli P 1990 Drop sizing by laser light scattring exploiting intensity angular oscillation in the mie regime *Particle & Particle Systems Characterization* **7** 221–225

[16] Leister R and Kriegseis J 2019 *Proceedings of the 13th International Symposium on Particle Image Velocimetry* ed Kähler C J, Hain R, Scharnowski S and Fuchs T URL `https://athene-forschung.unibw.de/doc/129141/129141.pdf`

[17] Spurk J H and Aksel N 2020 *Fluid Mechanics*, 8: Hydrodynamic Lubrication (Springer International Publishing) pp 251–283 URL `https://doi.org/10.1007/978-3-030-30259-7_8`

[18] Cierpka C, Rossi M, Segura R and Kähler C J 2011 On the calibration of astigmatism particle tracking velocimetry for microflows *Measurement Science and Technology* **22** 015401 URL http://stacks.iop.org/0957-0233/22/i=1/a=015401
[19] Rossi M and Kähler C J 2014 Optimization of astigmatic particle tracking velocimeters *Experiments in Fluids* **55** URL https://doi.org/10.1007/s00348-014-1809-2

One-dimensional calculation approach for gaseous clearance flows

T Jünemann[1] and A Brümmer[2]

[1]Flowserve Dortmund GmbH & Co KG, Dortmund, Germany

[2]Chair of Fluidics, Faculty of Mechanical Engineering, TU Dortmund University, Germany

E-mail: [1]`tjunemann@flowserve.com`, [2]`andreas.bruemmer@tu-dortmund.de`

Abstract. Clearance flows do strongly influence the efficiency of dry-running positive displacement machines. To model the operation behaviour of the machines an accurate prediction of the clearance mass flows is crucial. If a chamber model simulation is used, the clearance mass flows are commonly estimate with simple equation to keep the computational effort low. Therefore, regression functions or coefficient databases are built into simulation. However, finding such functions or coefficients is a challenging and expensive task since many boundary conditions and geometry parameters must be varied in experiment or simulation. To reduce the effort of clearance analysis this paper presents a one-dimensional approach to estimate the clearance mass flow in e.g. dry-running screw-type compressors, based on the well-known differential equation of A.H. Shapiro for compressible flow of ideal gases. The friction is modelled with respect to laminar and turbulent flow. The flow separation and the friction of laminar flow are modelled based on the solutions of the Jeffery-Hamel flow. The friction in the turbulent regime is modelled by the Blasius expression, while the flow separation criteria is assumed to be the same as for laminar flow. The moving clearance wall is modelled assuming a superposition of Couette and Poiseuille flow. In addition, the heat transfer between gas and clearance boundary as well as gas rarefaction can also be optionally modelled. Results of the one-dimensional approach are compared to measurements and 3D-CFD simulations from the literature, varying Reynolds number and wall velocity.

1. Introduction

For the simulation of positive displacement machines, accurate prediction of the clearance mass flows is crucial. Therefore, many investigations have focused on the determination of flow coefficients for different clearance types of screw machines.

Fister and Neumann [1] determined the flow coefficients of the housing clearance, intermesh clearance, front clearance, and blowhole using numerical steady-state compressible flow simulations. They varied the pressure ratio, gas temperature, and rotor speed for a given rotor geometry. Accordingly, the flow in the seal clearances can be considered laminar, while a Prandtl mixing turbulence model is used for the blowhole. Especially at high pressure ratios, an influence of the rotor rotation is found. A strong dependence on the Reynolds number is observed for the blowhole flow coefficient.

F.-J. Peveling [2] investigated in experiments a selection of general static clearance contours and varied the pressure ratio, the Reynolds number and the height. The Reynolds number and the pressure ratio are not varied separately, which complicates the interpretation. Increasing inlet pressure, and thus increasing Reynolds number, decreases the flow coefficient, although the opposite would be expected. Peveling's measurements show partial Saint-Venant and Wantzel flow coefficients greater than one. This can be explained by convergent-divergent clearance contours and the resulting improved flow.

The influence of rotor motion on the flow coefficient was studied by B. Dreißig [3]. In a generic experiment the mass flow rate in the clearance between a rotating cylinder and different concave counter contours was measured. With a dimensional analysis, Dreißig determined the dimensionless influence quantities for the flow coefficient. For an adiabatic case, these are the Reynolds number, the circumferential Mach number of the cylinder, the Mach number of the clearance flow, and the pressure ratio across the clearance. He then applied the results of the experiments to the static measurements of Peveling [2] and plotted them as a function of the angle of rotation for the housing and front clearance of a screw motor with SRM profile. A clear influence of rotor rotation can be seen in the results shown.

Z. Fan and Z. Chen [4] presented an approach for the calculation of the pressure distribution and its mass flow rate in a clearance of a compressor which is identical with the approach of the one-dimensional flow according to Shapiro [5]. However, this work seems to be unknown to the authors. They assumed an adiabatic flow and the friction is estimated laminar. As an example, the flow through a clearance between two eccentric cylinders is calculated and compared to experiments with good accuracy.

A quasi-one-dimensional calculation approach for adiabatic compressible flows in narrow clearances of large length is presented by Y. Huang [6]. He neglected the change of the state variables transverse to the flow direction as well as the change of the flow cross section in the mass conservation. The reduced partial differential equations for mass, momentum and energy are solved by an integral method. Information about the velocity profile and wall friction are directly part of the solution and do not have to be modelled as in the Shapiro approach [5]. For laminar, compressible flow this is an advantage. Huang compared the quasi-one-dimensional solution with a numerical two-dimensional solution of the Navier-Stokes equation, the deviation is between 15% for small clearance heights (0.1 mm) and 95% for very large clearance heights (2 mm).

Ishii et.al. [7] calculated the clearance mass flows in scroll compressors via an incompressible approach. The coefficient of friction is estimated according to Blasius. The density for the clearance flow is calculated via a polytropic relaxation at the inlet. For validation, an experiment is used which measures the time-dependent pressure drop in a container which relaxes against the atmosphere via a clearance contour. Simulation and measurements show good agreement. Finally, they compare the incompressible approach with a Fanno flow using the radial clearance as an example. The Fanno flow deviates clearly from the measurements since this does not represent the convergent-divergent course of the radial clearance contour. Ishii et.al. [8] extended the model by the influence of the surface roughness and investigated its influence on the efficiency of a scroll CO_2 compressor, where a higher surface roughness improves the volumetric efficiency and the overall efficiency of the compressor.

Müller [9] used an approach analogous to Ref. [5] and Ref. [4] for the calculation of the housing clearance of a screw vacuum pump. He neglected the cross-sectional variation and the moving boundary and instead considered the shape of the velocity profile in the conservation of momentum. He determined the friction and heat fluxes as a function of location in numerical flow simulations and used these as the data basis for the one-dimensional simulation. He developed his own models for transferring the incompressible data to compressible flow.

A detailed study of the clearance flows in a screw compressor is provided by M. Utri [10, 11, 12]. Starting from a dimensional analysis, he investigated the influence of various parameters on the flow coefficients of the housing and front clearance in experiment and numerical flow simulation. Based on the investigation, he developed correlation equations and implemented them into a chamber model simulation.

C. Huck [13, 14] developed a model for the clearance flows in vacuum pumps based on a one-dimensional Stokes-flow. He compared the model to measurements and DSMC simulations of a clearance between a rotating shaft and a plane counter plate.

T. Jünemann and A. Brümmer [15] combined the methods of Müller [9] and Huck [13] to develop a model that covers the entire range of gas rarefaction. They neglected the effects of undeveloped flow and flow shape and modelled Couette flow instead. The model works well for most gas rarefaction parameters, but needs to be improved for high Reynolds numbers.

This work advances the Shapiro-like calculation of clearance flow in screw compressors one step further. It follows the series of Ref. [5, 9, 15] and compares it's results with the screw clearance investigations of M. Utri [10, 11, 12]. The approach of Ref. [15] is extended by modelling flow profiles and possible flow separation in divergent clearance sections. Increased friction in laminar accelerated flow is modelled with respect to the solution of Jeffery-Hamel flow.

2. Theoretical Approach

The mass flow rate $\dot{m}$ of an ideal gas in a clearance of a screw machine is to be determined. The ideal gas is characterised by its gas constant R, its ratio of specific heats κ and its constant temperature-independent dynamic viscosity μ. The geometry of the clearance (see **Fig. 1**) is assumed to be a channel with varying rectangular cross section of constant width w and a height $h(x)$ varying with the spatial coordinate x. Typically, the height decreases until a minimum h_{min} is reached and then begins to increase. Here, the aspect ratio $\frac{h}{w} \ll 1$ is always assumed to be very small. The channel may have moving walls with velocity u_w, but the contour remains static. The channel connects an inlet (in) and an outlet (out) reservoir with different pressures. At the inlet, the temperature T_{in} is known and equal to the total temperature T_t of the entire flow, since the walls are assumed to be adiabatic and the work done by the moving walls should be negligible. To calculate the mass flow rate, the clearance is divided into three regions. The inlet and outlet region and the region of a one-dimensional flow between I and II.

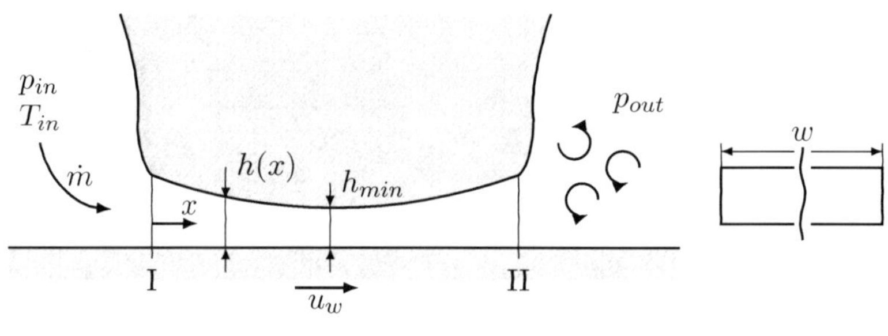

Figure 1. Sketch of clearance in rotatory displacement machine

The results will be given as the non-dimensional flow coefficient

$$\alpha = \frac{Re}{Re_{th}} \tag{1}$$

for varying Reynolds numbers

$$Re = \frac{2\dot{m}}{\mu w} \tag{2}$$

and for different circumferential Mach numbers of the wall

$$M_u = \frac{u_w}{(\kappa R T_{in})^{\frac{1}{2}}}, \tag{3}$$

whereby the theoretical Reynolds number Re_{th} is calculated according to Saint-Venant and Wantzel [16]

$$Re_{th} = Kn^{-1} \cdot (2\kappa\pi)^{\frac{1}{2}} \cdot \begin{cases} \left(\frac{2}{\kappa+1}\right)^{\frac{\kappa+1}{2(\kappa-1)}}, & \Pi \leq \left(\frac{2}{\kappa+1}\right)^{\frac{\kappa}{\kappa-1}} \\ \Pi^{\frac{1}{\kappa}} \left(\frac{2}{\kappa-1}\left(1 - \Pi^{\frac{\kappa-1}{\kappa}}\right)\right)^{\frac{1}{2}}, & \Pi > \left(\frac{2}{\kappa+1}\right)^{\frac{\kappa}{\kappa-1}} \end{cases} \tag{4}$$

depending on the pressure ratio

$$\Pi = \frac{p_{out}}{p_{in}} \tag{5}$$

and the Knudsen number

$$Kn = \left(\frac{\pi}{2} R T_{in}\right)^{\frac{1}{2}} \frac{\mu}{p_{in} \cdot h_{min}} \tag{6}$$

of the clearance.

2.1. Clearance Shape

To approximate the shape of the clearance contours, the approach of Utri [10, 11, 12] is used. According to this approach, each clearance contour can be described by two superellipses,

$$\left(\frac{x}{a}\right)^n + \left(\frac{y}{b}\right)^m = 1, \tag{7}$$

one on the inlet side and one on the outlet side, and a part with constant height h_{min} between them (see **Fig. 2**). Three different contours are discussed, an example of the male housing, the female housing, and the front clearance. For each clearance profile, the clearance height is varied while the other parameters are kept constant. The front clearance has a sharp wedge and therefore does not need to be described by superellipses, and the parameters for the other clearances are given in **Tab. 1**. The independent parameter for each contour is the ratio of the rotor pitch S to the crown circle diameter D.

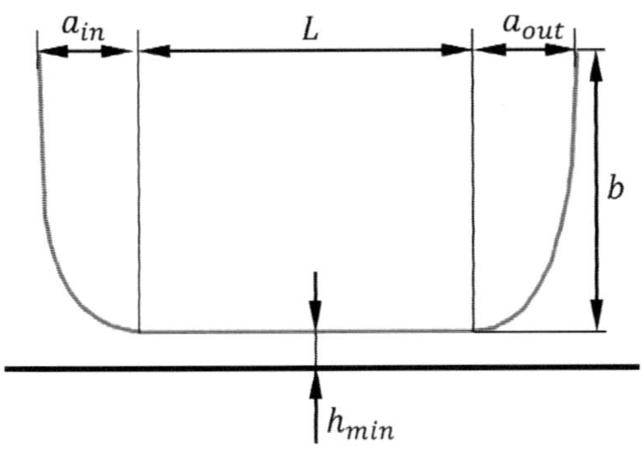

Figure 2. Example of a female rotor tooth contour with superellipses.

2.2. Modelling of the one-dimensional flow

To derive the equations for a one-dimensional compressible flow, the approach of Shapiro [5] is followed, but with the consideration of the averaged flow variables according to Ref. [9].

IOP Conf. Series: Materials Science and Engineering 1267 (2022) 012004 doi:10.1088/1757-899X/1267/1/012004

Table 1. Superellipse parameters for the discussed male and famle housing clearances [12].

	Male housing					Female housing					
	S/D	$a/D \cdot 72$	$b/D \cdot 72$	m	n	S/D_F	$a/D_F \cdot 72$	$b/D_F \cdot 72$	m	n	$L/D_F \cdot 72$
in	1.67	13.5	10	1.8	1	7.5	2.3	5	2.5	4.1	5.8
out		6.9		1	1		2.3		2.8	2.3	

As mentioned above, an ideal gas is assumed. In the first step, the conservation laws of fluid mechanics are applied to an incremental control volume within the clearance (see **Fig. 3**).

$$\oint \varrho \left(\boldsymbol{c} \cdot \boldsymbol{n}\right) \mathrm{d}\Omega = 0 \tag{8}$$

$$\oint \varrho \boldsymbol{c} \left(\boldsymbol{c} \cdot \boldsymbol{n}\right) \mathrm{d}\Omega + \oint p \boldsymbol{n} \mathrm{d}\Omega = \oint \boldsymbol{\tau} \cdot \boldsymbol{n} \mathrm{d}\Omega \tag{9}$$

$$\oint c_p \varrho T_t \left(\boldsymbol{c} \cdot \boldsymbol{n}\right) \mathrm{d}\Omega = \oint \boldsymbol{c} \cdot \left(\boldsymbol{\tau} \cdot \boldsymbol{n}\right) \mathrm{d}\Omega + \oint \dot{\boldsymbol{q}} \cdot \boldsymbol{n} \Omega \tag{10}$$

The density of the gas is represented by ϱ, the flow velocity by $\mathbf{c}$, the heat flux by $\dot{\boldsymbol{q}}$, the shear stress tensor by $\boldsymbol{\tau}$, and the specific isobaric heat capacity of the gas by c_p. In this integral form, Ω is the surface area of the control volume and $\boldsymbol{n}$ is the outward facing surface normal vector. All other flow variables have already been introduced.

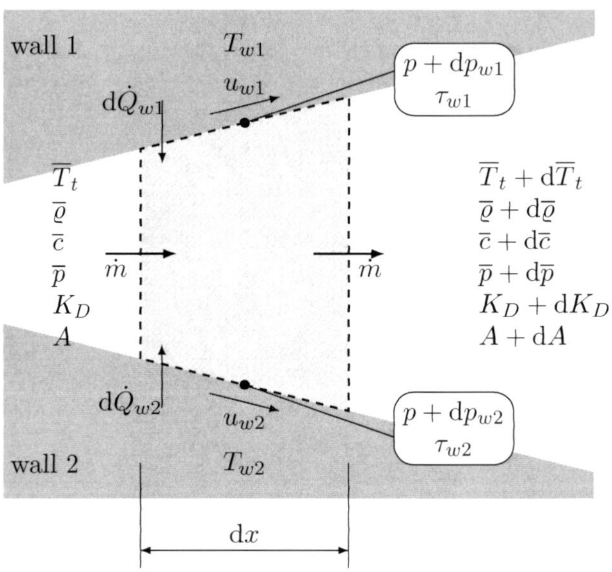

Figure 3. Incremental control volume in the clearance.

Since these equations are to be reduced to a one-dimensional form, the integrals must be replaced by mean values representing the flow in cross section A normal to the flow direction. The mean values are defined in the same way as in Ref. [9], starting with the mean surface pressure

$$\overline{p} = \frac{1}{A} \int_A p \mathrm{d}A. \tag{11}$$

IOP Conf. Series: Materials Science and Engineering 1267 (2022) 012004 doi:10.1088/1757-899X/1267/1/012004

The mean flow velocity is defined to satisfy the law of conservation of mass with the mean density $\overline{\varrho}$:

$$\overline{c} = \frac{1}{\overline{\varrho} A} \int_A \varrho c_x \mathrm{d}A. \tag{12}$$

Since this definition of mean flow velocity is not momentum conservative, a momentum shape factor is introduced:

$$K_D = \frac{1}{\overline{\varrho c}^2 A} \int_A \varrho c_x^2 \mathrm{d}A \tag{13}$$

The mean total temperature is intended to represent the convectively transported amount of energy of the flow and is therefore defined as follows:

$$\overline{T}_t = \frac{1}{\overline{\varrho c} A} \int_A \varrho c_x T_t \mathrm{d}A \tag{14}$$

In addition, the energy equation of gas dynamics should apply to the mean values, so that the mean static temperature is defined as follows

$$\overline{T} = \overline{T}_t \left(1 + \frac{\kappa - 1}{2} \overline{M}^2\right)^{-1}, \tag{15}$$

with the mean Mach number

$$\overline{M}^2 = \frac{\overline{c}^2}{\kappa R \overline{T}}. \tag{16}$$

Finally, the average density is defined with the help of the ideal gas equation

$$\overline{\varrho} = \frac{\overline{p}}{R\overline{T}}. \tag{17}$$

The local wall friction in the momentum equation is replaced by the Darcy friction factor f_x, which is a dimensionless friction force against the flow direction x. The full integral of the shear stress force is replaced by this definition, since the friction at the front and end faces of the control volume is usually negligible:

$$\begin{pmatrix} f_x \\ f_y \end{pmatrix} = -\frac{2}{\overline{\varrho c}^2} \frac{D_h}{\mathrm{d}x} \frac{1}{A} \oint \boldsymbol{\tau} \cdot \boldsymbol{n} \mathrm{d}\Omega \tag{18}$$

The local hydraulic diameter D_h is given as the ratio of four times the flow cross-sectional area A to the perimeter of the control volume. It can be calculated from the local height h and the channel width w

$$D_h = \frac{2h}{1 + \frac{h}{w}}. \tag{19}$$

In addition to wall friction, an additional loss mechanism can be identified, namely momentum loss due to a deformation of the velocity profile or, in other words, due to the change of the momentum shape factor. This additional dissipation is modeled by adding it to the wall friction, which leads to the definition of the apparent friction factor:

$$f_{x,app} = f_x + 2D_h \frac{\mathrm{d}K_D}{\mathrm{d}x} \tag{20}$$

Finally, the integrals of work and heat transfer in the law of conservation of energy are replaced by the specific work

$$w = \frac{1}{\dot{m}} \oint \boldsymbol{c} \cdot (\boldsymbol{\tau} \cdot \boldsymbol{n}) \, \mathrm{d}\Omega \tag{21}$$

and the specific wall heat flow

$$q = \frac{1}{\dot{m}} \oint \dot{\boldsymbol{q}} \cdot \boldsymbol{n} \mathrm{d}\Omega. \tag{22}$$

Applying all the definitions presented to the integral conservation laws of the incremental control volume leads to the three differential equations for compressible channel flows:

$$\frac{\mathrm{d}\overline{c}}{\overline{c}} + \frac{\mathrm{d}\overline{\varrho}}{\overline{\varrho}} + \frac{\mathrm{d}A}{A} = 0 \tag{23}$$

$$K_D \overline{\varrho}\overline{c}\mathrm{d}\overline{c} + \mathrm{d}\overline{p} = -f_{x,app} \frac{\overline{\varrho c}^2}{2} \frac{\mathrm{d}x}{D_h} \tag{24}$$

$$c_p \overline{T}_t = q + w \tag{25}$$

Using these equations and additionally the following differential forms of the ideal gas law, the definition of the Mach number of an ideal gas, the energy equation of compressible flow,

$$\frac{\mathrm{d}\overline{p}}{\overline{p}} = \frac{\mathrm{d}\overline{\varrho}}{\overline{\varrho}} + \frac{\mathrm{d}\overline{T}}{\overline{T}} \tag{26}$$

$$\frac{\mathrm{d}\overline{M}^2}{\overline{M}} = \frac{\mathrm{d}\overline{c}^2}{\overline{c}^2} - \frac{\mathrm{d}\overline{T}}{\overline{T}} \tag{27}$$

$$\frac{\mathrm{d}\overline{T}_t}{\overline{T}_t} = \frac{\mathrm{d}\overline{T}}{\overline{T}} + \frac{\frac{\kappa-1}{2}\overline{M}^2}{1+\frac{\kappa-1}{2}\overline{M}^2} \frac{\mathrm{d}\overline{M}^2}{\overline{M}^2} \tag{28}$$

and following the approach of Ref. [5] leads to a differential equation for the Mach number in the form

$$\frac{\mathrm{d}\overline{M}^2}{\overline{M}^2} = \frac{1+\frac{\kappa-1}{2}\overline{M}^2}{1-\frac{\overline{M}^2}{\overline{M}^2_{crit}}} G \tag{29}$$

with G defined as

$$G = \kappa \overline{M}^2 f_{x,app} \frac{\mathrm{d}x}{D_h} + \left(1 + K_D \kappa \overline{M}^2\right) \frac{\mathrm{d}\overline{T}_t}{\overline{T}_t} - 2\frac{\mathrm{d}A}{A} \tag{30}$$

and the critical Mach number defined as

$$\overline{M}^2_{crit} = \left(1 + (K_D - 1)\,\kappa\right)^{-1}. \tag{31}$$

2.2.1. Laminar flow (Re < 3000) - The Jeffery-Hamel solution The wall friction and the flow profile of a laminar flow are strongly influenced by whether it is accelerated or decelerated. Therefore, a model must be developed that takes this into account. The acceleration or deceleration of the flow can be due to either a divergent or convergent flow cross section or to the density change in a compressible flow. A good analogy is the Jeffery-Hamel flow, which is the solution of a fully formed laminar incompressible diffuser or nozzle flow. Moreover, the solution simultaneously provides a criterion for flow separation in the diffuser. In the following, a model for the friction coefficient and momentum shape factor is developed based on the JH solution.

As shown in **Fig. 4**, an incompressible flow in polar coordinates (r, φ) is considered, generated by a source or sink at the origin and solid walls as boundaries. Furthermore, the flow is assumed to be purely radial and fully developed, with no slip at the wall $(c\,(\varphi = \beta) = 0)$. Since the flow is symmetric, it is assumed to have a local maximum at the center $(c\,(\varphi = 0) = c_{max})$.

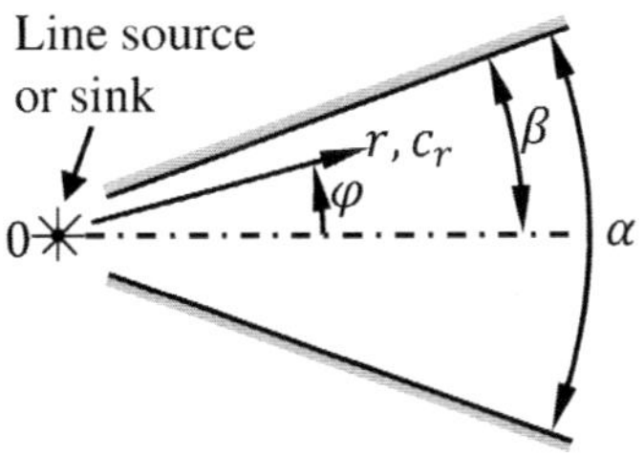

Figure 4. Geometry of the Jeffery-Hamel flow

Now the following dimensionless quantities are introduced

$$\eta = \frac{\varphi}{\beta}, t(\eta) = \frac{c_r}{c_{max}}, Re_{JH} = \frac{c_{max}\varrho r\beta}{\mu}, C = \left.\frac{\mathrm{d}t}{\mathrm{d}\eta}\right|^2_{\eta=1} \tag{32}$$

and, taking into account the conservation of mass and momentum, the governing equation for the flow profile is obtained

$$\eta(t, Re_{Jh}, \beta, C) = \int_t^1 \left((1-t)\left(\frac{3}{2}\beta Re_{JH}\left(t^2+t\right) + 4\beta^2 t + C\right)\right)^{-\frac{1}{2}} \mathrm{d}t, \tag{33}$$

where the boundary condition

$$\eta(t=0, Re_{Jh}, \beta, C) = 1 \tag{34}$$

specifies the constant C. [17]

The definition of Reynolds number for Jeffery-Hamel flow differs from the definition of this work (Eq. (2)), because the mass flow rate (or mean flow velocity) is not known until the problem is solved. To determine the mean flow velocity, the equation $t(\eta)$ must be inverted to a function of the form $t(\eta)$. The mean flow velocity is then given by

$$\frac{\overline{c}}{c_{max}} = \int_0^1 t(\eta, \beta, Re_{JH})\mathrm{d}\eta, \tag{35}$$

with which the Reynolds number of the JH flow can be converted into the Reynolds number introduced at the beginning for a channel flow:

$$Re = 4\frac{\overline{c}\varrho r\beta}{\mu} = 4\frac{\overline{c}}{c_{max}} \cdot Re_{JH} \tag{36}$$

If the velocity profile is known, the momentum form factor can also be calculated:

$$K_D = \left(\frac{c_{max}}{\overline{c}}\right)^2 \int_0^1 t^2(\eta, \beta, Re_{JH})\mathrm{d}\eta \tag{37}$$

The next step is to calculate the Poiseuille number from the JH solution,

$$Po = f \cdot Re = 8\frac{\tau_w}{\varrho\overline{c}^2}Re \tag{38}$$

and thus calculate the wall shear stress from the velocity gradient on the wall

$$\tau_w = \mu\frac{u_{max}}{r\beta}\left.\frac{\mathrm{d}t}{\mathrm{d}\eta}\right|_{\eta=1}, \tag{39}$$

leading to the relationship as a function of the constant C

$$Po = 32 \cdot \frac{\bar{c}}{c_{max}} C^{\frac{1}{2}}. \tag{40}$$

The Jeffery-Hamel flow is solved for the case $\beta \ll \beta Re_{JH}$, leading to a solution independent of the opening angle $\alpha = 2\beta$, shown in **Fig. 5**. Shown are the results of the momentum from the factor K_D and the Poiseuille number Po. Negative opening angles α represent the accelerated flow in a nozzle and positive opening angles represent the decelerated flow in the diffuser. For a plane flow ($\alpha Re = 0$), the results are identical to the Poiseuille flow in a plane channel. The Poiseuille number Po decreases with increasing opening angle and becomes zero at $\alpha Re = 37.66$, the point of flow separation. The momentum shape factor K_D decreases with increasing flow acceleration, but only slowly approaches a value of one.

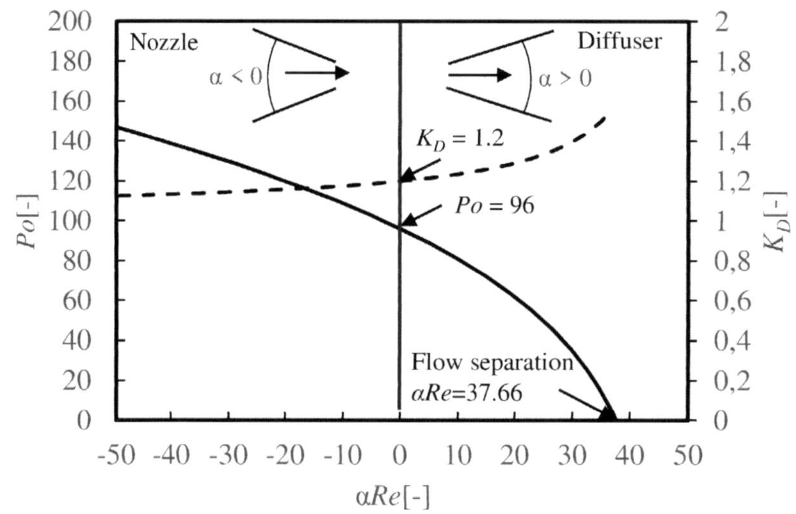

Figure 5. Results of Jeffery-Hamel flow for $\alpha Re \gg \alpha$

2.2.2. Laminar Couette-Poiseuille flow For a fully developed laminar flow in a plane channel, the friction and momentum shape factors of a Couette-Poiseuille flow can be determined directly from a superposition of the velocity fields of the partial solutions. An exact solution for Couette-Poiseuille flow in a divergent or convergent cross section analogous to the Jeffery-Hamel solution is not known and will not be derived in this article. Instead, an approximation for flows with moving boundary based on the Jeffery-Hamel solution is given. To determine the coefficient of friction, it is assumed to be analogous to the solution for plane channels given by the following equation:

$$f_x = \frac{Po}{Re}\left(1 - \frac{u_{w,x1} + u_{w,x2}}{2\bar{c}}\right) \tag{41}$$

For a pure Couette flow, the integral of the wall shear stresses is zero and so is the coefficient of friction. To determine the coefficient of friction of the Couette-Poiseuille flow, the friction is reduced by the fraction of the Couette flow caused by the moving wall ($u_{w1,x}, u_{w2,x}$). For the Poiseuille number Po, the Jeffery-Hamel solution as a function of the opening angle is still used. For the momentum shape factor, it is assumed to depend on the aperture angle α in the same way as for a pure Poiseuille flow. In the plane channel, it can be calculated as a function of the mean flow velocity and the wall velocity as follows:

$$K_{D,CP}(\alpha = 0) = 1.2 + 0.2\left(\frac{u_{w1,x}\cdot u_{w2,x}}{\bar{c}^2} - \frac{u_{w1,x}+u_{w2,x}}{\bar{c}}\right) + \frac{2}{15}\left(\frac{u_{w1,x}-u_{w2,x}}{\bar{c}}\right)^2 \tag{42}$$

The dependence on the opening angle is realised by relating the Jeffery-Hamel solution to its original value and then multiplying it by the Couette-Poiseuille value of the plane channel:

$$K_D = \frac{K_{D,JH}(\alpha Re)}{K_{D,JH}(\alpha Re = 0)} K_{D,CP}(\alpha = 0) \tag{43}$$

2.2.3. Laminar compressible flow In a compressible flow, a change in flow velocity can also be caused by a change in density. According to the literature, the friction in a compressible flow depends on the Mach number [18] or on the flow acceleration [9]. However, both papers only treat friction in the plane channel with generally decreasing density. Moreover, no model is presented to adjust the momentum shape factor. At this point, a correction analogous to Jeffery-Hamel flows for friction in laminar compressible flows is presented. For this purpose, the opening angle of the Jeffery-Hamel flow is converted into a flow acceleration by assuming a narrow channel with a small opening angle, so that this corresponds to the gradient of the height.

$$\alpha \approx \frac{\mathrm{d}h}{\mathrm{d}x} \tag{44}$$

If the incompressible conservation of mass is considered

$$\frac{\mathrm{d}\bar{c}}{\bar{c}} + \frac{\mathrm{d}h}{h} = 0, \tag{45}$$

the relationship between flow acceleration and opening angle can be obtained

$$\alpha = \frac{h}{\bar{c}}\frac{\mathrm{d}\bar{c}}{\mathrm{d}x}. \tag{46}$$

For the application of the Jeffery-Hamel solution for compressible flows, Eq. (27) and Eq. (28) are used to calculate the local flow acceleration from the Mach number, the local Mach number gradient, and the change in total temperature

$$\frac{\mathrm{d}\bar{c}}{\bar{c}} = \frac{1}{2}\left(\frac{\mathrm{d}M^2}{M^2}\left(\frac{1}{1+\frac{\kappa-1}{2}M^2}\right) + \frac{\mathrm{d}\overline{T}_t}{\overline{T}_t}\right). \tag{47}$$

This flow acceleration can be used to determine an equivalent opening angle according to Eq. (46) to access the data from Fig. 5.

At this point it should be mentioned that with this approach the Mach number gradient depends on itself with Eq. (29)-(31) and therefore has to be solved via a zero point search, e.g. with a bisection method. Therefore a calculation of the dissipation with this compressible approach has a much higher computational effort than with the incompressible values.

2.2.4. Turbulent flow ($Re > 8000$) The calculation in the turbulent case is much simpler, since according to Ref. [5] the friction coefficient of the incompressible solution can be adopted to the compressible flow. In addition, it is assumed that a superposition according to Eq.(41) is still approximately suitable. The turbulent Poiseuille number is calculated with the approximation according to Blasius:

$$Po = 0.316Re^{0.75} \tag{48}$$

The turbulent momentum shape factor is known to be close to one, since the flow profile is nearly rectangular. Müller [9] determined the value for a Poiseuille flow to be

$$K_D = 1.07. \tag{49}$$

It is assumed that it also has this value for a Couette-Poiseuille flow, since the moving wall is a subordinate effect for turbulent clearance flows in rotary displacement machines.

2.2.5. Transitional flow ($3000 < Re < 8000$) Transition flow is not studied in detail in this work, instead a linear interpolation is used for the friction coefficient and momentum shape factor.

2.2.6. Flow separation Especially for ducts with diverging outlet side it is important to define a criterion for the flow separation, otherwise the mass flow rate can be overestimated. To determine the exact point of flow separation, a precise knowledge of the boundary layer is required. However, since the boundary layer development strongly depends on the clearance geometry, it is not possible to find an exact, generally applicable criterion for the detachment point. An estimate can be made based on a fully formed flow. For this purpose, the Jeffery-Hamel solution is used. The criterion for the separation as a function of the Reynolds number and the opening angle is thus

$$(\alpha Re)_{FS} = 37.66, \tag{50}$$

although it only applies to laminar flows. Since turbulent flow is more stable than laminar flow, it follows that flow separation is overestimated in this case.

2.3. Inlet and outlet losses

The approach for inlet and outlet losses is taken from Ref. [9] and is merely recapitulated here. It is assumed that there is no exchange of heat and work between the inlet volume (in) and the entrance of the clearance channel (I) takes place (see Fig. 1). This means that the total temperature T_t is constant. Since the fluid is resting in the inlet volume, the total temperature $T_{t,in}$ is equal to the static temperature T_{in},

$$T_{t,in} = T_{in} = \overline{T}_{t,I}. \tag{51}$$

Therefore, the static temperature at the clearance entry (I) can be calculated from the Mach number M_I at the entry

$$\frac{\overline{T}_{in}}{\overline{T}_I} = 1 + \frac{\kappa - 1}{2} M_I^2. \tag{52}$$

The dissipation in the inflow is calculated with a polytropic approach that gives the pressure ratio at the inlet point

$$\Pi_{in} = \frac{\overline{p}_I}{p_{in}} = \left(\frac{T_{in}}{T_I}\right)^{\frac{n}{1-n}} \tag{53}$$

Reference [9] determines the polytropic coefficient from CFD simulations, which is almost constant for a wide range of Reynolds number:

$$n = 0.88 \cdot \kappa \tag{54}$$

Finally, for subsonic flows, it is assumed that the kinetic energy in the the outflow is completely dissipated and thus the pressure p_{II} in the outflow is equal to the pressure in the outlet volume p_{out}

$$p_{II} = p_{out}. \tag{55}$$

2.4. Solving the problem

Eq. (29) is an initial value problem which can be solved numerically, for example by Euler's method or more effectively by a Runge-Kutta method, provided it exhibits a defined behavior. The behavior of the differential equations becomes undefined when the denominator $(1 - \frac{M^2}{M^2_{crit}})$ becomes zero. An exceptional case is the condition for which G and the denominator become zero simultaneously. In this case, the derivative can be calculated using l'Hôpital's rule [5]. However, it turns out that this is not necessary in a numerical setup. Instead, during integration, a check is made to see if a particular behavior exists. If not, the integration is aborted and the input Mach number M_I is corrected downward. For a given Reynolds number and an assumed inlet Mach number M_I, the differential equation is integrated over x until either the end of the 1D region (II) is reached or until the flow separates. In addition, the change of the total temperature must be integrated to obtain the local value of the total temperature. However, it is not important for this work since the total temperature is assumed to be constant. With the progression of the Mach number and the total temperature, the pressure ratio of the channel can be calculated:

$$\frac{p_{end}}{p_I} = \frac{h_I}{h_{end}} \frac{M_I}{M_{end}} \left(\frac{1 + \frac{\kappa-1}{2} M_I^2}{1 + \frac{\kappa-1}{2} M_{end}^2} \right)^{\frac{1}{2}} \left(\frac{T_{t,end}}{T_{t,I}} \right)^{\frac{1}{2}} \tag{56}$$

The subscript $_{end}$ symbolises either the end of the channel (II) or the point of flow separation, whichever is the case. Combining this ratio with the pressure loss of the inlet and the outlet, the overall pressure ratio is given as follows:

$$\Pi = \frac{p_{out}}{p_{in}} \frac{p_{end}}{p_I} \frac{p_{out}}{p_{end}} \tag{57}$$

The guessed inlet Mach number is adjusted with a bisection method until this pressure ratio corresponds to the specified value. Once this is done, the theoretical Reynolds number and hence the flow coefficient can be calculated with the Knudsen number given by:

$$Kn = (2\kappa\pi)^{\frac{1}{2}} \frac{h(x=0)}{h_{min}} \frac{Ma_I}{Re} \left(1 + \frac{\kappa - 1}{2} M_I^2\right)^{\frac{n+1}{2-2n}} \tag{58}$$

3. Results

To verify the described approach, its results are compared with measurements and CFD simulations of the housing and front clearances of a screw compressor from Ref. [10, 11, 12]. As mentioned before, the clearance contours are mapped by superellipses. The parameters are given in Tab. 1. The results are compared for different Reynolds numbers at different clearance heights and a pressure ratio of $\Pi = 0.7$. In a second step, the influence of the moving wall is investigated using the front clearance as an example. Since no measurements are available for this, the comparison is made exclusively with the CFD results. All simulations shown were performed with the Spalart-Allmaras turbulence model [12].

First, the results for the front clearance space are discussed. **Fig. 6** shows the effect of Reynolds number on the flow coefficient for various height to length ratios. The markers represent the measurements, the dashed lines represent the CFD simulations, and the solid lines represent the 1D model. The highest clearance is shown in gray, the middle one in orange, and the smallest one in blue. Both the 1D and CFD simulations underestimated the measurements. However, the CFD and 1D simulations show a good agreement in the fully turbulent region. Especially in the laminar region, there is a good agreement between measurements and 1D simulations, which can be seen in the example of measurements at small passage heights. The strong deviations of the CFD simulations can be explained by the fact that the turbulence model used cannot represent re-laminarization. The transition from laminar to turbulent flow is represented in the 1D simulations by the collapse of the flow coefficient. This seems to be unphysical compared to the measurements. Comparison with the measurements at larger passage heights suggests that the assumption of laminar flows would lead to better agreement even at larger Reynolds numbers. Considering that larger passage heights represent relatively shorter distances, it can be concluded that the inlet flow is laminar dominated over a large distance before changing to turbulent flow at high Reynolds numbers.

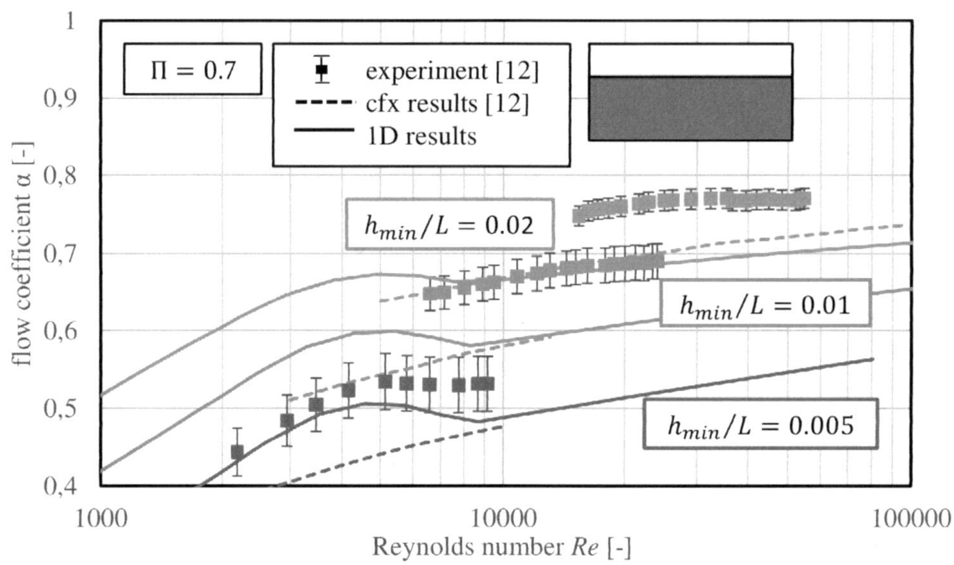

Figure 6. Flow coefficient of the front clearance. Comparison of the one-dimensional approach with measurements and CFD results of Ref. [12] using dry air as fluid. Results are shown for a pressure ratio of $\Pi = 0.7$.

A similar argumentation can be made for the female housing clearance (see **Fig. 7**). This is essentially a long clearance like the front clearance, but with rounded inlet and outlet edges. Thus, the inflow is much smoother and thus less turbulent with less dissipation. For the small clearance height, there is again good agreement for the laminar Reynolds number as already discussed for the front clearance. The flow coefficient estimated by the 1D model breaks down much more for higher Reynolds numbers than indicated by the measurements. The CFD simulation fits the measurements better than in the previous example, since due to the smoother flow at the clearance inlet, less turbulence is generated. Thus, the assumption is confirmed that the 1D model would be significantly improved by implementing an model for non-developed flow.

The last investigated contour is the housing clearance of the male rotor (see **Fig. 8**). Compared to the above investigated contours, this one is short and therefore is more dominated by inertia than friction. For this reason, modelling the momentum shape factor is very important. The 1D model overestimates the flow coefficient in the low Reynolds range while it slightly

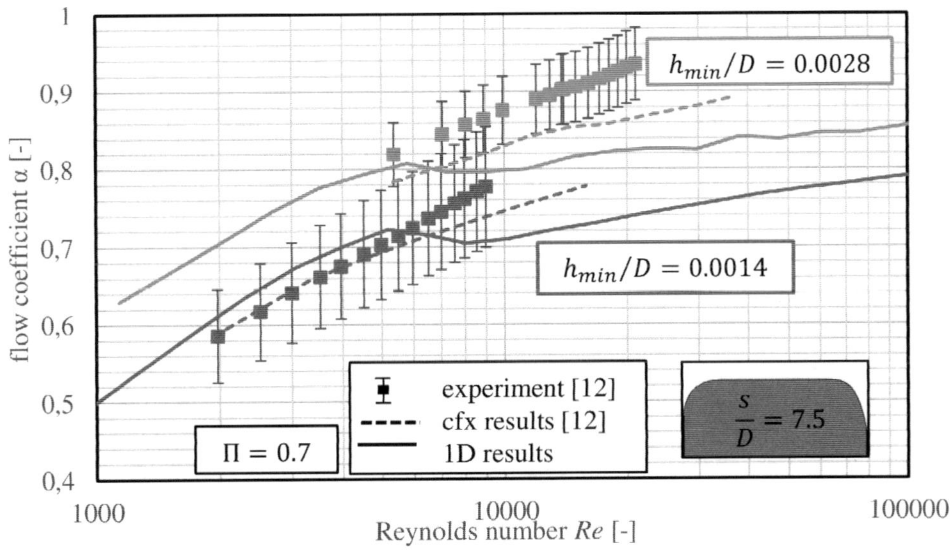

Figure 7. Flow coefficient of the female housing clearance. Comparison of the one-dimensional approach with measurements and CFD results of Ref. [12] using dry air as fluid. Results are shown for a pressure ratio of $\Pi = 0.7$.

underestimates it in the high Reynolds range. However, the quality is within the range of measurement accuracy. An improvement of the model can be possibly achieved if additionally dissipation in the non-developed flow is considered. In laminar flows, the friction factor in the inlet region is higher than in the fully developed flow and would therefore lead to a reduction in the flow coefficient at small Reynolds numbers.

Finally, the quality of the one-dimensional approach for moving boundaries has to be investigated. Here are no measured values available, instead only a qualitative comparison is done with CFD simulations of the front clearance (see **Fig. 9**). The case of a static wall as well as the cases of a positively and a negatively moving wall are shown. The wall velocities are given dimensionless as circumferential Mach numbers. Results for two different clearance heights are shown. The 1D model reproduces the influence of the moving boundary in a similar way as the CFD simulation. In the middle Reynolds number range are greater deviations, which can be attributed to the different modelling of the friction in the transition regime.

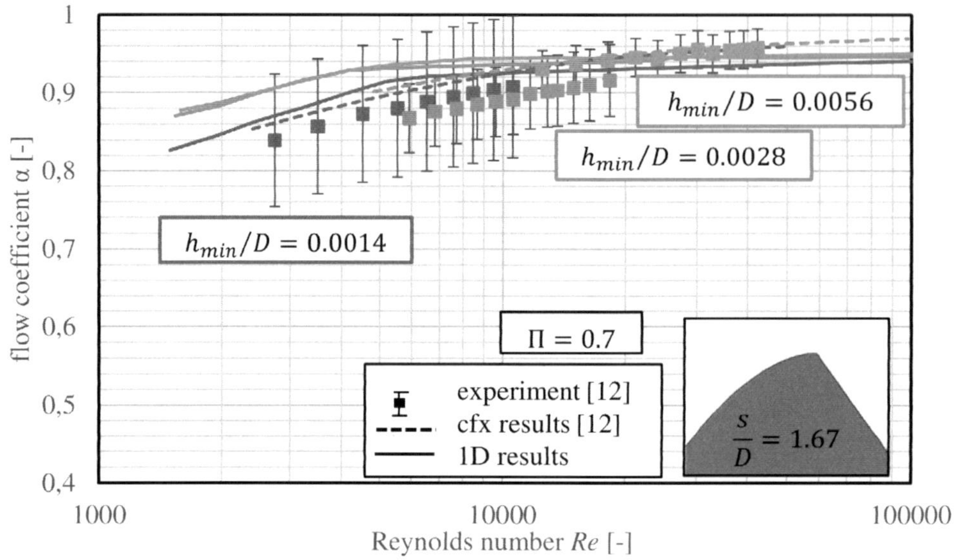

Figure 8. Flow coefficient of the male housing clearance. Comparison of the one-dimensional approach with measurements and CFD results of Ref. [12] using dry air as fluid. Results are shown for a pressure ratio of $\Pi = 0.7$.

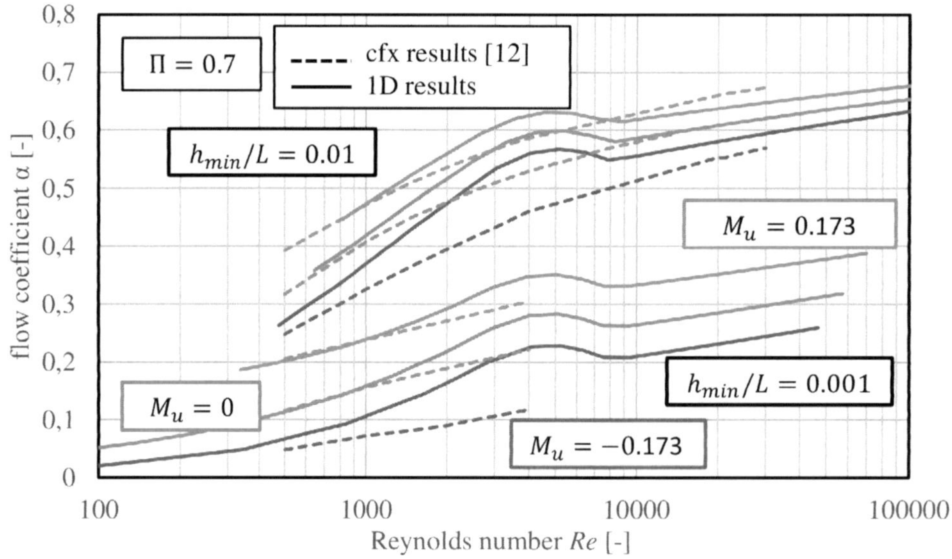

Figure 9. Flow coefficient of the front clearance with moving boundaries. Comparison of the one-dimensional approach with CFD results of Ref. [12] using dry air as fluid. Results are shown for a pressure ratio of $\Pi = 0.7$ and different circumferential Mach numbers.

4. Conclusion

In this paper, a one-dimensional approach for the calculation of clearance flows in dry-running rotating positive displacement machines is presented. The approach considers compressibility and friction as well as the flow profile. The modelling of the unknown quantities is based on the solution of the fully developed flow. The influence of the moving boundary is modelled by a superposition approach. The Jeffery-Hamel flow is used to approximate the effect of flow acceleration due to cross-sectional changes, but also due to density changes. The approach is tested assuming a constant total temperature and compared with reference measurements and simulations from the literature. In particular, a good agreement was found for laminar flows. In order to achieve an improvement for larger Reynolds numbers, a modelling of the flow inlet is recommended. A possible re-laminarization of the flow should also be considered. Furthermore, it should be mentioned that the approach is also suitable for modelling diabatic or rarefied flows, as described in Ref. [9, 15].

References

[1] Fister W and Neumann G 1984 *VDI-Berichte* **521** 299
[2] Peveling F J 1989 *Ein Beitrag zur Optimierung adiabater Schraubenmaschinen* Ph.D. thesis Dortmund University
[3] Dreißig B 1989 *Ein Beitrag zur Auslegung von trockenlaufenden Schraubenmotoren* Ph.D. thesis Dortmund University
[4] Fan Z and Chen Z 1994 *Int Compressor Eng Conf* 955
[5] Shapiro A H 1953 *The dynamics and Thermodynamics of Compressible Fluid Flow* vol 1 (Ronald Press)
[6] Huang Y 1994 *Int Compressor Eng Conf* 953
[7] Ishii N, Bird K, Sano K, Oono M and Iwamura S 1996 *Int Compressor Eng Conf* 1175
[8] Ishii N, Oku T and Sawai K A C W K K 2008 *Int Compressor Eng Conf* 1851
[9] Müller R 2013 *Spaltströmung mit Wärmeübertragung in Vakuumpumpen* Ph.D. thesis TU Kaiserslautern
[10] Utri M, Höckenkamp S and Brümmer A 2018 *IOP Conf Ser: Mater Sci Eng* **425** 012033
[11] Utri M and Brümmer A 2018 *Int Compressor Eng Conf* 2583
[12] Utri M 2021 *Potential von nicht-konstanter Rotorsteigung für Schraubenkompressoren* Ph.D. thesis TU Dortmund University
[13] Huck C, Pleskun H and Brümmer A 2018 *J Vac Sci Technol A* **36** 031606
[14] Huck C 2020 *Spaltströmungen im Vakuum* Ph.D. thesis TU Dortmund University
[15] Jünemann T and Brümmer A 2020 *J Vac Sci Technol B* **38** 044201
[16] Saint-Venant B and Wantzel L 1839 *J de l'Ecole Roy Poly* **16** 85
[17] White F H 1991 *Viscous Fluid Flow* (McGraw-Hill)
[18] Askao Y, Pi T, Turner S and Faghri M 2003 *Int J Heat Mass Trans* **46** 3041–3050

IOP Conf. Series: Materials Science and Engineering 1267 (2022) 012004 doi:10.1088/1757-899X/1267/1/012004

Nomenclature

Symbols

symbol	description	unit
a	coefficient super-ellipse	m
A	flow cross section area	m^2
b	coefficient super-ellipse	m
c	flow velocity	$m\,s^{-1}$
C	Constant of Jeffery-Hamel flow	-
c_p	specific heat capacity at constant pressure	$J\,kg^{-1}\,K^{-1}$
D	outer rotor diameter	m
D_h	hydraulic diameter of the channel	m
f	Darcy friction coefficient	–
G	characteristic term in the Mach number differential eq.	–
h	height of the clearance	m
Kn	Knudsen number	-
K_D	momentum form factor	-
n	exponent super-ellipse	-
n	polytropic exponent	-
$\boldsymbol{n}$	surface unit normal vector of the control volume (pointing outwards)	-
m	exponent super-ellipse	-
M	Mach number	-
p	pressure	Pa
Po	Poiseuille number	-
q	specific heat flow	$J\,kg^{-1}$
$\dot{q}$	heat flux	$W\,m^{-2}$
r	radial position Jeffery-Hamel flow (polar coordinates)	m
R	specific gas constant	$J\,kg^{-1}\,K^{-1}$
Re	Reynolds number	-
S	rotor lead	m
T	Temperature	K
t	non-dimensional flow velocity Jeffery-Hamel flow	-
u_w	velocity of the moving wall	$m\,s^{-1}$
w	width of the clearance	m
w	specific friction work of the wall	$J\,kg^{-1}$
x	position coordinate	m
y	position coordinate	m
α	flow coefficient	-
α	opening angle Jeffery-Hamel flow	rad
β	half opening angle Jeffery-Hamel flow	rad
η	non-dimensional angular position Jeffery-Hamel flow	-
κ	ratio of specific heats	-
μ	dynamic viscosity	Pa s
Π	pressure ratio of the clearance (outlet to inlet)	-
ϱ	gas density	$kg\,m^{-3}$
τ	shear stress	Pa
φ	angular position Jeffery-Hamel flow (polar coordinates)	rad
Ω	surface of the control volume	m^2

Acronyms, abbreviations and indices

symbol	description	symbol	description
app	apparently	min	minimal
crit	critical	max	maximal
CFD	computational fluid dynamics	out	outlet
DSMC	Direct Simulation Monte Carlo	r	radial
F	female rotor	t	total
I	entry of the channel	th	theoretical
II	exit of the channel	u	circumferential (german "Umfang")
in	inlet	$\bar{()}$	average flow value

Numerical investigation of contact loads of unsynchronised, dry-running screw machines

D Aurich[1], A Brümmer[1], A Wittig[2], D Stangier[2], W Tillmann[2], C-A Thomann[3], A Wittrock[3] and J Debus[3]

[1] TU Dortmund University, Chair of Fluidics, Leonhard-Euler-Str. 5, 44227 Dortmund, Germany

[2] TU Dortmund University, Institute of Materials Engineering, Leonhard-Euler-Str. 2, 44227 Dortmund, Germany

[3] TU Dortmund University, Experimental Physics 2, Otto-Hahn-Str. 4a, 44227 Dortmund, Germany

E-mail: daniel.aurich@tu-dortmund.de, andreas.bruemmer@tu-dortmund.de

Abstract. Unsynchronised, dry-running screw machines have advantages compared to conventional oil-injected and synchronised, dry-running screw machines. Because of the absence of an additional timing gear and no injected lubricant, the effort for manufacturing and assembly can be reduced and lubricant residues in the process gas are avoided. In this design the female rotor torque resulting from the gas pressure is transferred directly to the male rotor by the contacting rotor surfaces. To investigate this contact, a two-dimensional boundary element method contact algorithm is developed to calculate the loads in terms of contact tractions and sliding velocities. The contact kinematics of the screw rotors and the non-constant curvature of the rotor surfaces are considered. The movement between the screw rotors is transformed into a relative coordinate system fixed on the male rotor. Furthermore, the influence of manufacturing tolerances to the occurring contact loads are investigated.

1. Introduction

Screw machines are rotary positive displacement machines that are used in various applications of process gas technology [1]. Due to the compact design and the low-maintenance operation, screw machines are used in about half of all compressed air supply applications. Compared to common liquid-injected and common dry-running, synchronised designs a dry-running, unsynchronised screw machine has many advantages because of the absence of synchronisation gears, lubricants and the necessary peripheral devices. Due to high surface pressures and sliding velocities in the rotor contact, it is crucial to coat the screw rotors to minimize the contact wear and to ensure durability of the screw machine. The operation of dry-running unsynchronised screw machines has been tested in previous theoretical and experimental investigations [2, 3]. Various coatings were examined to reduce the wear due to dry friction [4–6], but until today only durabilities of 200 to 400 hours for unsynchronised, dry-running screw rotor pairs have been achieved [2]. It can be deduced that the optimization of the coating alone is not sufficient to ensure a continuous operation of an unsynchronised, dry-running screw machine. On the contrary, a detailed investigation of the mechanical parameters and the contact situation is required to improve the durability and wear properties of screw machines by means of mechanical and

geometrical optimization.
Until today, the mechanical properties of the contact area of screw machines have only been characterized by analytical models [4, 7, 8] that allow a simplified view of the contact situation. The Hertzian Contact is a common method for the calculation of contact stresses between two curved elastic bodies. Based on the curvature of the contacting surfaces and the applied force the pressure distribution can be directly calculated. Two important limitations of the Hertzian theory are that only frictionless contacts of surfaces with constant curvature can be considered. Furthermore, it is not possible to consider a surface coating in the contact area.
In order to take the coating into account semi-analytical solutions based on Airy stress functions and Fourier transformations can be used [4, 9, 10]. Using semi-analytical models, an elastic coating between two contacting elastic bodies can be taken into account. For two-dimensional problems, e.g. the contact between a cylindrical indenter and a coated half-plane, the contact tractions can be calculated. But even for simple problems the equations must be solved numerically. However, semi-analytical models are also not suitable for the calculations of contact loads for arbitrarily curved surfaces.
To gain more detailed modelling and investigation of contact tractions and displacements of realistically shaped, coated screw rotors, a numerical analysis is required. The aim of this work is to develop an efficient numerical method for the calculation of contact loads within unsynchronised, dry-running twin-screw machines. It is intended to form the basis for further extensions and investigations for coating modelling and wear prediction within screw machines. The finite element method (FEM) is a common method for structural analysis of engineering problems. By discretizing into small elements, an approximate solution of the displacement and stress field for an arbitrarily shaped body can be achieved numerically. The size of the contact area ($\sim \mu$m) between two curved bodies is usually several orders of magnitude smaller than the outer dimensions of the bodies ($\sim$mm...m) themselves. To solve the contact problem using the FEM, a large number of elements is required, because discretization is also required inside the body. This leads to very time-consuming solutions of large systems of equations.
The boundary element method (BEM) is another numerical computation method for solving partial differential equations. Compared to the FEM the BEM is quite unknown in the field of engineering. Using fundamental solutions of the underlying partial differential equation, the dimension of the appearing integrals are reduced by one. This results in several advantages of the BEM compared to the FEM for solving contact problems: For two-dimensional problems it is sufficient to discretize the boundary Γ of the solution domain Ω so that the number of elements can be reduced significantly. This leads to smaller systems of equations for solving the contact problem. Furthermore, the surface tractions and displacements are calculated by the BEM. These quantities can be used directly to solve the contact equations. Overall, the numerical handling is simplified. The BEM is a suitable method for solving contact problems of arbitrarily shaped elastic bodies, whereby only an approximation and a discretization of the surface geometry, the surface tractions and displacements is needed [11, 12].
In the following sections the operating principle and the kinematics of a dry-running, unsynchronized screw machine is presented. The rotor contact is defined in a rotating relative system that simplifies the numerical handling. The basic equations of the BEM are discussed and the resulting system of equations are transformed into a normal-tangential coordinate system that is suitable for the contact of arbitrary shaped surfaces. In the following, the implemented numerical method is compared to analytical solutions. Finally, the contact of screw rotors is calculated. For this, both ideal rotor profiles and measured geometries are used.

2. Kinematics of twin-screw machines

Twin-screw machines consist of two counterrotating rotors that are mounted in a tightly enclosing housing. Between the twisted rotors multiple chambers are separated, which

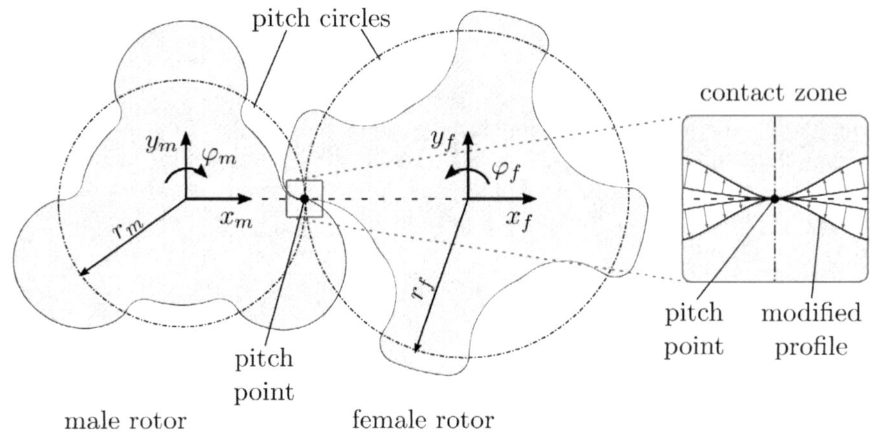

Figure 1. Cross section of a screw rotor pair and profile modification of the contact zone

periodically change their volume through rotation [1, 13, 14]. Due to the simple design of this type of machine and the absence of inertia forces, it is possible to operate at very high rotational speed. The kinematics of dry-running, unsynchronised screw rotors can be treated as that of helical gears. The cross sectional view of the rotor profiles considered in this work is shown in figure 1. The rotor profile is based on an asymmetric SRM rotor profile with three male rotor lobes and four female rotor lobes [1]. During operation the male rotor rotates with the angle φ_m around the origin of the $x_m y_m$-coordinate system, that is located at the absolute origin. The female rotor rotates counterwise around the origin of the $x_f y_f$-coordinate system with the angle φ_f. The ratio I of rotation between the male and female rotor is prescribed by the number of lobes per rotor Z_m and Z_f and by the radii of the pitch circles r_m and r_f:

$$I = \frac{\varphi_f}{\varphi_m} = \frac{Z_m}{Z_f} = \frac{r_m}{r_f}. \tag{1}$$

To ensure a defined contact, both rotor profiles are retracted by 0.25 mm in the direction of the surface normal, except for the part of the profiles where the intended contact zone is located. To minimize the occurrence of sliding velocities between the rotors, the contact zone is set to the pitch point. In the case of an ideal contact between two rigid bodies, the contact zone reduces to a single contact point and the sliding speed is calculated as zero [7].

Viewed in the absolute coordinate system, all points on both rotor surfaces move on a circular path around the origin of the respective coordinate system. For the numerical handling during the contact calculation it is beneficial to describe only the relative movement between both rotor surfaces. Figure 2 shows the relationships to describe the rotation of the female rotor from the point of view of a moving coordinate system fixed on the male rotor point $\underline{P}_m(\varphi_m)$. Due to the rotation φ_m around the absolute origin the male rotor point $\underline{P}_m^0$ and the base vectors $\underline{e}_{x,m}$ and $\underline{e}_{y,m}$ are transformed by the rotation matrix $\underline{\underline{R}}(\varphi_m)$

$$\underline{P}_m(\varphi_m) = \underline{\underline{R}}(\varphi_m) \cdot \underline{P}_m^0, \quad \underline{e}_x = \underline{\underline{R}}(\varphi_m) \cdot \underline{e}_{x,m} \quad \text{and} \quad \underline{e}_y = \underline{\underline{R}}(\varphi_m) \cdot \underline{e}_{y,m}. \tag{2}$$

For all further considerations the initial male rotor point and the base vectors are set to

$$\underline{P}_m^0 = \begin{bmatrix} r_m \\ 0 \end{bmatrix}, \quad \underline{e}_{x,m} = \begin{bmatrix} 1 \\ 0 \end{bmatrix} \quad \text{and} \quad \underline{e}_{y,m} = \begin{bmatrix} 0 \\ 1 \end{bmatrix}. \tag{3}$$

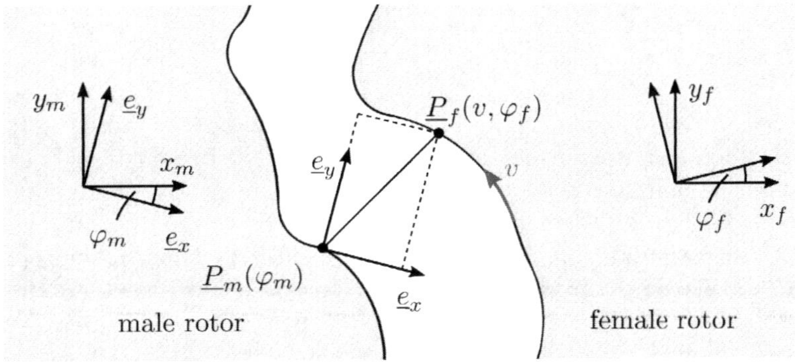

Figure 2. Relationship between the absolute and the moving male rotor coordinate system

The initial female rotor surface $\underline{P}_f^0(v)$ is interpolated by a B-Spline of degree two [15]. v denotes the internal spline parameter. The transformed female rotor surface is calculated via the rotation of the control points around the origin of the female rotor coordinate system

$$\underline{P}_f(v, \varphi_f) = \underline{\underline{R}}(\varphi_f) \cdot \underline{P}_f^0(v). \tag{4}$$

For the calculation of the relative coordinates $P'_{x,f}$ and $P'_{y,f}$ the relation between the absolute and the relative coordinate systems is used.

$$P_{x,f}\,\underline{e}_{x,m} + P_{y,f}\,\underline{e}_{y,m} = P'_{x,f}\,\underline{e}_x + P'_{y,f}\,\underline{e}_y + \underline{P}_m(\varphi_m) \tag{5}$$

Multiplying equation 5 with the transformed base vectors $\underline{e}_x$ and $\underline{e}_y$ gives

$$\underline{P}'_f(v, \varphi_m, \varphi_f) = \begin{bmatrix} \underline{e}_{x,m} \cdot \underline{e}_x & \underline{e}_{y,m} \cdot \underline{e}_x \\ \underline{e}_{x,m} \cdot \underline{e}_y & \underline{e}_{y,m} \cdot \underline{e}_y \end{bmatrix} \cdot \underline{P}_f(v, \varphi_f) - \begin{bmatrix} \underline{e}_x^t \\ \underline{e}_y^t \end{bmatrix} \cdot \underline{P}_m(\varphi_m). \tag{6}$$

According to equation 3 the base point $\underline{P}_m(\varphi_m)$ is located on the rotors pitch point for $\varphi_m = 0$. Furthermore, the rotation of the female rotor is defined to

$$\varphi_f = I\,\varphi_m + \delta. \tag{7}$$

The rotational offset δ is needed to describe the penetration between both rotor surfaces during the contact calculation of elastic bodies. Figure 3 shows the relative movement of the female rotor contact zone from the point of view of the moving xy-coordinate system with its base point

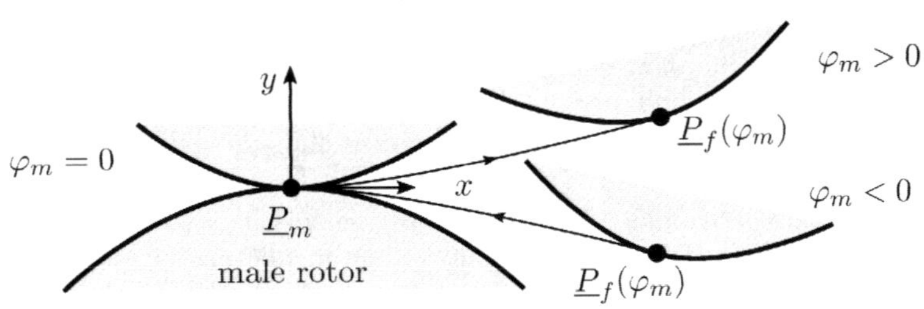

Figure 3. Relative movement of the female rotor ($\delta = 0$)

$\underline{P}_m$. If equation 7 is used and rigid rotors are assumed ($\delta = 0$), all female rotor points can be calculated only dependent on the male rotor rotation φ_m. In addition, figure 3 illustrates the trajectory of a female rotor point $\underline{P}_f(\varphi_f)$ for a fixed parameter v. At $\varphi_m = 0$ both rotor points touch at the pitch point. For small angles $|\varphi_m| \ll 1$ there is only a rolling movement between both rotor surfaces in the contact zone. Shortly before ($\varphi_m < 0$) and shortly after the contact ($\varphi_m > 0$) this rolling is dominated by a translational movement. At this point, the advantage of the relative view becomes clear: While in the absolute system all rotor points perform a circular motion, the relative perspective allows to separate the three mentioned phases of the rotor contact. In an elastic rotor contact, there is a penetration between the rotor surfaces and the rotational offset takes non-zero values $\delta > 0$. In this case, the penetration leads to a rotor contact beside the pitch point. In addition to the purely rolling contact between ideally rigid rotors, in the elastic case there is a sliding contact region beside the pitch point. For all further considerations only the relative xy-coordinate system between the rotor surfaces during the contact is used. As a result, the male rotor is considered to be a non-moving elastic body and the movement of the female rotor surface is described in the relative coordinates according to equation 6.

3. Boundary Element Method

As mentioned before, the BEM is a numerical computation method for solving partial differential equations, whereby the dimension of the underlying problem is reduced by one. The derivation of the resulting boundary integral equation for elastostatic problems is described in various literature sources [16–18]. The starting point of the derivation is the momentum balance in differential form

$$\mathrm{div}(\underline{\underline{\sigma}}) + \rho \underline{b} = \rho \underline{\ddot{u}}. \tag{8}$$

In the following the body forces $\underline{b}$ and the accelerations $\underline{\ddot{u}}$ are neglected. In general, both terms can be taken into account by the BEM. However, the contact stresses are of a higher order of magnitude, so that the influence of the body forces and the accelerations is only small. In addition, only the two-dimensional case is considered. To connect the stresses $\underline{\underline{\sigma}}$ and the displacement field $\underline{u}$ Hooke's material law and small strains are assumed. At the derivation of the FEM, the momentum balance in differential form is multiplied by an unknown test function and integrated over the solution domain. For the case of the BEM this first step is done in a similar way. The big difference is, that analytical solutions of the underlying partial differential equation - the fundamental solutions - are used for the test function [18]. Since the fundamental solutions are known, all volume integrals can be transformed to boundary integrals. Through this procedure a boundary integral equation can be formed that is only dependent on boundary variables

$$\underline{\underline{C}}(\underline{\tilde{x}}) \cdot \underline{u}(\underline{\tilde{x}}) = \int_{\Gamma} \underline{\underline{U}}^{\star}(\underline{\tilde{x}}, \underline{x}) \cdot \underline{t}(\underline{x}) - \underline{\underline{T}}^{\star}(\underline{\tilde{x}}, \underline{x}) \cdot \underline{u}(\underline{x}) \mathrm{d}\Gamma. \tag{9}$$

For the fundamental solutions $\underline{\underline{U}}^{\star}(\underline{\tilde{x}}, \underline{x})$ and $\underline{\underline{T}}^{\star}(\underline{\tilde{x}}, \underline{x})$ Kelvin's solution for plain strain is used, which is one of the most common fundamental solutions in the field of BEM [19]. With this solution the displacements and tractions at point $\underline{x}$ are calculated, if a single point force acts at source point $\underline{\tilde{x}}$ on an infinite elastic plane. The plain strain condition is used, because there is mostly line contact between the rotors and the strains in z-direction are expected to be small. $\underline{\underline{C}}(\underline{\tilde{x}})$ is an integral free term that only depends on shape of the boundary Γ at source point $\underline{\tilde{x}}$. For the solution of equation 9 the collocation method is commonly used in engineering. Thereby, the boundary integral equation is evaluated at a finite number of collocation points $\underline{\tilde{x}}_m$, that typically coincide with the discretized boundary nodes. To calculate an approximative solution the geometry $\underline{x}$, the traction field $\underline{t}$ and the displacement field $\underline{u}$ must be discretized

IOP Conf. Series: Materials Science and Engineering 1267 (2022) 012005 doi:10.1088/1757-899X/1267/1/012005

into elements.

$$\underline{x} \approx \sum_e^{n_{el}} \underline{x}^e \qquad \underline{t} \approx \sum_e^{n_{el}} \underline{t}^e \qquad \underline{u} \approx \sum_e^{n_{el}} \underline{u}^e \tag{10}$$

Within each element e a linear, isoparametric approach is used. This results in

$$\underline{x}^e(\eta) = \sum_{i=1}^{2} N_i(\eta)\, \underline{x}_i^e, \quad \underline{u}^e(\eta) = \sum_{i=1}^{2} N_i(\eta)\, \underline{u}_i^e \quad \text{and} \quad \underline{t}^e(\eta) = \sum_{i=1}^{2} N_i(\eta)\, \underline{t}_i^e. \tag{11}$$

$N_i(\eta)$ denotes the shape function for the linear approximation between the nodal values $\underline{x}_i^e$, $\underline{u}_i^e$ and $\underline{t}_i^e$ within each element e. The number of collocation points must correspond to the number of nodes. Inserting the approach of equation 10 and 11 into equation 9 and integrating over the surface Γ leads to the linear system of equation

$$\underline{\underline{U}} \cdot \underline{t} - \underline{\underline{T}} \cdot \underline{u} = \underline{0}. \tag{12}$$

The numerical integration is done elementwise over the parameter η using a 12 point Gaussian quadrature. It should be noted, that the fundamental solutions $\underline{\underline{U}}^\star(\tilde{\underline{x}}, \underline{x})$ and $\underline{\underline{T}}^\star(\tilde{\underline{x}}, \underline{x})$ show a weak and strong singular behavior if the field point $\underline{x}$ tends to the source point $\tilde{\underline{x}}$. For an accurate integration of the weak singular integrals a Gauss-Laguerre quadrature is used [16]. The strong singular integrals and the geometry dependent free term $\underline{\underline{C}}$ are indirectly calculated via the rigid body trick [19]. Since the number of collocation points coincides with the number of nodes, equation 12 contains twice as many unknowns as equations. To achieve a unique solution, boundary conditions must be applied to each node. If the displacement $\underline{u}$ is prescribed, the traction $\underline{t}$ remains as unknown variable and vice versa. The application of all boundary conditions leads to a system of linear equations

$$\underline{\underline{L}} \cdot \underline{X} = \underline{R}. \tag{13}$$

All prescribed boundary values are multiplied with the corresponding matrix entries and are sorted to the right hand side $\underline{R}$. The remaining unknown tractions and displacements and the corresponding matrix columns are assembled to the solution vector $\underline{X}$ and the matrix $\underline{\underline{L}}$.

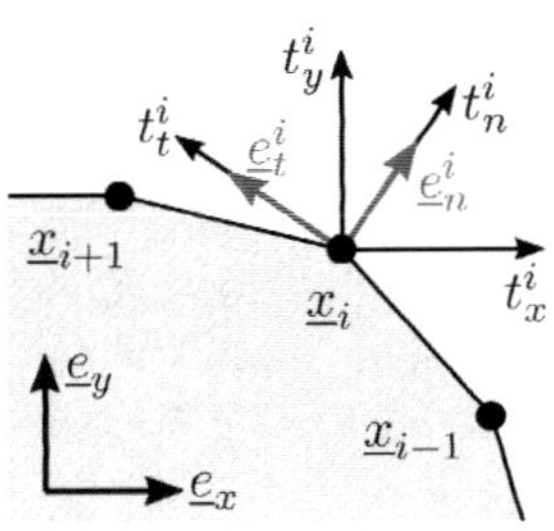

Figure 4. Traction in xy- and nt-coordinate system

For the calculation of contact problems of curved surfaces, it is beneficial to transform equation 12 into a local normal-tangential coordinate system. In this way the application of contact boundary conditions is simplified. Figure 4 illustrates the tractions in the xy- and the normal tangential nt-coordinate system at node $\underline{x}^i$.

$$\underline{t}_i = t_x^i \underline{e}_x + t_y^i \underline{e}_y = t_n^i \underline{e}_n^i + t_t^i \underline{e}_t^i \tag{14}$$

Equation 14 must be multiplied by the base vectors $\underline{e}_x$ and $\underline{e}_y$ at node i to get the transformation rule

$$\begin{bmatrix} t_x^i \\ t_y^i \end{bmatrix} = \begin{bmatrix} \underline{e}_n^i \cdot \underline{e}_x & \underline{e}_t^i \cdot \underline{e}_x \\ \underline{e}_n^i \cdot \underline{e}_y & \underline{e}_t^i \cdot \underline{e}_y \end{bmatrix} \cdot \begin{bmatrix} t_n^i \\ t_t^i \end{bmatrix}. \tag{15}$$

The transformation for the nodal displacements is done in the same way. The normal and tangential base vectors of node $\underline{x}_i$ are approximated by the neighbor nodes $\underline{x}_{i+1}$ and $\underline{x}_{i-1}$. According to figure 4 this leads for a counterclockwise node ordering to

$$\underline{e}_t^i = \frac{1}{\left|\underline{x}_{i+1} - \underline{x}_i\right|} \left[\underline{x}_{i+1} - \underline{x}_i\right] \quad \text{and} \quad \underline{e}_n^i = \begin{bmatrix} -e_{t,2}^i \\ e_{t,1}^i \end{bmatrix}, \tag{16}$$

whereby $e_{t,1}^i$ and $e_{t,2}^i$ denotes the first and second component of $\underline{e}_t^i$. For each collocation point m and for each node i the matrices $\underline{\underline{U}}$ and $\underline{\underline{T}}$ contain a 2x2-submatrix $\underline{\underline{U}}^{mi}$ and $\underline{\underline{T}}^{mi}$.

$$\begin{bmatrix} \ddots & & \ddots \\ & \begin{bmatrix} U_{xx}^{mi} & U_{xy}^{mi} \\ U_{yx}^{mi} & U_{yy}^{mi} \end{bmatrix} & \\ \ddots & & \ddots \end{bmatrix} \cdot \begin{bmatrix} \vdots \\ \begin{bmatrix} t_x^i \\ t_y^i \end{bmatrix} \\ \vdots \end{bmatrix} - \begin{bmatrix} \ddots & & \ddots \\ & \begin{bmatrix} T_{xx}^{mi} & T_{xy}^{mi} \\ T_{yx}^{mi} & T_{yy}^{mi} \end{bmatrix} & \\ \ddots & & \ddots \end{bmatrix} \cdot \begin{bmatrix} \vdots \\ \begin{bmatrix} u_x^i \\ u_y^i \end{bmatrix} \\ \vdots \end{bmatrix} = \underline{0} \tag{17}$$

Inserting of equation 15 leads to the transformed submatrix $\underline{\underline{U}}^{mi}$ in terms of the normal-tangential coordinate system at node i:

$$\begin{bmatrix} U_{xx}^{mi} & U_{xy}^{mi} \\ U_{yx}^{mi} & U_{yy}^{mi} \end{bmatrix} \cdot \begin{bmatrix} t_x^i \\ t_y^i \end{bmatrix} = \begin{bmatrix} U_{xx}^{mi} & U_{xy}^{mi} \\ U_{yx}^{mi} & U_{yy}^{mi} \end{bmatrix} \cdot \begin{bmatrix} \underline{e}_n^i \cdot \underline{e}_x & \underline{e}_t^i \cdot \underline{e}_x \\ \underline{e}_n^i \cdot \underline{e}_y & \underline{e}_t^i \cdot \underline{e}_y \end{bmatrix} \cdot \begin{bmatrix} t_n^i \\ t_t^i \end{bmatrix} = \underbrace{\begin{bmatrix} U_{nn}^{mi} & U_{nt}^{mi} \\ U_{tn}^{mi} & U_{tt}^{mi} \end{bmatrix}}_{\underline{\underline{U}}^{mi}} \cdot \begin{bmatrix} t_n^i \\ t_t^i \end{bmatrix} \tag{18}$$

The transformed submatrix $\underline{\underline{T}}^{mi}$ is determined analogously. The procedure must be applied columnwise in the $\underline{\underline{U}}$ and $\underline{\underline{T}}$ matrices for all submatrices corresponding to node i. The transformation of all columns of $\underline{\underline{U}}$ and $\underline{\underline{T}}$ results in a system of equations, whereby $\underline{u}$ and $\underline{t}$ contain the nodal displacement and traction values in terms of the nodal normal-tangential coordinate systems.

4. Contact Mechanics

The previous chapter outlines the basic equations of the BEM and the assembly to a linear system of equations, whereby the normal and tangential tractions and displacements represent the solution variables. In the following, the contact boundary conditions, that must be applied to solve the contact problem, are presented. Furthermore, an iterative contact algorithm based on the BEM, that is suitable to calculate the contact loads within unsynchronised screw machines, is derived.

Figure 5 shows parts of the male and the female rotor and the normal distances between the surfaces at load step k. The male rotor is assumed to be an elastic body with a Young's modulus E and a Poisson's ratio ν. Moreover, the female rotor is supposed to be a rigid body with a smooth, continuously differentiable surface. In addition, just the relative movement is considered, so that only the female rotor is moved. For a contact detection the normal distance $\underline{d}_i^k = d_i^k \underline{e}_{n,i}$ at each potential contact node $\underline{x}_i^k$ must be calculated.

$$\underline{x}_i^k + d_i^k \underline{e}_{n,i} = \underline{\Gamma}_f^k(v_i^k) \tag{19}$$

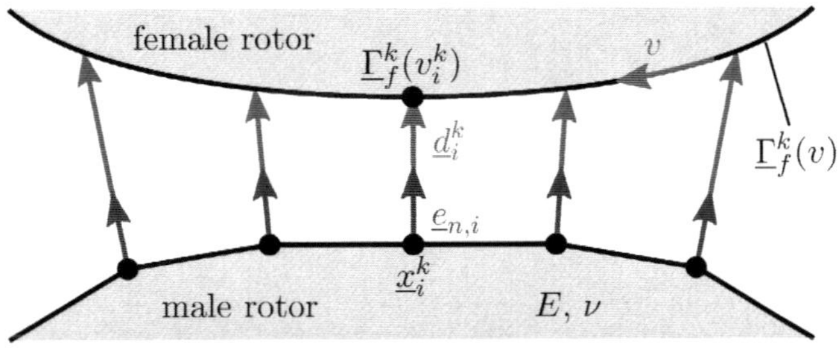

Figure 5. Normal distance between the rotor surfaces

To solve equation 19 iteratively for the unknown scalar valued normal distance d_i^k and the parameter v_i^k, Newton's method is used. If the normal distance is calculated to $d_i^k \leq 0$, the female rotor penetrates the male rotor, so that there is contact between both rotors. The penetration at the contacting nodes leads to an elastic deformation of the male rotor until both surfaces coincide. The resulting tractions and displacements at the contacting nodes depend on prescribed contact conditions. Three contact types are considered:

$$\text{free slip contact} \begin{cases} \Delta u_{n,i}^k = d_i^k \\ t_{t,i}^k = 0 \end{cases} \tag{20}$$

$$\text{stick contact} \begin{cases} \Delta u_{n,i}^k = d_i^k \\ \Delta u_{t,i}^k = -\delta_t^k \end{cases} \tag{21}$$

$$\text{frictional contact} \begin{cases} \Delta u_n^k = d_i^k \\ \Delta u_{t,i}^k = -\delta_t^k \quad \text{if} \quad (t_{t,i}^k \leq \mu t_{n,i}^k) \\ t_{t,i}^k = \mu t_{n,i}^k \quad \text{if} \quad (t_{t,i}^k > \mu t_{n,i}^k) \end{cases} \tag{22}$$

In general, the elastic contact of curved surfaces represents a non-linear problem. To handle this non-linearity, an incremental procedure is used to solve the contact equations. For that, the movement of the female rotor surface is divided into load steps. The displacement contact conditions are defined in an incremental form. During each load increment k, the distance between the moved female rotor and the deformed male rotor must be calculated. For a free slip contact, only the penetration in normal direction must be taken into account (equation 20). The tangential tractions $t_{t,i}^k$ must vanish and each node undergoes a tangential slip $\delta_{t,i}^k$ at each load step.

$$\delta_{t,i}^k = \left[\underline{\Gamma}_f^k(v_i^k) - \underline{\Gamma}_f^k(v_i^{k-1})\right] \cdot \underline{e}_{t,i} \tag{23}$$

To calculate this tangential slip, the parameter of the contact location on the female rotor of the previous load step v_i^{k-1} must be stored. For a stick contact node i must be pulled back into the tangential direction $\underline{e}_{t,i}$ to avoid a slip between both surfaces (equation 21). In the case of a frictional contact, Coulomb friction is assumed and node i sticks at the counter surface until the tangential traction exceeds the limit $t_{t,i}^k = \mu t_{n,i}^k$. μ denotes the frictional coefficient. In that case, the frictional traction must be prescribed (equation 22). It is important to avoid tensile tractions $t_{n,i}^k > 0$ at each load step, that occur during unloading. In this case, the contact node i must be released. The shown contact conditions and equations must be solved iteratively within each load step. Several algorithms using the BEM are presented in the literature, that solve the elastic contact iteratively in a comparable way [12, 17]. The basic idea of these algorithms

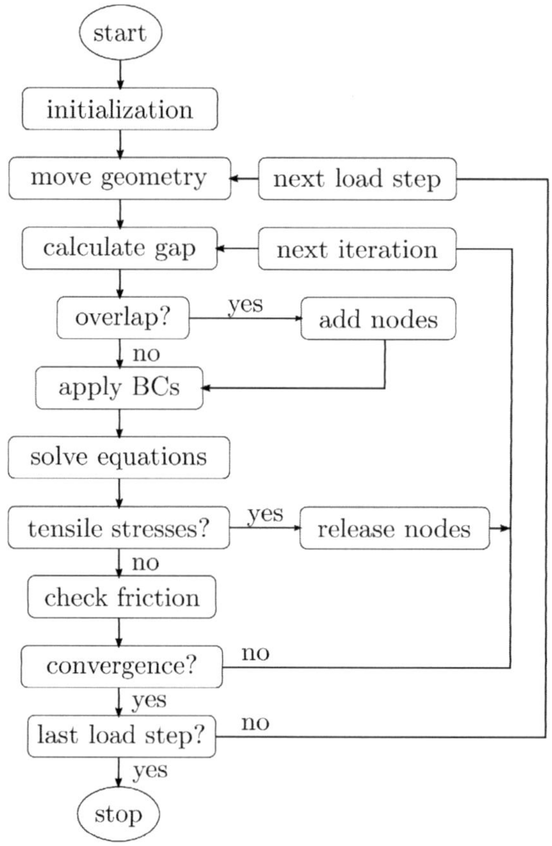

Figure 6. Contact algorithm

is adopted. Figure 6 illustrates the contact algorithm used for the further investigation of the screw rotor contact.

5. Validation

To validate the derived and implemented BEM contact algorithm, a test case is constructed. Figure 7 shows the geometry, the discretization and the static boundary conditions of the test case. The element size inside the potential contact zone is set to $1\,\mu$m, whereas the rest of the quadratic body is discretized by coarse elements with a size of 1 cm. The top surface is penetrated by a circular indenter with a prescribed indentation depth $\Delta u = 0.1\,\mu$m, which is divided into 40 load steps. This fine subdivision is necessary for both the frictionless and frictional contact to reduce the numerical error [20]. To compare the numerical results, the Hertzian contact is utilized [21–23]. In it's basic formulation, the Hertzian contact treats a frictionless elastic halfspace that is penetrated by a rigid, curved indenter. Furthermore, a substitution formula is given to calculate the contact of two elastic, curved bodies. If the material properties of both bodies are the same, this formula results in an equivalent Young's Modulus, which is only half of that for the case of a rigid indenter and an elastic halfspace. The implemented BEM contact algorithm is also based on an elastic body, which is penetrated by a rigid indenter. In order to treat the contact of two elastic bodies within the BEM, only half the Young's modulus is assumed for the elastic body. Figure 8 shows the numerically and analytically calculated normal

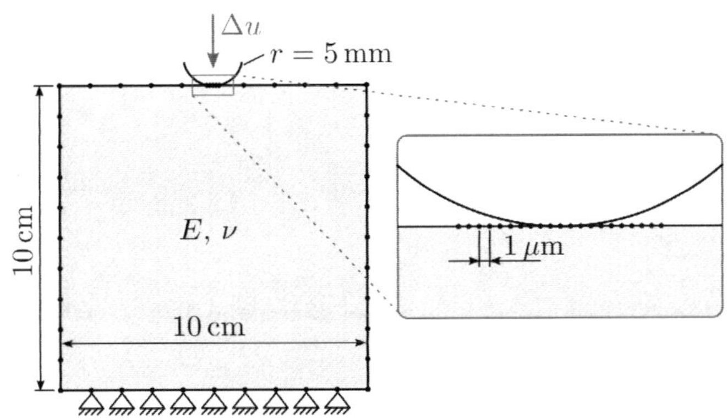

Figure 7. Test case for contact validation

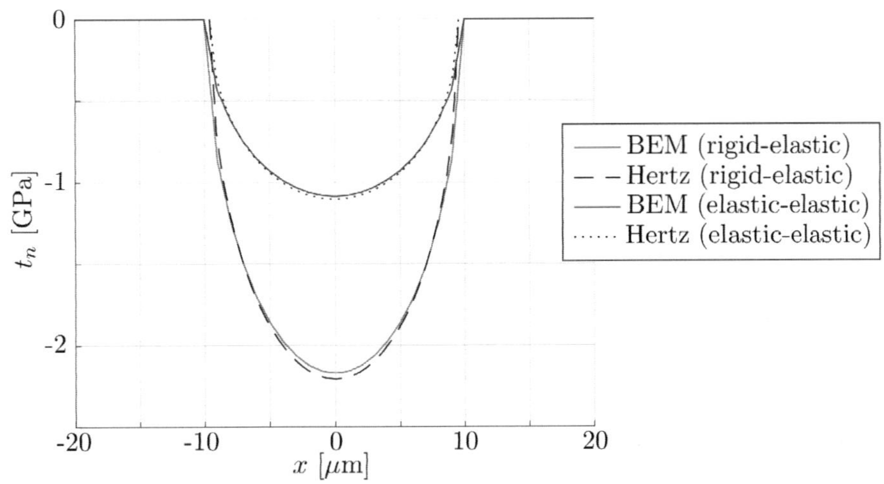

Figure 8. Normal tractions calculated by BEM (free slip, $\Delta u = 0.1\,\mu$m) and Hertzian contact

tractions within the contact zone for both the rigid and the elastic indenter. On the whole, the numerical and analytical results are in good agreement. However, the numerical maximum contact pressure in the center of the contact zone is slightly smaller, than the Hertzian pressure. In addition, the Hertzian contact width is smaller than the result of the BEM. The accuracy of the BEM regarding the contact width strongly depends on the discretization size of the elements. Because integration of both the numerical and analytical normal tractions leads to the same contact force, as well the maximum pressure is slightly influenced by the discretization size.

For the test case, the influence of friction on a normal contact is also calculated. Figure 9 demonstrates the normal tractions within the contact zone for different friction coefficients μ in comparison to the frictionless contact. The contact force is kept constant at $F = 1600\,\mathrm{Nm}^{-1}$. Regarding the contact width, there is no change within the discretization accuracy. But on the whole, the traction profile tends to narrow as friction increases. In addition, the maximum traction is significantly higher for the frictional contacts. Furthermore, figure 9 shows the tangential tractions t_t in relation to the local maximum tangential friction μt_t. Increasing the coefficient of friction leads to smaller sliding within the contact. While there is almost only

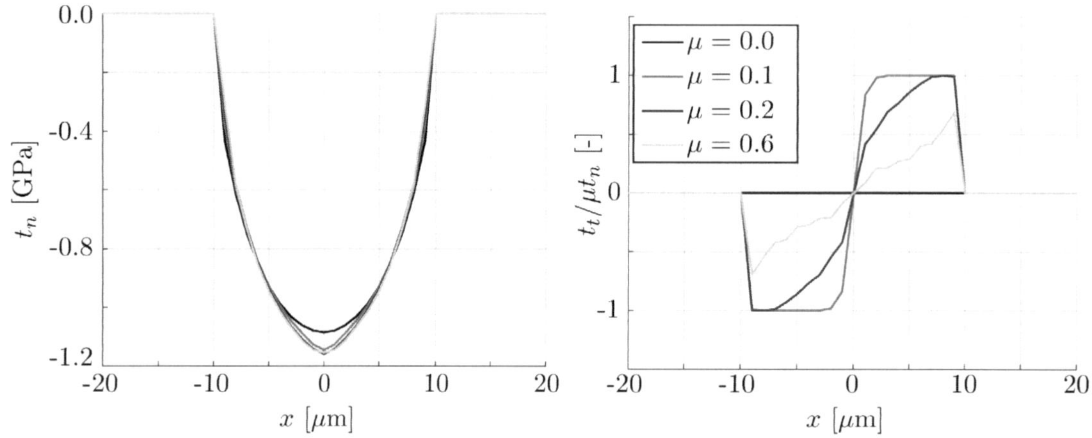

Figure 9. Normal and tangential tractions for different coefficients of friction μ ($F_y = 1600\,\mathrm{Nm^{-1}}$)

sliding for $\mu = 0.1$, the indenter completely sticks at the elastic body for $\mu = 0.6$. All in all, the accuracy of the calculated contact tractions agrees well with other results from the literature [11, 20, 24].

6. Rotor contact

The previous chapters describe the contact kinematics of unsynchronised twin-screw machines, the basics of the BEM and an iterative procedure to calculate the tractions and displacements within the contact of arbitrary shaped elastic bodies. In the following, these are applied to the contact of an unsynchronised twin-screw machine (figure 1). So far, only two-dimensional problems have been treated. If the kinematics are considered in the third spatial direction, the contact moves through the machine in z-direction during the rotor rotation. The speed of this z-movement is proportional to the rotational speed of the machine. Thus the change in the z-direction Δz only depends on an rotation increment.

$$z = \frac{s_m}{2\pi}\varphi_m \qquad \Delta z = \frac{s_m}{2\pi}\Delta\varphi_m \tag{24}$$

s_m denotes the male rotor lead and $\Delta\varphi_m$ corresponds to the rotational increment of the male rotor within each load step. From this, for each load step the position z is calculated. The three-dimensional contact area is assembled from the individual two-dimensional results of the load steps. This procedure is only valid for the case of a constantly moving, quasi-static contact, that only applies to the middle part of the rotor length.

The geometry of the rotor profiles is modified to ensure a unique contact in the pitch point (figure 1). Due to manufacturing tolerances when grinding the rotors, the real rotor geometries deviate from the ideal profile. To evaluate the contact loads of real twin-screw machines, the contact zone near the pitch needs to be measured. Figure 10 shows the measured contact geometry around the pitch point of all male and female lobes of a real rotor pair in comparison to the ideal profiles. The contact geometries were determined experimentally with a tactile coordinate measuring machine, whereby the same cross section at constant $z = 50$ mm was considered for all lobes. A stylus with a diameter of 1.5 mm was used. Since the roughness of the rotors of Rz 3.4 - 4.3 is much smaller than the stylus diameter, the measured profiles correspond to the envelope of the rotor lobes. Overall, the measured geometries of the male

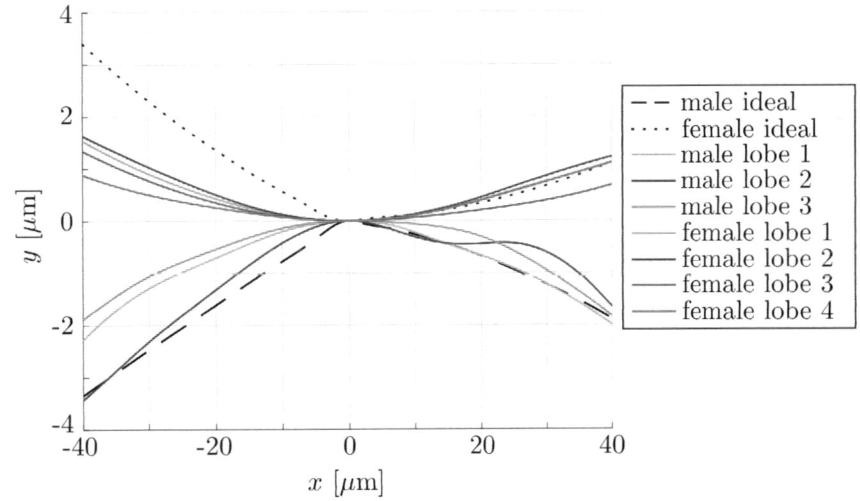

Figure 10. Ideal and measured rotor profiles near the pitch point

and female rotor show a smaller contact curvature than the ideal rotor profiles. In the area of $\pm 10\,\mu$m around the pitch point the real geometries deviate by about $1\,\mu$m. Furthermore, the contact geometries of the individual rotor lobes also differ significantly from each other. As a consequence, different contact loads are to be expected for the individual pairs of male and female lobes. To quantify the variation in contact load between the rotor lobes, only those measured geometries with the maximum and minimum contact curvature are considered. Compared to figure 10, lobe 2 of the male rotor and lobe 2 of the female rotor show the maximum contact curvature. In addition, the contact pair of lobe 3 of the male rotor and lobe 4 of the female rotor is investigated because of the minimum curvature around the pitch point.

For further numerical contact simulations the basic setup of figure 7 is used. Instead of the flat contact surface on the upper side of the elastic body, the male rotor profile is placed (figure 11). Furthermore, the female rotor profile is interpolated by a B-spline and used as rigid indenter, which moves relative to the elastic body. This relative motion is illustrated in figure 3 and is described by equation 6. In real operation of an unsynchronised screw machine, the contact load is imposed by the female rotor torque resulting from the gas forces. The introduced rotor

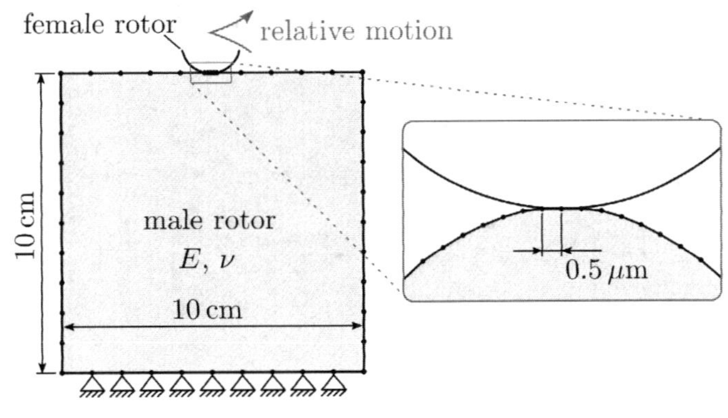

Figure 11. BEM discretization for rotor contact

kinematics from section 2 is displacement driven. Thereby, the contact load is applied by the rotational offset of the female rotor δ. It must be adjusted iteratively until the intended female rotor torque is achieved. The female rotor torque is equal to the sum off all nodal forces multiplied by the distance to the female rotor axis. For all further calculations a female rotor torque of 1 Nm is applied, that correspond to a normal contact force of 25 N. Furthermore, only free slip contact conditions are regarded for both the contact of the ideal and the real rotor geometries. The rotational increment for the male rotor is set to $\Delta\varphi_m = 0.0006250\,\text{rad}$, that correspond to a spatial discretization of $\Delta z = 26.857\,\mu\text{m}$.
Figure 12 shows the normal tractions t_n within the contact zone calculated for the ideal rotor geometry. The x-axis is about a hundred times finer resolved than the y-axis. This scaled view illustrates the curved contact area, that results for the non-linear elastic rotor contact. Since the z-coordinate is proportional to the male rotor rotation φ_m, the shown contact results can be interpreted both as the contact area that moves constantly into z-direction through the machine and as the contact at a fixed cross section for different rotational steps. Viewed from a fixed cross section, at $\varphi_m < 0$ the two rotors come into contact $53\,\mu\text{m}$ to the right of the pitch point ($x = 0\,\mu\text{m}$). As the male rotor rotation increases, the contact slides closer to the pitch point, where the x-coordinate remains nearly constant. In this contact area the rotors only perform a rolling motion. For $\varphi_m > 0$ the contact slides again slightly in the positive x-direction. The maximum normal tractions of $t_n = 4.31\,\text{GPa}$ are more than twice as large in the rolling area as in the sliding zones of the rotor contact. The whole contact area has a length of about 4 mm and a width of about $55\,\mu\text{m}$. Furthermore, the contact area is asymmetrical in both the x- and z-direction. Using the Hertzian contact theory for the present rotors and assuming the surface curvatures to be constant near the pitch point leads to a maximum normal traction of about 1 GPa. Thus, the Hertzian contact tractions are four times smaller than the results from the BEM. In addition, the Hertzian contact surface is calculated as an elliptical shape, that is symmetric to the x- and z-axis. These differences occur due to the assumption of constant curvature around the pitch point within the Hertzian theory. Compared to the Hertzian contact, the BEM results show that, at least for the present rotor geometry, it is not sufficient to assume

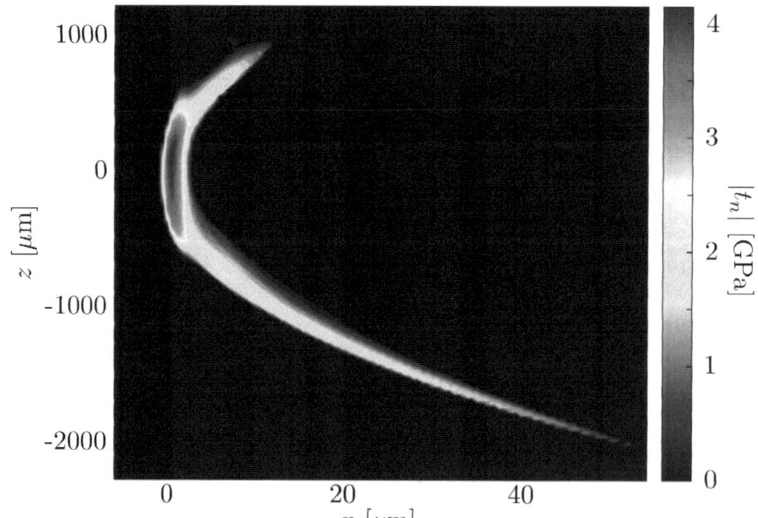

Figure 12. Ideal rotors: Normal tractions within contact zone

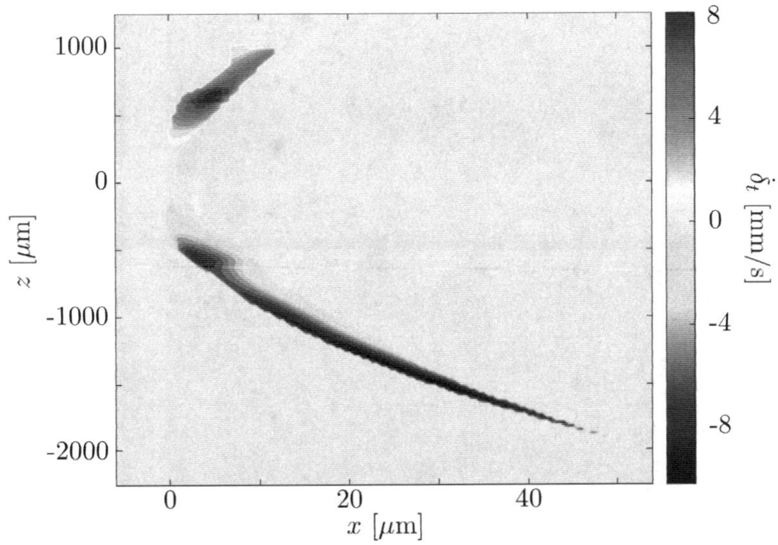

Figure 13. Ideal rotors: Sliding velocities within contact zone ($n = 5000\,\mathrm{min}^{-1}$)

a constant curvature within the contact zone of screw rotors. Due to the elastic deformation, the rotors also touch slightly beside the pitch point and areas with different curvatures come into contact. Figure 13 illustrates the calculated tangential sliding velocities $\dot{\delta}_t$ for the ideal rotor profiles. The pictured velocities are related to a rotational speed of $\dot{\varphi}_m = 5000\,\mathrm{min}^{-1}$. In contrast to the normal tractions, the maximum sliding velocities are located beside the pitch point. The sliding results from the kinematics, so that there is almost no tangential slide when

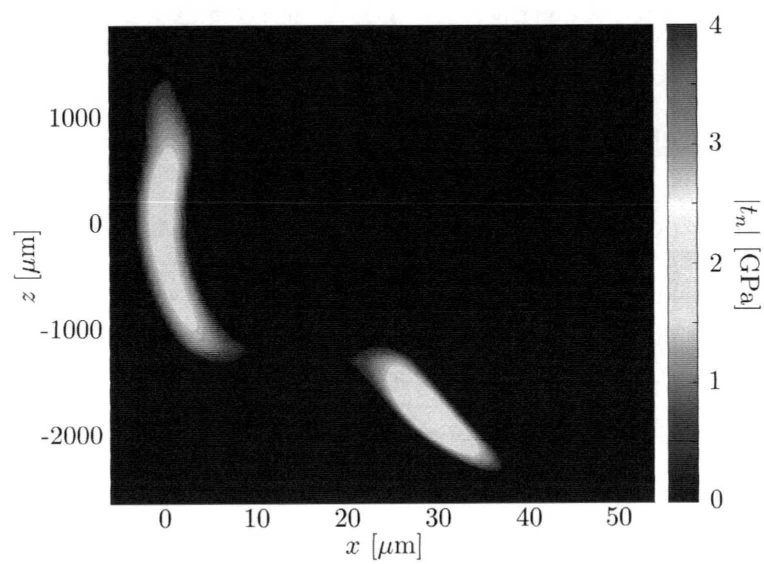

Figure 14. Measured rotors (m2-f2): Normal tractions within contact zone

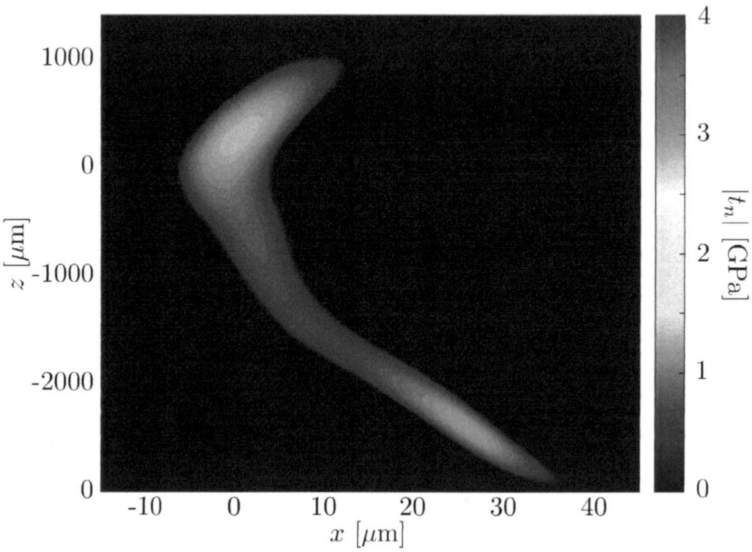

Figure 15. Measured rotors (m3-f4): Normal tractions within contact zone

the rotors roll at the pitch point. All in all, the magnitude of the sliding velocity of $10.89\,\mathrm{mm\,s^{-1}}$ is small compared to the maximum circumferential velocity of the rotors of about $24\,\mathrm{m\,s^{-1}}$. Figures 14 and 15 show the calculated normal tractions for the measured rotor profiles with the maximum and minimum curvature around the pitch point. In the case of the lobe pair with maximum curvature (m2-f2), the contact area is not continuous. The reason for this is the second hump of the male rotor lobe $30\,\mu$m beside the pitch point. As a consequence, the maximum normal traction of 2.24 GPa is located in the contact zone, where high sliding velocities are present. In contrast to the ideal rotors, the maximum contact load due to high normal tractions and sliding velocities is located beside the pitch point. In addition, due to the smaller curvatures compared to the ideal profiles, the maximum normal traction is almost half the size. For the minimum contact curvatures (lobe pair m3-f4), the calculated normal tractions are less than one third compared to the ideal contact. Overall, the local contact width is about twice as large as in the further results for the ideal contact.

Table 1. Comparison of contact results ($F_y = 25\,\mathrm{N}$)

profile	$\lvert\hat{t}_n\rvert$ [GPa]	$\lvert\bar{t}_n\rvert$ [GPa]	$\lvert\hat{\dot{\delta}}_t\rvert$ [$\mathrm{mm\,s^{-1}}$]	$\lvert\bar{\dot{\delta}}_t\rvert$ [$\mathrm{mm\,s^{-1}}$]	$F_{x,m}$ [N]
ideal	4.31	1.46	10.89	2.14	-0.67
m2 f2	2.24	1.20	15.16	0.48	-0.48
m3 f4	1.40	0.73	3.17	0.43	-0.04

Table 1 shows the maximum and average normal tractions and sliding velocities for the ideal rotor contact and the lobe pairs with maximum and minimum curvature. In general, smaller contact curvatures result in smaller normal tractions. In contrast, there is no direct correlation between the sliding velocity and the contact curvature. Furthermore, table 1 lists the x-component of the integral male rotor contact force retransformed into the fixed male rotor coordinate system. The calculations show that even without friction, a tangential force acts on the rotors that is perpendicular to the contact force. In the present cases, the calculated non-frictional tangential forces are 0.2-2.7 % of the prescribed normal force.

7. Summary and Outlook

The aim of this work is to evaluate the contact loads of unsynchronised, dry-running twin-screw machines. The BEM is an applicable method to calculate the tractions and sliding velocities within the elastic contact of the rotors. Several simplifications lead to an efficient contact algorithm, that enables the computation of a two-dimensional contact of arbitrary shaped, elastic bodies. Using the kinematics of the screw machine, the individual two-dimensional results are assembled to the three-dimensional contact zone. In addition, the individual two-dimensional results represent the contact progress depending on the rotor rotation at a specific cross section. Due to the relative coordinate system fixed on the male rotor and a local normal-tangential transformation, the contact evaluation and the definition of boundary conditions is simplified. The comparison between the BEM result and the Hertzian contact for a simple two-dimensional problem shows good agreement. The contact width and the pressure distribution is only slightly influenced by the discretization through the linear boundary elements. In the case of the screw rotor contact, the results from the Hertzian contact differ significantly from the BEM results. This is caused by the changing curvature over the contact length, which cannot be considered by the basic Hertzian contact formulation. Furthermore, the manufacturing tolerances of the rotor profiles show a significant influence on the loads within the contact zone. While the occurring normal tractions have a dependency on the contact curvature, there is no direct correlation for the sliding velocities. In addition, the BEM shows that, even for a non-frictional contact, non-vanishing tangential loads result from the elastic contact of the screw rotors. All in all, the results show a significant dependency of the contact loads within a twin-screw machine on the contact geometry. Even with a single rotor made in one manufacturing process, there is a large variation in the contact load between each lobe pair.
In further investigations the normal and tangential forces within the rotor contact will be determined experimentally and compared to the numerical results. The wear in the form of a change in the geometry of the contact zone will be quantified using a tactile measurement of the rotors after a defined number of loading cycles. The comparison to the numerical results will be used to investigate the wear mechanisms within the rotor contact. Furthermore, the influence of an elastic or elasto-plastic coating to the wear and friction within the rotor contact will be investigated numerically and experimentally. Therefore, the present BEM contact algorithm will be coupled with applicable wear and coating models.

References

[1] Rinder L 1979 *Schraubenverdichter* (Wien, New York: Springer-Verlag)
[2] Temming J 2007 *Stationärer und instationärer Betrieb eines unsynchronisierten Schraubenladers* Dissertation Universität Dortmund
[3] Hütker J and Brümmer A 2013 *8th International Conference on Compressors and their Systems* 407–416
[4] Dreischhoff U 1990 *Festkörperreibung in Schraubenmaschinen - Ein Beanspruchungsmodell für Werkstoffverbunde* Dissertation Universität Dortmund
[5] Dämgen U 1998 *Beanspruchung von verschleißbeständigen Werkstoffverbunden in Schraubenkompressoren* Dissertation Universität Dortmund
[6] Aurich D, Wittig A, Stangier D, Debus J, Thomann C-A, Tillmann W and Brümmer A 2021 *Vakuum in Forschung und Praxis* **33** 40–44
[7] Utri M, Aurich D, Brümmer A, Wittig A, Tillmann W, Moldenhauer H and Debus J 2019 *Process Technology Components*
[8] Wu Y-R and Tran V-T 2016 *International Journal of Refrigeration* **65** 111–128
[9] Vasu T S and Bhandakkar T K 2013 *International Conference on Computer Aided Engineering*

[10] Vasu T S and Bhandakkar T K 2015 *Sadhana* **40** 2221–2238
[11] Kosior F, Guyot N and Maurice G 1999 *Int. J. Numer. Meth. Engng.* **46** 65–82
[12] Leahy J G and Becker A A 1999 *The Journal of Strain Analysis for Engineering Design* **34** 235–251
[13] Radgen P and Blaustein E 2001 *Compressed Air Systems in the European Union. Energy, Emissions, Saving Potential and Policy Actions* (Stuttgart: LOG_X Verlag)
[14] Stosic N, Smith I and Kovacevic A 2005 *Screw compressors : mathematical modelling and performance calculation* (Berlin: Springer)
[15] Piegl L and Tiller W 2004 *The NURBS book* (Springer)
[16] Brebbia C A, Telles J C F and Wrobel L C 1984 *Boundary Element Techniques - Theory and Applications in Engineering* (Springer-Verlag)
[17] Becker A A 1992 *The Boundary Element Method in Engineering - A complete course* (McGRAW-HILL BOOK COMPANY)
[18] Gaul L and Fiedler L 2013 *Methode der Randelemente in Statik und Dynamik* (Springer Viewig)
[19] Beer G, Marussig B and Duenser C 2020 *The Isogeometric Boundary Element Method* (Springer)
[20] Takahashi S and Brebbia C A 1991 *Computer Methods in Applied Mechanics and Engineering* **93** 151–168
[21] Brezeanu L C 2015 *Procedia Technology* **19** 169–176
[22] Popov V L 2017 *Contact Mechanics and Friction* (Springer)
[23] Popov V L, Heß M and Willert E 2019 *Handbook of Contact Mechanics* (Springer)
[24] Takahashi S and Brebbia C A 1992 *Engineering Analysis with Boundary Elements* **9** 3–11

Contribution of modern rotor profiles to energy efficiency of screw compressors

S Patil[1,2], M Davis[1], N Stosic[1], A Kovacevic[1] and N Asati[2]

[1]Centre for Compressor Technology, City, University of London, Northampton Square, London EC1V 0HB, United Kingdom

[2]Department of Research & Development, Kirloskar Pneumatic Co. Ltd., Hadapsar, Pune 411 013, India

E-mail: sumit.patil@city.ac.uk

Abstract. Screw rotors are the heart of screw compressors. And the energy efficiency of industrial machines is a matter of tremendous significance now more than ever. Historically, rotor profile developments have played a key role in making screw compressors energy efficient and commercially viable. Further attention to manufacturing aspects of rotor profiles and the invention of the rack generated rotor profiles led to rotor profiles having good manufacturability. The principles of rotor profile generation and manufacturing are available in open literature since 1960's. But more and more literature on rotor profiling has been published since then. Modern screw rotor profiles (patented close to and in the 21st century) have all the principles of a good profile incorporated in their design. Hence the industry and profile designers at large are aware of the increasing difficulty to further make the twin screw compressor rotor profiles more energy efficient. This paper tries to quantify the contribution of rotor profile to energy efficiency of a typical twin screw compressor by comparing the most recent (hence modern) screw compressor rotor profiles. In order to fairly compare different rotor profiles, all are retrofitted to a single size, lobe combination, rotor length, and helix angle. Only the curves constituting the profile, as dictated in the patent documents of the respective profiles, have been changed. Tools such as SCORPATH and SCORG have been used to do the geometric and thermodynamic calculations on the profiles. Keeping the working conditions same for all the retrofitted but different profiles, a comparison has been made. This comparison sheds some light on how much is the energy efficiency of a particular twin screw compressor influenced by a mere change of profile. This analysis can be further extended to establish reasonable targets for twin screw compressor manufacturers to improve energy efficiency of their machines via improving their rotor profiles. This remains the future scope of this work.

1. Introduction

Twin screw compressors have evolved for almost a century since their invention in 1930's. Rotor profiles being an essential element of these machines, have gone through a similar evolution over these years. Advancements in rotor profile design and manufacturing played crucial role in putting the twin screw compressors on par with the other compression technologies. Energy efficiency of compressors could be talked about in terms of either their specific power which is power consumed to deliver a unit of compressed medium or the adiabatic or isothermal efficiencies of compression process evaluated for that machine. Early rotor profiles were not very energy efficient due to large leakage areas between the rotors. This scenario changed with the invention of asymmetric rotor profiles, first by Lysholm [1] and later with the SRM-A profile by Schibbye [2] which led to significant reduction of internal leakages with respect to total throughput of the rotors. The developments in manufacturing technology for making the helical rotors with precision also played a key role in making screw compressors energy efficient and hence commercially viable. Only by maintaining the tight form of rotor profiles in manufacturing, the internal clearances could be kept to minimum and thereby internal leakages were minimized too. There's a long and interesting history to how these principles of generating viable rotor profiles were discovered, patented and then re-discovered by other manufacturers. It is very well summarized by Stosic *et al.* [3]. By and large, most of the major screw compressor manufacturers had their own profile patents by 1980's [4] [5] [6] [7]. All these profiles were based on principles similar to SRM-A profile, wherein curves of choice were defined on either rotor to derive the conjugate curves on the other rotor using gearing condition. Efforts were put into choosing most suitable curves which would facilitate larger throughput area and minimize the leakage areas keeping manufacturability in consideration.

In the period that followed, the principle of rack generation was introduced in rotor profiling for screw compressors first by Menssen [8] followed by Rinder [9]. The principle of rack generation had several advantages over previous method such as better manufacturability by design, involute contact and a simplicity in the design. Still, the rack generated profiles had ample scope of improvement which was steadily and incrementally realized in later profiles such as Stosic [10] and as latest as Cavatorta and Tomei [11]. The rack generated profile by Rinder [9] though ingenious; had a problem of slightly larger blow hole area and poor sealing on the portion of rotors generated by the high pressure side of the rack. This particular shortcoming of the rack generated profile was overcome by Stosic and Hanjalic [12] through introduction of cycloids in this region of rack which could also be seen as a combined rotor-rack generation method. N-rotors are stronger yet lighter and also facilitate high throughput to leakage area ratio resulting in a higher energy efficiency of the machine.

Later and more recent contributions to the field of rotor profiling came through applications of finer design principles such as variable clearance distributions on rotors, controlling torque characteristics of the gate rotor [13] and an extensive work on optimisation techniques applied to rotor profile designs [14] [15] [16] [17]. The wealth of all this literature available in modern times has made the principles of good rotor profiling accessible to the majority of rotor profile designers.

The intent of this paper is to review the more recent or modern rotor profiles from last 25 years. These modern rotor profiles are based on the body of literature and knowledge briefly outlined hereby. The state of the art rotor profile patents owned by major compressor manufacturers are reviewed in light of their uniqueness and differences. Their contribution to energy efficiency of modern screw compressors is discussed. Based on the review and discussion, a general outline of what may lie in future for the screw rotor profiling is drawn.

2. Modern screw rotor profiles

Fu Sheng's [18], Gardner Denver's [11], City's [19] and Kaeser's [20] are some of the major rotor profile patents in last 20 years. They are chosen particularly to highlight the features of the state of the art rotor profiles. There are obviously more patents to be found on rotor profiles in this time period but these profiles are assumed to be representatives of good rotor profiling principles applied in the design of energy efficient screw compressors by the respective assignees.

Figures 1, 2 and 3 depict the mentioned modern rotor profiles as represented in their respective patent documents. The patent [19] is mainly about the N-profiled rotors designed for better torque characteristics and reduced noise. Geometrically, the profile curves in this patent are same as that of [10].

Two of the three profiles depicted herein (figures 1 and 2) are based on the principle of rotor-

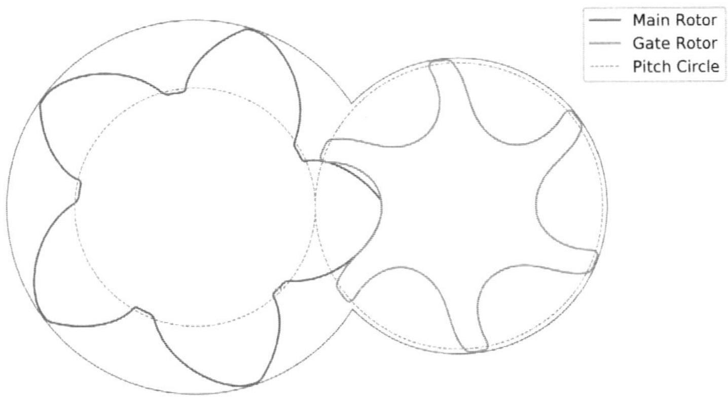

Figure 1. Fu Sheng Profile [18]

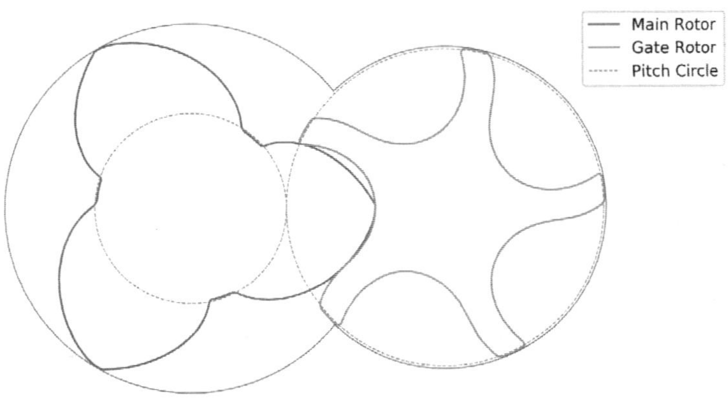

Figure 2. Gardner Denver Profile [11]

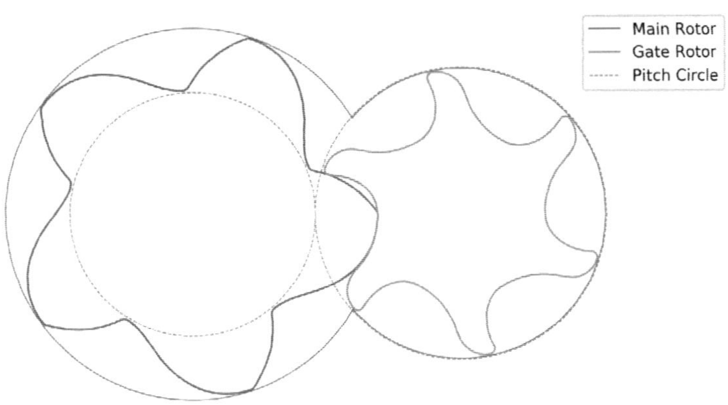

Figure 3. Kaeser Profile [20]

rack generation which is known have several advantages over classical rotor generation or rack generation principle. The racks used to define these profiles are very similar to the N-rack [10]. They all differ essentially by the curve used to define the low pressure side of the rack. Figure 4 depicts a general rack profile for rotor-rack generated profiles with nodes representing end points of the constituent curves. Table 1 then lists the individual curves to show how the patents are different. The respective conjugates on rotors are simply roulettes of the curves defined on rack calculated by solving the envelope gearing condition.

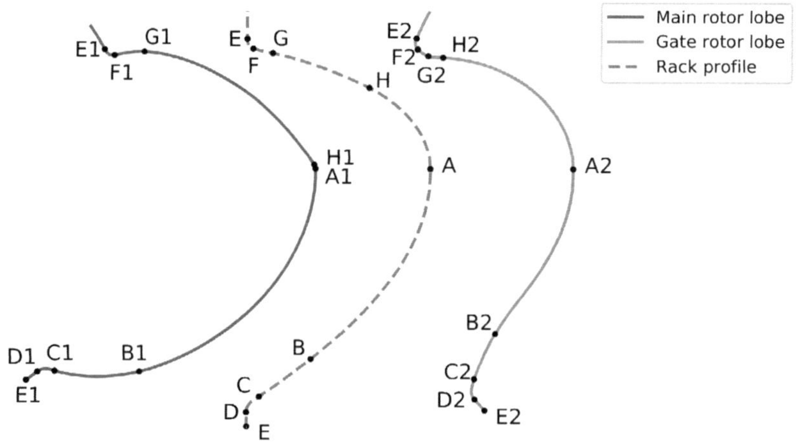

Figure 4. A general rack profile generating main and gate rotor profiles

Table 1. The curves constituting rack of the N-profile, Fu Sheng profile and the Gardner Denver profile

Rack curve	N-profile	Fu Sheng profile	Gardner Denver profile
D-E	Vertical straight line	Vertical straight line	Vertical straight line
E-F	Circular arc	Circular arc	Circular arc
F-G	Straight line	Straight line	-
G-H	Cycloid	Cycloid	Cycloid
H-A	Cycloid	Cycloid	Cycloid
A-B	**General arc***	**Elliptic arc**	**Hyperbolic arc**
B-C	Straight line	Straight line	Straight line
C-D	Circular arc	Circular arc	Circular arc

*Parabolic, Elliptic or Hyperbolic arc, depending upon the arc exponents [10]

The key change in the three similar profiles in table 1 is in the curve *A-B*. This particular portion of the rack determines what would be main and gate rotor lobe thicknesses which influence the total combined throughput area of the rotor pair. It also affects the sealing line length of the meshing rotor pair. Hence, the changes in arc to get various conic sections such as ellipse and hyperbola are probably done for striking a good balance between total throughput area and the sealing line leakage area. Curve *A-B* does not affect high pressure side blow-hole area but all the curves above horizontal axis do. In this context, it is interesting to note that the straight line F-G is omitted in the Gardner Denver profile. This slight change of directly connecting the cycloid with circular arc on high pressure side helps marginally reduce the blow hole area.

Only the Kaeser profile out of most of the modern profiles is based on the principle of rotor generation. Majority of this profile's curves are defined on the gate rotor and the remaining on the main rotor. It is largely based on its predecessor SIGMA profile by Bammert [4]. The latest profile has made amendments with the targets of improving gate rotor torque characteristics, minimizing blow-hole area, strengthening gate rotors and optimizing the profiles with respect to multi-objective criteria. Especially the profile version of Kaeser's new profile depicted in figure 3 has up to four times smaller blow-hole than an equivalent rack-generated profile due to slightly negative gate rotor addendum combined with a well designed sharp gate rotor tip. This however comes at a price of relatively lesser throughput area due to smaller profile depth. Such trade-offs are ubiquitous in rotor profile design. Therefore, one has to weigh in the role of operating conditions and application of the screw compressor to make the best trade-offs.

3. Contribution of modern profiles to the energy efficiency of screw compressors

The literature has several clues and reportings of the improvement in energy efficiency of screw compressors with the onset of rotor-rack generation method along with the application of finer design principles such as variable clearance distributions and multi-variate optimisation to rotor profiles. That is to say, contribution of the modern screw rotor profiles replacing the older profiles of late 20th century to the energy efficiency of screw compressors can be pinned down with evidence. Stosic *et al.* [21] presents a case of retrofitting old SRM-A profiled rotors in an oil flooded twin screw compressor with the N-profiled rotors and thereby improving the energy efficiency of the machine by 2.5%.

From a purely geometric point of view, profile does not solely contribute to the energy efficiency of screw machine. Other characteristics of the rotors such as lobe combination, diameter, length, helix angle, speed and also the suction and discharge ports affect the performance to a significant degree. Hence, only a holistic approach to optimum choice of all the geometric and operating parameters would lead to substantially efficient compressors. The combined profile and geometry optimisation with Kaeser's new profile [20] is claimed to have improved upon the old SIGMA profile [4] by up to 3%. Similarly, Fu Sheng's rack generated profile (figure 1) is claimed to have improved on their own old rotor generated profile [22] by 1.32% in terms of the energy efficiency [17].

Nevertheless, all three of the presented examples arise from a comparison of the modern profiles with old profiles. But, if one wishes to gauge the contributions of only modern rotor profiles into improvements of energy efficiency of screw compressors in last 25 years, only the modern profiles must be compared starting from N-profile [10]. As elaborated in the previous section, majority of the modern screw rotor profiles have incorporated the principles of designing good profiles. This could be the reason for minimal differences among their designs. But to what extent do these minimal changes affect the energy efficiency of machines? To answer this, Fu-Sheng profile [18] and the Gardner Denver profile [11] can be compared through retrofitting and numerical evaluation of their geometric and thermodynamic properties. The profile with 5/6 lobe combination of rotors as depicted in figure 1 true to Fu Sheng's patent description is generated and taken as a reference for calculation. The Gardner Denver profile's patent depiction (figure 2) is in 3/5 lobe combination of rotors. In order to retrofit with Fu Sheng profile, a Gardner Denver profile with 5/6 lobe combination is generated based on the patent description of curves and other details of the profile. The two profiles are set at the same centre distance of approximately 100 mm and have a 140 mm main rotor outer diameter. The length to diameter ratio is set to 1.55 and the wrap angle is set to 300° for both rotor pairs. Figure 5 is a representation of these two retrofitted profiles.

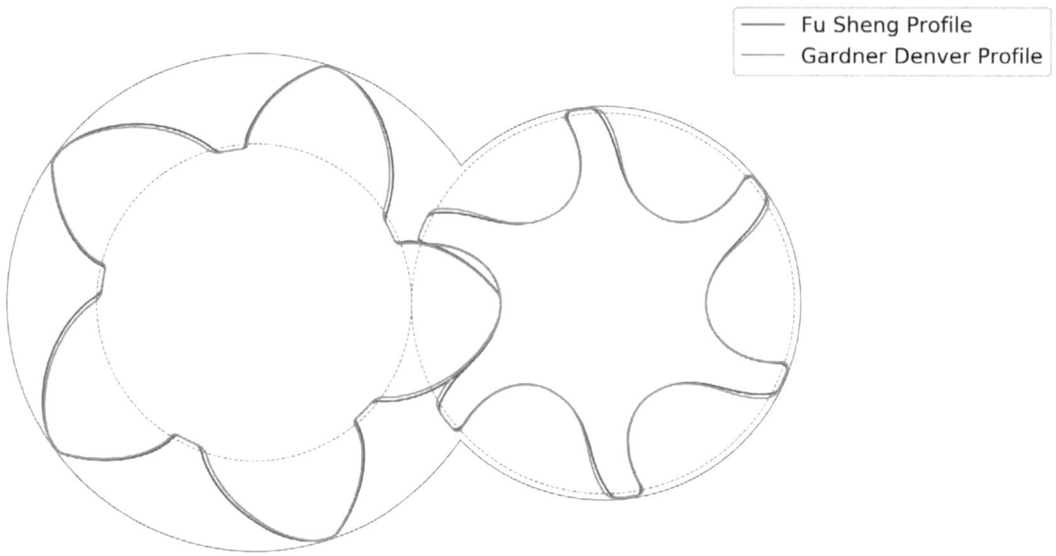

Figure 5. Retrofitted rotor profiles with 5/6 lobe combination-[18] and [11]

As elaborated earlier, the high pressure side of these profiles is identical except the absence of straight line in Gardner Denver profile. A closer look at the leading edge tip of the gate rotors of Gardner Denver profile shows that it would have a smaller blow hole area compared to Fu Sheng profile on account of this small change. The low pressure sides of these profiles have Ellipse and a Hyperbola respectively. That makes them look slightly different on leading edge of the main rotor. The throughput area and the leakage areas for these two profiles are calculated and given in table 2. It is worth noting that both, FuSheng and Gardner Denver profiles can be generated through the N rotor form by choosing coefficients and exponents in its general curve to resemble circles, ellipses, parabolas or hyperbolas.

Table 2. Geometric profile characteristics for retrofitted Fu Sheng and Gardner Denver profiles

Profile Characteristic	Fu Sheng Profile	Gardner Denver Profile
Throughput area (mm^2)	8507	8621
Interlobe sealing area (mm^2)	4.693	4.876
Blow hole area (mm^2)	1.852	1.191

The Gardner Denver profile has a slight advantage in throughput area as well as blow hole area on account of the finer details elaborated in previous section. If these profiles are subjected to a thermodynamic simulation for an oil injected screw compressor air application, the difference in energy efficiency of these machines is observed to be of the order of 0.3% to 0.5% only. The simulation was done in SCORG software within the envelope of operating conditions - tip speed 10 m/s to 40 m/s, suction pressure 1 bar absolute (bara) and discharge pressure 6.5 bara to 12.5 bara. The oil injection parameters and other peripheral details were set to practically reasonable values and same for both setups to avoid their influence on evaluation. The observed order of difference shows that the modern profiles which are more or less equivalent in application of good design principles do not make a big difference in terms of energy efficiency. Even the gap of 0.3% could be minimized if the profile with larger leakage area is allowed to compensate using other characteristics such as rotor size, sharper tips or rotor speeds, etc. If an N-profile were designed for similar size, it wouldn't be any different because structurally both the profiles in figure 5 are equivalent to it. Either FuSheng or Gardner Denver profiles can be generated through the N rotor patent definition by changing the curve *A-B* exponent *n* [10]. No more than 0.5% of difference in energy efficiency would be observed by only interchanging curves such as-parabola, ellipse and a hyperbola.

4. Conclusion & Future scope

Most of the known features of a good rotor profiles are already incorporated in all the modern rotor profiles. Hence, an advantage by a mere change of curves or their placement in a profile is practically not possible at least in modern profiles. Unlike the leap in energy efficiency of screw compressors first by introduction of asymmetric profiles in 1960's and then through the principle of rotor-rack generation in 1990's, modern profiles are unlikely to come by such leap through only change of profile curves and shapes.

Screw compressor manufacturers already in possession of a good modern profile should expect improvements from any new profiles only in the order of 1% at maximum. More improvements are only going to come from either innovations in optimisation techniques or a scientific disruption in the methods of profile generation and manufacturing. One example of such an innovation which could improve rotor profiles if combined with good optimisation framework is application of topology inspired by path homotopy in profiling [23].

Since all modern profiles are more or less equally efficient, the weight of designing more efficient screw compressors lies with the ability of profile designer. Use of the state of the art optimisation algorithms, understanding of the machine application and underlying physics of compression is paramount if one wishes to fine tune any good profile for the desired duty. Energy efficiency of not only operation but also during manufacturing of screw compressors until the end of complete product life-cycle should be the focus for future of rotor profiling.

References

[1] Lysholm A 1967 Screw rotor machine US Patent 3,314,598
[2] Schibbye L B 1969 Screw rotor machine and rotors therefor US Patent 3,423,017
[3] Stosic N, Smith I K and Kovacevic A 2003 *Proceedings of the Institution of Mechanical Engineers, Part E: Journal of Process Mechanical Engineering* **217** 157–170
[4] Bammert K 1982 Intermeshing screw rotor machine with specific thread profile US Patent 4,350,480
[5] Bowman J L 1983 Helical screw rotor profiles US Patent 4,412,796
[6] Astberg A 1984 Screw rotor machine and rotor profile therefor US Patent 4,435,139
[7] Hough D, Morris S J and Barber A D 1987 Screw rotor machines with specific tooth profiles US Patent 4,636,156
[8] Menssen E 1977 Screw compressor with involute profiled teeth US Patent 4,028,026
[9] Rinder L 1987 Screw rotor profile and method for generating US Patent 4,643,654
[10] Stosic N R 1997 Plural screw positive displacement machines US Patent 6,296,461
[11] Cavatorta P and Tomei U 2014 Screw compressor having male and female rotors with profiles generated by enveloping a rack profile US Patent 8,702,409
[12] Stosic N and Hanjalic K 1997 *Journal of Fluids Engineering* **119** 659–664
[13] Stosic N, Smith I and Kovacevic A 2005 *Screw compressors: mathematical modelling and performance calculation* (Springer Science & Business Media)

[14] Su S H and Tseng C H 2000 *Journal of mechanical design* **122** 543–552

[15] Kauder K, Helpertz M, Reusch B and Berlik S 2002 *VDI BERICHTE* **1715** 29–50

[16] Stosic N, Smith I K and Kovacevic A 2003 *Applied Thermal Engineering* **23** 1177–1195

[17] Wu Y R and Fong Z H 2009 *Mechanism and Machine Theory* **44** 66–82

[18] Lee H T, Fong Z H and Wu Y R 2006 Mechanism of the screw rotor US Patent App. 10/961,056

[19] Stosic N R 2017 Reduced noise screw machines US Patent 9,714,572

[20] Weih G 2019 Rotor pair for a compression block of a screw machine US Patent 10,400,769

[21] Stosic N, Smith I, Kovacevic A and Venumadhav K 2000 Retrofit 'n' rotors for efficient oil-flooded screw compressors *Proc. of the 2000 International Compressor Engineering Conference at Purdue*

[22] Lee H T 1990 Screw-rotor machine with an ellipse as a part of its male rotor US Patent 4,890,992

[23] Patil S, Stosic N, Kovacevic A, Smith I and Asati N 2021 Application of path homotopy in twin screw compressor rotor profile design *IOP Conference Series: Materials Science and Engineering* vol 1180 (IOP Publishing) p 012004

Acknowledgments
Authors would like to thank Kirloskar Pneumatic Company Limited, India for financially supporting this research.

Investigation of radial gap size change under load and the impact on performance for a twin screw compressor using numerical simulation

R Andres, J Hesse and A Spille

CFX Berlin Software GmbH, Karl-Marx-Allee 90a, D-10243 Berlin, Germany

E-mail: rainer.andres@cfx-berlin.de

Abstract. Computational Fluid Dynamics (CFD) and the Finite Element Method (FEM) are common and validated tools in research and industry for solving fluid flow or performing structural analysis respectively. Despite being rather challenging, also the numerical simulation of screw compressors benefits from development and enhancement of numerical models, tools and methodologies. Differences between the real machine and the numerical model are inevitable, yet increasing accuracy and prognosis quality are desired by refining existing models and accounting for specific phenomena.

Clearances within the working chamber (mainly intermesh clearance between rotors and axial and radial gaps between rotors and housing) strongly affect the flow characteristics, efficiency and thus the overall compressor performance. These clearances change under operating conditions because of thermal expansion or due to forces acting on the structure, whereas the clearance size is often only known for the reference state out of operation. The extent of deformation and the quantification of the resulting clearances is therefore of high interest in order to get the numerical model closer to real operating conditions.

In this paper numerical simulation was used to determine the change in radial gap size between rotors and housing when the compressor is running and exposed to thermal and pressure loads. For a specific operating point, a 3D-CFD simulation for a 4-6 twin screw compressor with SRM profiles was conducted for the undeformed reference compressor geometry to calculate the flow field within the machine. Also, the temperature field within the housing and rotor solids was computed using the Conjugated Heat Transfer (CHT) method. The temperature and pressure fields were then submitted to the FEM solver to compute the deformation of rotors and housing. The results of this structural analysis served as input for a modification of geometry and numerical grids to account for the change in radial gap size. A re-run of the flow simulation in the deformed state enabled to quantify the impact on the machine performance and specific flow quantities. It could be shown that taking deformation into account, the volumetric efficiency is clearly affected.

Commercial solvers from Ansys were utilized for solving fluid and structure. The spatial discretization of the fluid volume around the rotors was realized by employing pre-generated grids using the mesh generator TwinMesh. Here, the rotor deformation can be defined in dependence of the axial position. The mesh of the housing maintained unchanged for all CFD simulations as a resulting gap size was determined by considering housing and rotor deformation relative to each other.

1. Introduction

Screw compressors are widely used in the industry, e. g. for pneumatic power tools or in the automotive industry. In this paper an oil free compressor is modeled, which is often used when purity of the air is critical. Within oil injected compressors, there is the risk of contaminated air which has an adverse effect of downstream processes and production. However, oil free compressors lack the sealing and cooling effect of the oil, thus in general the achieved pressure ratios are lower compared to oil flooded compressors. Furthermore, the lack of sealing increases leakage losses, and the lack of cooling increases solid temperatures and therefore expansion.

In the CFD simulation axial and radial clearances are fully resolved to account for gas flow between the rotors and between rotor and housing. These clearances lead to reverse flows through the gaps from the chamber with higher pressure to the one with lower pressure and therefore losses inside the machine. The initial gaps will change under operating conditions because of thermal expansion or due to forces acting on the structure. The impact on calculation results when taking the gap change into account due to thermal deformation is demonstrated in [1] and [2].

For the manufacturer, reliable tools to evaluate overall machine performance such as compression ratio and power consumption as well as to calculate losses and efficiency are of great desire. Also, the impact of design changes should be evaluated before prototypes are built so that developing costs can be reduced.

The compressor investigated in this work consists of a generic housing geometry and rotors with SRM profiles. Table 1 lists the essential parameters for the undeformed geometry.

Table 1. Geometry parameters of the investigated screw compressor

	Male rotor	Female rotor	
Number of lobes	4	6	-
Length	168.1		mm
Tip diameter	101.9	101.1	mm
Root diameter	58.7	57.9	mm
Rotor Wrap angle	300	200	deg
Center distance	80		mm
Inner volume ratio	2.2		-

2. Rotor chamber modeling

The rotor chambers of the screw compressor consist of volumes, which deform over time, depending on the rotor position. The chambers are connected by gaps that allow the rotation of the rotors in the static housing. Due to the intermeshing of the two rotors, the chambers strongly deform. For most CFD codes, several approaches are possible to account for this deformation. For a systematic introduction to 3D CFD simulations of screw compressors see [3]. Various methodologies for chamber modeling and the corresponding advantages and disadvantages are shown in [3]. Common approaches are e. g. automatic remeshing, immersed solid methods or customized grid generation.

The present analysis uses an automated mesh generation by the commercial software TwinMesh. At a fixed angle increment, the working chamber volumes around both rotors are discretized for the different rotor positions with a hexahedral 3D mesh consisting of two O-type grids. Here, the rotational periodicity of the rotor pair is considered, so that the set of grids is only generated for one pitch angle, i. e. 90° (360° divided by the number of male rotor lobes). This meshing approach and the CFD models were validated for a 2-stage screw compressor [5], a screw refrigeration compressor [6], and a screw expander [7], and in several confidential projects for clients.

Within the intermesh region, the two O-type grids are 1:1 connected to each other. Inflation layers near rotor and housing walls are present. Also, as part of the working chamber grid, radial, axial and intermesh clearances are taken into account. A 2D slice of the working chamber grid is shown in Figure 1.

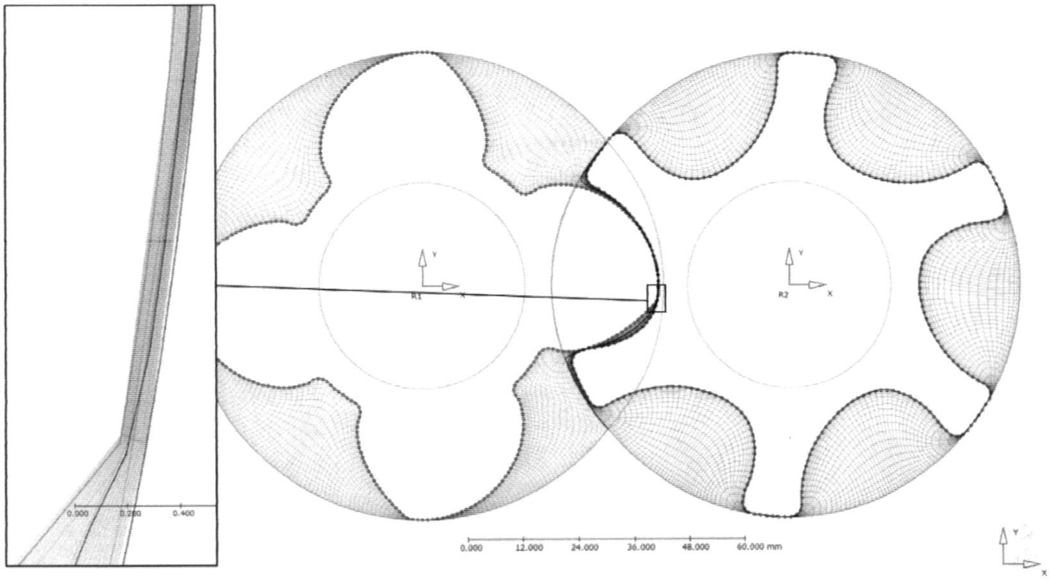

Figure 1. 2D cross section through the working chamber of the compressor

During the transient simulation, the location of each working chamber grid node is updated depending on the rotation angle via a routine at the start of each time step. The mesh topology is retained (node numbers and node connectivity), therefore no interpolation between individual time steps is necessary. An exception from this rule is the start of a new pitch angle range (i. e. every 90° of male rotation angle) because the set of pre-generated grids is re-used and results are interpolated from the last to the first grid in the set.

3. Simulation setup

3.1. Methodology of fluid-structure interaction

The aim of this paper is to evaluate the change of the radial clearance sizes of the screw compressor due to thermal and pressure load and its effect on compressor performance. Three simulation steps are done sequentially:

1. Transient CFD simulation of the screw compressor with straight rotors and uniform radial gap size of 50 µm; conjugate heat transfer (CHT) is included to account for solid heating of rotors and housing
2. Steady-state structural simulation of rotors and housing with loads from step 1 (temperatures of rotors and housing are almost constant, pressure load is taken arbitrarily from last time step)
3. Transient CFD simulation of the screw compressor with deformed rotors; the deformation is taken from step 2; since TwinMesh does not support housing deformation (this will be included in the next release) but only shaft deflection and rotor expansion, these are chosen so that the radial gap sizes of step 2 are met best.

3.2. Fluid setup

The simulations are set up and performed in the commercial solver Ansys CFX. Figure 2 shows the assembly of the setup which consists of separate grids for fluid and solid parts. The fluid grids discretize the working chambers and the stator volumes, the solid grids represent the solid parts for rotors and casing. Table 2 gives an overview on the mesh resolution of the different parts of the CFD simulation.

Table 2. Number of vertices and elements in the meshes for the CFD simulation

	Vertices	Elements
Fluid around male rotor	830,700	776,000 hexahedrons
Fluid around female rotor	877,500	816,000 hexahedrons
Fluid in stator	263,981	217,042 tetrahedrons; 91,719 prisms; 166,434 hexahedrons
Solid of male rotor	22,145	83,138 tetrahedrons
Solid of female rotor	23,788	88,223 tetrahedrons
Solid of casing	390,120	923,058 tetrahedrons; 263,697 prisms; 47,200 hexahedrons

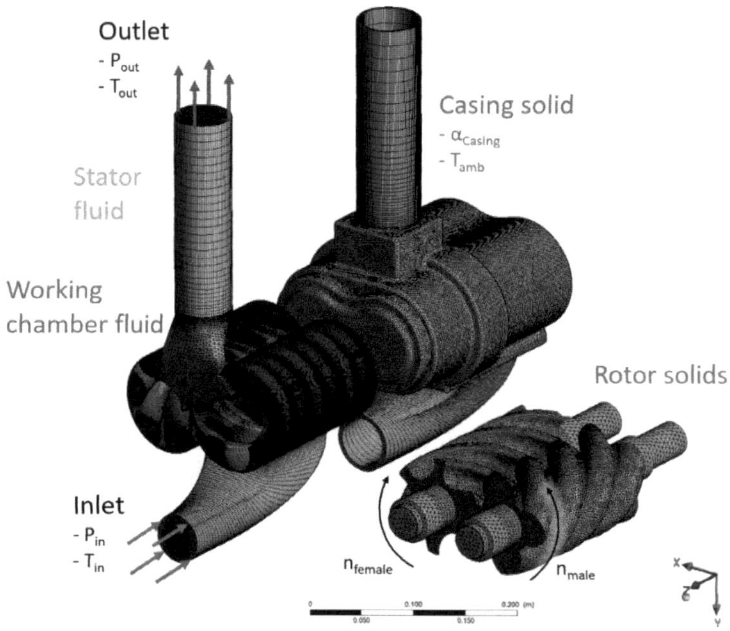

Figure 2. Boundary conditions and numerical grids for fluid and solid parts: Top right shows static solid for casing and pipe walls, bottom right the two rotating solids for the rotors; left side shows the fluid volumes consisting of static parts in suction and pressure ports and connected pipes (light blue) and deforming parts around both rotors (dark blue). The inlet is at bottom, the outlet at top side.

For the conservative flux of mass, momentum and energy between individual fluid volumes, i. e. stator volume and deforming working chamber, fluid-fluid interfaces are specified. In addition, heat flux between fluid and solids as well as temperatures within the structure are calculated using fluid-solid interfaces and the CHT method. Here, convection boundary conditions are set for the outer walls of the structure. At the outer casing wall, an ambient temperature of 20°C is assumed with an estimated heat transfer coefficient of 10 W/(m² K) for quiescent air [8]. The rotor shafts are usually located within an enclosure and experience heat dissipation from the bearings. Since these parts are not considered in the model, for the outer walls of the rotor shafts an estimated and fixed temperature of 70°C is specified.

The operating point for all calculations is defined by a rotational speed of the male rotor of 12 333 rev/min and a pressure ratio between inlet and outlet of 1:3. The fluid is air as an ideal gas, all solids are from steel. The boundary conditions are summarized in Table 3. Inlet and outlet are specified as openings via non-reflective boundary conditions; outlet temperature is only needed in case of backflow. The radial, axial and intermesh clearances for the initial run without deformation are shown in Table 4. For this investigation, synchronized rotors are considered, thus the intermesh clearance between male and female rotor is constant.

Table 3. Boundary conditions of the initial run

Angle increment (male rotor)	$\Delta\theta$	1	deg
Time step	Δt	13.51	µs
Rotation speed male	n_{male}	12333	rev/min
Rotation speed female	n_{female}	8222	rev/min
Inlet pressure (total)	p_{in}	1	bar(a)
Outlet pressure (static)	p_{out}	3	bar(a)
Inlet temperature	T_{in}	20	C
Outlet temperature	T_{out}	160	C
Rotor shaft temperature	T_{shaft}	70	C
Ambient temperature	T_{amb}	20	C
Heat transfer coefficient for outer casing walls	α_{casing}	10	W/(m² K)

Table 4. Clearances for the initial simulation run

Radial male (uniform)	50	µm
Radial female (uniform)	50	µm
Axial male (equal for pressure and suction side)	100	µm
Axial female (equal for pressure and suction side)	100	µm
Intermesh	100	µm

As shown above, the required order of magnitude for time step to calculate the fluid field is approx.14 µs leading to (depending on mesh resolution and used hardware) computational times in the range of hours for one revolution. Opposed to this timescale, the required time to heat up the structure is significantly larger in the range of 100,000 revolutions. Based on experience, it may take 10-15 min until the rotor and casing solids reach a certain operating temperature. It is not applicable to cover this rather large time span, so that a transient simulation of the full heating up cannot be done. Here, we are only interested in the periodic state after the heat up phase, i. e. chamber pressures and temperatures change periodically due to rotation, and solid temperatures are almost constant with only very small periodic changes. Once this temperature level is reached it is assumed that for a fixed point within the rotor or housing, the local temperature should be almost constant, independent of the rotation angle.

To overcome this challenge, the timescale for the energy equation within the solid volumes is increased by a factor up to 10,000 at simulation start so that solid temperatures increase fast. This approach is similar to decreasing the heat capacity of the solid material. Once the temperature monitors within the solids approach a periodic state, the timescale factor is decreased sequentially down to a value of 100. This reduction is necessary because the larger timescales artificially increase the fluctuation at the monitored points and lead to inhomogeneous temperature distributions, especially in circumferential direction, where the rotor lobe which lies within a pocket of hot gas heats up faster than in reality due to this acceleration approach.

The position of the temperature probes in the male rotor solid near the pressure side can be seen in Figure 3. As it can be seen in Figure 4, their temperatures show still differences between individual lobes and trends regarding the mean value; especially timescale factor 1000 should have been used for a longer time. Moreover, timescale factors lower than 100 are not considered here to limit the simulation effort required for this analysis. It must be noted that the temperature probes for the individual rotor lobes do not share the exact same radial and circumferential position, since they are only attached to the nearest grid node relative to this, whereas the mesh is not periodic in circumferential direction.

To improve these results, simulation time should be extended as well as the monitoring methodology to reduce errors introduced by these factors. Still, as a compromise between simulation time and accuracy, the outlined methodology is considered acceptable, where general feasibility of the workflow and the significance of taking deformations into account are in focus.

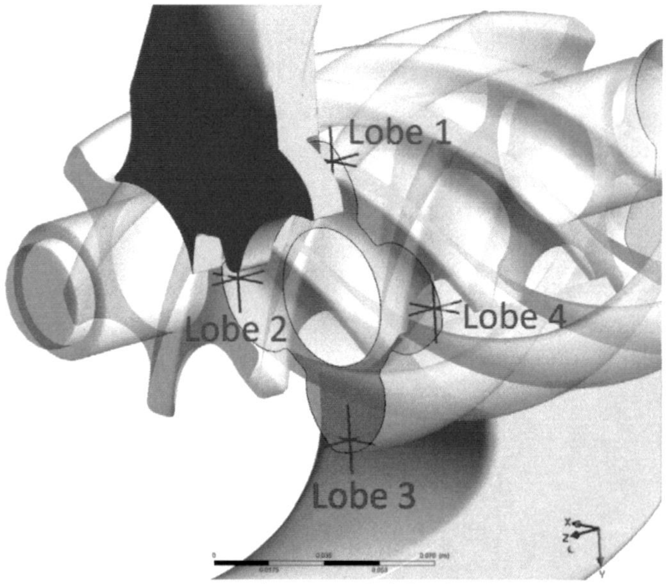

Figure 3. Temperature probes at each lobe for the male rotor

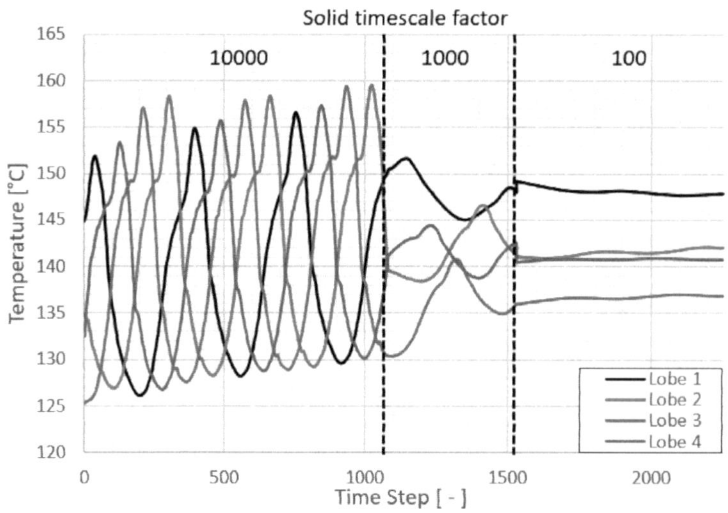

Figure 4. Temperature at male rotor probes for 3 different solid timescale factors (10000, 1000, 100)

3.3. Structural setup

The structural simulation for the screw compressor is done in the commercial solver Ansys Mechanical (Mechanical APDL). The structural model includes the geometry of both rotors and the whole casing, from inlet to outlet. A fixed and arbitrary rotor position is used for the steady-state structural simulation. The influence of the rotational position is not investigated and might be a topic for further work.

A cut through the CAD model is shown in Figure 5, with the male rotor in blue, the female rotor in orange and the housing in green. Highest impact on the deformation of the bodies has the temperature, which leads to thermal strains from temperature gradients and overall thermal expansion. The thermal and pressure loads for the static structural simulation are taken from the last CFD time step. Due to the high heat capacity of the steel, the temperature inside the solids remains nearly constant. Any influence due to rotor dynamic effects, like gyroscopic forces and interferences due to eigenfrequencies, were neglected.

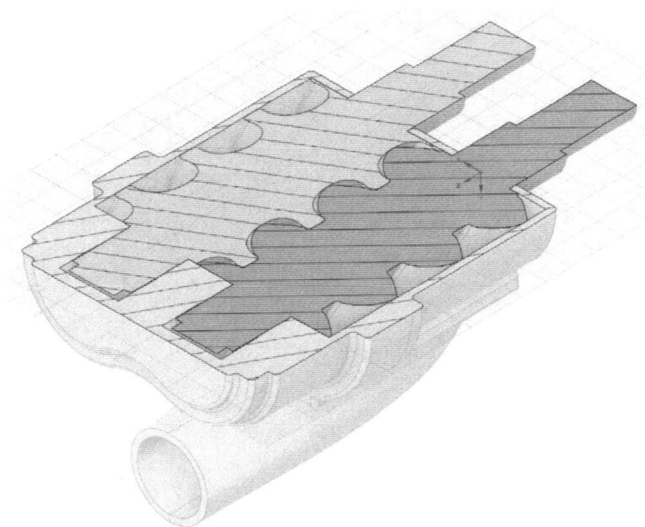

Figure 5. Cut through the CAD model with the male rotor in blue, the female rotor in orange and the housing in green

The goal is to determine the displacement of the housing and the rotors due to gravitational force, centrifugal force, operating pressure and operating temperature. From the displacements the changes in radial and axial clearances between both rotors and the rotors and the housing are derived. The structural and thermal material properties of the structural steel used for the rotors and casing are given in the following table 3.

Table 5. Material properties of structural steel

	Steel
Density	7850 kg/m^3
Young's modulus	200 GPa
Poissons ratio	0.3
Coefficient of thermal expansion	1.2e-5/K

On the left side both rotors are supported on the second shaft shoulder with a floating bearing to a fixed reference point in space (not moving), see Figure 6. This side represents the motor. On the right side on the second shaft shoulder, they are supported by a fixed bearing to the inner casing. For the fixed bearing the axial displacement is fixed, with no axial stiffness defined. This is necessary to keep the axial gap on the discharge side less depended on the thermal expansion of the rotor. For both bearings a radial stiffness of 500,000 N/mm is considered. The casing is supported on two fixed points and a frictionless support on the bottom horizontal face of the outlet, to let the housing deform as freely as possible due to temperature changes.

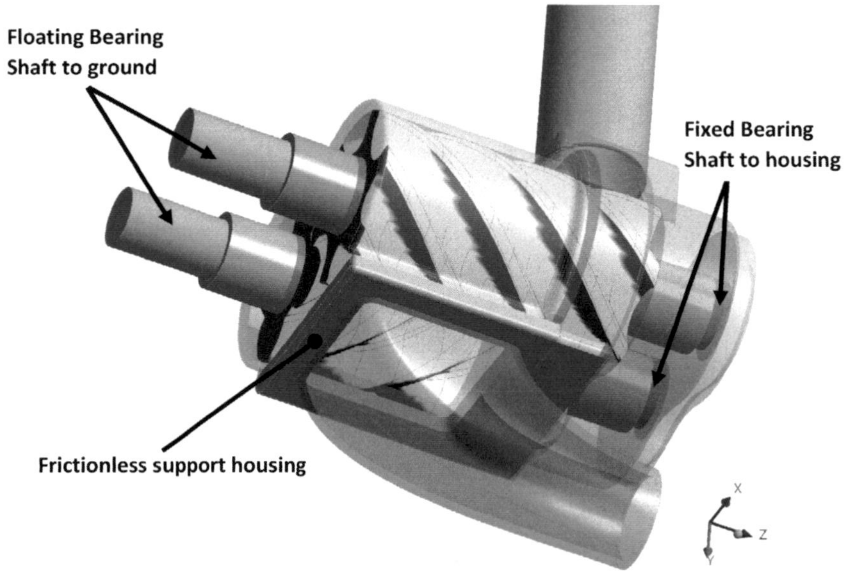

Figure 6. Bearing conditions

Between the rotating shafts and the housing no contacts are defined. The initial radial clearances are 50 µm and the initial axial clearances are 100 µm. If contact between bodies is assumed, due to higher deformations, frictional contacts can be used. To assure a minimal clearance between both shafts or the shafts and the housing for the CFD simulation, a minimum contact distance or clearance can be defined. This minimum clearance cannot be interpenetrated.

Both the rotors and housing are meshed with Ansys Meshing. The shafts are meshed with high quality hexahedral elements. Due to the geometric complexity of the housing, it is meshed with tetrahedrons. The CFD results of the distributed pressure and the temperature on the inside of the rotors and the inner casing are interpolated on the FEM mesh, for one time step.

4. Results

4.1. CFD results for initial geometry

The screw compressor compresses the air from 1 bar to 3 bar, see pressure distribution in Figure 7.

Due to compression the temperature increases towards the discharge port, see temperature distribution in Figure 1. The figure shows the temperature on the rotor walls and on a slice plane through each rotor axis to see the temperature also in the casing solid.

Velocities are shown in Figure 9. The high velocities are located in the clearances. Depending on the pressure change and the clearance size backflow results which reduces the efficiency of the machine.

The mass flow is nearly constant at the inlet and fluctuating at the outlet depending on the size of over-compression at the discharge port. Averaged over the last calculated male rotor revolution, the mass flow is 0.146 kg/s which equals an inlet volume flow of 0.126 m³/s. The theoretical volume flow rate, derived from chamber volumes and rotational speed, is 0.177 m³/s, i. e. volume efficiency is 71.1%. The averaged power within the same interval is 24.1 kW. The time resolved values for mass flow are shown in Figure 17.

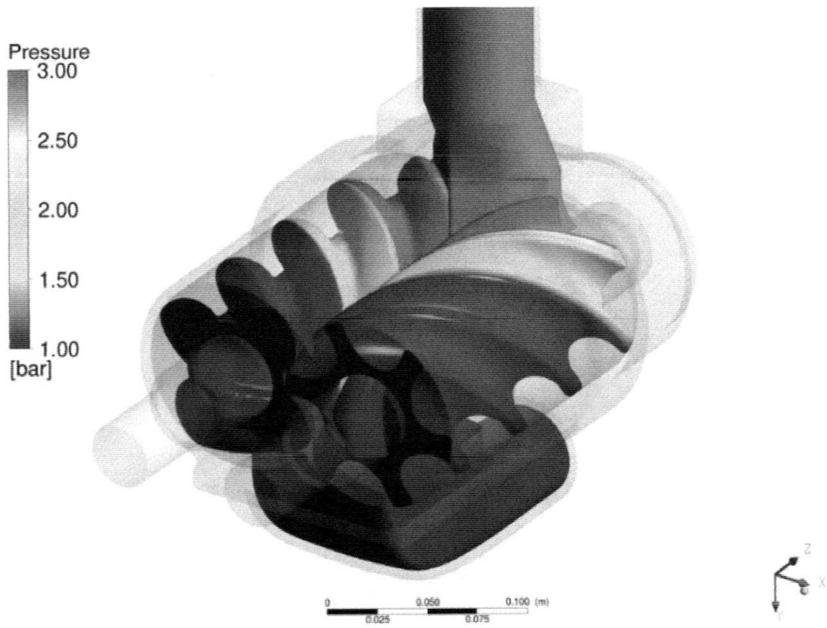

Figure 7. Absolute pressure (initial geometry)

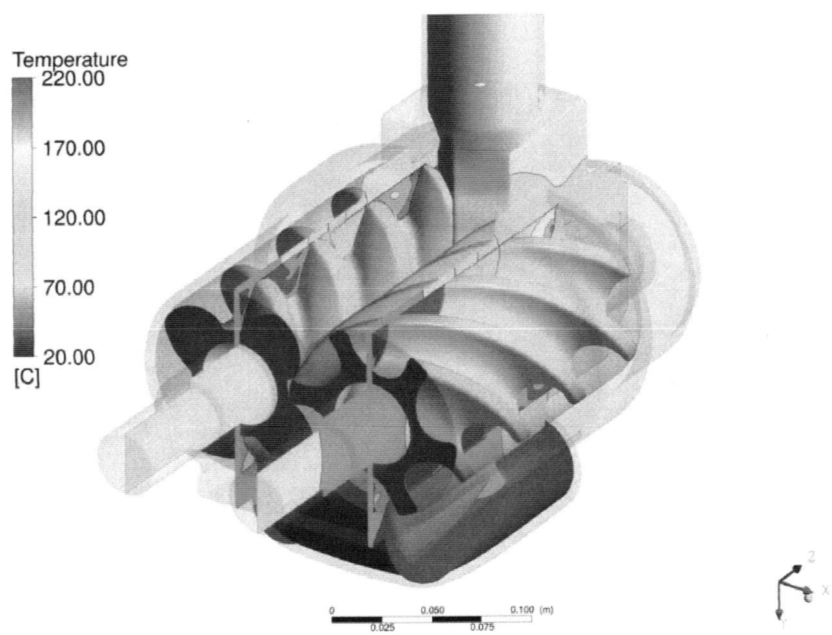

Figure 8. Temperature (initial geometry)

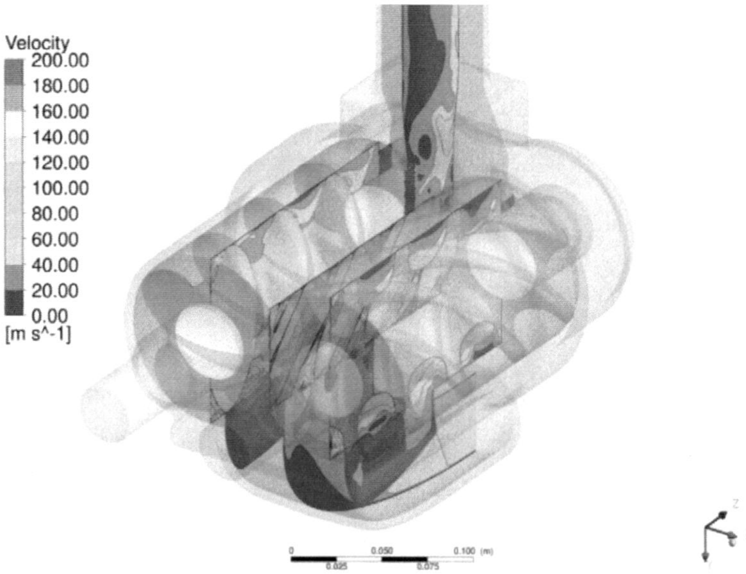

Figure 9. Velocity (initial geometry)

4.2. Structural results

For the structural results the deformations and therefore changes in radial and axial clearance are of most interest. Figure 10 shows the scaled deformation of the rotors. Because of the increasing temperature towards the discharge port the outer diameter of the rotors is also increasing. The housing is deformed more at the top side (discharge pipe), because a frictionless support is used on the bottom side and the temperature at the outlet pipe is much higher, see Figure 11.

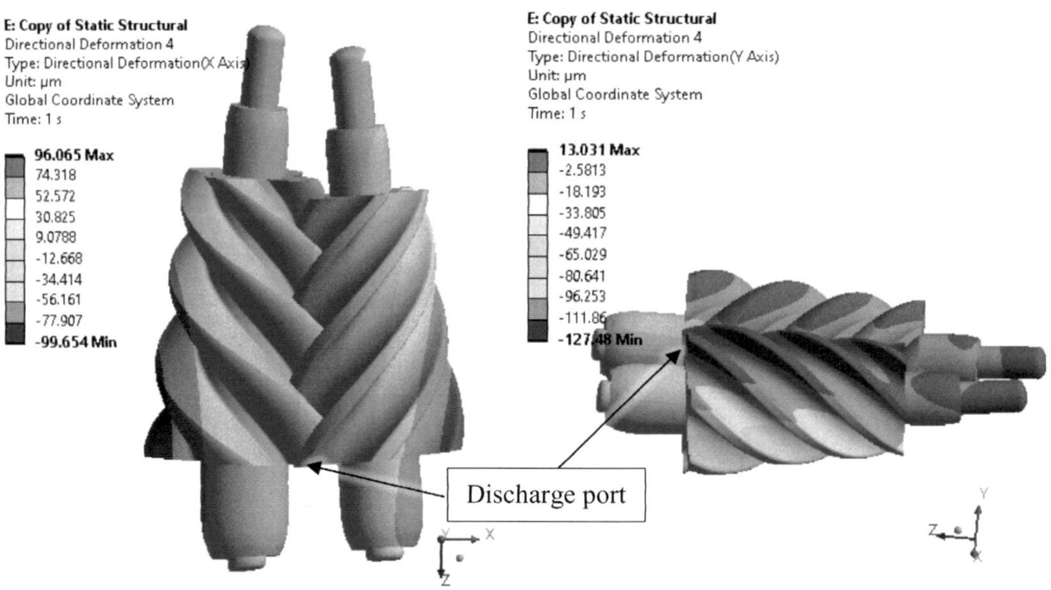

Figure 10. Rotor deformation scaled by a factor of 250 (left: x-deformation, right: y-deformation)

The deformation of both rotor shafts and the housing comes mainly from the temperature gradients inside the bodies. The structural results could be used to adapt the flow volume depending on the deformation of the rotors and the casing. In comparison to the whole model, the deformation is small but it has a large influence on the flow through the clearances.

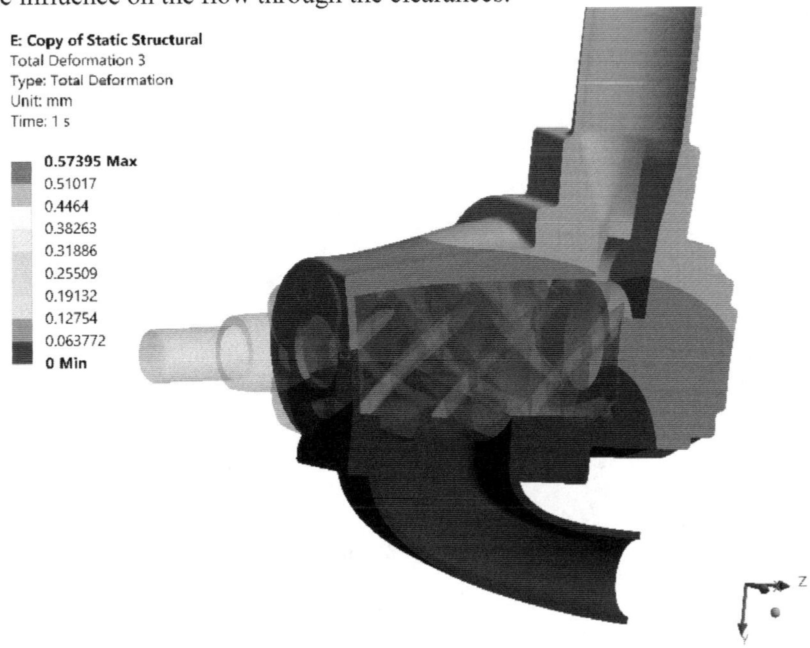

Figure 11. Casing deformation scaled by a factor of 250

4.3. Transferring clearance changes into CFD geometry

TwinMesh has the possibility to modify the radial clearance size depending on the position along the rotor axis by bending and scaling of the rotors. TwinMesh does not support housing deformation yet, this will be included in the next release. Depending on the deformations calculates by the structural solver, it may not possible to reach the exact deformation by using the available options.

To get a simplified definition paths are created for each rotor in x and y-direction, see Figure 12. Along these paths the deformation values are exported for the rotors and the casing. For missing values in between, where no solid is located along the path, the values are linearly interpolated.

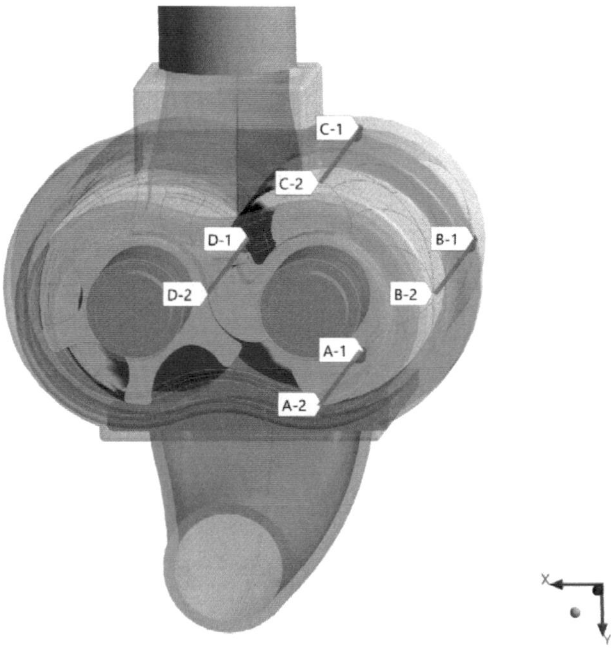

Figure 12. Paths to get deformation values for the main rotor

The variation of clearance size can be calculated from the deformation values, called "target" values here. Figure 13 (dashed lines) shows that the intermesh clearance and the clearance in positive y-direction is decreasing towards the discharge port depending on the increasing rotor (mainly thermal deformation). On the negative y-direction the clearance increases because the deformation of the casing is larger than the thermal expansion of the rotor. On the female rotor the same tendency exists (Figure 14).

The modified radial clearances could be realized in TwinMesh by specifying the axis position and the scaling offset (positive value means a shrinking rotor normal to the rotor surface) depending on the position along the rotor axis, see Figure 15 and Figure 16. In addition to that the casing radius is reduced along the x-axis by 50 μm (slightly oval curvature).

Figure 13 and Figure 14 show the realized deformations (solid lines). The intermesh clearance and the radial clearances in y-direction are good in comparison to the structural results (dashed lines). Only the x-direction gap to the casing is different.

The change of the axial clearance is not taken into account in the following CFD simulation because only a constant axial clearance size could be defined in TwinMesh for both rotors. The axial clearance for male rotor on the suction side changes from 100 μm to 46 μm and on the pressure side to 107 μm, for the female rotor on the suction side from 100 μm to 71 μm and on the pressure side to 106 μm.

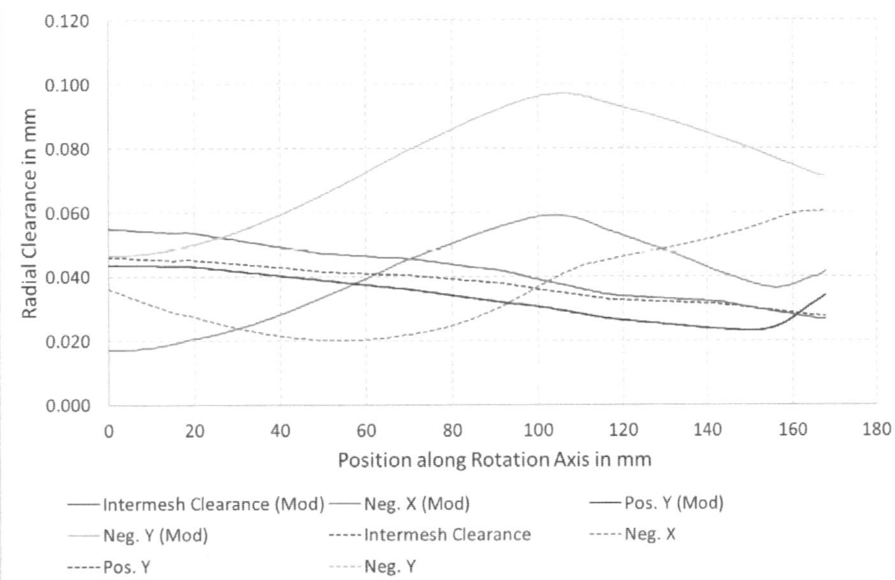

Figure 13. Target (dashed) and realized (solid) radial clearances of the male rotor

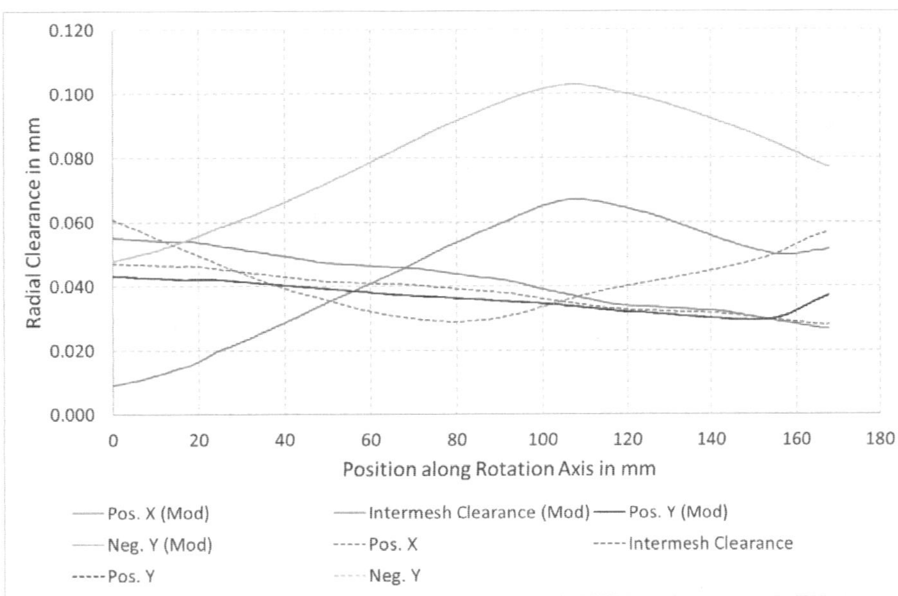

Figure 14. Target (dashed) and realized (solid) radial clearances of the female rotor

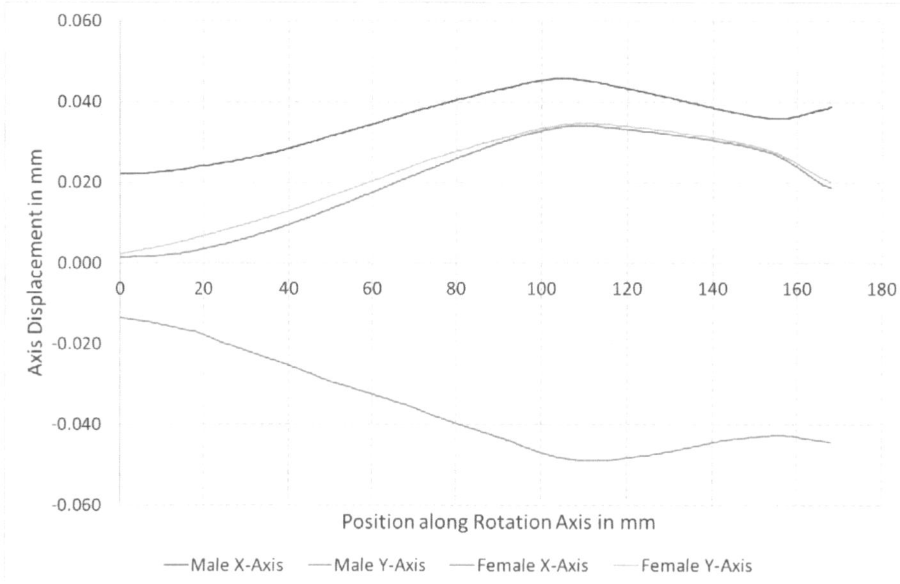

Figure 15. Axis displacement

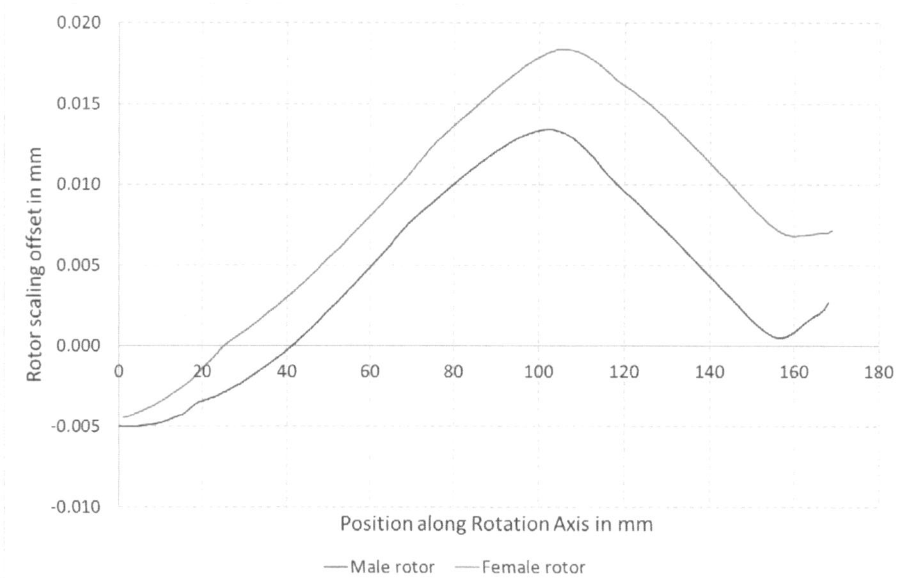

Figure 16. Rotor scale offset

4.4. CFD results for adapted geometry

The flow field looks nearly the same as without modified clearances. The influence of modified clearances (mainly reduced clearance size) can be seen by the mass flow and the power. As expected, mass flow and the power increase after adapting deformation. The mass flow is now 0.151 kg/s (+2.9%) and the power 24.7 kW (+2.5%). The volume efficiency rises to 73.1%.

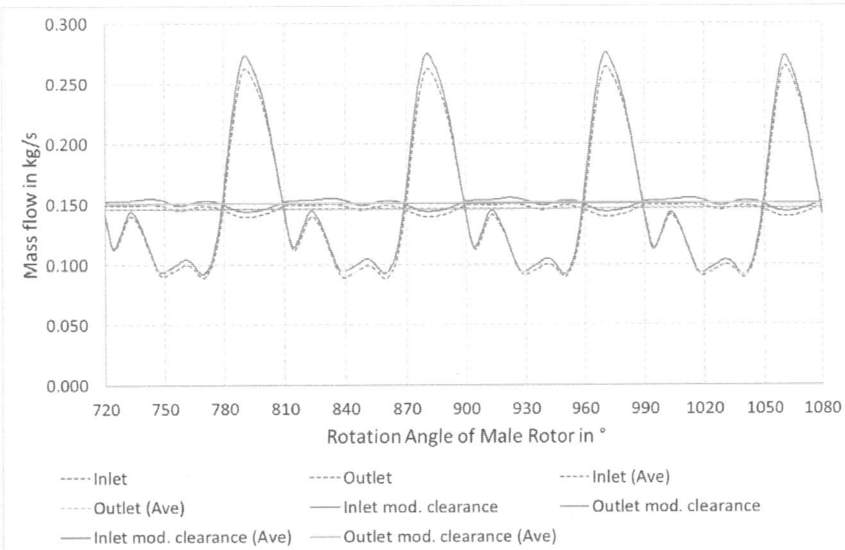

Figure 17. Mass flow for initial (dashed) and adapted (solid) geometry over rotation angle

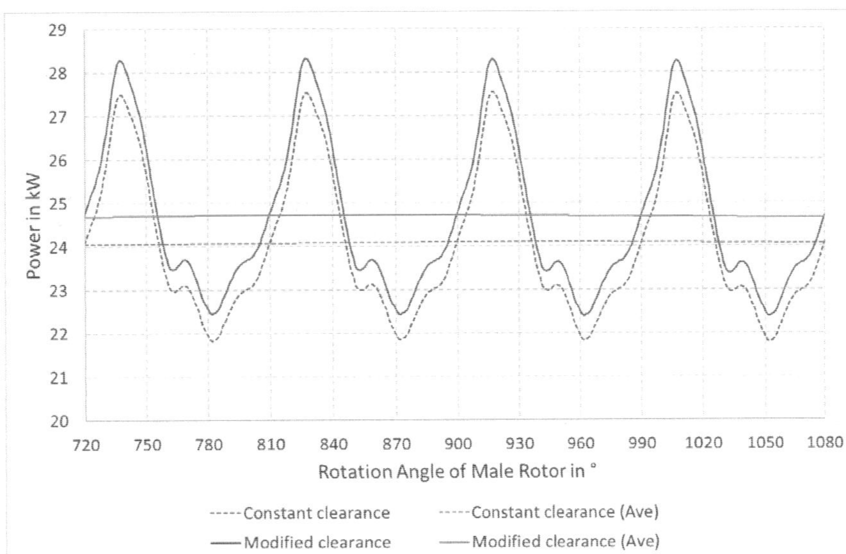

Figure 18. Power for initial (dashed) and adapted (solid) geometry over rotation angle

5. Summary and outlook

This paper shows one workflow to adapt the deformation results of a structural simulation to a CFD setup. It starts with a CFD simulation (including CHT for the rotor and housing solids) with constant radial clearances using TwinMesh to realize the necessary mesh deformation inside the working chamber and Ansys CFX as the CFD solver. In a next step the temperature of the housing and the rotor solids and the pressure on the fluid sides are mapped on a structural mechanical setup solved by Ansys Mechanical. The change of the radial clearances depending on the deformation of the solids is used as an input for the final CFD simulation using an adapted mesh for the working chambers.

A simplified method to take the change of the radial clearance sizes along the rotor axis into account was shown. It can be used to see the impact on the machine behaviour in more detail and get more realistic results for the real operating condition.

The next step will be to develop a direct method to modify the mesh of the working chambers for each location (radial and axial clearances) depending in the structural deformation values. The challenge is to keep the mesh quality high by considering all degrees of freedom of the deformation.

References

[1] Nikolov A, Huck C and Brümmer A 2012 Influence of Thermal Deformation on the Characteristic Diagram of a Screw Expander in Automotive Application of Exhaust Heat Recovery *21st International Compressor Engineering Conference* Purdue

[2] Buckney D, Kovacevic A and Stosic N 2014 Accounting for Local Therman Distortions in a Chamber Model for Twin Screw Compressors *22nd International Compressor Engineering Conference* Purdue

[3] Kovacevic A, Stosic N and Smith I 2007 Screw compressors: three dimensional computational fluid dynamics and solid fluid interaction vol 46 (Springer Science & Business Media)

[4] Spille-Kohoff A, Hesse J and El Shorbagy A 2015 CFD simulation of a screw compressor including leakage flows and rotor heating *9th International conference on compressors and their systems (ICCS)* London

[5] Andres R., Hesse J and Low D 2018 CFD Simulation Of A Two Stage Twin Screw Compressor Including Leakage Flows And Comparison With Experimental Data *24th International Compressor Engineering Conference* Purdue

[6] Wu H, Huang H, Zhang B, Xiong B and Lin K 2019 CFD Simulation and Experimental Study of Working Process of Screw Refrigeration Compressor with R134a *Energies* Vol. 12 (issue 11)

[7] Andres R, Hesse J, Salecker U, Spille-Kohoff A, Nikolov A and Brümmer A 2016 CFD Simulation of a Twin Screw Expander including Leakage Flows *23rd International Compressor Engineering Conference* Purdue

[8] Stöcker H 1998 *Taschenbuch der Physik* p 691 (Verlag Harri Deutsch)

Analysis of the flow through the blowhole of twin-screw machines with different rotor profiles using dimensionless numbers

M Heselmann and A Brümmer

TU Dortmund University, Chair of Fluidics, Leonhard-Euler-Str. 5, 44227 Dortmund, Germany

E-mail: `matthias.heselmann@tu-dortmund.de`

Abstract. To ensure safe operation, gaps in a screw machine are essential. Some are relatively easy to adjust, like front and housing gaps, while it is more difficult, if they depend directly on the selected rotor profile shape, such as the inter-lobe clearance and the blowhole. Consideration of mass flow rates through those clearances is crucial to predict operation conditions or design screw machines. When using chamber model simulations, it is common to calculate idealized, theoretical mass flow rates and to combine them with a flow coefficient which represents the effect of e.g. friction losses. This paper deals with the flow coefficient of the blowhole, which is formed by the two rotors and the housing cusp. Therefore, a set of dimensionless numbers is used, including physical and geometrical influences on the flow coefficient. The real mass flow rate of an ideal gas (e.g. air) through the blowhole is calculated by CFD-simulations. The dimensionless numbers are systematically varied to identify their specific influences on the gained mass flow rates.

1. Introduction

Twin-screw machines consist of two counter wise wounded rotors which are mounted in a tight housing. The simple design of this type of machine causes that no free mass forces have to be balanced. Therefore it is possible to operate at very high RPMs, which leads to a great energy density and a leading role in the use of this type of machine as a compressor in the compressed air and refrigeration industry [1–3]. To ensure safe operation there is a need for clearances between the rotors and the housing. In general there are four clearance types namely the front gap, the housing gap, (both between the rotors and the housing in axial and radial direction), the inter-lobe clearance where the rotors intersect each other and the blowhole between the rotors and the cusp of the housing (Figure 1). Clearances represent a decisive factor in the machine design since they significantly influence the operation behavior. Thermodynamic calculations of this machine type are performed using computational fluid dynamics (CFD) [4, 5] or chamber model simulation [6]. The latter are based on conservation of mass and energy and require much less computational effort. Therefore chamber model simulation are more suitable when many configurations need to be simulated. However, to predict the machine behavior accurate, the knowledge of clearance mass flow rates is crucial. Within the chamber model simulation mass flow rates between connected chambers or ports are usually calculated as pure pressure induced

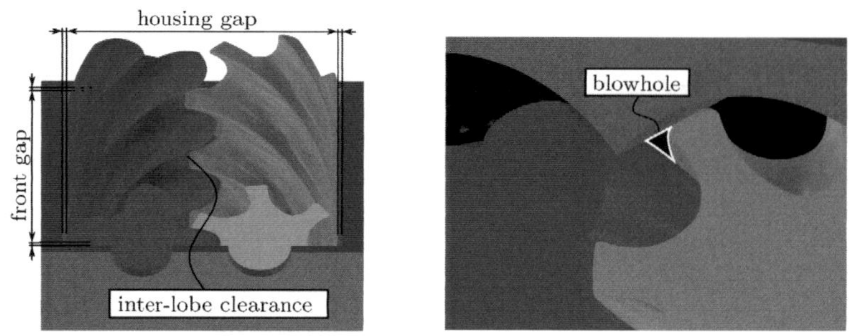

Figure 1. Sketch of the clearances occurring within a twin-screw machine (significantly enlarged).

flow using an isentropic nozzle flow [7]:

$$\dot{m}_{th} = p_{hp} \cdot A_c \cdot \sqrt{\frac{2 \cdot \kappa}{(\kappa - 1) \cdot R_s \cdot T_{hp}} \cdot \left(\Pi_c^{\frac{2}{\kappa}} - \Pi_c^{\frac{\kappa+1}{\kappa}} \right)} \quad \text{for} \quad \Pi_c = \frac{p_{lp}}{p_{hp}} > \Pi_{crit} \tag{1}$$

Equation 1 is valid for an adiabatic, frictionless flow from a cavity with high pressure (index hp) through a clearance with minimum connection area A_c to a cavity with low pressure (index lp) as long as the pressure ratio is subcritical $\Pi_c > \Pi_{crit}$. R_s is the specific gas constant, κ the isentropic exponent, and T the Temperature. If the pressure ratio is critical $\Pi_c < \Pi_{crit}$ the mass flow rate is choked in the minimum area and independent of the low pressure conditions:

$$\dot{m}_{th} = p_{hp} \cdot A_c \cdot \sqrt{\frac{2 \cdot \kappa}{(\kappa + 1) \cdot R_s \cdot T_{hp}}} \cdot \left(\frac{2}{\kappa + 1} \right)^{\frac{1}{\kappa-1}} \quad \text{for} \quad \Pi_c = \frac{p_{lp}}{p_{hp}} < \Pi_{crit} \tag{2}$$

Since the isentropic nozzle flow does not consider influences of friction, a flow coefficient α is used which rates the actual mass flow rate to the theoretic mass flow rate:

$$\alpha = \frac{\dot{m}}{\dot{m}_{th}} \tag{3}$$

This procedure is already a proven concept for front and housing gaps [8], and is usually adopted to every connection. The crucial aspect is to determine or guess the flow coefficient in an appropriate way.

However, the definition of the flow coefficient is not unique. For orifices the definition can be found in [9]. The estimation of mass flow rates through seals is given in [10]. The determination of flow coefficients through the housing gap of twin-screw compressors are examined among others in [11] as well as the flow through the front gap in [12]. The flow coefficient for those connections is in the range of $\alpha = 0.01 \ldots 0.8$ depending on the boundary conditions. Dominating dimensionless numbers are found, so that the flow coefficient is calculated as a result of a regression functional [8]. Focusing on the blowhole, the examination of the flow coefficient becomes more difficult since the shape of the narrowest cross section and the flow direction have a three-dimensional shape. Therefore, several suggestions exists to determine the narrowest cross section [2, 13]. The experimental examination of the flow through the blowhole of a twin-screw machine is very difficult due to additional connections through the inter-lobe clearance and the housing gap. Patel and Lakhera examined the flow through the blowhole using an iterative method to estimate the mass flow rate and compare their model with experiments [14]. Here the blowhole is approximated by an equilateral triangle.

In this study the mass flow rate through the blowhole is calculated using the software Ansys CFX© and is related to the isentropic nozzle flow. The narrowest cross section is calculated using the algorithmic method derived by Rinder respectively Grafinger [2, 15]. Boundary conditions are varied by means of dimensionless numbers to yield comparability to other physical conditions or rotor geometries containing profile and rotor lead variations. The aim of this study is to quantify the flow coefficient for the blowhole of a twin-screw machine and to identify the main influence parameters by varying single characteristic numbers. The results will be used by the authors in the future to increase the accuracy of chamber model simulations.

2. Dimensionless numbers for mass flow rate through blowhole

To determine independent dimensionless numbers for the flow through the blowhole of a screw machine the Π-theorem is used [16]. The advantage of this method is that just the independent influences must be found. Comparing the blowhole to other clearances, like front and housing gaps, it is affected by more geometric parameters. In sum, it depends on:

$$\dot{m} = F(p_{hp}, p_{lp}, c_p, c_v, \eta, T_{hp}, z_m, z_f, D_m, D_f, a, l, n_m) \tag{4}$$

Assumptions for the calculation of these parameters are that the fluid properties (isobaric heat capacity c_p, isochoric heat capacity c_v and the dynamic viscosity η) do not vary with the flow from the high to the low pressure side. Further, the system is treated as adiabatic. Special attention is given to the theoretic mass flow rate (equations 1 and 2), respectively the narrowest cross section. According to Rinder the cross section is found in the front section of the rotor profile and depends on the shape of the female rotor, the female housing bore, the male rotor shape and the position of the highest inter-section point of the contact line (figure 2). Due to

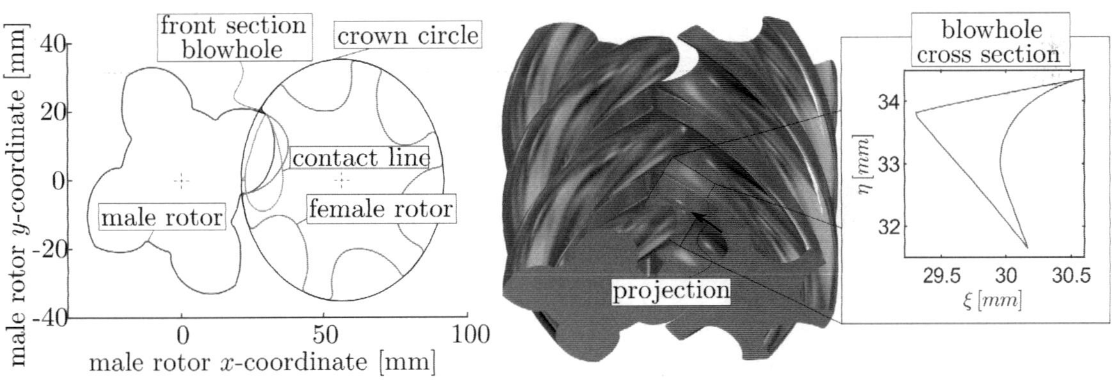

Figure 2. Sketch of the determination of the blowhole in the front section and project it along the female rotor lead.

the fact, that Rinder derived his method from the analytical profile developed by Svenska Rotor Maskines (SRM-A), the identification of the influence parameters for the effective blowhole cross section is possible. Those are the lobe numbers z_m, z_f and crown circle diameters D_m, D_f from male and female rotor, the axis distance a and the male rotor lead l. Therefore, all necessary parameters for the calculation of the theoretic mass flow rate are independent, so that the theoretic mass flow rate can substitute the temperature T_{hp}. Calculating the theoretical

Reynolds number requires the determination of a characteristic length. As for other non circular shapes it is common to use the hydraulic diameter:

$$d_{hyd} = \frac{4 \cdot A_c}{U_c} \tag{5}$$

where U_c is the perimeter of the blowhole, which can be calculated directly since the parameters influencing the cross section area are known. Since the purpose of this paper is to vary the rotor profile, the hydraulic diameter substitute the rotor lead. Performing the dimensional analysis, the following characteristic numbers are obtained:

$$\begin{aligned} &\Pi_1 = \frac{p_{lp}}{p_{hp}} = \Pi_c,\ \Pi_2 = \frac{c_p}{c_v} = \kappa,\ \Pi_3 = \frac{\dot{m}_{th} \cdot d_{hyd}}{A_c \cdot \eta} = Re_{th},\ \Pi_4 = \frac{\dot{m}}{\dot{m}_{th}} = \alpha,\ \Pi_5 = z_m, \\ &\Pi_6 = \frac{z_m}{z_f} = i,\ \Pi_7 = \frac{D_f}{D_m},\ \Pi_8 = \frac{a}{D_m},\ \Pi_9 = \frac{d_{hyd}}{D_m},\ \Pi_{10} = \frac{u_m}{a_{hp}} = Ma_u \end{aligned} \tag{6}$$

3. CFD simulation

Since the measurement of the mass flow rate through the blowhole is hardly possible due to occurring housing gaps and inter-lobe clearance CFD-simulation is used. This results in a more realistic mass flow rate than the theoretical mass flow rate, because on the one hand the influence of friction, flow contraction and separation is taken into account and on the other hand no flow direction through the blowhole is specified. Thus the flow cross section is the result of the simulation. The software used is Ansys CFX© which uses a finite volume method to solve the set of partial differential equations. As simulation domain one high pressure and one low pressure chamber of the screw machine is chosen. It is to mention, that there are no housing gaps or inter-lobe clearances. Figure 3 visualizes the simulation domain. On the left, beside the high pressure side, the whole low pressure chamber is highlighted. However, a mesh study shows, that the low pressure chamber size can be reduced significantly without influencing the results. This leads finally to the simulation domain shown on the right. The mesh is created automatically and consists mainly of tetrahedrons. In a second step prisms layers are added to reduce the y+-value. The inlet and outlet surfaces are defined as opening, allowing mass flow rates in- and

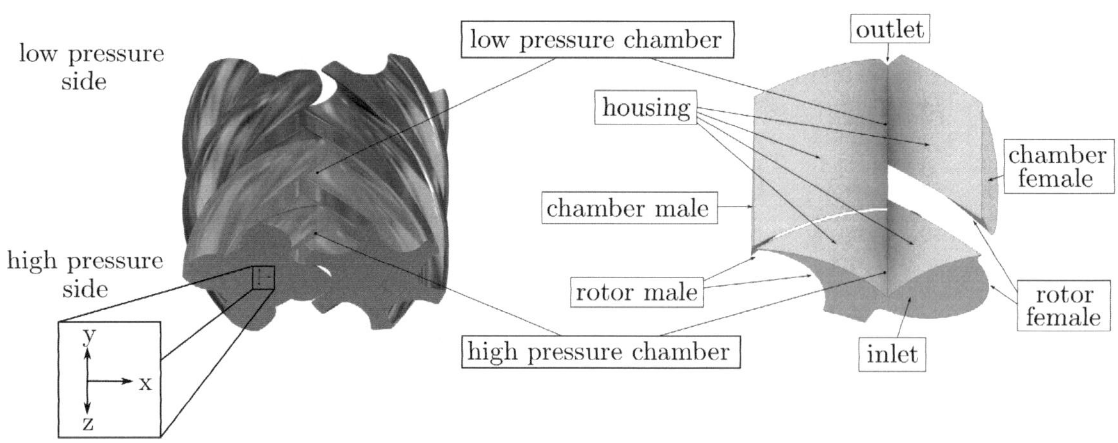

Figure 3. Extracting the fluid domain (working chambers) as negative of the rotors and the housing bore (not shown).

outwards the domain. During the mesh study the inlet is defined as mass flow inlet for better stability of the simulation. The surfaces of the male and female chamber are defined as free slip walls, since they are within a working chamber. All other surfaces are defined as moving, no slip walls to consider the rotation of the rotors. However, in this study the RPMs are set to zero. The mesh study is performed according to Richardson extrapolation, knowing that this method is only valid for structured meshes, is gives an idea how far an extrapolated solution is though [17]. The mesh is investigated for a SRM-A type profile with a lobe combination of 4+6 and a male rotor crown diameter of 72 mm while the ratio of crown diameters is $\Pi_7 = 0.98$. The axis distance is set to a = 56.52 mm and the hydraulic to crown circle diameter to $\Pi_9 = 0.007$. The results of the Richardson extrapolation are shown in figure 4 on the left hand side. The pressure in the low pressure chamber is set to 2 bar. While the calculated pressure at the inlet is very close ($\Delta p/p < 0.1\%$) to the extrapolated value, the y+value is quite high ($y+ > 180$).

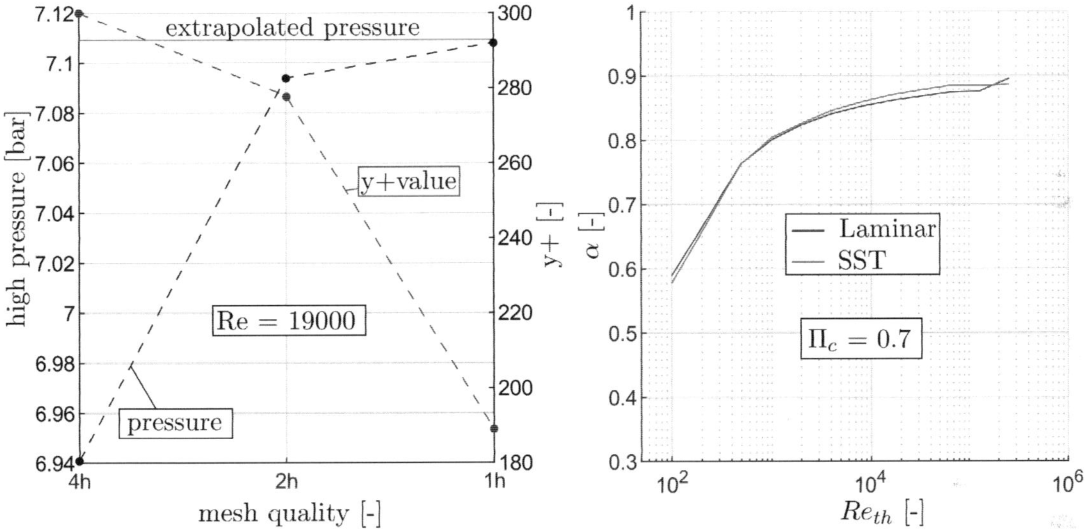

Figure 4. Evaluation of the mesh quality according to Richardson extrapolation and the influence of turbulence modeling.

Therefore prism layers are added on housing, rotor male and rotor female (refer figure 3). Leading to a y+value < 5 for the investigated conditions (not shown). The introduced prism layers affect the pressure about 1%, but the results should be more trustworthy with respect to the boundary layer. The final mesh information are given in table 1. Bad elements in terms of maximum angle occur in corners where the rotors reaches the crown circles. The worst elements

Table 1. Mesh information.

parameter	value/range	portion
prism layer [-]	20	housing, male and female rotor
maximum angle [°]	63 - 126	91% of all tetrahedrons
maximum angle [°]	90 - 144	91% of all pyramids
maximum angle [°]	90 - 126	90% of all pentagons
aspect ratio [-]	0.55 - 1	88% of all elements
mesh expansion factor [-]	0 - 306	99% of all elements

with respect to the aspect ratio and expansion factor are found at the transition from prism layers to tetrahedrons.

To choose an appropriate turbulence model the shear stress transport (SST) model is compared to a laminar flow simulation where the turbulent viscosity is set to zero. The results show, that the influence of the turbulence model on the flow coefficient is neglectable (figure 4 - right hand site). Similar to the male rotor housing gap, the blowhole flow is not affected by turbulence [8]. Since the numerical effort for the turbulent simulation is reasonable and to cover occurring influences while varying geometric and physical properties, the SST-model is used for all further simulations. An overview over all boundary conditions remaining constant during this investigation is given in table 2.

Table 2. Constant boundary conditions.

	parameter	**value**
boundaries	inlet, outlet	opening
	chamber male, female	free slip wall
	housing	no slip wall
	rotor male, female	no slip wall
	RPMs	0
fluid	specific gas constant [J/kg/K]	287.058
	isentropic exponent [-]	1.4
simulation	analysis type	stationary
	heat transfer fluid - solid	adiabatic
	turbulence model	SST
	wall roughness	smooth wall

4. Variation of pressure ratio

Results of the variation of the pressure ratio are shown in figure 5. To hold the theoretical Reynolds number constant the absolute pressures and/or the temperature on the high pressure side are varied, since with constant theoretical Reynolds number the theoretical mass flow rate remains constant. In general the flow coefficient increases slightly with decreasing pressure ratio. In the range around the critical pressure ratio the influence on the flow coefficient reaches a tipping point. From pressure ratio about $\Pi_c < 0.4$ the flow coefficient remains nearly at constant level since the flow is choked. The influence of the pressure ratio is dependent on the theoretical Reynolds number. The flow coefficient varies about 7% for $Re_{th} = 100000$ and more than triples to 23% for $Re_{th} = 100$. Furthermore, the trend of the flow coefficient over the pressure ratio is similar to that of the front and housing gaps, which are proven by measurements [8]. The flow through the blowhole is mainly influenced by geometric induced flow separation at the housing cusp. The assumption that the flow is tangential to the female rotor is especially true for high pressure ratios (figure 6). As the pressure ratio decreases, the flow separates from the female rotor and follows the contour of the male rotor. As soon as this happens around $\Pi_c < 0.6$, the flow pattern indicates that the separation vortex acts like a flow guidance. It reduces the pressure in the narrowest cross section and therefore increases the flow coefficient.

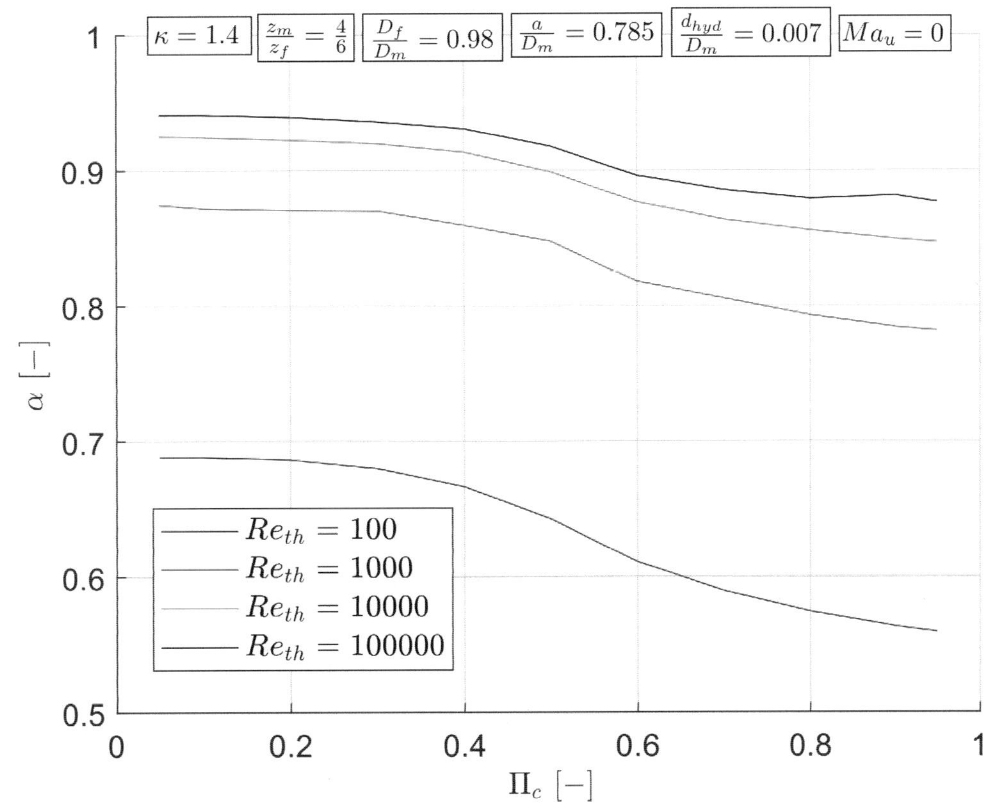

Figure 5. Flow coefficient as a function of the pressure ratio for different theoretic Reynolds numbers for SRM-A profile with lobe combination 4+6.

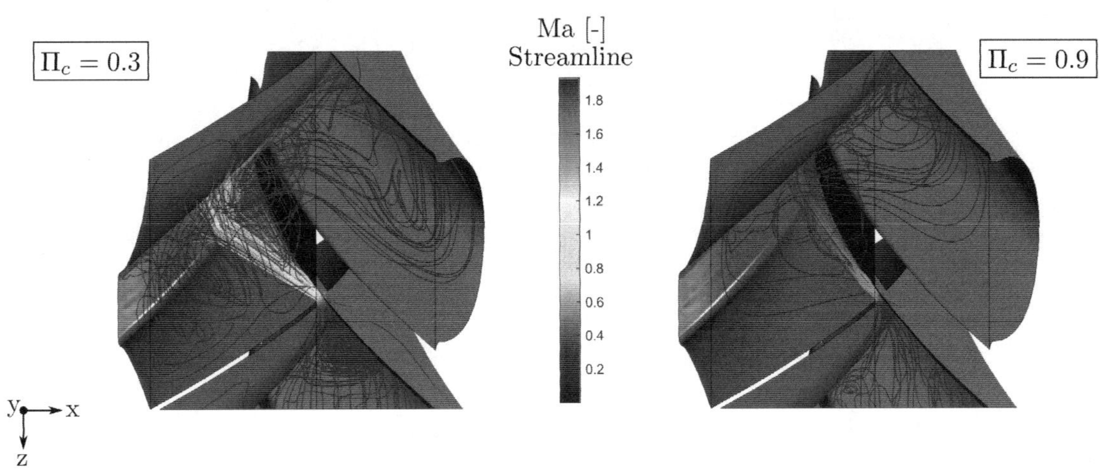

Figure 6. Streamlines through the blowhole for different pressure ratios at $Re_{th} = 100000$ for SRM-A profile with lobe combination 4+6.

5. Variation of Reynolds number

Although the flow through the blowhole is not dominated by friction (due to the neglegtable length of the clearance), the theoretical Reynolds number has a noticeable influence. As the Reynolds number increases, the inertia forces exceed the viscous forces increasingly. Figure 7 depicts the Mach number, that is defined as the ratio of velocity to the speed of sound, in a cross section along the housing cusp[1].

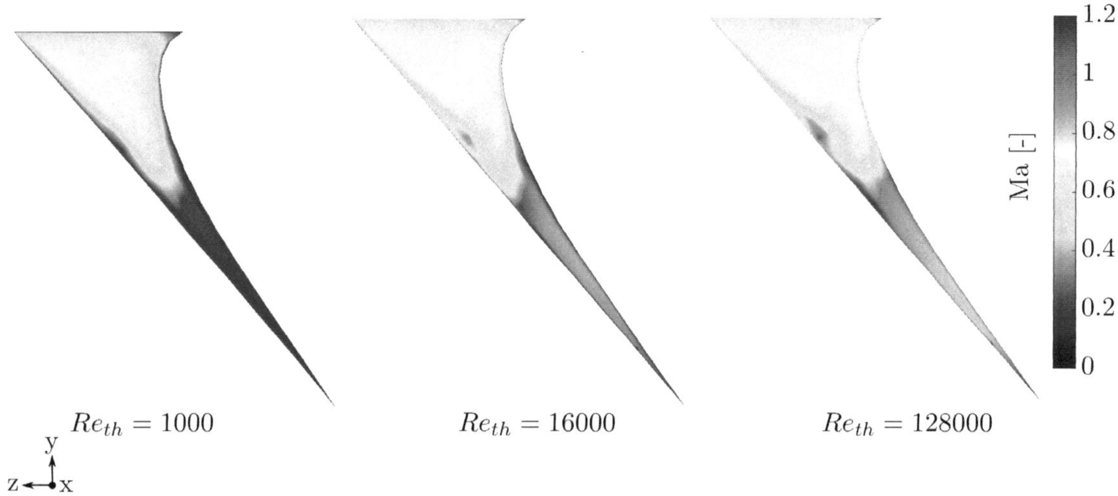

Figure 7. Mach number in a plane along the housing cusp to visualize the velocity gradient for pressure ratio of $\Pi_c = 0.7$.

The flow velocity in the center of the cross section is primarily dependent on the pressure ratio. With increasing Reynolds number, e.g. by increasing the absolute pressures, the mass flow rate increases and the velocity gradient at the wall becomes steeper. This is illustrated by the faster color change. For large Reynolds numbers ($Re_{th} \geq 16000$), no significant difference is observed. Results, in terms of the flow coefficient, for the variation of the theoretical Reynolds number are presented in figure 8. In general the curse of the flow coefficient for the blowhole is very similar to that of the male rotor housing gap, but it is a bit higher [11]. For small theoretical Reynolds numbers friction causes a reduced velocity through the clearance and reduces the flow coefficient. The higher the Reynolds number, the steeper is the velocity gradient at the wall. It reaches a limiting point and the flow coefficient remains constant for further increased mass flow rates, theoretical Reynolds number respectively. Beside the theoretical Reynolds number, the ratio of hydraulic diameter to male rotor crown circle diameter is varied. The hydraulic diameter is varied by changing the rotor lead at constant male rotor crown circle diameter. The flow coefficient decreases as the ratio d_{hyd}/D_m decreases. This is quite intuitive because the flow through a smaller cross section is more influenced by friction. This is validated for convergent nozzles by Bragg and front and housing gaps by Utri [8, 18].

[1] The plane of minimal cross section according to Rinder is not useful here, because this cross section is not surrounded by walls at the same time.

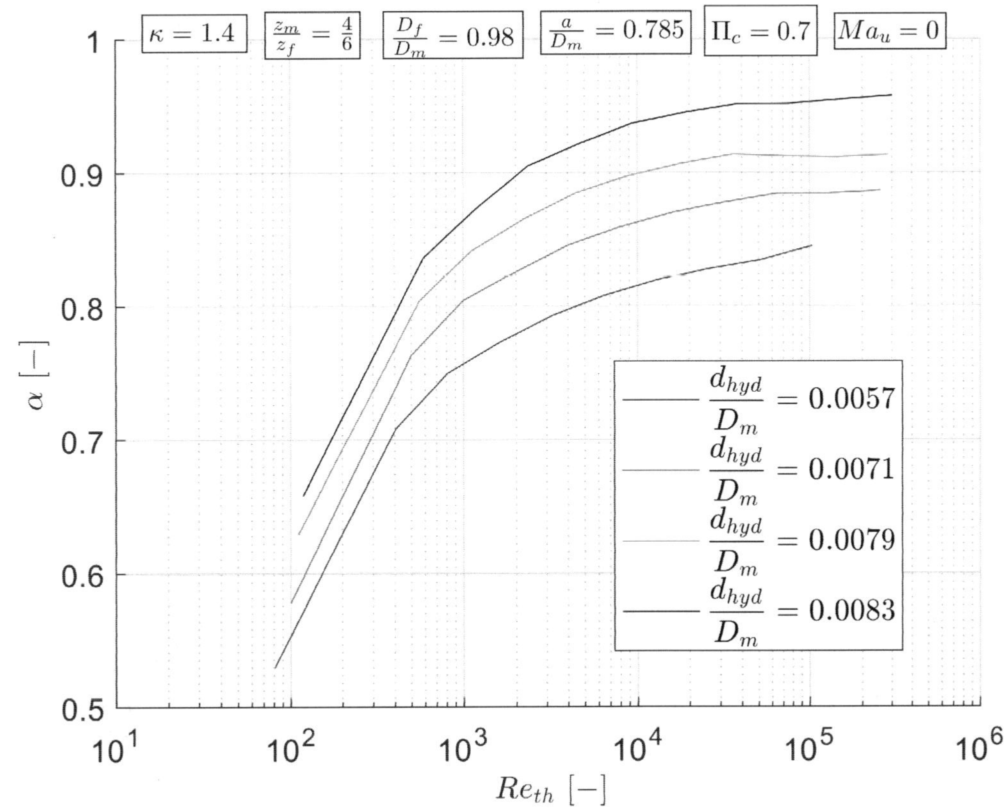

Figure 8. Flow coefficient as a function of the theoretical Reynolds number for different ratios of d_{hyd}/D_m for SRM-A profile with lobe combination 4+6.

6. Variation of hydraulic to crown circle diameter ratio

In this section the hydraulic to crown circle diameter ratio is varied for a SRM-A profile with lobe combination 4+6. Since all other dimensionless numbers should be kept constant, it is convenient to simply vary the hydraulic diameter via the lead of the rotors. The results are shown in figure 9. The flow coefficient shows an almost linear curse with the ratio of hydraulic to crown circle diameter which is explained by the shrinking influence of friction. Beside the variation of the ratio of hydraulic to crown circle diameter the absolute value of the crown circle diameter and thus all dimensionless numbers containing the crown circle diameter has been changed for two different Reynolds numbers. First thing observed is, that the gradient of the flow coefficient with respect to the ratio of hydraulic to crown circle diameter is independent of the Reynolds number. This supports the statement that the flow through the blowhole is dominated by an impulse loss and not by friction. Second, that the flow coefficient do not changes ($\Delta\alpha/\alpha < 1\%$), as long as the dimensionless numbers remain constant. Therefore the system of independent, dimensionless numbers, to describe the flow through the blowhole is valid.

IOP Conf. Series: Materials Science and Engineering 1267 (2022) 012008 doi:10.1088/1757-899X/1267/1/012008

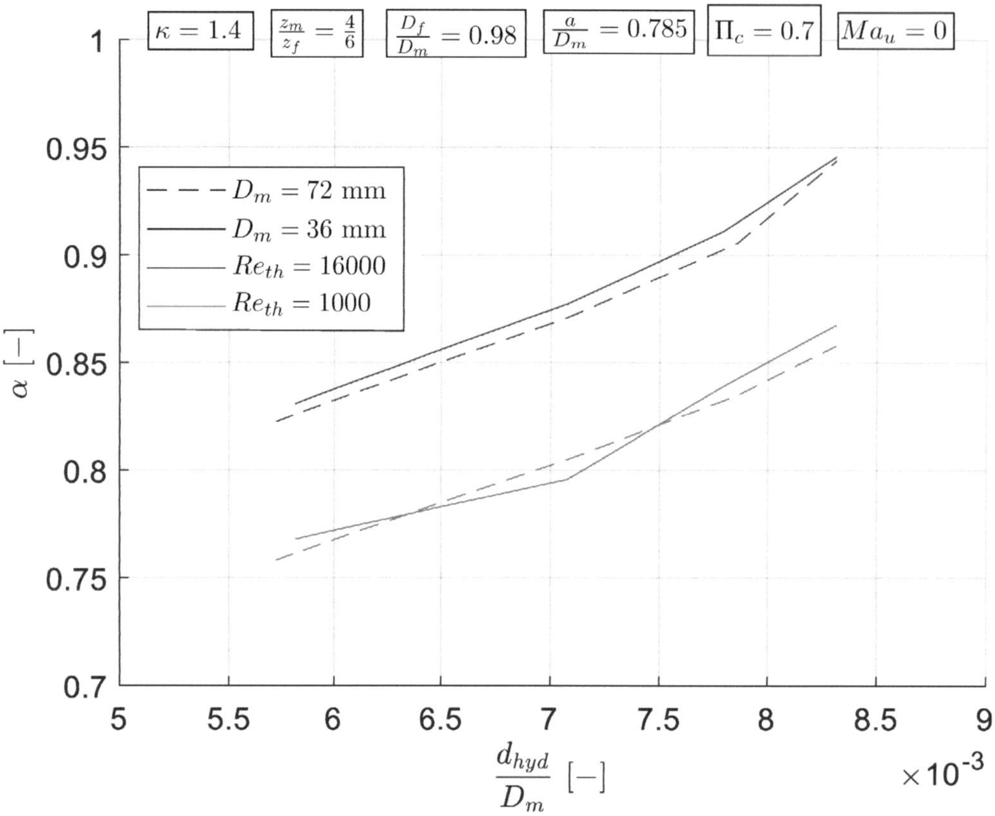

Figure 9. Flow coefficient as a function of d_{hyd}/D_m for different crown circle diameters but equal dimensionless numbers and SRM-A profile with lobe combination 4+6.

7. Profile variation

It is shown in advance that the system of dimensionless numbers is valid for the SRM-A profile with a lobe combination of 4 male and 6 female lobes. While the flow coefficient for a variation of the physical characteristic numbers shows a similar course as the male rotor housing gap, these are kept constant in the following and the geometric characteristic numbers are examined instead.

7.1. Note on the lobe number

Investigating the flow coefficient using dimensionless numbers is particularly useful when only one number is varied at a time to determine its isolated influence. A systematic variation of the geometric characteristic numbers is hardly possible with regard to the rotors, because their numbers of lobes can only be varied as integers. Furthermore, the kinematic conditions of the gearing law creates dependencies between crown circle diameters, axis distance and lobe numbers, which makes it even more difficult to vary only one parameter. At this point, the geometry of the front section of two profiles is considered, so that it is possible to vary only one parameter. First things first, the significance of the profiles considered here is not taken into account. The focus at this point should be on the geometric interpretation of the lobe number of the male rotor. Figure 10 shows the SRM-A profile for only a variation of the lobe number of the male rotor, namely with 2 and 4 rotor lobes. It is obvious that the geometric characteristic numbers i, a/D_m and D_f/D_m, which are kept constant, define the lobe contour. With the absolute lobe number, the lobe contour is simply arranged around the center axis according to the tooth pitch angle ($\varphi = 360°/z_m$). The blowhole area of the two profiles that is decisive here is therefore identical.

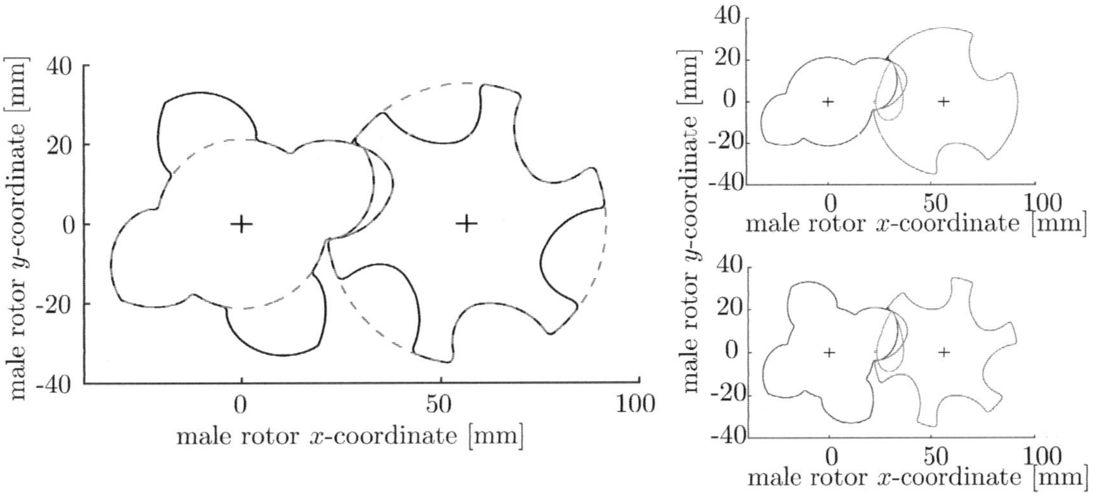

Figure 10. Variation of male rotor lobe number for constant geometric ratios.

As the preceding investigations have shown, the flow coefficient is very much influenced by the physical characteristic numbers and by the ratio of the hydraulic to the crown circle diameter. Now, these ratios are sensibly kept constant and a variation of the number of lobes on the male rotor is performed. According to figure 10 this does not even affect the blowhole area in the front section. Therefore, there is no reason to assume at this point that the flow coefficient would change by varying the number of lobes.

7.2. Variation of lobe number ratio

The variation of the lobe number ratio is performed in this section. The fact, that this is not possible without varying the number of lobes, is mentioned in the previous section and it was explained that the number of lobes should not have any effect on the flow coefficient. The results are presented in figure 11. Shown is the flow coefficient plotted against the ratio of hydraulic to crown circle diameter for varying lobe combinations. It should be noted that the limits of the curves shown are a function of the blowhole geometry. The lower limit results from a very low rotor lead, which strongly distorts the blowhole, making the hydraulic diameter small. The upper limit results from the hydraulic diameter in the front section. Two things can be observed: First, the flow coefficient decrease as the lobe number ratio increase. The reason for this is the ratio of the hydraulic diameter of the blowhole to its area. This means that for a constant hydraulic diameter, the area of the blowhole increases with the lobe number ratio and thus the theoretical mass flow rate. The simulated mass flow rate, on the other hand, remains approximately constant and the flow coefficient is correspondingly lower. Second, the gradient of the flow coefficient with respect to d_{hyd}/D_m decreases with increasing lobe number ratio. The reason for that is also the increasing blowhole area, which requires very large differences in the male rotor lead. For example: To achieve $d_{hyd}/D_m \approx 0.008$ with the lobe combination of 4+6 and a male rotor crown diameter of $D_m = 36$ mm (which is the highest shown blue value in figure 11) a male rotor lead of $l_m = 100$ mm is required. Changing the lobe combination to 5+7 requires a male rotor lead of only $l_m = 41$ mm. Therefore, the blowhole area increases much more than the blowhole perimeter with increasing lobe number ratio.

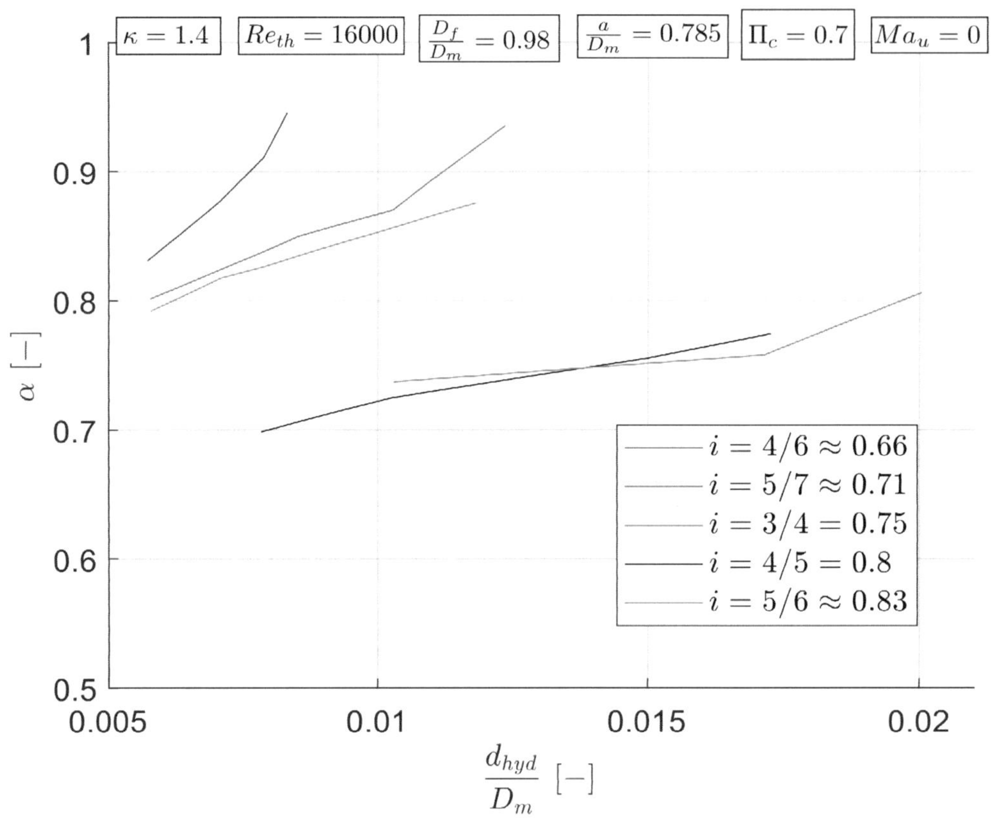

Figure 11. Flow coefficient as a function of d_{hyd}/D_m for different lobe combinations of SRM-A profile.

7.3. Variation of profile family

So far it is shown, that the system of dimensionless numbers is valid for SRM-A type profiles and the influence of some important ratios is examined. Last but not least, the question arises whether the system of characteristic numbers is also valid for other profile families. This is examined on the basis of the SLF profile [19]. At first, it is investigated weather the system of dimensionles numbers is valid for the profile family itself. Second, if a variation of the profile family is permissible within the dimensionless system. The results are shown in figure 12. The flow coefficient is plotted against d_{hyd}/D_m for SRM-A and SLF profile. Considering the SLF profile first (red and blue), the absolute geometric values are varied for constant dimensionless numbers. It is observed, that those do not affect the flow coefficient. This in turn means that the established system of dimensionless numbers should also be valid for the SLF profile. The comparison of the two profile families shows in a second step that the curves of the flow coefficient over the ratio of hydraulic to crown circle diameter are similar but differ in magnitude. The comparison between the two profiles is shown for two different Reynolds numbers of $Re_{th} = 1000$ and $Re_{th} = 16000$. The difference between the flow coefficients, is apparently independent of the Reynolds number, ranges from 6 % for small d_{hyd}/D_m to 2 % for larger d_{hyd}/D_m.

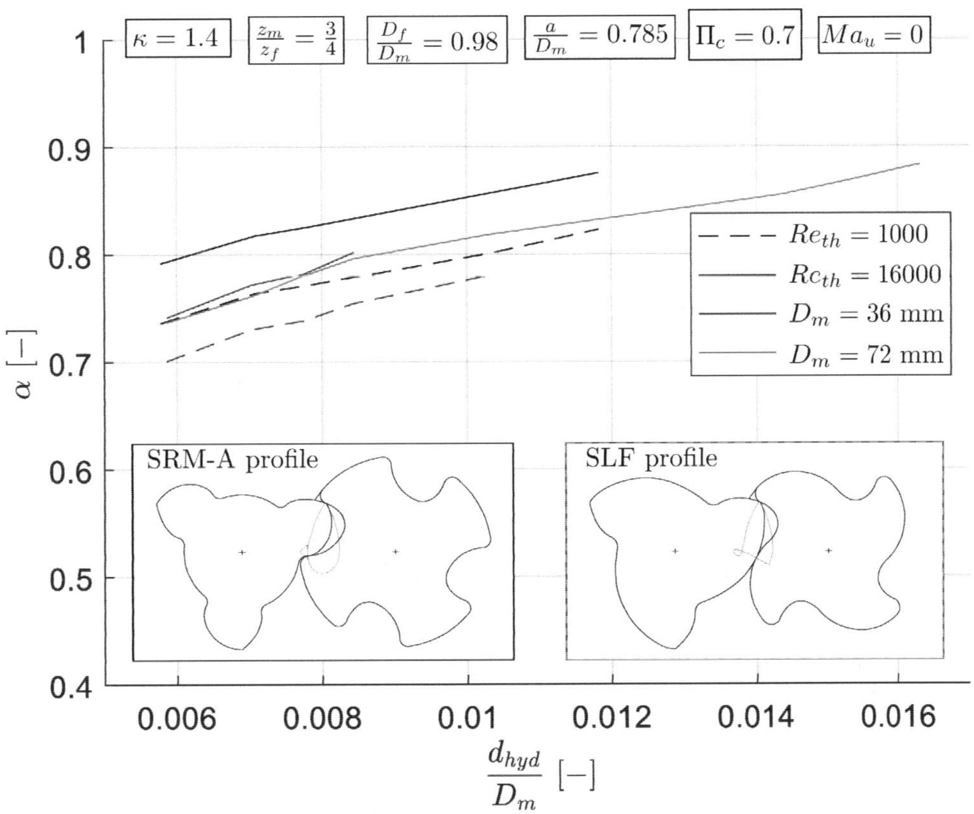

Figure 12. Flow coefficient as a function of d_{hyd}/D_m for different rotor profiles, absolute dimensions and theoretical Reynolds number.

If the shape of the profiles is considered in the front section, it should be mentioned that the geometries differ in relation to the blowhole mainly due to the female rotor tooth flank. This means that the area is larger, but the perimeter is significantly smaller. In order to make the flow coefficient additionally comparable for different profile families, it is necessary to find another influencing variable.

8. Conclusion and outlook

This paper investigates the 3D flow through the blowhole of twin-screw machines using a system of dimensionless numbers. The flow coefficient is determined using CFD-simulations. Pressure ratio and Reynolds number as physical and lobe number as well as the lobe number ratio and the ratio of hydraulic to crown circle diameter as geometrical parameters are varied. The investigated physical numbers have a significant influence on the flow coefficient. From geometric view the lobe number does not affect the flow coefficient. In opposite to the lobe number ratio, which has a significant influence. The influence of the hydraulic to crown circle diameter is strongly dependent on the lobe number ratio. Although the system of dimensionless numbers is established from SRM-A type profile, it is found to be valid for other profile families as the SLF profile too. A comparison between the two profiles shows deviations of about 5 %, which means that for full comparability of different profile families further influencing variables have to be identified. It is worth mentioning that the influence of the considered ratios is significantly

IOP Conf. Series: Materials Science and Engineering 1267 (2022) 012008 doi:10.1088/1757-899X/1267/1/012008

larger than 5 %. Further investigations will focus on the influence of the remaining parameters which were kept constant in this study: isentropic exponent, circumferential Mach number, ratio of axis distance to crown circle diameter and ratio of crown circle diameters. Furthermore, the gained results have to be summarized in a functional and implement in a chamber model simulation.

Nomenclature			**Subscript**	
A	area	$[m^2]$	c	clearance
a	axis distance	[m]	crit	critical
c_p	isobaric heat capacity	[J/kg/K]	f	female
c_v	isochoric heat capacity	[J/kg/K]	hp	high pressure
D	diameter	[m]	hyd	hydraulic
d	diameter	[m]	lp	low pressure
i	lobe number ratio	[-]	m	male
l	rotor lead	[m]	th	theoretical
$\dot{m}$	mass flow rate	[kg/s]		
Ma	Mach number	[-]		
Ma_U	circumferential Mach number	[-]		
n	RPMs	$[min^{-1}]$		
p	pressure	[Pa]		
Re	Reynolds number	[-]		
R_s	specific gas constant	[J/kg/K]		
T	temperature	[K]		
U	perimeter	[m]		
u	circumferential speed	[m/s]		
x,y	coordinates	[m]		
z	lobe number	[-]		
α	flow coefficient	[-]		
η	dynamic viscosity	$[Pa \cdot s]$		
κ	isentropic exponent	[-]		
Π	pressure ratio	[-]		
Π	characteristic number	[-]		
ρ	density	$[kg/m^3]$		
φ	tooth pitch angle	[°]		

References

[1] Radgen P and Blaustein E 2001 *Compressed Air Systems in the European Union. Energy, Emissions, Saving Potential and Policy Actions* (Stuttgart: LOG_X Verlag GmbH)

[2] Rinder L 1979 *Schraubenverdichter* (Wien New York: Springer-Verlag)

[3] Stosic N, Smith I and Kovacevic A 2005 *Screw compressors. Mathematical modelling and performance calculation* (Springer-Verlag)

[4] Rane S, Kovacevic A and Stosic N 2015 *IOP Conf. Ser.: Mater. Sci. Eng.* vol 90 p 012008

[5] John B, Voorde V and Vierendeels J 2005 *17th AIAA Computational Fluid Dynamics Conference* p 4701

[6] Kauder K, Janicki M, Rohe A, Kliem B and Temming J 2002 *VDI-Berichte* vol 1715 (VDI Verlag, Düsseldorf) pp 1–16

[7] Saint-Venant B and Wantzel L 1839 *J. de l'ecole Roy. Poly.* vol 16 p 85

[8] Utri M 2021 *Potenzial von nicht-konstanter Rotorsteigung für Schraubenkompressoren* (Berlin: Logos Verlag Berlin)

[9] DIN EN ISO 5167-2 2004 Measurement of fluid flow by means of pressure differential devices inserted in circular-cross section conduits running full - part 2: Orifice plates

[10] Trutnovsky K and Komotori K 1981 *Berührungsfreie Dichtungen* (Düsseldorf: VDI Verlag)
[11] Utri M, Höckenkamp S and Brümmer A 2018 *IOP Conf. Ser.: Mater. Sci. Eng.* vol 425 p 012033
[12] Utri M and Brümmer A 2018 *International Compressor Engineering Conference* p 2583
[13] Singh P J and Bowman J L 1988 *International Compressor Engineering Conference* p 786
[14] Patel H H and Lakhera V J 2021 *Engineering Research Express* vol 3 p 015024
[15] Grafinger M 2010 *Die computerunterstützte Entwicklung der Flankenprofile für Sonderverzahnungen von Schraubenkompressoren* (Shaker Verlag)
[16] Buckingham E 1914 *Physical review* vol 4 pp 345–376
[17] Ferziger J H, Peric M and Street R L 2019 *Computational Methods for Fluid Dynamics* (Springer International Publishing)
[18] Bragg S L 1960 *J. Mech. Sci.* vol 2 pp 35–44
[19] Buthmann P 1985 *Rechnergestützte Schraubenverdichterprofilentwicklung und -rotorauslegung unter Berücksichtigung der betriebsbedingten Verformungen* (Dissertation, Ruhr-Universität Bochum)

IOP Conf. Series: Materials Science and Engineering 1267 (2022) 012009 doi:10.1088/1757-899X/1267/1/012009

Design parameter study to extend the capacity range of dry twin screw compressors

J-G Persson [1,2]
[1] JGP Teknikkonsult, Åkersberga, Sweden
[2] Machine Design Department, KTH-Royal Institute of Technology, Stockholm, Sweden

E-mail: jgp@kth.se

Abstract. The possibility for extending the capacity range for oil-free dry twin screw compressors has been studied. Basic geometry parameters influence on compressor performance has been derived. The two parameters shaft speed and rotor diameter are critical. The lower capacity limit for dry screw compressors depends on maximum acceptable shaft speed, considering bearings, gearbox or power electronics. The higher capacity limit depends on rotor size, considering machining costs and tolerances for large rotor diameters. Thermodynamic efficiency and optimal rotor tip speed are dependent on internal throttling and leakage losses. Simplified analytical modelling has been applied, neglecting detailed geometry with minor influence on compressor performance. Basic kinematic relations give shaft speed as dependent on screw geometry non-dimensional specific displacement, on rotor tip speed (moderately dependent of machine size), and on capacity. Basic kinematic relations also define the rotor diameter dependence on these three parameters, specific displacement; tip speed, and capacity.
The rotor profile geometry, i.e. number of lobes, rotor lobe depth/diameter, male tip width/diameter; as well as rotor length/diameter ratio and wrap angle determine the specific displacement. The specific displacement is non-dimensional and hence independent of the machine size or capacity. Also the specific internal leakage path lengths in the rotor mesh and between rotors and casing, as well as the specific outlet port size will be determined by the rotor profile major geometry parameters. Here specific means in relation to the male rotor diameter. A fixed internal pressure ratio (volume ratio) has been assumed.
A simplified computerized model was developed to calculate the necessary geometry data and in particular the specific displacement and the optimal tip speed, from basic geometry parameters. For the case of constant specific clearances (non-dimensional), the optimal tip speed is constant, independent of compressor size and capacity. For the case of constant absolute clearances, the optimal tip speed is slightly dependent on the compressor size and capacity. A simplified, non-female addendum, point generated, asymmetric profile was assumed. Rotor profile details with minor influence on performance have then been omitted. A cylindrical part of the male tip would allow a bigger outlet port while maintaining the built-in volume ratio.
The theoretical results from this study indicate that optimal compressor data should primarily be dependent on the number of male lobes and the lobe depth, i.e. the specific displacement. A screw profile with a lower number of lobes will result in a lower optimal shaft speed and should hence enable the design of dry screw compressors for lower capacity than the commonly used 4:6 lobe configuration, without exceeding the drive train high speed constraint. A screw profile with a higher number of lobes would result in a higher optimal shaft speed and in particular reduced rotor size (rotor diameter). It should hence enable the design of dry screw compressors for higher capacity than for the commonly used 4:6 lobe configuration, without having the machining problems of too large a rotor diameter.
Considering both the potential capacity range extension and the efficiency, low capacity dry screw compressors with fewer rotor lobes, e.g. 2:4, might be the most promising option.

1. Introduction

The objective of this study is to find possible ways to extend the capacity range for oil-free dry twin screw compressors while maintaining, or possibly improving, the thermodynamic efficiency. The shaft speed for lower capacity units should still be within an acceptable range, considering rolling element bearings, gearbox or power electronics. The rotor diameter for higher capacity units should still be acceptable considering machining costs and tolerances. This paper presents a theoretical study to estimate the influence of basic rotor geometry parameters on compressor performance. In particular, the male rotor tip speed (male rotor peripheral velocity) would have significant influence. The optimal tip speed depends primarily on the rotor geometry, e.g. on the number of rotor lobes, considering pressure ratio, wrap angle, clearances etc kept constant for all cases compared.

General capacity – shaft speed relations for fluid machinery of both turbo and positive displacement types, have been analysed in [1]. For a positive displacement compressor, basic kinematics gives relations for male rotor shaft speed n and male rotor diameter D as functions of capacity (inlet suction volume flow) $\dot{V}$ and parameters male rotor tip speed u and the non-dimensional specific displacement (V/D^3) which is independent of rotor size for a specific rotor geometry.

$$u = \pi\, n\, D \tag{1}$$

$$\dot{V} = n\, V \tag{2}$$

Equations (1), (2) give:

$$n = \sqrt{\frac{1}{\pi^3} \cdot \frac{(V/D^3)\, u^3}{\dot{V}}} \tag{3}$$

$$D = \sqrt{\pi \cdot \frac{\dot{V}}{(V/D^3)\, u}} \tag{4}$$

The generic relations in equations (3), (4) have been illustrated in figure 1, for the case of specific displacement $V/D^3 = 0.75$ representative for a twin screw compressor. The influence of compressor tip speed on the feasible capacity range is clearly illustrated.

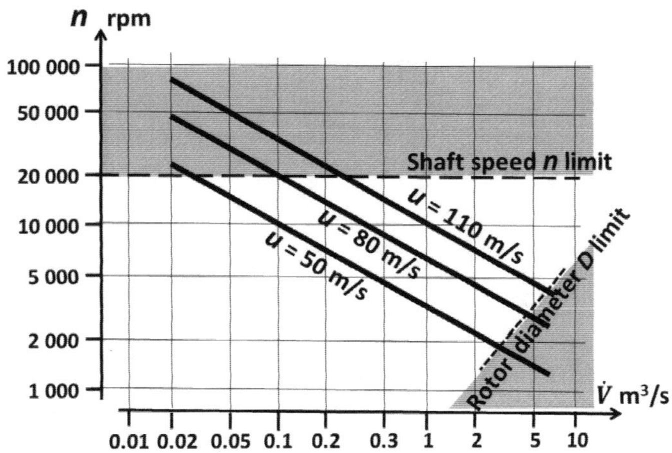

Figure 1. Shaft speed n as function of capacity $\dot{V}$ for twin screw compressors at different values of the male rotor tip speed u. All compressors in the diagram have the specific displacement V/D^3=0.75.

It has been assumed that the compressor variants to be studied operate at their optimal shaft speed which is in fact dependent on the optimal tip speed. According to the theory in [2], when assuming constant absolute clearances independent of rotor size, the optimal tip speed for a specific rotor geometry will be moderately dependent on capacity and size [2], [3].

2. Modelling of twin screw machines

There have since long been many approaches to analysing twin screw compressor performance [4], [5]. Computer simulations based on a lumped chamber model (i.e. at every time step the fluid state is approximated as constant over a chamber volume) and numerical integration have been used to study internal leakage, throttling in ports and isentropic efficiency for screw compressors and expanders [6], [7], [8], [9]. During the last decades, CFD modelling has been introduced for simulation. Further development of the CFD models would however be required for improved accuracy [10], [11].

On the contrary to detailed numerical simulations, an important feature of simplified analytical modelling is that you could easily see the influence of major design and operational parameters. Conclusions can then be drawn without carrying out a great number of time-consuming simulations with detailed models. This approach has been applied in the following. One critical parameter would be the optimal male tip speed. With a rotor profile design having a lower optimal tip speed, the compressor size and capacity could be decreased without violating the shaft speed constraint, as can be seen in figure 1. With a rotor profile design having a higher optimal tip speed, the compressor capacity could be increased without violating the rotor diameter constraint, also seen in figure 1.

The optimal tip speed is defined by two speed-dependent thermodynamic loss components; internal leakage, and throttling losses at inlet and outlet ports [2]. These loss components can approximately be estimated as functions of design parameters for the specific rotor geometry. Hence the optimal tip speed for a specific rotor geometry could also be estimated as a function of major design parameters, e.g. number of rotor lobes.

3. Rotor geometry

3.1. Rotor profile geometry

A simplified, non-female addendum asymmetric profile with approximately no blow-hole on the high pressure side has been assumed and geometry details with minor influence on performance have not been considered. Major design parameters are: number of male and female lobes, lobe depth, male lobe tip width, length/diameter ratio and wrap angle. By introduction of a cylindrical male tip section, a larger outlet port should be possible without decreasing the built-in volume ratio (built-in pressure ratio). A cylindrical male tip should also be favourable concerning the male tip inter-lobe leakage. The profile geometry to be analysed is presented in figure 2. The equations for generating the profiles are well known, e.g. [5], and have not been included in this paper.

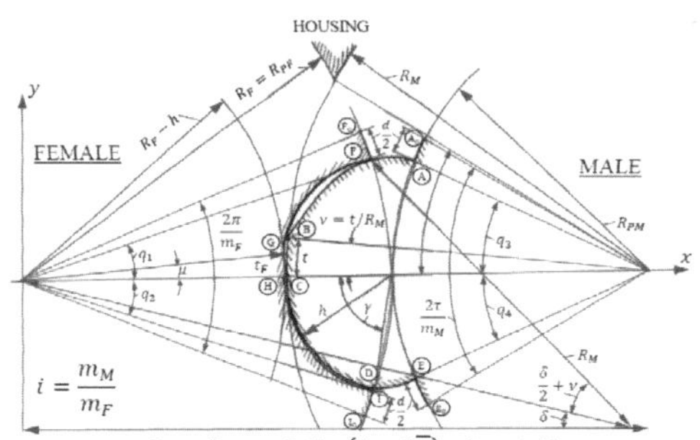

$$R_M = D/2$$
$$R_{PM} = D/2 - h$$
$$R_F = R_{PM}/i$$
$$v = t/R_M = 2 \cdot \overline{t}$$
$$t_F = \mu \cdot (R_F - h)$$
$$\mu = \frac{t \cdot i}{R_F - h} = \frac{\overline{t} \cdot i}{(1/2 - \overline{h})/(i - \overline{h})}$$
$$q_1 = \kappa - \delta$$
$$q_2 = \arcsin\frac{h \cdot \sin(\gamma)}{R_F}$$
$$q_3 = q_1/i$$
$$q_4 = q_2/i$$
$$\delta = i[\arccos\frac{c^2 + R_M^2 - R_F^2}{2 \cdot c \cdot R_M} - v]$$
$$\kappa = \arccos\frac{c^2 + R_F^2 - R_M^2}{2 \cdot c \cdot R_F}$$
$$\gamma = \frac{\pi}{2} - \arcsin(\frac{h}{2 \cdot R_F})$$

Figure 2. Profile geometry. The detailed geometry model has not been included in this paper, but profile parameters and equations have been indicated in this figure.

The screw profiles studied have been illustrated in figure 3.

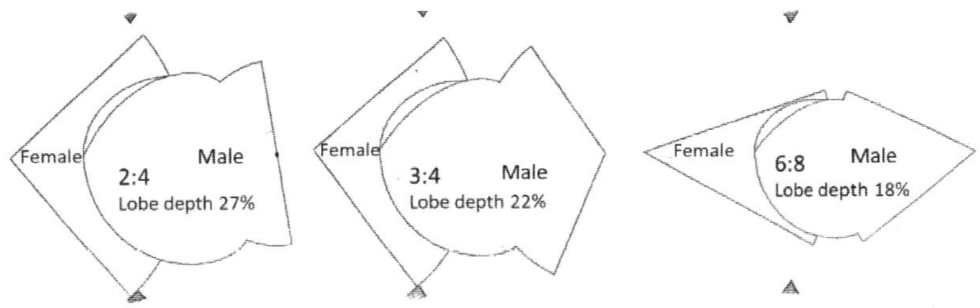

Figure 3. Screw profiles: 2:4, lobe depth 27%; 3:4, lobe depth 22%; and 6:8, lobe depth 18%.

3.2 Rotor and casing geometry

A computer program was used to calculate geometry data from the input design parameters:

- m_M number of male lobes
- m_F number of female lobes
- h/D relative lobe depth
- t/D relative male tip width
- L/D relative rotor length
- ω wrap angle

Calculated data: specific displacement V/D^3, male and female lead S_M, S_F, maximum relative speed male/female, relative speed in the groove bottom, and sealing line lengths rotors/casing and in the rotor mesh. The male and female profile groove cross section areas A_M and A_F were calculated by numerical integration from the profile coordinates. Neglecting the rotor mesh influence on the displaced volume, the displaced volume per revolution will approximately be:

$$V \approx m_M \cdot (A_M + A_F) \cdot L \tag{5}$$

3.2.1 Leakage path lengths. Rotor/casing sealing line lengths can easily be obtained from the "herring-bone" diagram i.e. the plan view of the unwrapped compressor envelope, figure 4.

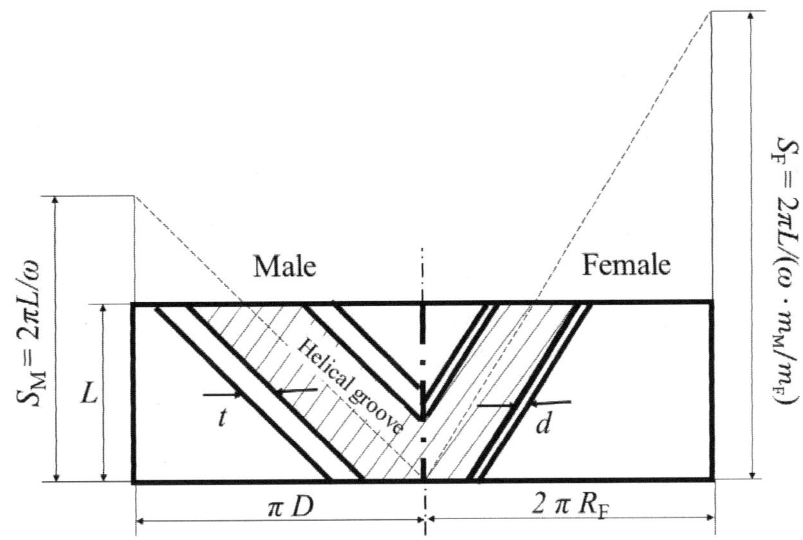

Figure 4. Compressor "herring-bone" diagram.

The different sections of the rotor mesh sealing line illustrated in figure 5, have been calculated.

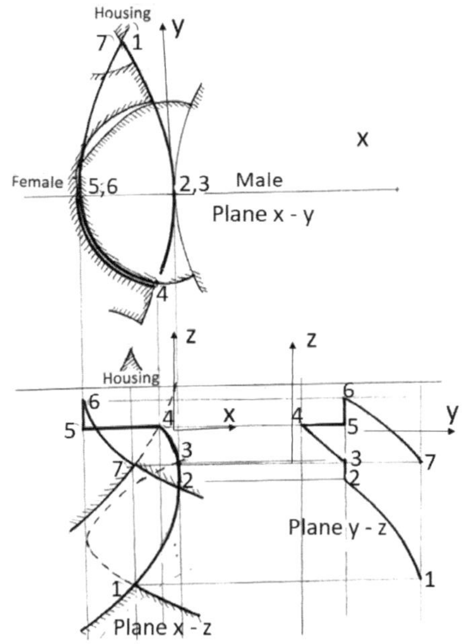

Figure 5. Sections of the rotor mesh sealing line.

3.2.2 Outlet port area and internal compression. A required internal pressure ratio $\pi_\mathrm{i} = 3$ has been assumed. κ is the isentropic exponent (1.4 for air). Internal pressure ratio and internal volume ratio:

$$\pi_\mathrm{i} = \left(\frac{V_\mathrm{max}}{V_\mathrm{min}}\right)^{\kappa} \approx \left(\frac{L}{a+e/2}\right)^{\kappa} \tag{6}$$

The radial outlet port geometry with parameters and geometry relations is illustrated in figure 6.

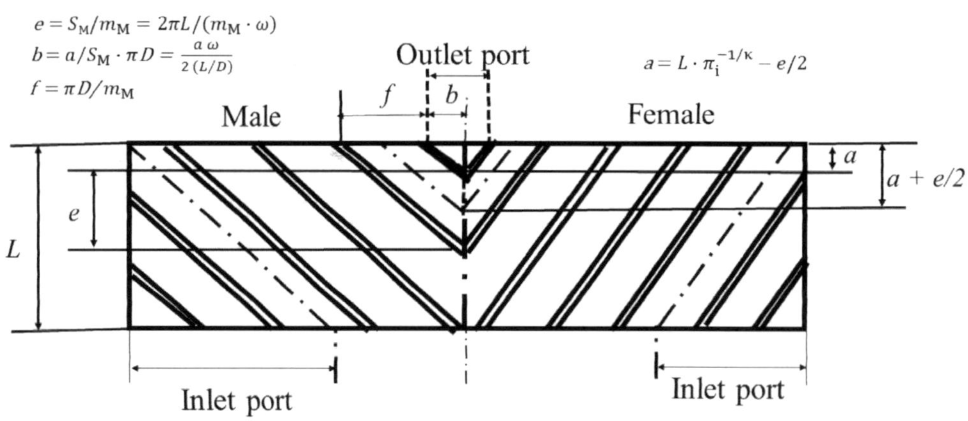

Figure 6. Outlet port radial part with geometry parameter relations.

The radial outlet port area is:

$$A_\mathrm{out,rad} \approx \frac{a \cdot b}{2} \cdot \left(1 + \frac{m_\mathrm{M}}{m_\mathrm{F}}\right) \tag{7}$$

The axial outlet port area will approximately be:

$$A_{\text{out,ax}} \approx \frac{b}{f}(A_{\text{M}} + A_{\text{F}}) \tag{8}$$

For the case of a modified rotor profile with a cylindrical portion at the male rotor tip, the outlet port area might be increased as compared to the case with a male apex tip with the same built-in pressure ratio and the same number of rotor lobes. This possible modification of the outlet port is illustrated in figure 7. This optional profile modification has however not been applied in the following analysis.

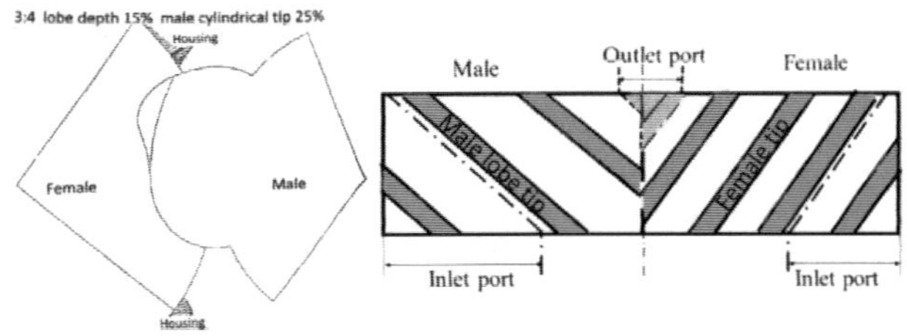

Figure 7. Possible enlargement of the outlet port, for a rotor profile with a cylindrical male tip.

4. Leakage and throttling losses, optimisation

4.1 Optimal tip speed criteria

The optimal tip speed u_{opt} and the optimal shaft speed n_{opt} for a dry twin screw compressor with rolling element bearings, will be determined by the two speed-dependent thermodynamic loss components internal leakage and throttling in ports. According to the theory for positive displacement rotary compressors [2], the speed-dependent relative loss λ due to recompression of internal leakage will be:

$$\lambda = A/u \tag{9}$$

where for constant relative clearances δ_i/D, the constant A is dependent on the specific leakage path lengths L_i/D, the relative clearances δ_i/D, number of male lobes m_{M} and specific displacement V/D^3.

$$A \propto \frac{m_{\text{M}} \cdot \Sigma_i[(L_i/D) \cdot (\delta_i/D)]}{V/D^3} \tag{10}$$

The speed-dependent relative loss ξ due to throttling in ports [2] will be:

$$\xi = B \cdot u^2 \tag{11}$$

where the constant B is dependent on the internal pressure ratio π_{i}, the relative port areas A_{in}/D^2, A_{out}/D^2, the throttling coefficients z_{in}, z_{out}, and the specific displacement V/D^3. The throttling coefficients z have been adjusted to account for the fluid velocity in inlet and outlet ports, with factors for these velocities related to the tip speed. The inlet port velocity c_{in} is approximated to the axial velocity through the machine:

$$c_{\text{in}} \approx \frac{2\pi}{\omega} \cdot L \cdot n = \frac{2(L/D)}{\omega} \cdot u \tag{12}$$

The outlet port velocity c_{out} will approximately be:

$$c_{\text{out}} \approx \frac{\dot{V} \cdot \pi_{\text{i}}^{-1/\kappa}}{A_{\text{out}}} = \frac{1}{\pi} \cdot \frac{(V/D^3) \cdot \pi_{\text{i}}^{-1/\kappa}}{A_{\text{out}}/D^2} \cdot u \tag{13}$$

The sensitivity to design parameters for the throttling loss constant B can then be established, according to equation (14) where c_{in} and c_{out} are embedded in the throttling coefficients z_{in}, z_{out}.:

$$B \propto (V/D^3)^2 \cdot \left[\frac{z_{\rm in} \cdot \pi_{\rm i}^{\frac{\kappa-1}{\kappa}}}{(A_{\rm in}/D^2)^2} + \frac{z_{\rm out} \cdot \pi_{\rm i}^{-2/\kappa}}{(A_{\rm out}/D^2)^2}\right] \tag{14}$$

Optimal tip speed is determined [2] by:

$$\frac{\partial}{\partial u}(\lambda + \xi) = 0 \tag{15}$$

and from equations (9), (11), (15) the optimal tip speed will be [2]:

$$u_{\rm opt} = \sqrt[3]{\frac{A}{2B}} \tag{16}$$

4.2 Optimal conditions for the case of constant relative clearances δ_i/D

Insertion in equation (16) of the relations from equations (10), (14) gives:

$$u_{\rm opt} \propto \frac{1}{(V/D^3)} \cdot \left[\frac{m_{\rm M} \cdot \Sigma_i[(L_i/D)\cdot(\delta_i/D)]}{\left[\frac{z_{\rm in} \cdot \pi_{\rm i}^{\frac{\kappa-1}{\kappa}}}{(A_{\rm in}/D^2)^2} + \frac{z_{\rm out} \cdot \pi_{\rm i}^{-2/\kappa}}{(A_{\rm out}/D^2)^2}\right]}\right]^{1/3} = const. \tag{17}$$

Absolute values would not be necessary for a comparative study, just the sensitivity to design parameters should be sufficient. Equations (3), (17) give the optimal shaft speed – capacity relation, for the case of constant relative clearances δ_i/D:

$$n_{\rm opt} \propto \dot{V}^{-1/2} \cdot \frac{1}{(V/D^3)} \cdot \left[\frac{m_{\rm M} \cdot \Sigma_i[(L_i/D)\cdot(\delta_i/D)]}{\left[\frac{z_{\rm in} \cdot \pi_{\rm i}^{\frac{\kappa-1}{\kappa}}}{(A_{\rm in}/D^2)^2} + \frac{z_{\rm out} \cdot \pi_{\rm i}^{-2/\kappa}}{(A_{\rm out}/D^2)^2}\right]}\right]^{1/2} \propto \dot{V}^{-1/2} \tag{18}$$

Similarly, equations (4), (17) give the optimal rotor diameter – capacity relation, for the case of constant relative clearances δ_i/D:

$$D_{\rm opt} \propto \dot{V}^{1/2} \cdot \left[\frac{\left[\frac{z_{\rm in} \cdot \pi_{\rm i}^{\frac{\kappa-1}{\kappa}}}{(A_{\rm in}/D^2)^2} + \frac{z_{\rm out} \cdot \pi_{\rm i}^{-2/\kappa}}{(A_{\rm out}/D^2)^2}\right]}{m_{\rm M} \cdot \Sigma_i[(L_i/D)\cdot(\delta_i/D)]}\right]^{1/6} \propto \dot{V}^{1/2} \tag{19}$$

The loss factor $f_{\rm opt}$ at optimal conditions will be [2]:

$$f_{\rm opt} = \frac{1}{\eta_{\rm opt}} - 1 = 3\sqrt[3]{A^2 B / 4} \tag{20}$$

It should be noted that the efficiency $\eta_{\rm opt}$ is based only on the two speed-dependent loss components λ and ξ, so this efficiency is not representing the total isentropic efficiency for the compressor. It has been assumed that all other loss components would be the same for all cases. Hence they are not represented in this analysis. For the case of constant relative clearances δ_i/D, the loss factor will be:

$$f_{\rm opt} \propto \{m_{\rm M} \cdot \Sigma_i[(L_i/D)\cdot(\delta_i/D)]\}^{2/3} \cdot \left[\frac{z_{\rm in} \cdot \pi_{\rm i}^{\frac{\kappa-1}{\kappa}}}{(A_{\rm in}/D^2)^2} + \frac{z_{\rm out} \cdot \pi_{\rm i}^{-2/\kappa}}{(A_{\rm out}/D^2)^2}\right]^{1/3} = const. \tag{21}$$

4.3 Optimal conditions for the case of constant absolute clearances δ

With the assumption of constant absolute clearances that was applied for the cases studied, the optimal tip speed will instead be slightly dependent of the compressor capacity (size). By replacing δ_i/D by δ

in equation (17) and insertion of the rotor diameter D from equation (4), the relation for optimal tip speed will then be modified according to equation (22):

$$u_{\text{opt}} \propto \dot{V}^{-1/5} \cdot \frac{1}{(V/D^3)} \cdot \left[\frac{m_M \cdot \sum_i (L_i/D) \cdot \delta}{\left[\frac{z_{\text{in}} \cdot \pi_i^{\frac{\kappa-1}{\kappa}}}{(A_{\text{in}}/D^2)^2} + \frac{z_{\text{out}} \cdot \pi_i^{-2/\kappa}}{(A_{\text{out}}/D^2)^2} \right]} \right]^{2/5} \propto \dot{V}^{-1/5} \qquad (22)$$

Equations (3), (22) give the optimal shaft speed – capacity relation for the case of constant absolute clearances δ:

$$n_{\text{opt}} \propto \dot{V}^{-4/5} \cdot \frac{1}{(V/D^3)} \cdot \left[\frac{m_M \cdot \sum_i (L_i/D) \cdot \delta}{\left[\frac{z_{\text{in}} \cdot \pi_i^{\frac{\kappa-1}{\kappa}}}{(A_{\text{in}}/D^2)^2} + \frac{z_{\text{out}} \cdot \pi_i^{-2/\kappa}}{(A_{\text{out}}/D^2)^2} \right]} \right]^{3/5} \propto \dot{V}^{-4/5} \qquad (23)$$

Equations (4), (22) give the optimal rotor diameter – capacity relation for the case of constant absolute clearances δ:

$$D_{\text{opt}} \propto \dot{V}^{3/5} \cdot \left[\frac{\left[\frac{z_{\text{in}} \cdot \pi_i^{\frac{\kappa-1}{\kappa}}}{(A_{\text{in}}/D^2)^2} + \frac{z_{\text{out}} \cdot \pi_i^{-2/\kappa}}{(A_{\text{out}}/D^2)^2} \right]}{m_M \cdot \sum_i (L_i/D) \cdot \delta} \right]^{1/5} \propto \dot{V}^{3/5} \qquad (24)$$

For the case of constant absolute clearances δ the loss factor will be:

$$f_{\text{opt}} \propto \dot{V}^{-2/5} \cdot \left[m_M \cdot \sum_i (L_i/D) \cdot \delta \right]^{4/5} \cdot \left[\frac{z_{\text{in}} \cdot \pi_i^{\frac{\kappa-1}{\kappa}}}{(A_{\text{in}}/D^2)^2} + \frac{z_{\text{out}} \cdot \pi_i^{-2/\kappa}}{(A_{\text{out}}/D^2)^2} \right]^{1/5} \propto \dot{V}^{-2/5} \qquad (25)$$

4.4 Optimal design dependency on the scale influence for clearances

The lines for n_{opt} and D_{opt} have different slope in a logarithmic capacity – speed diagram, dependent on whether constant relative clearances or constant absolute clearances have been assumed, according to equations (18), (19), (23), (24). For constant absolute clearances the leakage loss will be lower for large capacity units with larger rotor diameter, equation (25).

5. Parameter influence

Equations (22), (23), (24), (25) have been used to analyse compressor performance sensitivity to major design parameters, and in particular to the number of rotor lobes. For all cases studied the absolute value of clearances was considered constant. The external pressure ratio and the built-in pressure ratio has been set to 3 for all cases. The absolute values of the operational parameters at optimal conditions have not been calculated, as these equations only give proportionalities, for comparison of alternative designs. Calculations of the specific displacement V/D^3 and optimal data have been carried out for two rotor configurations, 2:4 and 6:8 (not having the cylindrical male tip according to figure 7). Optimal data n_{opt}, u_{opt}, D_{opt} and f_{opt} have been calculated as relative values compared to the corresponding parameters for a standard compressor with 4:6 lobes as reference case. The geometry parameters related to leakage and throttling, embedded in factors A and B, were calculated and compared to the reference 4:6 compressor. All other data including the absolute value δ of the clearances were considered constant and the same for all compressors studied, independent of their rotor size. The absolute values of the losses were then not necessary to calculate. The 4:6 reference case capacity was set to 100%. Performance data were calculated for two cases:

- 2:4 compressor with a shaft speed 100% as compared to the 4:6 reference case. The capacity for this 2:4 compressor geometry was found to be 30% as compared to the 4:6 reference.

Optimal tip speed was 65%. The efficiency was found to be slightly improved, as compared to a 4:6 compressor with the same capacity.

- 6:8 compressor with a male rotor diameter 100% as compared to the 4:6 reference case. The capacity for this 6:8 compressor was found to be 123% as compared to the 4:6 reference. Optimal tip speed was 137%. The efficiency was found to be slightly lower.

Results have been summarised in table 1 and table 2.

Table 1. 2:4 compressor. Shaft speed, tip speed, rotor diameter and speed-dependent losses at optimal conditions, as compared to the reference case 4:6.

	Reference case 4:6 $V/D^3 = 0.75$	----- 2:4 compressor ----- $V/D^3 = 0.85$	
Capacity $\dot{V}$	100%	**100%**	**30%**
Optimal shaft speed n_{opt}	**100%**	37%	**100%**
Optimal tip speed u_{opt}	100%	40%	65%
Optimal male rotor diameter D_{opt}	100%	145%	65%
Speed-dependent losses f_{opt}	100%	**78%**	110%

Table 2. 6:8 compressor. Shaft speed, tip speed, rotor diameter and speed-dependent losses at optimal conditions, as compared to the reference case 4:6.

	Reference case 4:6 $V/D^3 = 0.75$	----- 6:8 compressor ----- $V/D^3 = 0.67$	
Capacity $\dot{V}$	100%	**100%**	**123%**
Optimal shaft speed n_{opt}	100%	155%	133%
Optimal tip speed u_{opt}	100%	145%	137%
Optimal male rotor diameter D_{opt}	**100%**	90%	**100%**
Speed-dependent losses f_{opt}	100%	130%	**115%**

These results have been illustrated in figure 8. Shaft speed – capacity relations for the optimal 2:4 and 6:8 compressors analysed are compared to the shaft speed – capacity relations for compressors with constant tip speed. The extended capacity regions for the optimised compressors have been indicated.

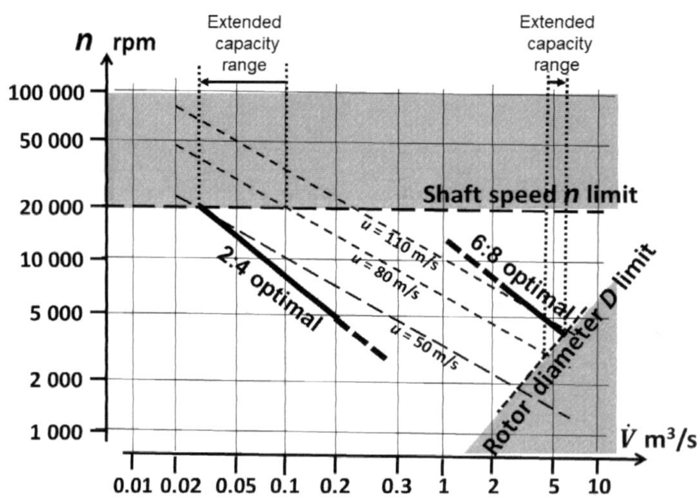

Figure 8. Shaft speed – capacity relations plotted for the optimal 2:4 and 6:8 compressors analysed, with the extended capacity regions indicated. Constant absolute values of clearances have been assumed. The dashed lines represent 4:6 compressors with specific displacement $V/D^3 = 0.75$ and constant tip speed.

6. Conclusions

From this study it was confirmed that the capacity range for dry twin screw compressors might be widened, without violating the shaft speed or the rotor diameter constraints. Considering the possible increase of the capacity range for the respective machine, as well as the efficiency, a low capacity dry screw compressor with fewer rotor lobes, e.g. 2:4, might be the most promising option.

Conclusions based on this theoretical study have however not been verified e.g. by detailed numerical simulations, nor validated experimentally by prototype tests. These results should then rather be seen as an indication of possible ways to extend the capacity range by means of rotor geometry modifications. Further research based e.g. on CFD simulations and possibly prototype tests would be required, for verification and validation.

For the case of an industrial implementation, a general and efficient product development process as follows is proposed:

1 Simplified analytical approach for analysis of design parameter influence and sensitivity
2 Verification by detailed numerical simulation, e.g. by CFD modelling
3 Validation by prototype manufacturing and tests

7. References

[1] Persson J-G 2000 Turbo vs´ displacement type fluid machinery - characteristics and operating ranges *ICME '2000 First International Conference on Mechanical Engineering* (Shanghai)

[2] Persson J-G 1990 Performance mapping versus design parameters for screw compressors and other displacement compressor types *VDI-Berichte 859* (Düsseldorf: VDI-Verlag) pp 15-31

[3] Persson J-G 1990 Performance evaluation of fluid machinery during conceptual design *CIRP Annals Manufacturing Technology 12/1990* (Paris: CIRP, Elsevier) pp 137-140

[4] Golovincov A G, Rumyantsev V A, Ardashev V I, Peshti Yu V, Plastinin P I, Suslov A D, Frolov E S and Yaminsky V V 1964 Ротационные компрессоры (Rotational compressors) *Иэдательство машиностроение (Educational engineering)* (Moscow: Иэдательство машиностроение) Винтовые компрессоры (Screw compressors) pp 248 - 290

[5] Rinder L 1979 Schraubenverdichter *Springer-Verlag* (Wien: Springer-Verlag ISBN 3-211-81535-X)

[6] Jonsson S 1987 Performance simulations of twin-screw compressors for refrigeration purpose *VDI-Berichte 640* (Düsseldorf: VDI-Verlag) pp 187-207

[7] Platell P 1993 Displacement expanders for small scale cogeneration *KTH Machine Design Licentiate thesis* (Stockholm: NUTEK-VK-93-16; KTH TRITA-MAK-LA-93-1) Appendix 2

[8] Sjöstedt C-J 2005 On the modular modelling for dynamical simulation with application to fluid systems *KTH Machine Design Licentiate thesis* (Stockholm: KTH TRITA – MMK 2005:30 ISSN 1400-1179 ISRN/KTH/MMK/R-05/30-SE)

[9] Sjöstedt C-J, Persson J-G and Chen D-J 2006 Automotive fuel cell system simulation, component and compressor modelling *VDI-Berichte 1932* (Düsseldorf: VDI-Verlag) pp 217-235

[10] Kovacevic A, Rane S and Stosic N 2016 Computational fluid dynamics in rotary positive displacement screw machines *16th International Symposium on Transport Phenomena and Dynamics of Rotating Machinery* (Honolulu, USA: hal-01879361)

[11] Andres R, Hesse J, Hetze F and Low D 2018 CFD Simulation of a two stage twin screw compressor including leakage flows and comparison with experimental data *24th International Compressor Engineering Conference at Purdue* (West Lafayette, IN, USA: Purdue e-Pubs)

Acknowledgements

The author would like to thank Atlas Copco Airpower Belgium, for our previously fruitful collaboration in compressor development.

IOP Conf. Series: Materials Science and Engineering 1267 (2022) 012010 doi:10.1088/1757-899X/1267/1/012010

Validation of inhomogeneous chamber states in rotary positive displacement vacuum pumps

H Pleskun[1], **T Jünemann**[1] **and A Brümmer**[1]

[1]TU Dortmund University, Chair of Fluidics, Leonhard-Euler-Str. 5, 44227 Dortmund, Germany

E-mail: heiko.pleskun@tu-dortmund.de

Abstract. Chamber model simulation is a common approach to simulate rotary positive displacement vacuum pumps. Therefore the pump is abstracted into working chambers and connecting clearances, whereby the clearance leakages can be identified as the major loss mechanism in such machines. The clearance mass flow rates are calculated with respect to the thermodynamic states in the adjacent chambers, which are inhomogeneous for rarefied gases due to the movement of the rotors which causes a pressure gradient within the chamber. This effect increases with higher Knudsen numbers, because of the increasingly dominant friction. These inhomogeneous chamber states are assumed to be quasi-static in case that the chamber volume is constant with time. Therefore the chamber must not have a connection to the suction or discharge port. This can be modelled with a one-dimensional approach for geometrically abstracted chambers. In order to validate the one-dimensional characteristics in circumferential direction three-dimensional steady state simulations of a working chamber are performed using a Computational Fluid Dynamics (CFD) solver. To improve the accuracy for rarefied gases Maxwell velocity slip boundary conditions are applied. It is shown, that the inhomogeneous chamber states can be approximated by a regression analysis of a dimensionless number. Furthermore the housing clearance and the radial clearance mass flow rates for given boundary conditions are geometrically abstracted and calculated using a one-dimensional model. The new clearance models and the inhomogeneous chamber states are implemented in a chamber model simulation software and results of a test machine are compared to measurements and to previous simulations.

1. Introduction

Dry running vacuum pumps have become an important part of vacuum systems due to their oil-free working principle. The need for a clean vacuum, for instance in the semiconductor industry, has pushed the development of dry running positive displacement pumps, such as screw and roots pumps (see reference [1]). Due to contactless operation of the pistons the sealing between the working chambers is realized by small clearances. Regarding Figure 1 the effective mass flow rate of a screw machine can be calculated by the delivered mass flow rate $\dot{m}_d$ of a working chamber driven by the rotation of the rotors and reduced by the clearance mass flows $\dot{m}_{cl}$. Clearances are located between both rotors like the radial clearance, the inter-lobe clearance and the blow hole and between the rotors and the housing. In order to calculate the thermodynamic operation of these pumps, knowledge of these clearance mass flow rates, which represent the main loss mechanism, is essential. A common approach to calculate the thermodynamic operation of vacuum pumps is chamber model simulation where the pump is divided in a network of chambers that transport gas from low pressure port to high pressure port and connecting clearances which cause a mass flow rate in opposite direction [2][3][4].

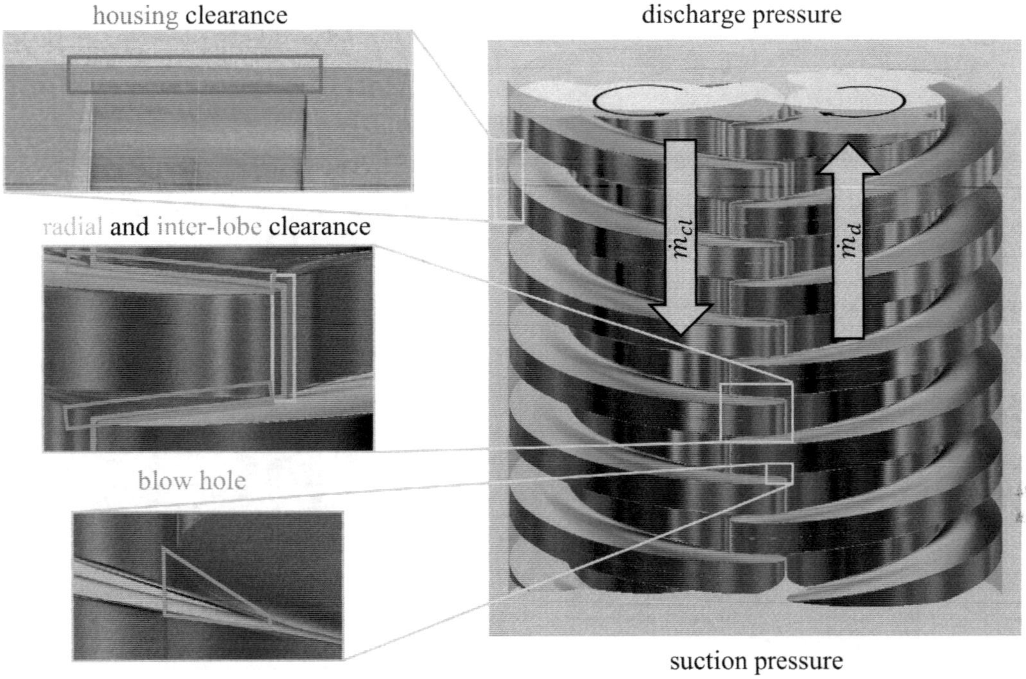

Figure 1. Sketch of an SSVP with 2-lobed cycloid profile and the corresponding clearances.

Typically in chamber model simulation the thermodynamic state of the gas within the chambers is considered to be homogenous and the clearance boundary conditions are taken from mean values within the adjacent chambers. The calculation of clearance flows for rarefied gases including moving boundaries has previously been investigated in references [3],[4] and [5] who created a data base for different clearance shapes and by [6], [7], [8] and [9] who derived and validated a 1D model in order to calculate the clearance mass flow rates. In this paper the different clearances are analysed. The different modelling approaches are discussed and the impact on the simulated suction speed curve is shown. The simulation results are compared to measurements of a test machine.

In a previous paper it was shown that the chamber states are not homogeneous for rarefied gases but that a pressure gradient occurs in tangential direction of the rotor lobe [10]. This is again highlighted in Figure 2 which shows the chamber model abstraction for a screw spindle vacuum pump (SSVP) with the sketched network of chambers and connecting clearances. In this representation pressure p related to the chamber's mean pressure p_{ch} is sketched. The inhomogeneous pressure distribution causes the clearances to not only connect two adjacent chambers, but the high pressure side and a low pressure side of adjacent chambers. Thus the boundary conditions of the clearances have to be corrected for better agreement of simulation results to the pump's operation performance. In a previous work a one-dimensional model was presented to calculate the pressure distribution within the chamber. Although computational fluid dynamics (CFD) solver based on the Navier-Stokes equations fail if the whole machine is calculated due to the gas rarefaction effects within the clearances, in the chambers the Knudsen numbers are lower due to the characteristic length scales. Therefore CFD simulations including Maxwell slip boundary conditions as presented in reference [11] can be applied within a wide range. In this paper the one dimensional model is validated with 3D CFD simulations. Furthermore the impact of different materials of the rotors or housing on the pressure distribution within the chambers is investigated. A dimensional analysis is performed and a regression functional is provided which depicts the main impact of the pressure distribution within the chamber to reduce the computation time. This is implemented in a chamber model and compared to the results where the differential equation is solved.

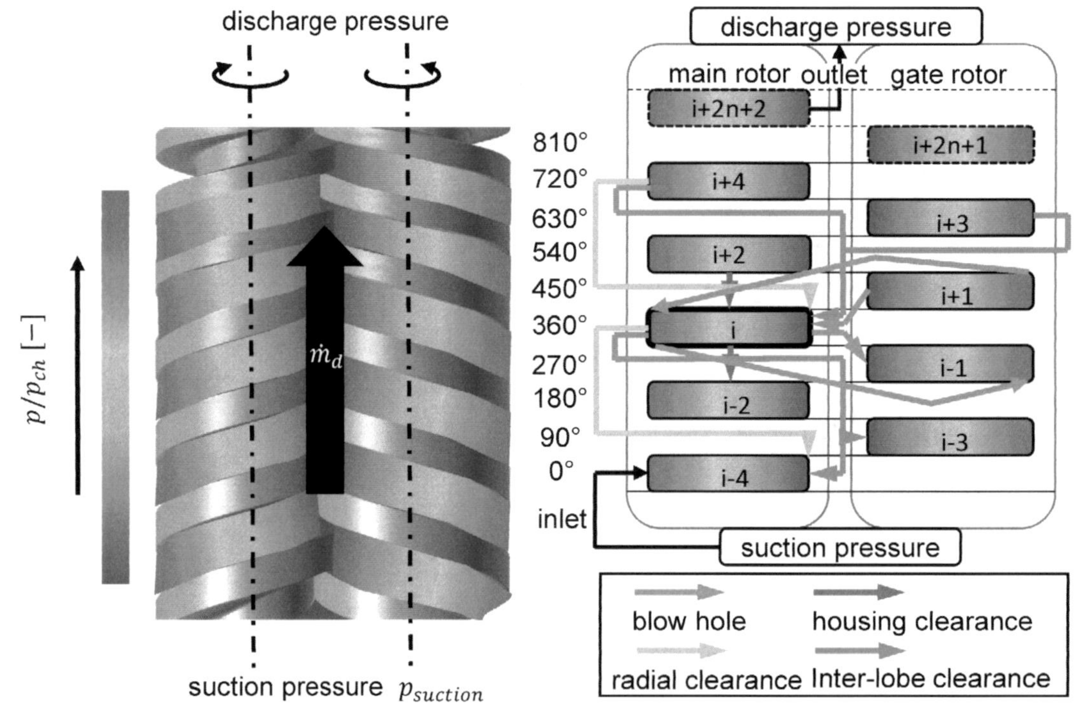

Figure 2. Sketch of an SSVP and the investigated impact of inhomogeneous chamber states on the boundaries of adjacent clearances. Clearance connections and models are similar to [10].

2. Inhomogeneous chamber states

This section deals with the calculation of inhomogeneous chamber states in SSVPs. At first the theoretic approach is presented which leads to a simplified 1D model. In the following 3D CFD simulations are performed in order to validate the simplified model and finally regression functionals are given that can accurately predict the pressure ratios within a working chamber of SSVPs.

2.1. Kinematics

In order to show why the pressure gradient occurs within the chambers it is necessary to take the kinematics into account. Figure 3 shows the sketch of an unwound cylinder cut of one rotor of an SSVP, where the chamber and the profile are alternately placed along the rotor axis. The rotor lead and the piston's rotation causes a chamber to move from low pressure port to high pressure port along the rotor axis with a velocity

$$U_{ax} = s \cdot f \tag{1}$$

where f is the rotational speed and s is the rotor lead. U_{ax} is the velocity of an encapsulated mass package, that is carried in axial direction. Furthermore, each point on the rotor has a movement in circumferential direction

$$U_{circ} = \pi \cdot D \cdot f \tag{2}$$

which is the absolute velocity, where D is twice the distance from the rotor axis to the considered rotor point. The relation between the absolute velocities of the rotor points due to the rotation and the movement of the mass package in axial direction can be obtained by the angle φ.

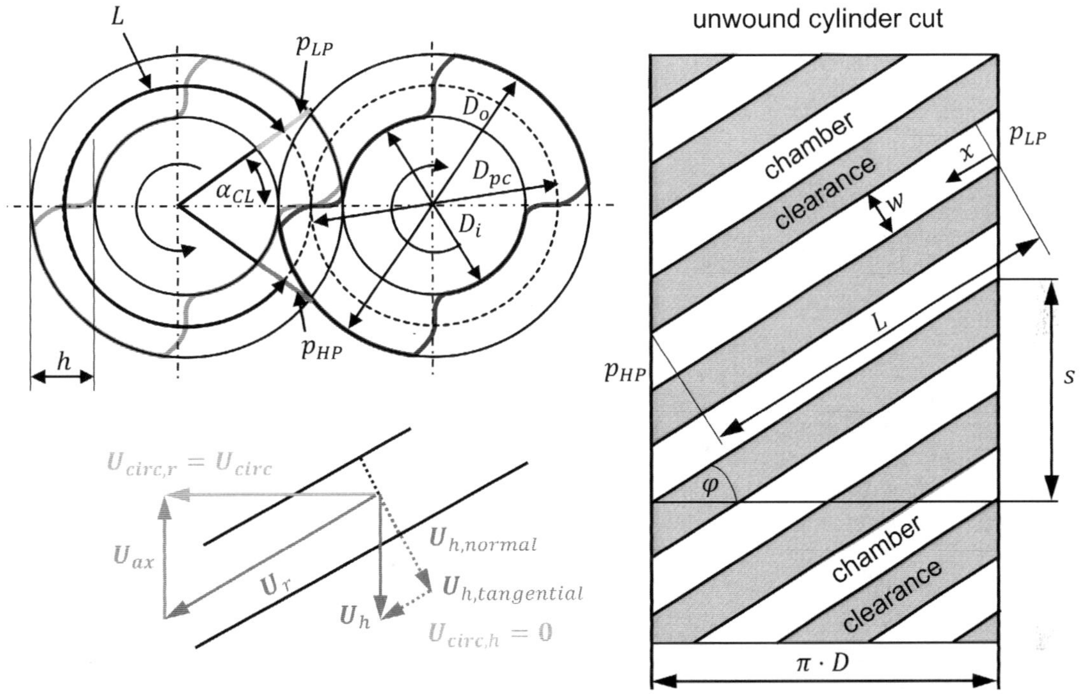

Figure 3. Geometry of a two-lobed cycloid profile and boundary velocities in the relative system of screw vacuum pumps.

Therefore the magnitude of the relative movement U_r of rotor points within the chamber's fixed coordinate system can be obtained by

$$U_r = \frac{\pi D \cdot f}{\cos(\varphi)} \qquad \varphi(D) = \tan^{-1}\left(\frac{s}{\pi D}\right) \qquad D_i \leq D \leq D_o \tag{3}$$

where D_o is the vacuum pump's outer diameter and D_i the inner diameter respectively.

The housing (h) does not move, so the housing's relative velocity $U_h = -U_{ax}$ is negative to the movement of the relative system. This can be divided into a part tangential and normal to the rotor lobe:

$$U_{h,tangential} = s \cdot f \cdot \sin(\varphi(D_o)) \tag{4}$$
$$U_{h,normal} = s \cdot f \cdot \cos(\varphi(D_o)) \tag{5}$$

In each cross section of a chamber there are four bounding surfaces causing a Couette flow in the same direction tangential to the rotor lobe and the housing which additionally causes a flow in direction normal to the rotor lobe. Because of the chamber's encapsulation due to the other rotor and the surrounding housing this flow is dammed up. This leads to a pressure gradient which results in a Poiseuille flow in opposite direction of the Couette flow caused by the moving boundaries.

Geometrically the chamber can be abstracted as a channel with rectangular cross section with height

$$h = \frac{D_o - D_i}{2} \tag{6}$$

and the channel width

$$w = \frac{s}{2z} \cos\left(\varphi(D_{pc})\right) \tag{7}$$

which represents the linearised mean distance between two lobes. D_{pc} is the pitch circle diameter and mean value of inner and outer diameter. The chamber length is defined as

$$L = (\pi - \alpha_{CL}) \cdot D_{pc} \cdot \sqrt{1 + \left(\frac{s}{\pi D_{pc}}\right)^2} \tag{8}$$

with the angle of the cut lens

$$\alpha_{CL} = \tan^{-1}\left(\frac{\sqrt{D_o^2 - D_{pc}^2}}{D_{pc}}\right) \quad \text{(see Figure 3)} \tag{9}$$

This is the length of the helix, at the pitch circle. The end chamber geometry is defined between the intersection lines of rotor axis and the cusps of the cut lens which is the intersection area of the outer diameters of both rotors (see the yellow and red lines in Figure 3). The mean pressure in each of these intersection lines is defined as the chamber's high pressure p_{HP} and low pressure p_{LP} respectively, which is used as inlet and outlet boundary values of adjacent clearances. Thus the chamber is defined with constant cross section area for constant rotor lead s. The impact of a non-constant rotor lead on the pressure distribution within the chamber is neglected in this work.

2.2. Dimensional analysis

Assuming isothermal conditions and a constant gas mixture, that can be related to single gas parameters of a perfect gas, the following thirteen parameters influence the pressure distribution within an SSVP's chamber with the above geometric constraints: The specific gas constant R, the mean temperature within the chamber T, the dynamic viscosity μ, the chamber's mass averaged pressure

$$p_{ch} = \frac{m_t \cdot R \cdot T}{V_t} \tag{10}$$

where m_t is the total mass of the gas within the chamber and V_t is the chamber's total volume. Furthermore, a mass flow rate $\dot{m}_{cl,t}$ within the chamber due to adjacent clearances tangential to the lobe is considered. Further parameters are the rotational speed f and the geometric constraints: The rotor lead s, the outer and inner rotor diameter D_o and D_i and the number of lobes z. Two more parameters are the tangential momentum accommodation coefficients (TMAC) of the housing α_h and the rotor α_r. These parameters are dimensionless numbers $0 \leq \alpha \leq 2$ that describe the gas surface interaction in rarefied gases and depend on the wall material and the gas properties and can be obtained by measurements in the literature, see for example reference [12]. If the rotor is made of a different material than the housing (for example due to coating), different TMACs are needed for the certain walls in the model. This also changes if the gas is exchanged for example instead of air, helium or hydrogen is evacuated.

According to Buckingham's theorem [13] this set of thirteen parameters leads to the following set of nine dimensionless parameters:

The pressure ratio

$$\Pi = \frac{p(x)}{p_{ch}} \tag{11}$$

for an arbitrary position x within the chamber, the gas rarefaction parameter δ which is inverse to the Knudsen number related to the model of hard sphere molecules [14]

$$\delta = \frac{\min(h, w) \cdot p_{ch}}{\mu \cdot c_m} \tag{12}$$

with $c_m = \sqrt{2RT}$ the molecule's most probable speed, the speed ratio

$$u_0 = \frac{U_{circ,o}}{c_m} = \frac{\pi D_o \cdot f}{\sqrt{2RT}} \tag{13}$$

the dimensionless mass flow rate

$$C_0 = \frac{\dot{m}_{cl,t}}{\dot{m}_{eff}} = \frac{\dot{m}_{cl,t} \cdot \sqrt{\pi} c_m}{p_{ch} \cdot hw} \tag{14}$$

with $\dot{m}_{eff}$ the chamber's effusion mass flow rate and the geometric relations

$$g_1 = \frac{D_o}{D_i} \quad g_2 = \frac{D_o}{s} \tag{15}$$

and the already dimensionless number of lobes z which scales the channel width and the two TMACs α_h and α_r.

2.3. *One-dimensional model*

In order to calculate the pressure distribution within the chamber the approach of an isothermal channel flow model is used as presented in reference [15]. Thus in steady state the mass flow rate $\dot{m}$ in a certain direction can be expressed as the sum of the mass flow rate $\dot{m}_C$ caused by a Couette flow and the mass flow rate $\dot{m}_P$ caused by a Poiseuille flow:

$$\dot{m} = \dot{m}_C + \dot{m}_P \tag{16}$$

The mass flow rate caused by a Couette flow in a rectangular channel is the sum of the mass flow rate caused by each moving wall:

$$\dot{m}_C = \rho A \cdot \sum_{j=1}^{4} G_{C,j}(\delta, \eta, \alpha_h, \alpha_r) \cdot U_{w,j} \tag{17}$$

ρ is the density, $A = hw$ the cross section area, $U_{w,i}$ is the mean velocity of the respective wall and $G_{C,i}$ is the shape factor depending on the geometric ratio $\eta = \tilde{w}^2/A$, with $\tilde{w}$ the width of the respective moving wall within the cross section. The shape factors for diffuse wall scattering are derived in a previous work in reference [15]. This has been extended to depict the influence of different wall materials in reference [16].

The mass flow rate of the Poiseuille flow reads as follows

$$\dot{m}_P = -\frac{\min(h, w) \cdot A}{c_m} \cdot G_P(\delta, \eta, \alpha_h, \alpha_r) \cdot \frac{\partial p}{\partial x} \tag{18}$$

and is related to the pressure gradient in flow direction (in this case the tangential lobe direction, see the local coordinate system in Figure 3). The dimensionless flow rates G_P used in this model are described in a previous work [6]. For different wall materials the flow rates of the Poiseuille flow can be taken from reference [16].

Inserting equations 17 and 18 in equation 16 leads to a differential equation in tangential lobe direction which is solved by a 5th order Dormand-Prince-Method with adaptive step size with an accuracy of 10^{-9} from the boost library [17]:

$$\frac{\partial p}{\partial x} = \frac{2}{G_P \min(h,w) c_m} \left(p \sum_{i=1}^{4} G_{C,i} \cdot U_{w,i} - \frac{\dot{m}_{cl,t} RT}{hw} \right) \tag{19}$$

Using the dimensionless numbers of section 2.2 equation 19 can be normalized as follows:

$$\frac{L}{p} \frac{\partial p}{\partial x} = \frac{L}{\min(h,w) G_P} \left(\frac{2 \sum_{i=1}^{4} G_{C,i} U_i}{c_m} - \frac{C_0}{\sqrt{\pi}} \right) \tag{20}$$

If the mass averaged pressure within the chamber is known a bisection method is used to find the low pressure in a way that the resulting pressure distribution conserves the mass within the chamber under isothermal conditions

$$m_{1D} = \frac{A}{RT} \int_0^L p \, \mathrm{d}x \approx \frac{A}{RT} \sum_{i=1}^{N} \bar{p}_i \Delta x_i \tag{21}$$

where the trapezoidal rule is used. Δx_i is the step size between two pressure values p_i and p_{i+1} of the discretized differential equation 19 with variable step size and $\bar{p}_i = (p_i + p_{i+1})/2$ is the average pressure value in this increment. As the trapezoidal method is exact for linear curves and the step size is calculated adaptively to provide small step sizes in case of non linear curves, the error provided by this method is negligible. If the pressure in the middle of the chamber is known, equation 19 can be solved starting in the middle of the chamber once with a positive sign to find the high pressure and once with a negative sign to find the low pressure. Therefore no bisection method is needed.

2.4. Three-dimensional simulation

In order to validate the used one-dimensional model, Computational Fluid Dynamics simulations (CFD) using ANSYS CFX 2021 R2 are performed where the Navier-Stokes-Fourier equations (NSF) are solved numerically [18]. The used mesh including the boundaries are shown in Figure 4. The setup is designed to be similar to an experimental test rig at the Chair of Fluidics of TU Dortmund University. Therefore one vacuum chamber with zero rotor lead but a constant width of $w = 0.015$ m is simulated. Similar to the real machine, the chamber is encapsulated due to the other rotor which forms a radial clearance. In order to vary the clearance mass flow this radial clearance is replaced by an inlet and an outlet where the same mass flow rate is prescribed with a uniform velocity profile. The distance between inlet and outlet is 0.02 m. The system pressure is prescribed with the pressure opening. The inner diameter and the side wall are rotating around the rotation axis. The opposite boundary of the side wall is a symmetry plane. The other walls are static. For all walls Maxwell slip boundary conditions with accommodation coefficients $\alpha = 1$ as described and validated in reference [11] are used in order to improve the results for the slip regime $\delta \geq 10$. For lower gas rarefaction parameters the first order slip boundary condition does not provide reliable results any more [19][20]. In order to compare the results to the one-dimensional model the chamber's high pressure p_{HP} and low pressure p_{LP} are evaluated as an area average as shown in Figure 4.

The domain's outer diameter D_o is 0.17 m and the inner diameter D_i is 0.1 m. The opening size is $0.02 \times 0.02 \times 0.0075$ m and is placed in the middle of the chamber on the opposite side of the radial clearance. Due to the symmetry plane the domain width is half the chamber width. The mass flow rate through the opening in steady state is numerically zero. The temperature at the walls is fixed to 298.15 K and is equal to the temperature at the opening and the inlet.

A Richardson extrapolation according to references [21] and [22] with two finer meshes and a constant refinement factor of 1.5 in all directions for the case $\delta = 10$ and $\delta = 100$, $u_0 = 0.12$ and $C_0 = 0.01$ has been performed. For $\delta = 10$ the numerical error of the used mesh to the theoretically exact value for the pressure ratio is $\frac{\Delta(p_{HP}/p_{ch})}{p_{HP}/p_{ch}} \approx 1\%$ and $\frac{\Delta(p_{HP}/p_{LP})}{p_{HP}/p_{LP}} \approx 2.8\%$. For $\delta = 100$ the numerical errors for the pressure ratio are $\frac{\Delta(p_{HP}/p_{ch})}{p_{HP}/p_{ch}} \approx 0.3\%$ and $\frac{\Delta(p_{HP}/p_{LP})}{p_{HP}/p_{LP}} \approx 0.8\%$.

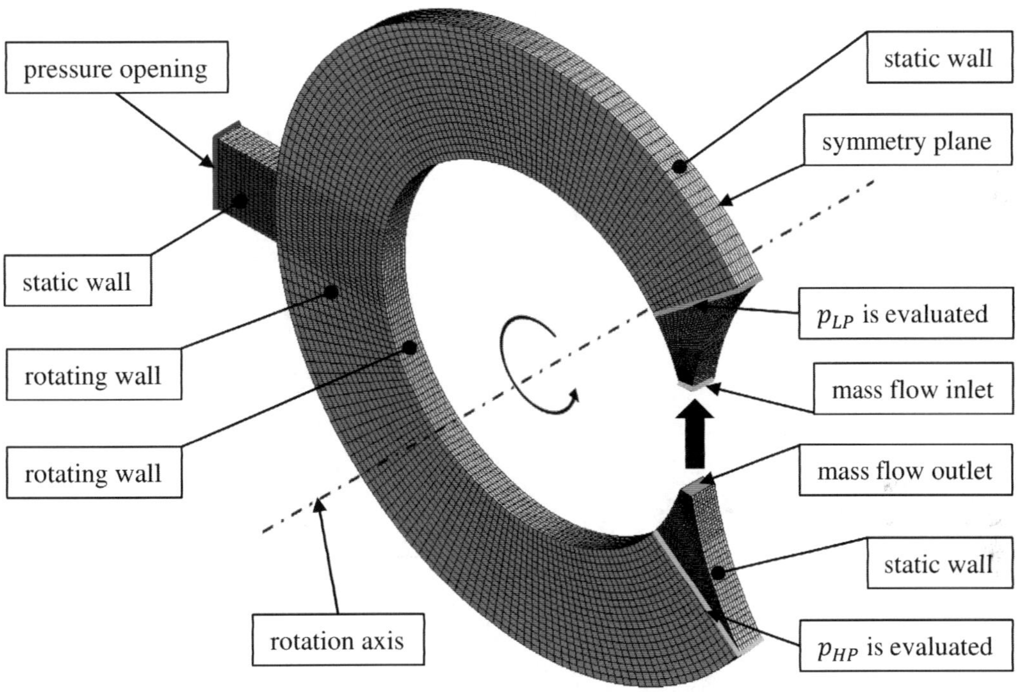

Figure 4. 3D-Domain in Ansys CFX.

2.5. Validation

In order to compare the results of the one-dimensional model to those of the CFD simulations for equal boundary conditions the procedure in in section 2.3 is applied to find the midpoint pressure p_{ch} in the middle of the chamber at $x = L/2$. Figure 5 shows the pressure ratio p_{HP}/p_{ch} as a function of the gas rarefaction parameter $\delta(p_{ch})$ for different speed ratios, with p_{ch} the opening pressure (CFD) or midpoint pressure (1D). The normalized mass flow rate is $C_0 = 0$. For this purpose in CFD the inlet and outlet are replaced by walls. The results of the 1D model show a maximum of the pressure ratio at $\delta \approx 0.5$. For $\delta < 0.5$ the pressure ratio tends to an asymptotic value for the free molecular flow. For $\delta > 0.5$ the pressure ratio tends to unity thus for the hydrodynamic regime $\delta > 1000$ a homogeneous chamber state is obtained.

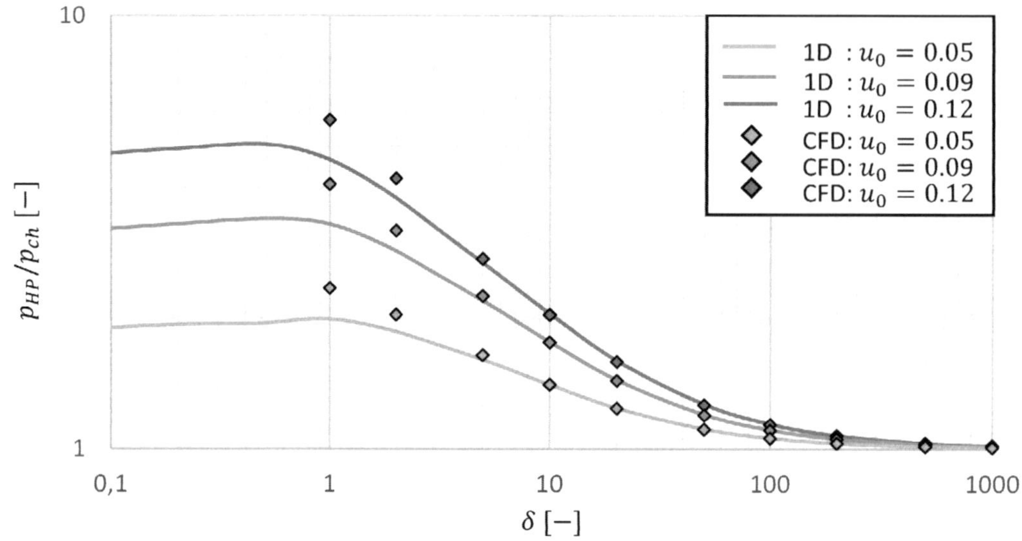

Figure 5. Comparison of the pressure ratio p_{HP}/p_{ch} between 1D model and CFD for different speed ratios u_0 for $C_0 = 0$.

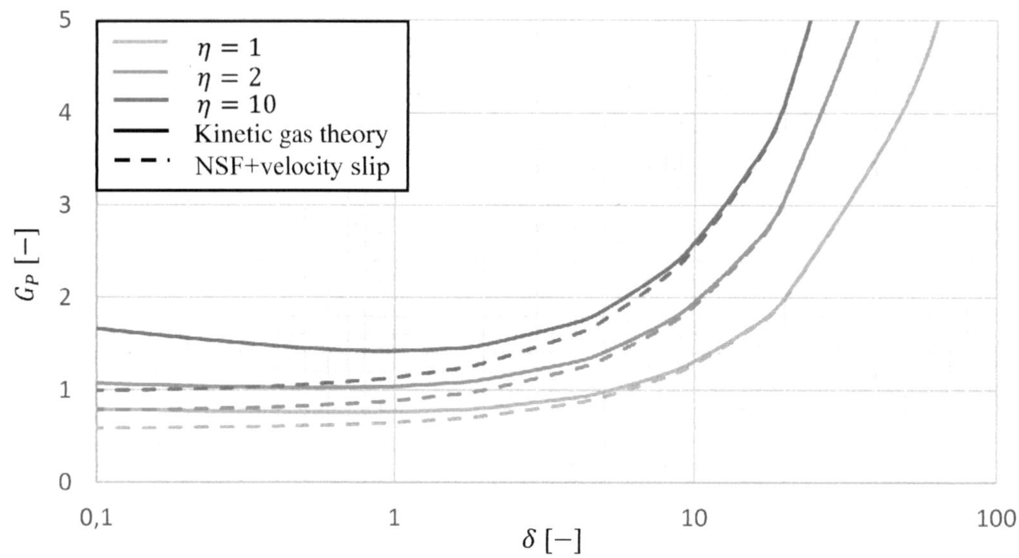

Figure 6. Comparison of the normalized Poiseuille flow as a function of the gas rarefaction parameter δ for different width to heigh ratios η.

Generally the pressure ratio increases with a higher speed ratio u_0. The results of the 1D model and of the CFD simulations are in good agreement for $\delta > 10$ as the slip velocity boundary condition is applicable there. For $\delta < 10$ the Boltzmann equation based on the kinetic theory of gases has to be solved. In this regime the pressure ratio obtained by CFD is higher than that of the 1D model. This can be explained with the behaviour of the normalized Poiseuille flow

$G_P(\delta)$ shown in Figure 6 for different width to height ratios η. The values for the diagram can be derived from the data given in reference [19]. The results of the kinetic gas theory deliver the well-known Knudsen minimum [23] whereas the NSF equations with velocity slip boundary conditions do not show this minimum but tend monotonically to an asymptotic value lower than that of the kinetic theory. This behaviour is seen in the results of the pressure ratio. As the Couette flow is not as sensitive to the gas rarefaction as the Poiseuille flow, the pressure ratio has its maximum at a similar rarefaction parameter where the normalized Poiseuille flow has its minimum. On the other side for large values of δ a very small pressure gradient is sufficient to compensate the mass flow rate caused by the Couette flow. As the Knudsen minimum is not depicted by the NSF equations, the pressure ratio for small values of δ is higher than that by the 1D model. And thus the 1D model provides even more physically accurate results. The same behaviour is obtained for the pressure ratio p_{HP}/p_{LP} shown in Figure 7 but on a much higher level. Note that a higher temperature level has the same effect on the pressure distribution as a lower rotational speed.

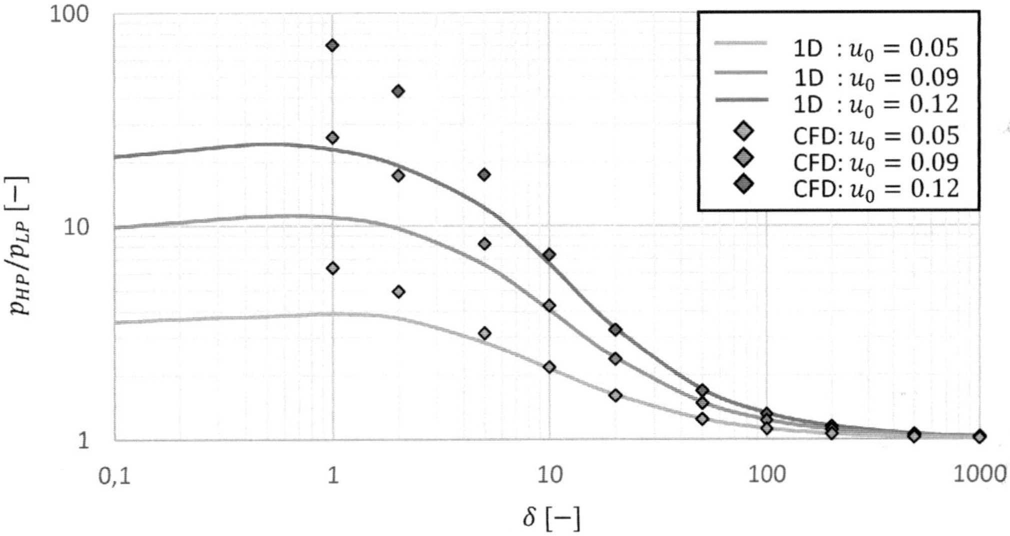

Figure 7. Comparison of the pressure ratio p_{HP}/p_{LP} between 1D model and CFD for different speed ratios u_0 for $C_0 = 0$.

Figure 8 shows the normalized pressure distribution as a function of the normalized chamber length for a speed ratio $u_0 = 0.12$ and a mass flow ratio $C_0 = 0$. The plot for the CFD results is the pressure distribution along the pitch circle between the planes where high and low pressure are evaluated. Within the slip regime for $\delta = 20$ the pressure distributions of 1D and CFD results match perfectly and the pressure distribution is nearly linear. In the transition regime $\delta = 2$ the pressure distribution is not linear any more and according to the deviations in the previous plots the CFD simulation shows a higher gradient than the 1D model. The non linearity in the pressure distribution implies that there is a deviation between the midpoint pressure and the mean pressure thus in order to assure conservation of mass in chamber model simulation the mass averaged mean pressure has to fulfil the total mass within the chamber and it is not sufficient to take the midpoint pressure.

Figures 9 and 10 show the impact of a steady clearance mass flow carried along the chamber. Such a mass flow rate slightly reduces the impact of the Couette flow within the chamber

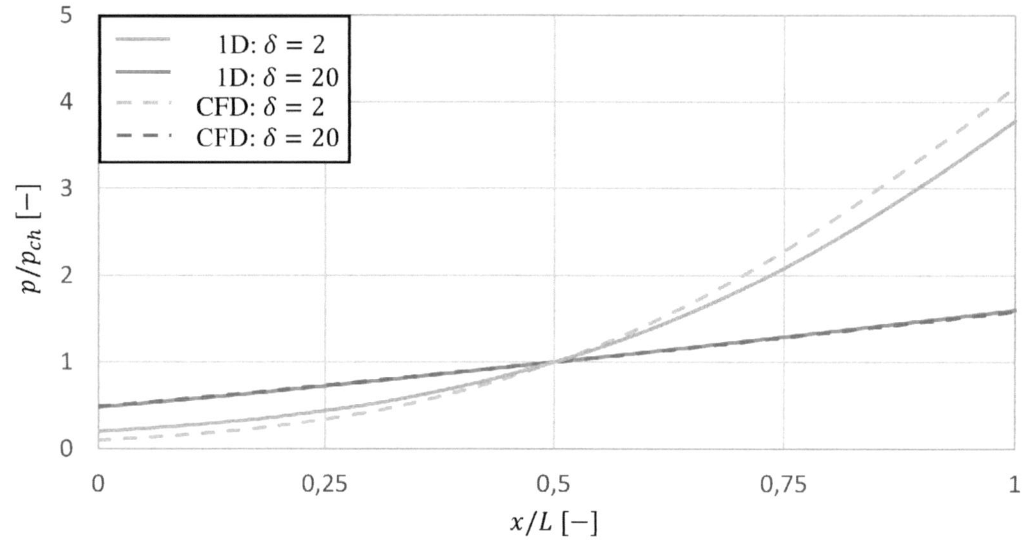

Figure 8. Comparison of the normalized pressure distribution between 1D model and CFD for $u_0 = 0.12$ and $C_0 = 0$

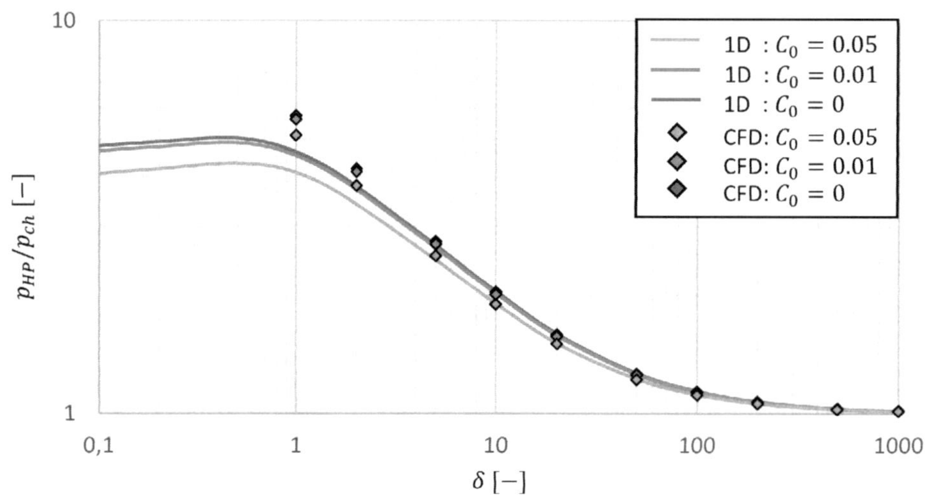

Figure 9. Comparison of the pressure ratio p_{HP}/p_{ch} between 1D model and CFD for different mass flow ratios C_0 for $u_0 = 0.12$.

and as a result the pressure ratio shrinks. Again CFD and 1D model are in good agreement for $\delta > 10$. Although the NSF equations are not exact in the transition regime $\delta < 10$ the qualitative behaviour of both models is the same. Note that a pretty high normalized mass flow rate is needed to have a significant impact on the pressure distribution. The measurements in references [6] and [24] show that the typical mass flow ratios for clearances including positive wall movement in this regime are $0.01 < C_{0,cl} < 0.4$ with $C_{0,cl}$ related to the minimum clearance cross section area which is typically more than factor 100 smaller than the chamber's cross

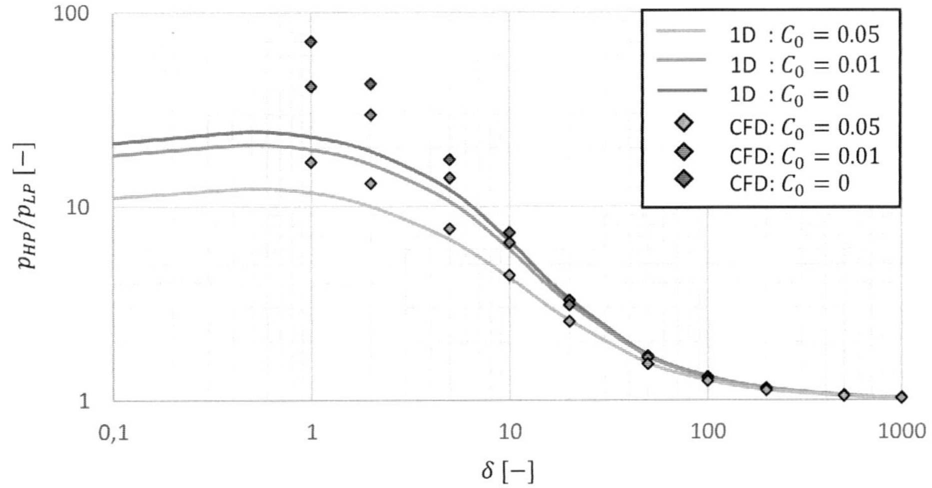

Figure 10. Comparison of the pressure ratio p_{HP}/p_{LP} between 1D model and CFD for different mass flow ratios C_0 for $u_0 = 0.12$.

section area. So there can be an impact, if the clearance heights and pressure ratios are large.

Figures 11 and 12 show the impact of a steady normalized housing clearance mass flow $C_{0,h}$ carried along the chamber with an additional constant tangential mass flow rate $C_0 = 0.01$ for $u_0 = 0.12$. The housing mass flow rate is modelled as mass source coming from the housing with the assumption, that it is equally distributed along the chamber length. To reach a steady state, the sum of $C_0 + C_{0,h}$ is modelled as the mass flow outlet. Therefore the mass flow rate increases with a linear function from p_{LP} to p_{HP}. This procedure is adapted for the 1D model and therefore the mass flow rate $\dot{m}_{cl,t}$ is not a constant, if the housing clearance is taken into account. Again a good agreement between CFD and 1D model is achieved for $\delta \geq 10$ with the rising deviation for $\delta < 10$. It needs to be noted, that typically there is an incoming housing clearance mass flow rate and an outflowing housing clearance mass flow rate within a chamber, so this mass source model only takes the difference of both mass flow rates into account.

Finally the impact of the rotor material is investigated and shown in figures 13 and 14. Typically for technical applications the assumption of diffuse wall scattering with a TMAC $\alpha = 1$ is made as mostly air collides with steel walls. In the literature these values are obtained by measurements in the range of $0.86 \leq \alpha \leq 1.01$ (see reference [12]). So next to the diffuse scattering ($\alpha = 1$) the case is investigated, that the rotor has a TMAC $\alpha = 0.9$. For air molecules colliding with aluminum walls a TMAC $\alpha = 0.72$ is obtained [12]. It is assumed, that the housing has a TMAC $\alpha_h = 1$ and the rotor material is exchanged. It can be seen, that the gas surface interaction has a much higher impact on the pressure distribution within the chamber than the corresponding mass flow rate and the pressure ratio within the chamber significantly shrinks with a shrinking TMAC. Both the 1D model and the CFD simulations are in perfect agreement for $\delta > 10$ with the previously mentioned higher deviation in the transitional regime. This indicates, that the choice of the rotor material for example due to coating can have a significant impact on the operation performance regarding the ultimate pressure. It needs to be noticed, that a positive effect regarding the inhomogeneous chamber state can be contrary to the effect on the clearances which are also effected by the pair of the TMACs. The integral impact of the TMACs on the machine's performance should be investigated further.

IOP Conf. Series: Materials Science and Engineering 1267 (2022) 012010 doi:10.1088/1757-899X/1267/1/012010

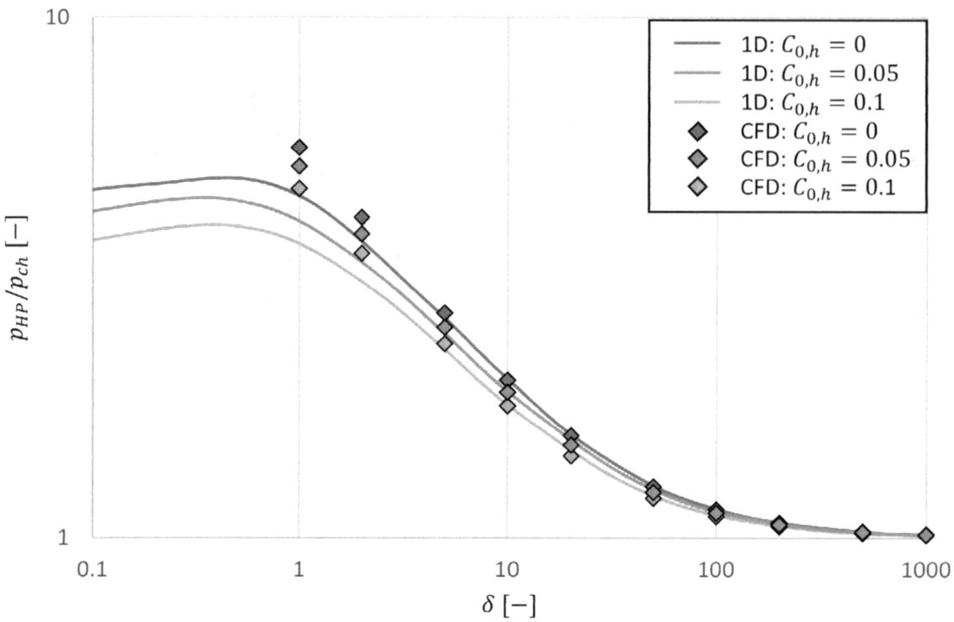

Figure 11. Comparison of the pressure ratio p_{HP}/p_{ch} between 1D model and CFD for different mass flow ratios of the housing clearance $C_{0,h}$ for $u_0 = 0.12$ and $C_0 = 0.01$.

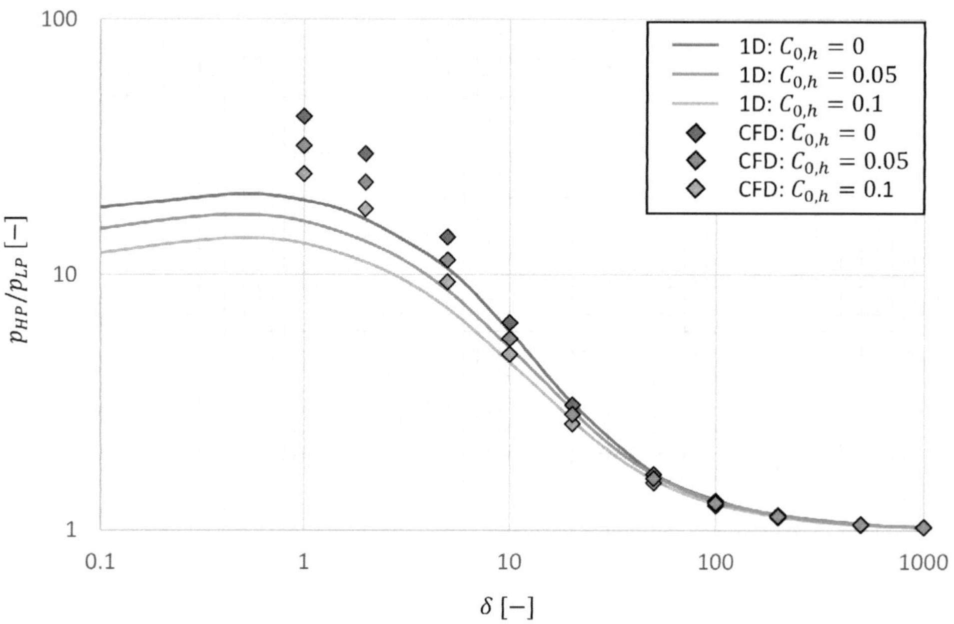

Figure 12. Comparison of the pressure ratio p_{HP}/p_{LP} between 1D model and CFD for different mass flow ratios of the housing clearance $C_{0,h}$ for $u_0 = 0.12$ and $C_0 = 0.01$.

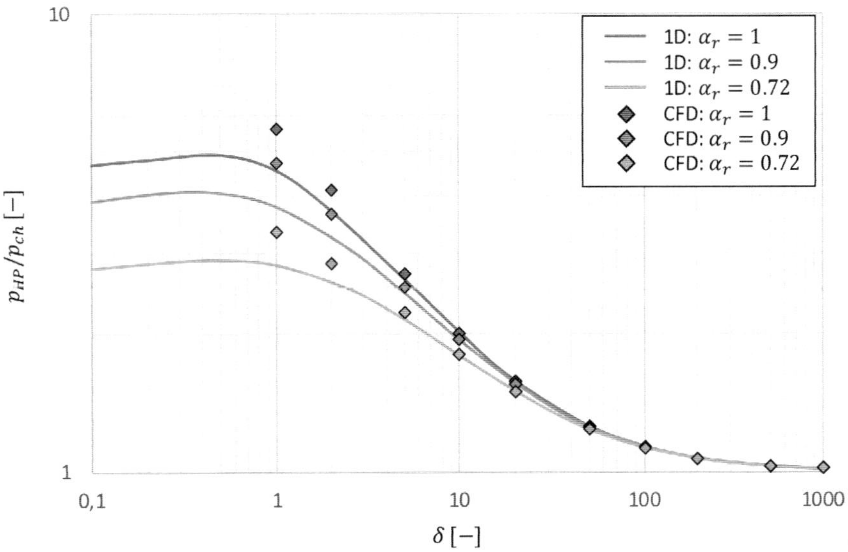

Figure 13. Comparison of the pressure ratio p_{HP}/p_{ch} between 1D model and CFD for different α_r with $\alpha_h = 1$, $C_0 = 0.01$ and $u_0 = 0.12$.

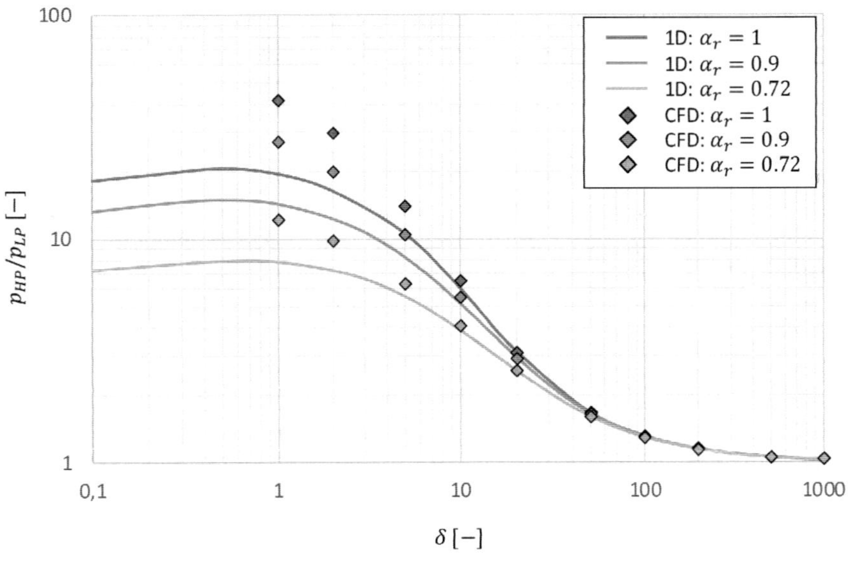

Figure 14. Comparison of the pressure ratio p_{HP}/p_{LP} between 1D model and CFD for different α_r with $\alpha_h = 1$, $C_0 = 0.01$ and $u_0 = 0.12$.

2.6. Regression Analysis

After the geometrical abstraction of the 1D model is validated, it is ready to use for chamber model simulation, but the computational cost of solving differential equations every time the mass within a chamber changes is non-negligible. Therefore the pressure distribution is analysed. Following Figure 2, in the context of chamber model simulation the average pressure, the low

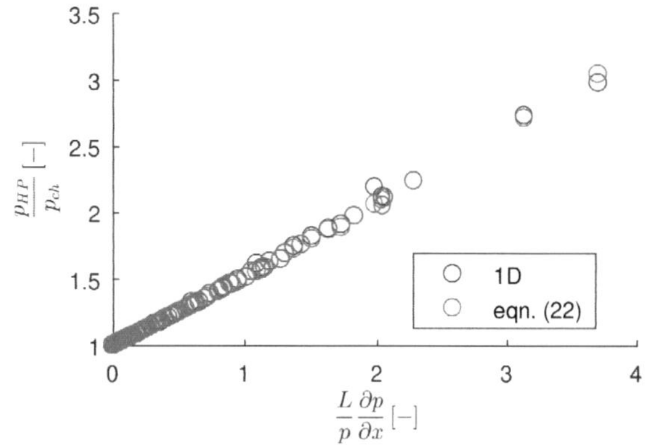

a) $10 \leq \delta \leq 10^5$

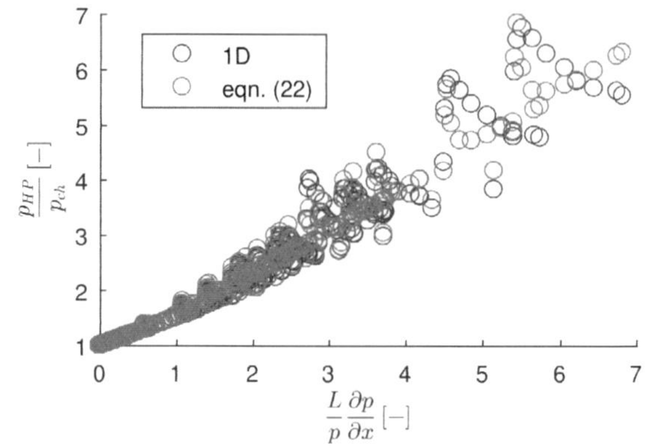

b) $0.001 \leq \delta \leq 10^5$

Figure 15. Results of the 1D-Model compared to equation 22.

pressure and the high pressure of a chamber is needed to correct the boundary conditions, while the mass averaged pressure is known. In the following cases the dimensionless numbers of section 2.2 are varied as summarized in Table 1. Therefore the geometrical ratios, the number of lobes z, the velocity ratio u_0 and the mass flow ratio C_0 are varied in common ranges. Therefore both the housing clearance and the other clearances are taken into account, but while for the 1D model the mass source model is used, for the dimensionless pressure gradient only half of the effective housing clearance mass flow rate is added to calculate C_0. Therefore it is assumed, that the linear increasing mass flow rate of the housing clearance acts like a constant mass flow rate which is only half of that value. Different TMACs for the rotor are investigated in case, that the rotor surface material is exchanged. Furthermore the gas is exchanged to hydrogen with housing and rotor made of steel is simulated, where the TMAC is changed to $\alpha \approx 0.83$ [25]. The 1D model is used for the simulation results and the pressure ratio p_{HP}/p_{ch} is given as a function of the dimensionless pressure gradient given in equation 20 which is evaluated with the known mass averaged pressure p_{ch} and the length L given in equation 8. The gas rarefaction parameter is varied in a range $0.001 \leq \delta \leq 100000$ with three values per decade. Regarding Figure 15a it is obtained that the pressure ratio p_{HP}/p_{ch} is a linear function of the dimensionless pressure gradient $\frac{L}{p}\frac{\partial p}{\partial x}$ as long as the gas rarefaction parameter is $\delta \geq 10$. For $\delta < 10$ the pressure ratio differs from the linear function with higher deviations for larger values of the dimensionless pressure gradient. This is taken into account to form the functional in equation 22 with two constants k_1 and k_2 that are found with a Levenberg-Marquardt algorithm. The overall agreement between the 1D model and equation 22 is very good for $\delta \geq 10$ with a mean deviation of 0.15 % and depicts the main behaviour in the whole range of the gas rarefaction with a mean deviation of 4.3 % for $\delta < 10$ (see Figure 15b). Especially for a very large dimensionless pressure gradient there are some theoretical operation points that have a deviation between both models of about 20 %, but these boundary conditions will not be reached in a fore vacuum pump.

$$\frac{p_{HP}}{p_{ch}} \approx 1 + \frac{L}{p_{ch}}\frac{\partial p}{\partial x} \cdot \left(\frac{1}{2} + k_1 \cdot \sqrt{\frac{L}{p_{ch}}\frac{\partial p}{\partial x}} \cdot \exp\left(-\left|\ln(\delta + k_2)\right|\right)\right) \tag{22}$$

$$k_1 = 0.238 \qquad k_2 = 0.4226$$

Figure 16a shows the pressure ration p_{ch}/p_{LP} as a function of the dimensionless pressure

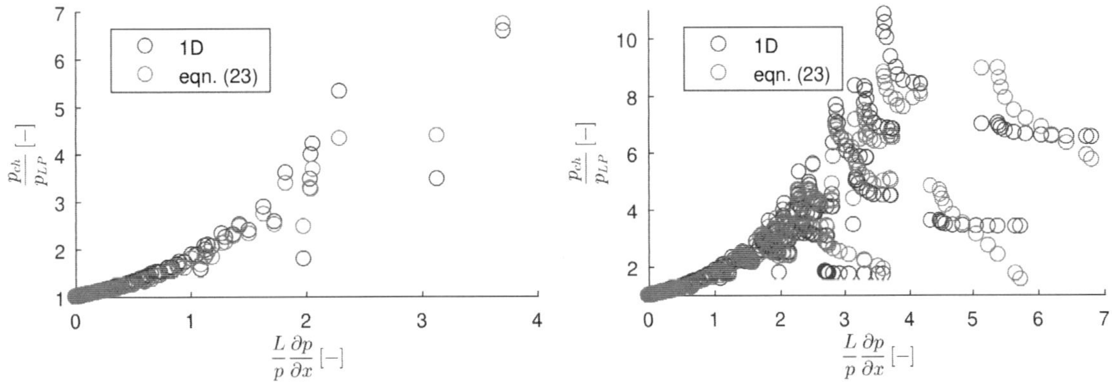

a) $10 \leq \delta \leq 10^5$ **b)** $0.001 \leq \delta \leq 10^5$

Figure 16. Results of the 1D-Model compared to equation 23.

gradient for $\delta \geq 10$. It is obtained, that this is already a non linear function which starts to spread at $\frac{L}{p}\frac{\partial p}{\partial x} > 2$. But a main trend can be seen. The pressure ratio even spreads more from the main trend for low values of the gas rarefaction parameter, which is highlighted in Figure 16b. The main deviations occur in the right hand side of the diagram, which mark the large values of C_0 at $\delta < 10$. As explained before in Figures 9 and 10, a mass flow rate within the chamber has a much larger impact on the low pressure than on the high pressure. Therefore this relation has to be added to the functional in equation 23 which fits the pressure ratio p_{ch}/p_{LP} as a function of the dimensionless pressure gradient $\frac{L}{p}\frac{\partial p}{\partial x}$, δ and u_0/C_0 with seven constants $g_1 - g_7$ which are found with Levenberg-Marquardt algorithm. For $\delta \geq 10$ the mean deviation is about 1.9 %. For $\delta < 10$ the mean deviation is about 9.5 %. Although this deviation seems pretty large, it needs to be noted, that the low pressure has a much smaller impact on the adjacent clearances than the high pressure. As the high pressure side marks the inlet pressure for the clearances, the resulting mass flow rate scales with this value. The low pressure side of the chamber only has an impact on the pressure ratio of the certain clearance. But this has only a small impact on the mass flow rate which can be seen in references [6, 7, 24]. Moreover as before most operation points are within the scope of smaller values of the dimensionless pressure gradient. For more stability within the scope of chamber model simulation it is convenient to fix the pressure ratio in the range $1 \leq p_{ch}/p_{LP} \leq 10$.

$$\frac{p_{ch}}{p_{LP}} \approx 1 + \left[g_1 \cdot \left(\frac{L}{p_{ch}} \frac{\partial p}{\partial x} \right)^{g_2} + \left(\frac{L}{p_{ch}} \frac{\partial p}{\partial x} \right)^{g_3} \right]$$
$$\times \left(g_4 + g_5 \cdot \exp\left(-\frac{g_6 \cdot \sqrt{\delta}}{\frac{L}{p_{ch}} \frac{\partial p}{\partial x}} \right) \right) \cdot \left(1 - g_7 \cdot \frac{L}{p_{ch}} \frac{\partial p}{\partial x} \sqrt{\frac{C_0}{u_0}} \right) \tag{23}$$
$$g_1 = 1.923 \quad g_2 = 2.23 \quad g_3 = 0.66 \quad g_4 = 0.27$$
$$g_5 = -0.095 \quad g_6 = 8.56 \quad g_7 = 0.2$$

3. Clearance flows

In the previous sections the impact of the homogeneous chamber states on the boundary conditions of the clearances have been discussed in detail. In reference [10] the clearance mass flow rates for given boundary conditions have been determined by interpolation of measurement

Table 1. Different configurations for the regression analysis

D_o/D_i	D_o/s	u_0	C_0	z	α_h	α_r	gas
1.7	2	0.12	0	1	1	1	air
1.7	2	0.12	0	2	1	1	air
1.7	2.8	0.12	0	1	1	1	air
1.7	2.8	0.12	0	2	1	1	air
1.7	2.8	0.12	0.05	2	1	1	air
1.7	2.8	0.12	0.1	2	1	1	air
1.7	2.8	0.155	0.01	2	1	1	air
1.7	2.8	0.09	0.01	2	1	1	air
1.5	2.8	0.12	0.01	2	1	1	air
1.9	2.8	0.12	0.01	1	1	1	air
1.7	2.8	0.12	0.01	2	1	0.72	air
1.7	2.8	0.12	0.01	2	1	0.9	air
1.7	2.8	0.06	0.01	2	0.83	0.83	H2
1.7	2.8	0.034	0.01	2	0.83	0.83	H2
1.7	2.8	0.155	0.05	2	1	1	air
1.571	5.641	0.14	0.05	2	1	1	air
1.571	5.641	0.14	0.1	2	1	1	air
1.571	5.641	0.14	0.2	2	1	1	air

data of Wenderott [5] for static walls with simulation data for moving walls in the free molecular regime by Stratmann [3]. This procedure was questioned in recent works as large deviations can occur for clearances in the transitional and slip regime ($0.1 < \delta < 100$) that are dominated by Couette flow.[24] It was shown, that the differential equation equation 19 can also be used to calculate clearance mass flow rates even for non constant cross sections as long as the Reynolds and Mach numbers are low [6, 26], but it fails for high Mach and Reynolds numbers. This problem was solved in a previous work by Jünemann [7] extending the differential equation of Shapiro [27]. The extended 1D model is validated with measurements for different flow types and channel shapes [7, 8]. This is now used to improve the calculation of the mass flow rates for given boundary conditions in the context of chamber model simulation. Regarding figures 1 and 2 there are four different clearance types in SSVPs. The housing clearance is geometrically modelled as two parallel plates with constant clearance height. The width is the helix of the outer diameter between the two cusps of the cut lens and the length L_h is the shortest distance between two chambers along the outer lateral surface of the rotor. As the machine is calculated isothermally with air and rotor lead and rotor frequency are constant, the reduced flow rate of the housing clearance $C_{0,h}$ is only a function of the gas rarefaction parameter δ and the pressure ratio

$$\Pi_{cl} = \frac{p_{out}}{p_{in}}. \tag{24}$$

The reduced mass flow rate of a clearance is defined as

$$C_{0,cl} = \frac{\dot{m}_{cl} \cdot \sqrt{\pi} c_m}{p_{in} \cdot h_{cl} \cdot w_{cl}} \tag{25}$$

with the index h being the dimensions of the housing clearance. The movement of the housing in flow direction can be calculated with equation 5. Figure 17 shows the dimensionless mass flow rate $C_{0,h}$ of the test machine's housing clearance as a function of the gas rarefaction parameter

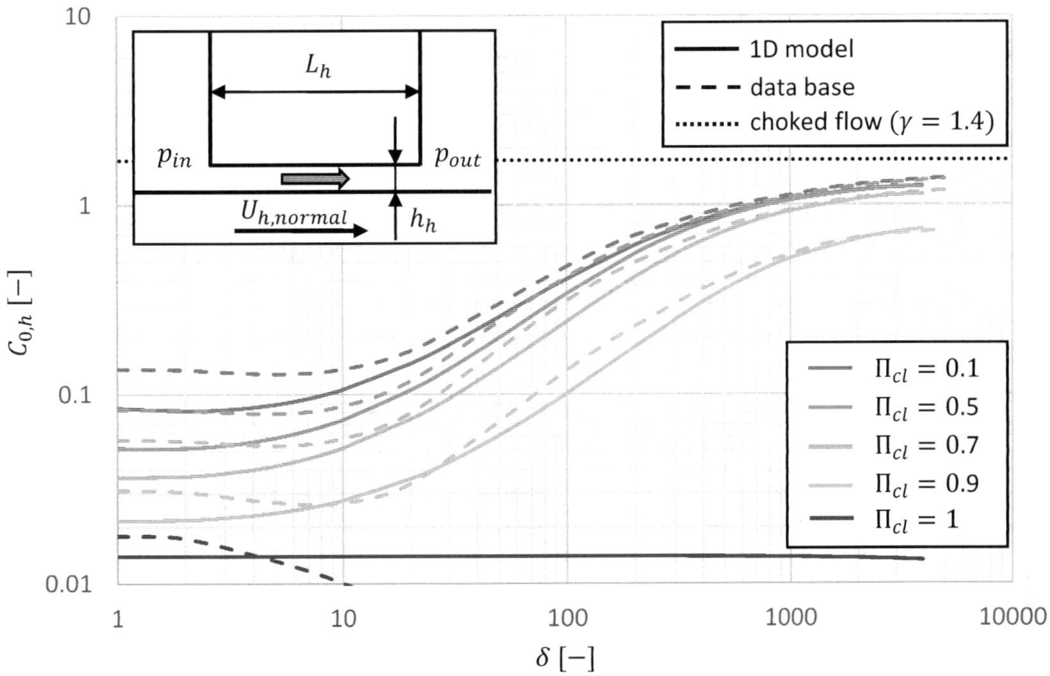

Figure 17. Reduced mass flow rates $C_{0,h}$ for the test machine's housing clearance as a function of the gas rarefaction parameter δ for different pressure ratios Π_{cl}.

for different pressure ratios Π_{cl} for the 1D model and the used data base. In addition a sketch of the abstracted clearance model is given and the line of the choked flow for a heat capacity ratio $\gamma = 1.4$ is added. For the housing clearance mass flow rate the data base is principally in good agreement with the results of the 1D model and both are well below the choked flow, which is a theoretical limit. A log-log plot is used to emphasize the differences for low gas rarefaction parameters δ. Major deviations occur for a pressure ratio $\Pi_{cl} = 1$ where the data base neglects the impact of the Couette flow for $\delta > 10$. Furthermore the data base does not depict the Knudsen minimum as a constant value is set at $\delta \approx 2$. Due to the interpolation process in the data base a overall slightly higher mass flow rate is calculated.

The radial clearance is placed between the outer and inner diameter of the two rotors and forms a converging-diverging channel shape along the length direction as sketched in Figure 18. The radial clearance height h_{rc} is the shortest distance between the two rotating cylinders and the width w_{rc} is similar to the housing clearance length. Both cylinders are rotating with the same rotor frequency and for the 1D model the cosine fraction in flow direction is used. Again due to the isothermal simulation of air and the constant rotor lead and rotor frequency, the reduced flow rate $C_{0,rc}$ of the test machine's radial clearance is only a function of the gas rarefaction parameter δ and the pressure ratio Π_{cl} which is shown in Figure 18 for the 1D model and the data base respectively as well as the limit of the choked flow for a heat capacity ratio $\gamma = 1.4$. This time only a semi log plot is used, as this shows the impact better than the log-log plot. It is obtained, that the data base underestimates the impact of the Couette flow in the range of $10 < \delta < 1000$, as the flow rate drops with higher values of the pressure ratio compared to the 1D model. Again at $\delta = 100$ the mass flow rate of the data base is zero for $\Pi_{cl} = 1$ although the 1D model indicates, that still a pretty large amount of mass is carried due to the

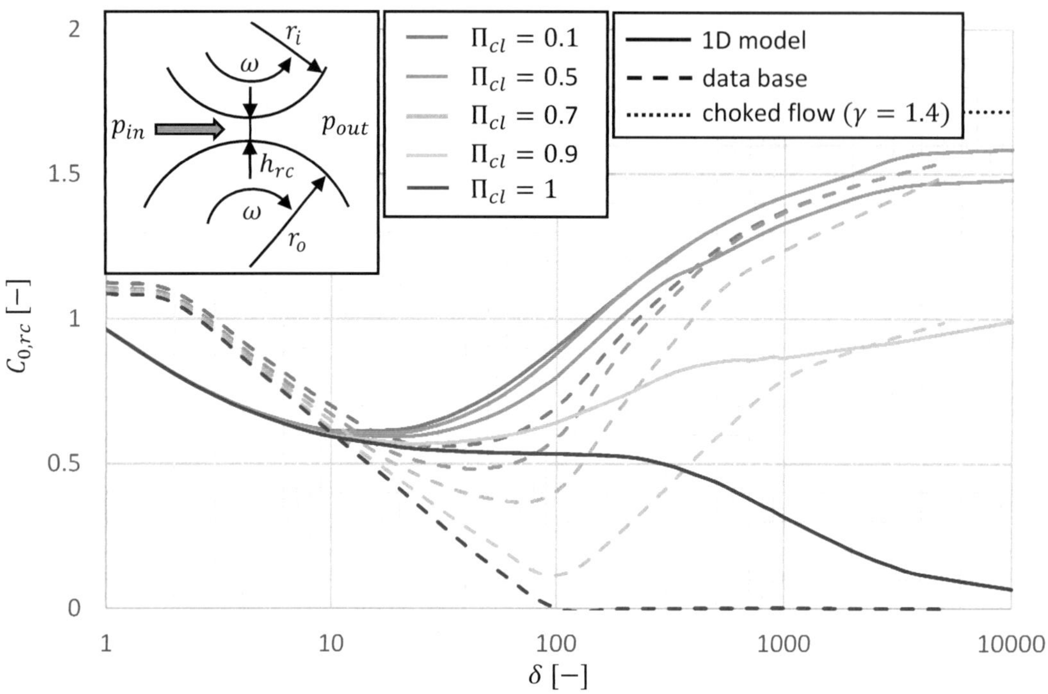

Figure 18. Reduced mass flow rates $C_{0,rc}$ for the test machine's radial clearance as a function of the gas rarefaction parameter δ for different pressure ratios Π_{cl}.

moving walls. For low gas rarefaction parameters the data base overestimates the reduced flow rates, as the molecular limit is set at $\delta \approx 2$.

The blow hole is nearly a slit which at this state is not validated with the 1D-model. Therefore the measurements of Wenderott are used for this clearance. The inter-lobe clearance is not well studied yet in detail. Theoretically more than two chambers are connected due to this clearance type, but only binary connections are considered. So this is handled a bit pragmatically by dividing the whole clearance area equally on the connections sketched in figure 2. The contact line forms an "S"-shape within the cut lens, while the wall velocity is assumed to be at average the velocity at the pitch circle. Therefore this model is a simplification of the stream lines that will actually occur in the different flow regimes of the inter-lobe clearance. Due to this simplification in the geometric model itself, we did not calculate this with the 1D-model, but kept the data base. For a better model more research has to be done on the abstraction of the inter-lobe clearance taking the different flow regimes of the Couette-Poiseuille flow into account.

For the housing clearance and the radial clearance a data base has been created using the 1D model for a fixed geometry and fixed dimensionless wall velocities. The gas rarefaction parameter is discretized with the equally distributed steps per decades in the range of $0.01 \leq \delta \leq 10000$. The pressure ratio is discretized for each gas rarefaction parameter with ten steps from $0.1 \leq \Pi_{cl} \leq 1$. For intermediate states a bilinear interpolation method is applied. A second data base has been created for negative wall velocities in case that the flow direction changes.

4. Chamber model simulation

In this section the different models presented above are included in a chamber model simulations and the impact of each model is compared to measurements of a test machine.

4.1. Geometric and thermodynamic model

As sketched in Figure 2 for the chamber model the vacuum pump is divided into chambers that transport mass from low pressure port to high pressure port and connecting clearances. Therefore the geometric properties of the chambers and clearances as a function of the rotation angle are needed. The geometric model used in this paper is the same as in reference [10]. Therefore the clearances are not simply connected with two chambers, but with a certain chamber position. The blow hole connects the high pressure side of two chambers and the low pressure side of two chambers, while the radial and inter-lobe clearance connect the high pressure side of the chamber at higher rotation angle with the low pressure side of the chamber with lower rotation angle considering the rotation angle to start at the low pressure port. The boundary conditions of the housing clearance are taken from mass averaged mean values of the certain chamber. The mass flow rates for given geometric and thermodynamic boundary conditions are obtained by tabulated data of measurements and simulations of the specific clearance shapes investigated and described in references [3] and [5].

Within the chamber model simulation the total chamber volume with the given chamber mass for every rotation angle is available. But due to the chamber's geometric abstraction for the inhomogeneous model, there is a discrepancy between the model's chamber volume $V_{mod} = L(t)hw$ and the total chamber volume V_t, as the volume within the cut lens is neglected. Thus for chambers, that are not connected to the high or low pressure port, the total chamber volume is divided as follows

$$V_t = V_s + V_{mod} + V_e \tag{26}$$

where $V_s = V_e$ are the chamber's start and end volume respectively. This classification is performed as the total mass m_t within the chamber is divided on these three volumes with the linearisation

$$m_s + m_e = \int \rho \, \mathrm{d}V_s + \int \rho \, \mathrm{d}V_e \approx \frac{p_{ch}}{RT} \left(V_s + V_t\right) \tag{27}$$

which assumes a linear pressure distribution with the chamber. As either the start and end volumes are small compared to the total chamber volume or the deviation from the linear pressure distribution is small in case, that V_{mod} is small, the total error of this assumption is small. Therefore the current chamber length for the model can be calculated by

$$L(t) = \frac{V_t(t) - V_s - V_e}{hw} \tag{28}$$

without changing the mass averaged pressure for the model or violating conservation of mass. Now a case distinction is done whether the chamber is open or closed. As the test machine has a constant rotor lead, the machine is closed when the gradient of the volume curve $\partial V_t/\partial t = 0$. Otherwise the chamber has a connection to the low or high pressure port. Therefore equations 22 and 23 are used to calculate p_{HP} and p_{LP} if $L(t) > 0$ and $\partial V_t/\partial t \neq 0$.

In order to consider the impact of the mass flow rate of adjacent clearances on the pressure distribution within the chamber, the whole mass flow rate $\dot{m}_{cl,t}$ to calculate C_0 in equation 14 the sum of the mass flow rates on the chamber's low pressure side $\sum \dot{m}_{LP}$, the sum of the mass flow rates on the chamber's high pressure side $\sum \dot{m}_{HP}$ and the sum of inflowing and outflowing housing clearance $\sum \dot{m}_h$ of the certain chamber are evaluated for each time step explicitly. Thus in order to calculate the pressure distribution within the chamber the mass flow rates of the last time step are used. As for closed chambers a stationary approach is used, the mass flow that enters on the low pressure side needs to flow out on the chamber's high pressure side. Which in chamber model simulation typically is not the case. Therefore a simplification

$$\dot{m}_{cl,t} \approx \min\left(\left|\sum \dot{m}_{LP}\right|, \left|\sum \dot{m}_{HP}\right|\right) + \frac{\sum \dot{m}_h}{2} \tag{29}$$

is done in order to improve the model accuracy. For open chambers the $\min -$ function is not needed, as only one side has adjacent clearances. The reason why only half of the sum of the housing clearance mass flow rates of each chamber is used is explained in section 2.6. Furthermore due to convergence problems the maximum mass flow rate has been capped thus $0 \leq C_0/\pi \leq \sum_i G_{C,i} U_i/c_m$ holds. The new model is implemented in the chamber simulation software *KaSim* developed at the Chair of Fluidics of TU Dortmund University described in reference [28].

4.2. Impact of the inhomogeneous chamber states

In order to work out the impact of the new investigated model for inhomogeneous chamber states, simulations for a test machine are performed and compared to measurements. This machine is an isochoric two-lobed cycloid profiled SSVP. The dimensionless data is given in Table 2, where L_R is the rotor length, h_h, h_{il} and h_{rc} are clearance heights of the housing clearance, the inter-lobe clearance and the radial clearance respectively and ψ_t is the machines total wrap angle. As the machine is cooled it is assumed to be fully isothermal at room temperature $T = 293K$. The machine neither has a radial inlet area nor a pressure relief valve.

Table 2. Data of the test machine

D_o/D_i	D_o/s	D_o/L_R	D_o/h_h	D_o/h_{im}	D_o/h_{rc}	ψ_t	z
1.571	0.667	5.641	880	733.33	733.33	3474°	2

Figure 19 shows the ratio of the suction speed

$$\dot{V} = \frac{\dot{m}_{effective} \cdot R \cdot T}{p_{suction}} \tag{30}$$

to the theoretical suction speed

$$\dot{V}_{th} = 2 \cdot V_{ch,max} \cdot f \cdot z \tag{31}$$

as a function of the suction pressure $p_{suction}$. $\dot{m}_{effective}$ is the effective mass flow rate of the machine. The discharge pressure is always 10^5 Pa. Comparing the results of *KaSim* without the new model (red line) to the measurements, a good agreement is obtained for large values of the suction pressure. But with the decrease of the suction pressure those two curves diverge. The measurements show that the machine cannot evacuate the low pressure volume lower than about 6 Pa at the given rotational speed, while the simulated suction speed is still at a high value. Regarding the inhomogeneous model from reference [10] (blue line) the curve is not affected in the high pressure regime, but collapses at low suction pressures in qualitatively good agreement with the measurements. But the new model in the current implementation slightly seems to overrate the effect, as the results indicate a worse suction speed than the measurements. In these simulations the differential equation 19 was solved directly within *KaSim* and the impact of the mass flow rates of adjacent clearances on the pressure distribution within the chambers was neglected. In order to validate the functionals in equations 22 and 23, the same boundary conditions are applied, but without directly solving the differential equation and just evaluating the explicit expressions (green line). It is obtained, that the ultimate pressure is slightly different, but the overall suction curve is mainly the same. This small error is justified by just a small fraction of computation time which is needed with the new functionals. Finally the orange line shows the results when the impact of the mass flow rate on the pressure distribution is considered. Apparently the ultimate pressure is almost matched exactly with the measurements, while still a small deviation is shown regarding the nominal suction speed.

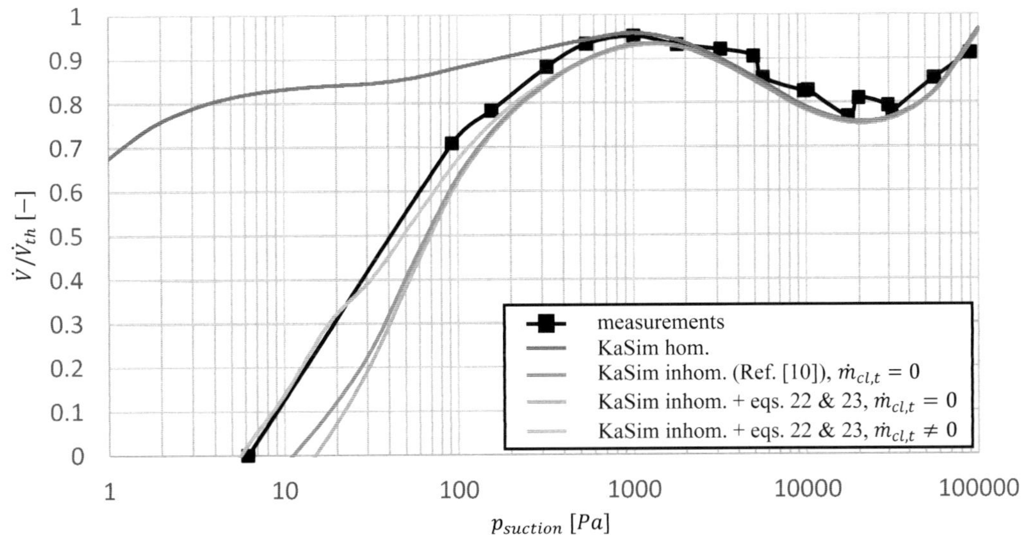

Figure 19. Comparison of the suction speed curves of measurements and chamber model simulation with homogeneous and inhomogeneous pressure distribution within the chambers for a rotational speed $f = 83.33$ Hz at constant temperature $T = 293$ K.

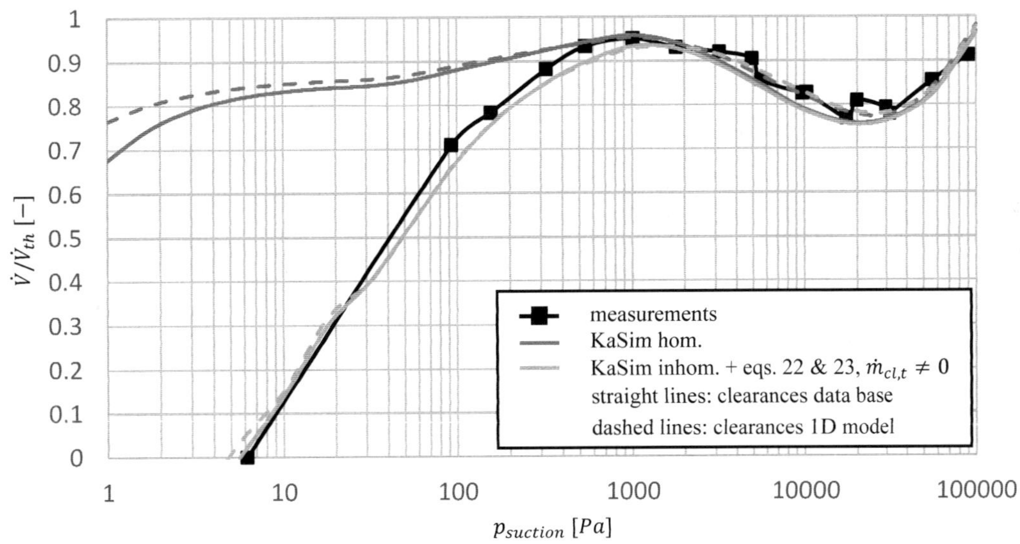

Figure 20. Comparison of the suction speed curves of measurements and chamber model simulation with homogeneous and inhomogeneous pressure distribution within the chambers for a rotational speed $f = 83.33$ Hz at constant temperature $T = 293$ K.

4.3. Impact of the clearance flow models

Figure 20 shows a comparison of the suction speed curve similar to Figure 19. The straight lines represent the simulation using the data base of Wenderott and Stratmann to calculate the clearance mass flow rates. The dashed lines represent the respective simulations using the 1D model in order to calculate the mass flow rate of the housing clearance and the radial clearance. For $p_{suction} > 2000\,\mathrm{Pa}$ the 1D model leads to a slightly higher suction speed which is in good agreement with the measurements as the simulation underrated the machine's performance in this regime. This might be explained by the impact of the housing clearance which has the largest cross section area. For 100 Pa $\leq p_{suction} \leq$ 2000 Pa the calculation of the clearance mass flow rates seem to have no impact on the integral machine performance. This can be a coincidence due to the fact, that the 1D model predicts a much larger mass flow rate for the radial clearance than the data base, but a slightly lower mass flow rate for the housing clearance. Thus the two impacts are cancelled out. For lower suction pressures the 1D model again leads to a larger suction speed. For the inhomogeneous chambers, the difference is small because the suction speed already tends to zero.

5. Conclusion

In this work a one-dimensional model in order to calculate the pressure distribution within the chambers of an SSVP has been validated with CFD. The results of both approaches are in good agreement for gas rarefaction parameters $\delta > 10$. It has been shown, that for $\delta < 10$ the results of the one-dimensional model are even more physically correct than the CFD simulations because it depicts the gas rarefaction effects of the kinetic theory of gases more precisely. The results show that the maximum of the inhomogeneous pressure distribution within the chamber occurs at $\delta \approx 0.5$ which correlates with the well-known Knudsen minimum. Furthermore, it has been shown that with increasing speed ratio and decreasing mass flow ratio of adjacent clearances, the pressure ratio within the chambers increases. The results indicate that the gas surface interaction resulting in the medium and wall material have a large impact on the inhomogeneity within the chambers. This will also affect the clearances, so whether the integral effect is positive or negative should be further investigated.

The 1D model is analysed and two functionals are derived in order to evaluate the high pressure and low pressure of the chamber directly. This is implemented into the chamber model simulation software *KaSim* and the results of a test machine are compared to previous simulations, where the differential equation of the 1D model was calculated explicitly. The results are in good agreement between the simplified regression functionals and the direct solution of the differential equation, but with much less computation time. Furthermore, the impact of the mass flow rate of adjacent clearances on the pressure distribution within the chambers is investigated which provides a good agreement with measurements regarding the ultimate pressure. The radial clearance and the housing clearance are analysed regarding the calculation of the mass flow rates for given boundary conditions. An even better accuracy is expected if a better model for the inter-lobe clearance is provided and results of transient simulations regarding the chamber filling process should be taken into account which is shown in reference [29].

References

[1] Jousten K 2018 *Wutz - Handbuch der Vakuumtechnik* vol 12 (Springer Vieweg)
[2] Huck C, Brümmer A and Sachs R 2014 Thermodynamical operation behaviour of screw vacuum pumps with cycloid- and quimby-toothed rotors *9th Int. Conf. on Screw Machines: VDI Bericht 2228* pp 331–346
[3] Stratmann D 2010 *Kennfeldverhalten zweiwelliger Rotationsverdrängervakuumpumpen* Dissertation Technische Universität Dortmund
[4] Rohe A 2005 *Wärmehaushalt von Schraubenspindel-Vakuumpumpen* Dissertation Technische Universität Dortmund
[5] Wenderott D 2001 *Spaltströmungen im Vakuum* Dissertation Technische Universität Dortmund
[6] Huck C, Pleskun H and Brümmer A 2018 *J. Vac. Sci. Technol. A* **36** 031606
[7] Jünemann T and Brümmer A 2020 *J. Vac. Sci. Technol. B* **38** 044201
[8] Jünemann T and Brümmer A 2022 One-dimensional calculation approach for gaseous clearance flows iOP Conf. Ser.: Mater. Sci. Eng.

[9] Müller R 2013 *Spaltströmung mit Wärmeübertragung in Vakuumpumpen* Dissertation Universität Kaiserslautern URL `http://d-nb.info/1044891858`
[10] Pleskun H, Jünemann T, Bode T and Brümmer A 2021 *IOP Conf. Ser.: Mater. Sci. Eng* **1180** 012009
[11] Jünemann T, Pleskun H and Brümmer A 2021 *IOP Conf. Ser.: Mater. Sci. Eng* **1180** 012037
[12] Acharya T, Falgoust J, Schoegl I and Martin M J 2019 *J. Thermophy. Heat Trans.* **33** 733
[13] Buckingham E 1914 *Phys. Rev.* **4** 345
[14] Sharipov F 2016 *Rarefied Gas Dynamics - Fundamentals for Research and Practice* (Wiley-VCH)
[15] Pleskun H, Bode T and Brümmer A 2022 *Physics of Fluids* **34** 032004
[16] Pleskun H and Brümmer A 2022 *Physics of Fluids* **34** 082009
[17] Boost 2021 Boost C++ Libraries version 1.77.0 `http://www.boost.org/` last accessed 2022-07-19
[18] CFX 2021 Modeling Guide
[19] Sharipov F 1999 *J. Vac. Sci. Technol., A* **17** 3062
[20] Bird G A 1994 *Molecular Gas Dynamics and the Direct Simulation of Gas Flows* (Clarendon Press, Oxford)
[21] Richardson L F and Gaunt J A 1927 *Philosophical Transactions of the Royal Society of London. Series A, Containing Papers of a Mathematical or Physical Character* **226** 299–361
[22] Ferziger J H and Peric M 2002 *Computational Methods for Fluid Dynamics* vol 3 (Springer)
[23] Knudsen M 1909 *Annalen der Physik* **333(1)** 75–130
[24] Huck C 2020 *Verdünnte Strömung in Spalten von Vakuumpumpen mit bewegter Berandung* Dissertation Technische Universität Dortmund URL `http://d-nb.info/121831334X`
[25] Veltzke T 2013 *On gaseous Microflows under isothermal Conditions* Dissertation Universität Bremen
[26] Huck C and Brümmer A 2018 *IOP Conf. Ser.: Mater. Sci. Eng.* **425** 012027
[27] Shapiro A H 1953 *The Dynamics and Thermodynamics of Compressible Fluid Flow* vol 1 (Ronald, New York)
[28] Janicki M 2007 *Modellierung und Simulation von Rotationsverdrängermaschinen* Dissertation Technische Universität Dortmund
[29] Pleskun H, Syring A and Brümmer A 2022 Transient chamber filling in rotary positive displacement vacuum pumps iop Conf. Ser.: Mater. Sci. Eng

Table 3. List of most important symbols

symbol	explanation	unit	symbol	explanation	unit
A	cross section area	m^2	c_m	most probably speed	m/s
C_0	reduced mass flow rate	–	D	diameter	m
f	rotational speed	Hz	G	reduced flow rate	–
h	chamber/clearance height	m	L	chamber length	m
m	mass	kg	$\dot{m}$	mass flow rate	kg/s
p	pressure	Pa	R	specific gas constant	J/(kg K)
s	rotor lead	m	T	temperature	K
u_0	reduced velocity	–	U	velocity	m/s
V	volume	m^3	w	chamber width	m
z	number of lobes	–	δ	gas rarefaction parameter	–
η	cross section ratio	–	μ	dynamic viscosity	Pa s
ρ	density	kg/m^3	ψ	wrap angle	–

Simulating the dynamic behavior of a Triple Screw Pump without CFD

S Kosmann[1], O Trossmann[1], M-J Fikacek[1] and Dr H Tielbörger[2]

[1]Leistritz Pumpen GmbH, Nürnberg, Germany

[2]Siemens Industry Software GmbH, München, Germany

skosmann@leistritz.com

otrossmann@leistritz.com

Abstract. Pressure pulsations in screw pumps lead to noise, vibration and system strain. In order to increase the amount of dissolved and undissolved gas in the fluid while keeping noise, vibration and stress on the system at an acceptable level, the pump behavior must be understood in detail. Physical tests and CFD are common methods in modern pump design. However, both methods are not able to assess numerous design changes and machine integration challenges quickly. The system simulation approach tackles both challenges efficiently and allows to evaluate machine dynamics.
Siemens Simcenter Amesim has been used to model a *Leistritz Flexcore 45/90* triple screw pump. The model, containing components of a multi-physics library, calculates the movement of the chambers along the spindle package length by having virtual chambers alternately change their volume depending on the spindle rotation. Fluid properties, pressure dependent aeration and inertia effects are considered. The generation and displacement of the chamber volumes cause the suction and discharge of the fluid. The overlay of the single chambers results in the total volume flow of the pump. Internal leakage between the chambers is also considered, turning the model into a realistic digital twin of the screw pump. For the overall model and each individual chamber, the model gives volume flow rates, discharge pressure, pressure pulsation and internal pressure evolution. The air content has a significant influence on the pressure pulsations. To verify the model, measurements of operating points on a real pump have been carried out. Simulated volume flows and pressure pulsation show very good agreement with measurements. This new method for modeling screw pumps dramatically increases the speed of simulating pump behavior and opens the possibility of integrating the models into higher-level systems. The reproduction of dynamic effects and system responses for control tasks is also possible. The existing model can be extended towards further physics like friction modeling or additional system components like an electrical power unit.

1. Introduction

Screw pumps are rotary positive displacement pumps used for a wide field of applications, for example oil, naval or chemical industry. For a lot of use cases high efficiency and low pulsations are inevitable. In some applications it is necessary to convey fluids with high gas volume fractions as well as solid particles. Screw pumps in different varieties and designs can handle all these cases. Increasing the amount of gas in the fluid usually leads to higher pressure pulsation which again causes vibration, noise and stress on the system. Besides the gas content there are many other factors that influence the dynamic behavior of the pump. To optimally adapt the pump design to the system it is being implemented in, the internal behavior must be studied precisely.

A common method in pump design is to run physical tests, which show reliable outcomes, but are very expensive and time consuming. Another contemporary method is the use of CFD, which allows evaluation of the pump design prior to physical fabrication. This numerical approach is working well for a deep and accurate analysis of the flow characteristics for consistent operating points and leads to illustrative results. The downside, though, is that the system boundaries are very limited and investigations on the impact of the surrounding system lead to a significant increase in calculation time.

The aim of this paper is to show a method different from numerical simulation for emulating the dynamic behavior of a screw pump. Therefore, the simulating software *Siemens Simcenter Amesim* has been used. It provides the opportunity to create simulation models from components of different physical domains. For dynamic investigations of a screw pump, the combination of mechanics and hydraulics is required. The design for the model is based on a triple screw pump *Leistritz Flexcore 45/90*.

2. Methodology

2.1. Conceptual design

Triple screw pumps contain a drive spindle and two running spindles. The intermeshing spindles and the housing wall form enclosed chambers. They continuously move linearly due to the spindle rotation. Geimer [1] describes the progression of the chamber movement in five steps, as shown in Figure 1.

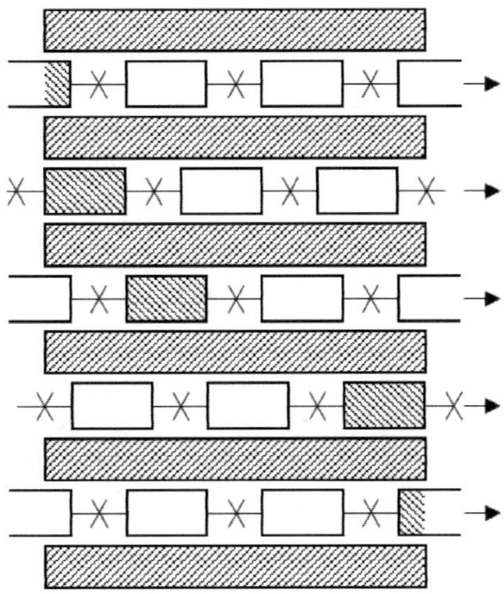

Figure 1. Conveying phases of a screw pump by Geimer [1].

At the beginning of the chamber development, the corresponding volume increases, depending on the angle of rotation of the spindles, and extracts fluid from the suction area. Then, it travels along the length of the spindle with a constant volume until it opens towards the pressure side and expels the fluid. During this process, several chambers are present simultaneously. Based on this principle, the approach shown in Figure 2 has been elaborated. When a chamber is fully expelled at the end of the spindle, it can be reused to again create a nascent chamber at the beginning of the spindle. Since the sequence of chambers along the length of the spindle constantly changes in this procedure, a parallel arrangement of different variable volumes is appropriate. The suction and ejection processes of the individual chambers together result in the total theoretical constant volume flow of the pump.

A *Leistritz Flexcore* screw pump has three different types of leakages. The running spindle leakage and the throttle leakage can be implemented as a return pipe with an aperture and the profile leakage can be reproduced with connections between the individual chambers. As described by Wincek [2], the leakage causes a linear increase of the chamber pressure along the spindle length. For the pressurization at the pump outlet, an adjustable throttle can be set to simulate the system pressure. A design sketch for this approach with an exemplary number of three chambers can be seen in Figure 2. Because of the cyclic reuse of the chambers to represent a rotating spindle in the sketch, the sequence of chambers alternates as they would be located in the spindle profile of the real pump. For this reason, chambers 1 and 3 are adjacent phase-wise and must also have a profile leakage connection.

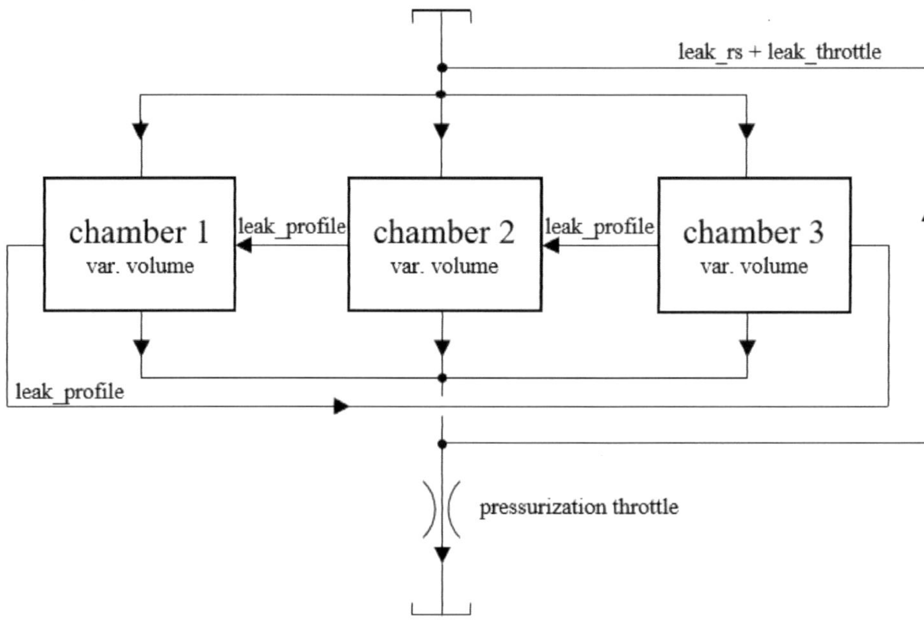

Figure 2. Conceptual design sketch of the model with three chambers.

2.2. Modelling

Every chamber needs the information about its current volume as a function of the angular position of the driving spindle. For this purpose, a chamber model in CAD has been constructed based on the spindle geometry data, which can be seen in Figure 3. The chamber volume has been determined by measuring the cross-sectional area for every 1° spindle rotation, which corresponds to 0.25 mm of linear displacement for a pitch of 90 mm per rotation. Figure 3 shows the resulting characteristic curve along the spindle package length. When a chamber is just beginning to be formed, the spindle must rotate 402° until the chamber is completely shaped and enclosed. At 527°, the chamber reaches the end of the spindle package and opens on the discharge side. From this point on, the fluid is pushed out of the chamber by reducing the chamber volume until it no longer exists after 928°.

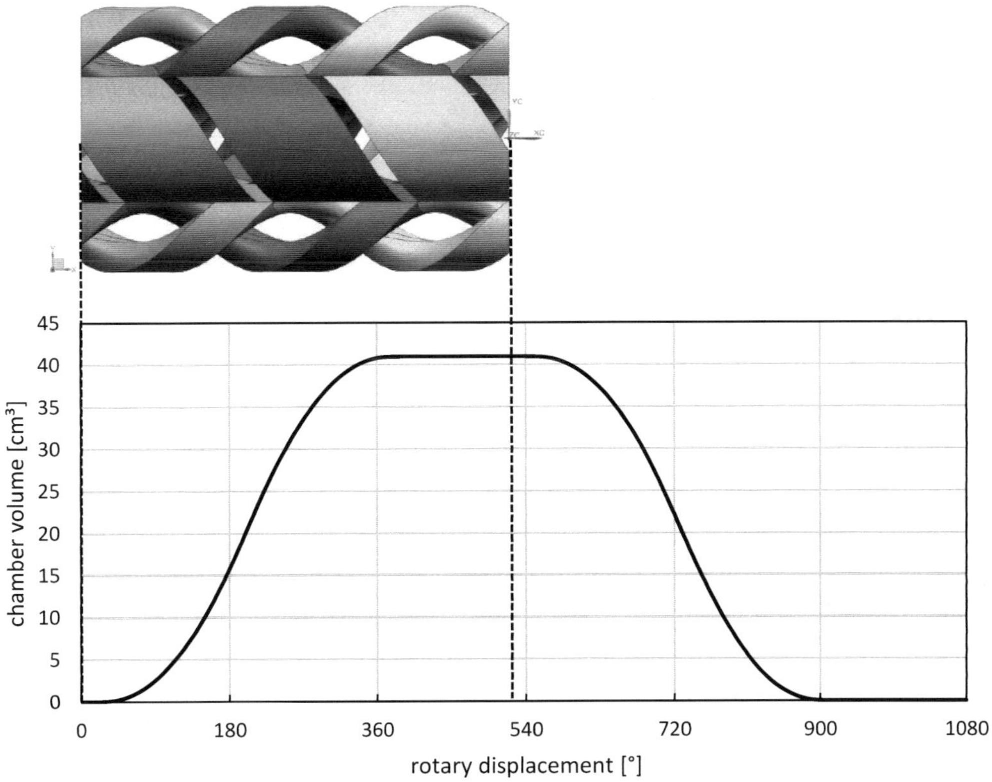

Figure 3. CAD chamber model and chamber volume of one chamber over rotary displacement.

Figure 4 shows the *Simcenter Amesim* sketch of a single chamber. The blue lines represent the hydraulic connection of the individual components and transmit the information about the pressure and the volume flow between the components. The black colored modules are called supercomponents (*sc*), which include different subcomponents. This has the advantage that the overall sketch is more compact. The structures of the *sc*'s are shown below under points 1-4. The signal "*W*" gives the momentary spindle angle. In *sc* 1, the brown colored modules provide the information about the chamber volume. A phase shift "*k*" depending on the chamber number is added to the spindle angle incoming from port 4, since each chamber is offset from the previous one by 180°. Additionally, the torque that is created by the pressurization of the fluid is read out. It is calculated from the pressure present in the chamber and the change in volume per change in angle of rotation as follows:

$$T = p \cdot \frac{dV_{ch}}{dW} \tag{1}$$

Above and below *sc* 1 there is an inlet and an outlet throttle. They are intended for the angle dependent opening and closing of the chamber to the suction and pressure ports. While the chamber volume is growing, the inlet throttle needs to be open and the outlet throttle closed. During the stationary shifting both orifices must be closed and while the volume decreases, the outlet is open. This process is controlled by the *sc*'s 2 and 3.

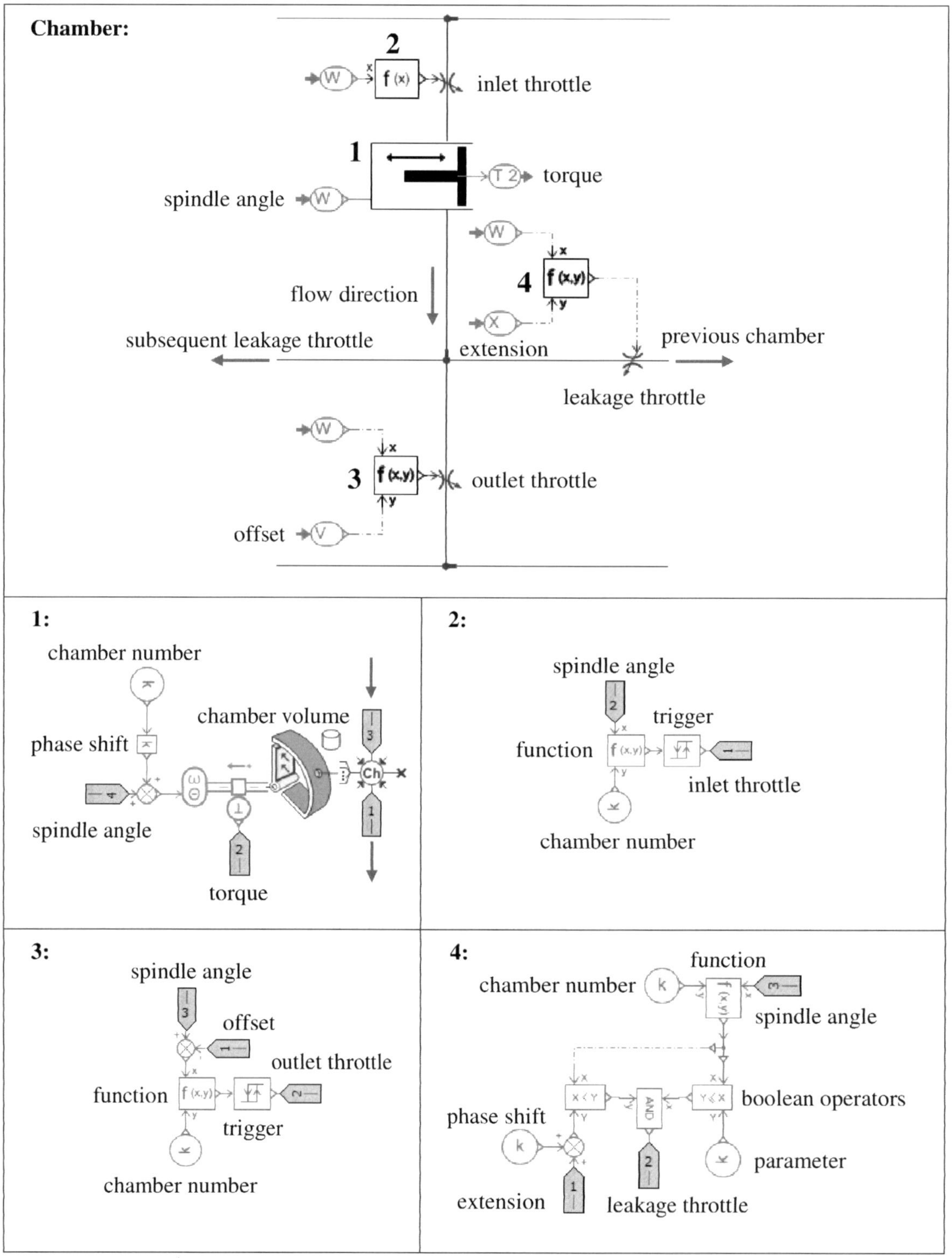

Figure 4. Model structure of one chamber in *Simcenter Amesim*.

The *sc* 2 calculates the opening signal "$signal_{in}$" from the spindle angle and the chamber number "k_n" with the modulo operation

$$signal_{in} = (W + k_n \cdot 180°)\ \%\ \varphi_{period} \tag{2}$$

which resets the value after a period of $\varphi_{period} = 1080°$ and shifts it depending on the chamber number. Thereafter a "trigger" component converts the signal into a binary information for "1 = open" if $0 < signal_{in} < 402°$ and "0 = closed" if $402° < signal_{in}$. In *sc* 3 the spindle angle signal for the outlet throttle receives an additional offset "*V*". It is calculated with the formula:

$$V = \varphi_{ex} - 402° - 540° \tag{3}$$

Between the chambers the profile leakage is integrated as an adjustable throttle. The differential pressure dependent volume flow of the throttle is given by a table. The characteristic must be calculated for every individual pump design based on the geometrically determined flow dynamics. For this model, a company-internal tool was used to determine the profile leakage flow. The profile leakage can only be present while both involved chambers are existing, and the associated throttles get addressed by *sc* 4. By applying a modulo operation, the function component converts the angle signal into a repeating sequence, which is shifted by the chamber number. The boolean operators convert it into a binary signal "1 = open" and "0 = closed". The leakage begins as soon as the chamber is fully developed, which is the parameter of 402°. This value is the same for every *Leistritz Flexcore* pump and has been determined from CAD models. The duration of the leakage depends on the pump size and is adjusted by the extension "*X*" at port 1. The value of "*X*" is obtained by the following formula:

$$X = \varphi_{ex} - 402° \tag{4}$$

Therefore, the existence angle φ_{ex} is required. This is the total angle of rotation during which the chamber exists both open and closed. The *Leistritz Flexcore 45/90* has a spindle length $l = 131.5\ mm$ and a pitch $h = 90\ mm$. From the formula

$$\varphi_{ex} = \frac{(l + h \cdot 1.119)}{h} \cdot 360° \tag{5}$$

the existence angle of $\varphi_{ex} = 928°$ is obtained for the entire existence time of a chamber. A chamber is slightly longer than its pitch due to the complex spindle geometry. For this reason, the pitch must be multiplied by a factor of 1.119 to get the chamber length. This was determined by measuring the CAD data and is the same for all pump sizes. Rounded up to whole rotations, the period angle for a repeating period corresponds to $\varphi_{period} = 1080°$. Since the spindles are double threaded, the phase shift of the chambers is $\varphi_{phase} = 180°$. The total number of chambers $n_{total} = 6$ follows from:

$$n_{total} = \frac{\varphi_{period}}{180°} \tag{6}$$

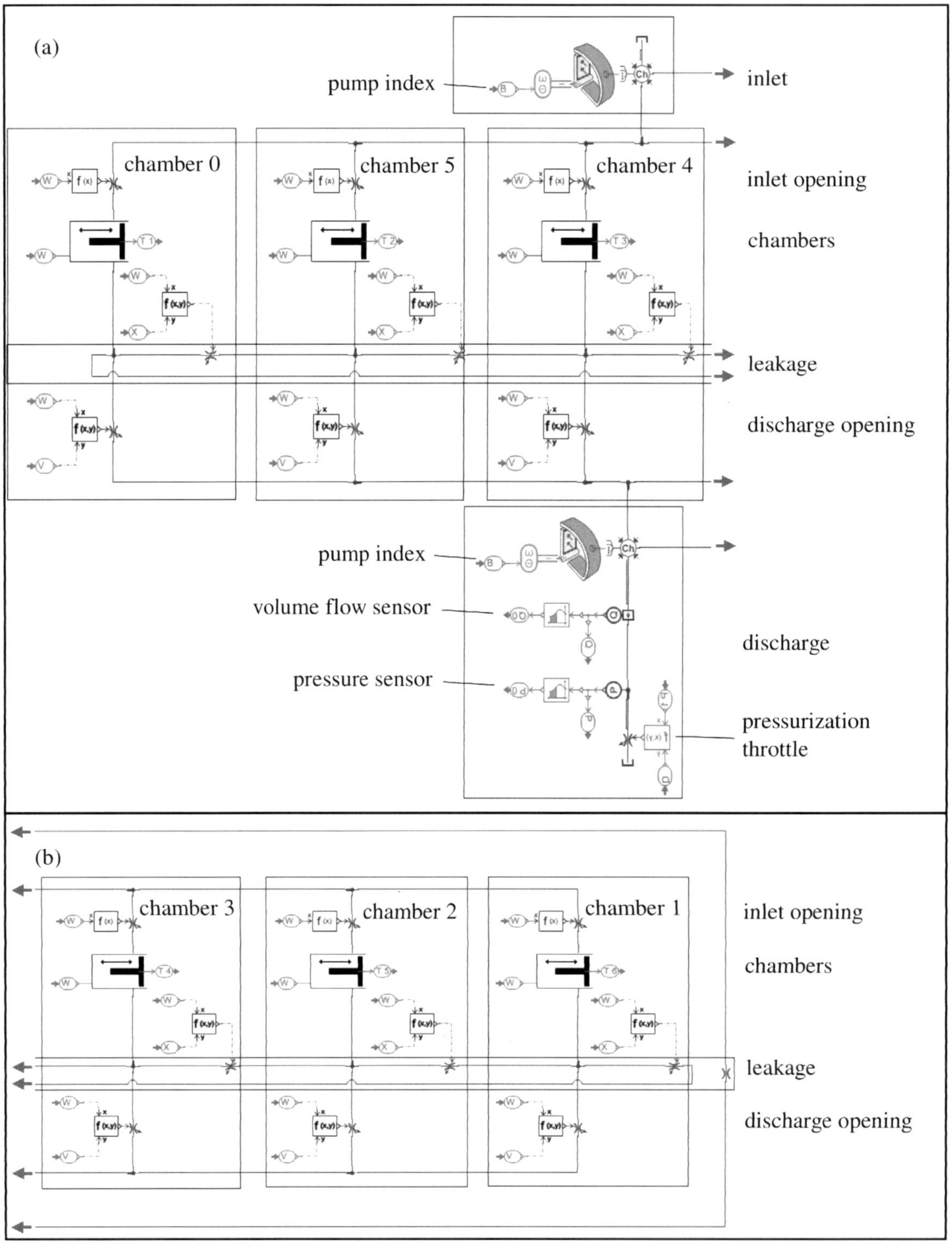

Figure 5. Composite model sketch of a *Leistritz Flexcore 45/90* in *Simcenter Amesim*. a) left part and b) right part.

The composite model can be seen in Figure 5. The running spindle leakage and the throttle leakage are combined into one pipe around the spindle package and its characteristic is included in the properties of a fixed orifice. As with the leakage profile, a table is stored in this module which characterizes the pressure-dependent leakage volume flow. A company-internal tool was also used to create this characteristic curve. The system pressure gets emulated by conveying against an adjustable throttle at the outlet. It has a circular cross-section of 790 mm² at maximum opening and a maximum flow coefficient $C_p = 0.7$. The generated pressure depends on the one hand on the opening ratio, which can be selected between 0 and 1, and on the other hand on how large the volume flow is that goes through it. In this example with pump size 45/90, a function component was used to automatically adjust the orifice geometry. By plotting different operating points, the formula for calculating the opening ratio can be determined. If another pump size is used, the formula must always be redetermined or the opening ratio has to be set manually. For a more appropriate solution, further developments still need to be done. Sensors for measuring the dynamic volume flow and the pressure over time are built in before the pressure throttle. To ensure that the model is flexible for modifications to different pump sizes, the suction and pressure chamber volumes can be adjusted.

Figure 6 contains the input parameters to be set by the user. The fluid properties can be specified in the blue drop symbol. The hydraulic fluid model is computed as a slightly compressible oil and it is assumed, that it always contains a certain amount of air or gas. The gas content is defined as the volume percentage of the total fluid volume. The liquid phase of the fluid has a constant bulk modulus and a constant absolute viscosity which in this example are set at $K = 1{,}7\ GPa$ and $\nu = 2{,}7087 \cdot 10^{-3}\ Pa \cdot s$. The density of the liquid must be given for atmospheric pressure. For other pressures, the component calculates the density to be consistent with the bulk modulus for mass conservation. The user can set a saturation pressure, at which the gas phase is totally dissolved and the fluid is only influenced by the properties of the liquid. Beneath the saturation pressure, the gas exists as bubbles in the liquid and modifies the bulk modulus and the absolute viscosity of the fluid. For simplification, the saturation pressure is constantly set at 100 MPa.

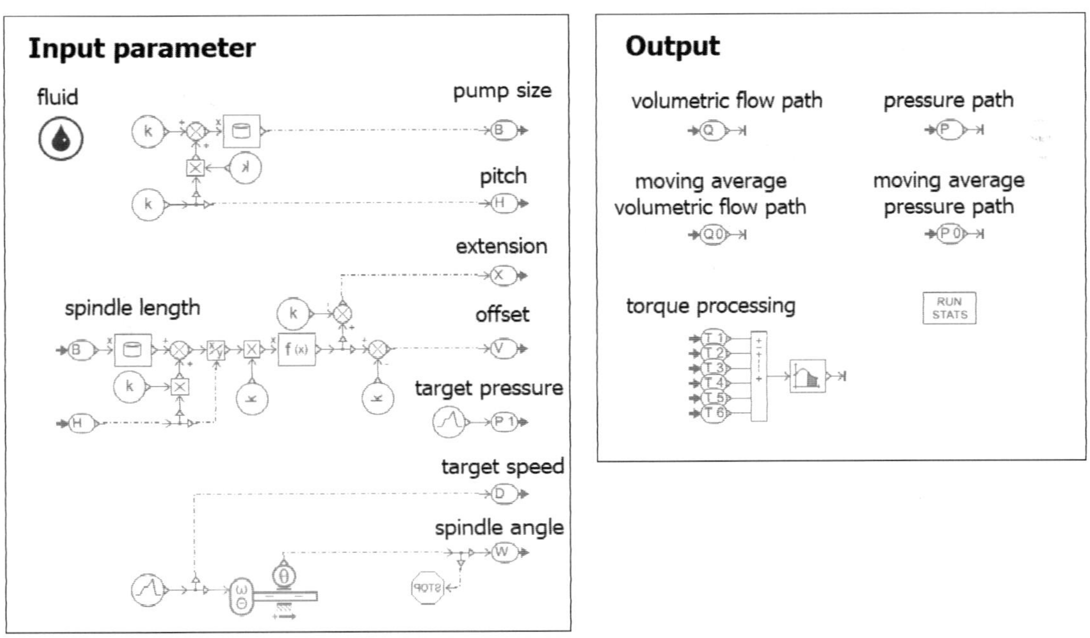

Figure 6. Input parameters and output of the simulation model in *Simcenter Amesim*.

The pump size and pitch refer to a pump index from a deposited table. Another table returns the corresponding spindle length. From spindle length and pitch the extension "*X*" and the offset "*V*" are calculated. Finally, the operating data in terms of target pressure and target speed must be entered. These can be variable over time. By integrating the rotary velocity, the momentary spindle angle "*W*" is calculated. After reaching a defined angular position, the simulation stops automatically.

The output area provides the most important simulation results. At the top the volumetric flow curve and the pressure curve over time can be displayed. In order to make the absolute values easier to identify, high frequencies are smoothed by a moving average. The hydraulic torque of each individual chamber caused by the pressurization of the fluid is summed up to the total torque acting on the driving spindle. The "run stats" component is used to determine some statistics about the current simulation run.

Further settings can be made before running the simulation. It is recommended to select a resolution ("Print Interval") of $t_{print} = 0{,}0001\ s$ under "Run Parameters". This way, the resolution of the results is high enough to clearly display and examine the pulsations. If operating points are simulated where the pulsation is insignificant, the resolution can be decreased in order to minimize the result file. However, such a decrease does not affect simulation accuracy since the solver automatically adjusts its local time step to the model dynamics.

3. Measurement Setup

For the sake of validating the simulation results, measurements of the *Leistritz Flexcore 45/90* on the test bench are necessary. Figure 7 shows the hydraulic circuit diagram of the experimental setup. A temperature sensor "TR2432" with a range from -40 °C to 300 °C is used to measure the temperature of the fluid. It has a resolution of 0.1 K and an accuracy of ±0.3 K. The pressure is measured before and after the pump. The sensor at the suction side is a "PN009A" with a measuring range from -100 kPa to 0 kPa, a resolution of 0.5 kPa and an accuracy of ±1%. At the discharge side a "PN2023" is used with a range from -0.1 MPa to 2.5 MPa and a resolution of 0.05 kPa. The accuracy is ±1%. To get a higher resolution of the outlet pressure, a pressure pulsation sensor of type "4065A200" is attached at the pump outlet as well. It has a sampling rate of 10000 Hz, reaches from 0 to 200 bar, has a resolution of 0.1 Pa and the accuracy is ±0.5%. The pressure is applied by means of a hand wheel adjustable valve. A screw type flow meter "OMG68.70" with an accuracy of ±0.1% and a range up to 1050 l/min determines the flow rate of the pump. The power shaft contains a digital torque meter "T12" with a range up to 2 kNm and an accuracy of ±0.01%. The "T12" also contains a speed sensor with a resolution of 0.1 min^{-1} and a maximum deviation of 1.5 min^{-1}.

The fluid used in the test bench is the lubricating oil *Magna 2*, which at a temperature $\theta = 25\ °C$ has a density $\rho = 812\ \frac{kg}{m^3}$ and an absolute viscosity $\nu = 2{,}7087 \cdot 10^{-3}\ Pa \cdot s$. While the tests were running, it was ensured that the temperature was steadily at about 25°C.

The measurements were carried out at operating points of 750, 1500 and 2100 rpm and differential pressures of 2, 5, 10, 12 and 15 bar. The operating point at 750 rpm and 15 bar was omitted because it exceeds the application limits. Pressure pulsation was recorded at 1500 rpm at absolute pressures of 5, 10 and 15 bar.

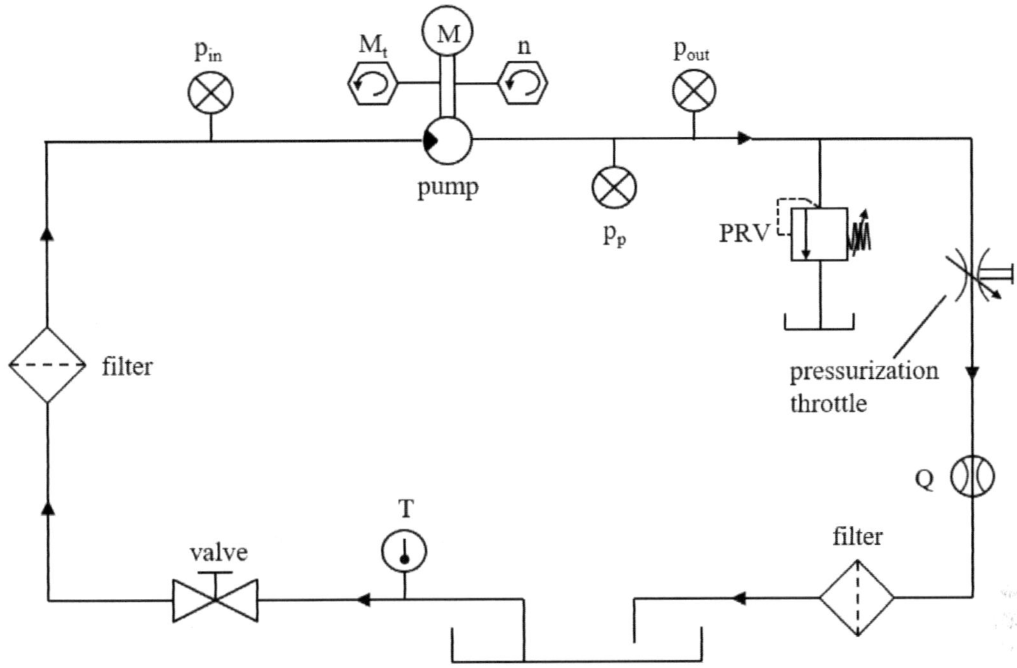

Figure 7. Hydraulic circuit diagram of the test bench.

4. Results and discussion

4.1. Stationary Operating Points

For validation, the measurement results of the stationary operating points are compared with the simulation results. Table 1 summarizes the relative deviations of the volume flow Q and the strain torque M_t.

Table 1. Relative deviations of the volume flow Q and the strain torque M_t between measurement and simulation results.

n [rpm]	750		1500		2100	
Δp [bar]	**dev. Q**	**dev. M_t**	**dev. Q**	**dev. M_t**	**dev. Q**	**dev. M_t**
2	-2.50%	5.63%	1.53%	4.66%	1.45%	0.00%
5	-0.47%	4.26%	2.15%	4.25%	1.89%	2.93%
10	1.62%	5.00%	3.41%	4.26%	2.78%	4.09%
12	2.54%	4.86%	3.79%	4.70%	3.10%	4.13%
15	-	-	4.54%	4.66%	3.56%	4.34%

The volume flow measurement shows the tendency that the relative difference from the simulation increases with rising pressure. It can be concluded that the deviations result from the leakage flow. As the differential pressure increases, the influence of leakage on the total volumetric flow becomes larger, which is why deviations become more significant. In addition, the calculation of the leakage characteristics assumes minimal component tolerances and ideal conditions, which means that the simulated leakage is lower than in practice. For this reason, it would be desirable to detach the model from a manual characteristic input, so that the behavior of the leakage restrictors is calculated based on physical conditions in the simulation itself. This requires a detailed investigation of the gap flows.

The deviation of the torque shows a slight decrease at higher rotational speed. It is mostly between 4% and 5%, except for lower pressures at 1500 rpm where the deviation is lower. In the simulation the total torque is caused by the pressurization of the fluid, while some other aspects like friction are not considered. This challenge could be solved by adding mechanical components such as bearings to the model to take friction into account.

4.2. Pressure Pulsation

Figure 8 shows the pressure pulsation during 0.1 seconds for 15, 10 and 5 bar at 1500 rpm. The black graph shows the measurement recording, while the red graph represents the simulation result. For the 15 bar simulation the gas content was set at 1% and for the 10 and 5 bar simulation it was decreased to 0.5%. The gas content in the test rig fluid was not known during the tests, so it had to be estimated. Through iterative adjustments, the best matches of the pressure curves were achieved with the above values.

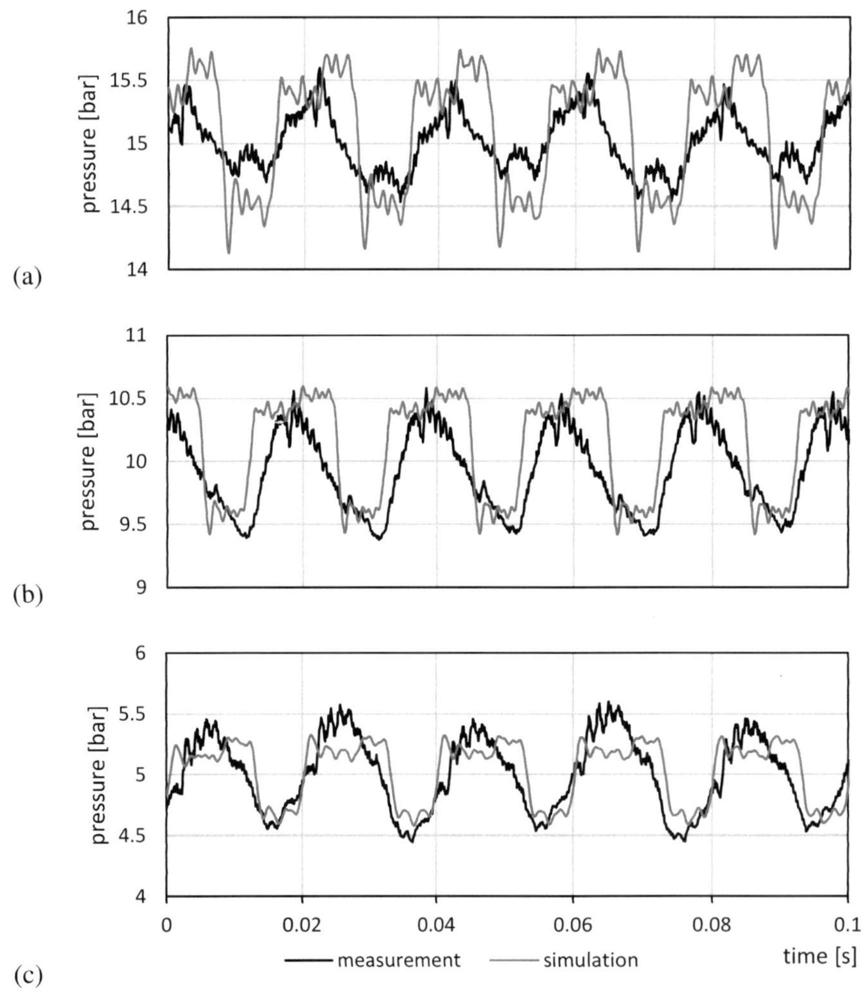

Figure 8. Measured and simulated pressure pulsation curves at 1500 rpm and a) 15 bar; b) 10 bar; c) 5 bar.

The main frequency of the pulsation measurement coincides with the pressure pulsation in the simulation at 1500 rpm with 50 Hz. This corresponds to the chamber opening frequency, which can be calculated from:

$$f = \frac{1500 \frac{1}{min}}{60 \frac{s}{min}} \cdot 2 = 50\ s^{-1} \tag{7}$$

This effect has been explained by Geimer [1], according to which, due to the compressibility of the medium, compensating currents takes place when a chamber opens to the pressure side, which generates a pressure pulsation of 1-3% of the set pressure. The level of pulsation is significantly higher than this value, both in the measurements and in the simulation results, but Geimer does not take the influence of the gas content into account here. This effect can be quite large, but for a more precise statement, further investigations on the impact of the gas content must be conducted. The gas content gives the fluid increased compressibility. The piping between the chambers and the pressurization throttle is capable of taking into account the inertia of the fluid and thus the volumetric flow rate along the length of the pipe in a location-differentiated manner. Due to reflections of the pulses, the observable high frequencies in the pressure pulsation occur.

The overall shape of the pulsations is similar. Both measurements and simulation results, show increasing oscillations at higher pressure. There is a jump in the pressure peaks that becomes more significant when the target pressure is higher, in which measurement and simulation coincide. A noticeable difference in shape is that the simulated curves are more rectangular than the measured curves. A reason for the different rates of rise and fall could be the inertia of the measurement setup. Another difference occurs in the amplitudes. The amplitude of the model is about 0.5 bar larger than the real pump at the 15 bar setting, while it is about 0.3 bar smaller at the 5 bar setting and matches at the 10 bar setting. The differences could also be caused by frequencies in the driving unit or minor unbalances of the spindles, which change the curve by means of a superimposition.

4.3. Internal pressure buildup

Figure 9 shows the pressure buildup of the single chambers along the spindle package over the rotary angle at a constant speed of 1500 rpm and at about 15 bar. The plots depict the chambers of the standard pump size 45/90. The process is explained using the red chamber 0 as an example: Up to 1480° the chamber is opened to the suction side of the pump, so the pressure equals zero. From 1480° to 1610° the chamber is closed to both sides and the pressure becomes 7.5 bar, which is half the discharge pressure. When the chamber opens to the outlet at 1610°, the complete adjustment to the discharge pressure takes place. Since chamber 5 is not completely closed at 1660°, there is no fully enclosed chamber along the spindle package present. Chamber 0 ejects the containing fluid with shrinking volume until it gets annihilated at around 2010°. For software logical reasons, the chamber volume can only become almost zero, which is why this small chamber is enclosed at the time of the actual extinction and a constant pressure prevails. This remains until the chamber of the model is used again at 2160° to form a new chamber at the start of the spindle. How the problem can be solved that a prevailing pressure is displayed here although the chamber is no longer present has yet to be investigated. Nevertheless, this does not affect the rest of the pressure curve and can therefore be ignored. The pressure increase in the enclosed chambers does not occur abruptly but shows a damped rise due to the throttling behavior of the leakage gaps. For chamber 0 this can be seen at about 1480-1490°. When the chamber is opened at 1610°, on the other hand, the pressure rises immediately.

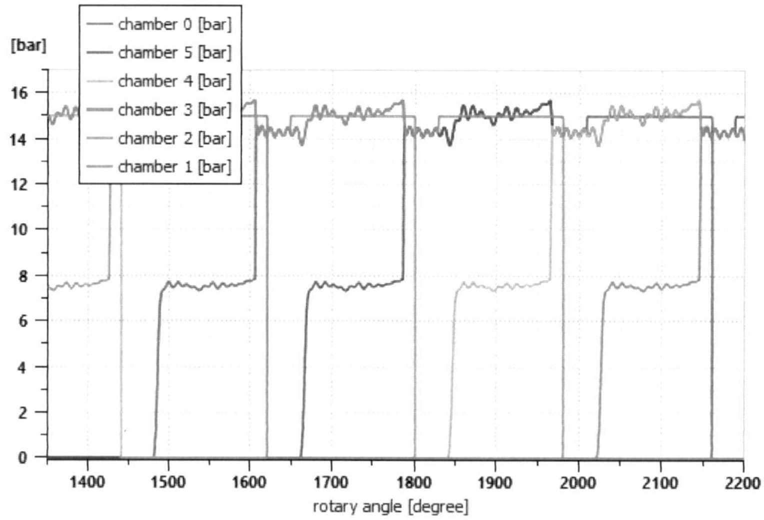

Figure 9. Chamber pressure buildup over the rotary angle at 1500 rpm and 15 bar for the *Leistritz Flexcore 45/90.*

If a chamber is open to the discharge side, it could be assumed that the curve must correspond to the outlet pressure pulsation. However, a slight difference from Figure 8 a) can be observed, which can be justified on the basis of the inertial properties of the fluid in the pipes. Figure 9 shows that the main frequency of the pulsation is caused by the opening of the chambers. As described before, according to Geimer [1], when a chamber opens to the discharge side, compensating currents lead to this effect. This simulation plot, instead, suggests a different explanation: In sections with one enclosed chamber, there are two gaps to be passed by the profile leakage. The pressure step across a single chamber boundary is in this case 7.5 bar. In a section where only one boundary must be crossed by the profile leakage, the pressure step is 15 bar in comparison. If the pressure step is greater, the resulting profile leakage increases, which in turn reduces the total volume flow. With a larger volume flow, a larger discharge pressure is created at the pressurization throttle.

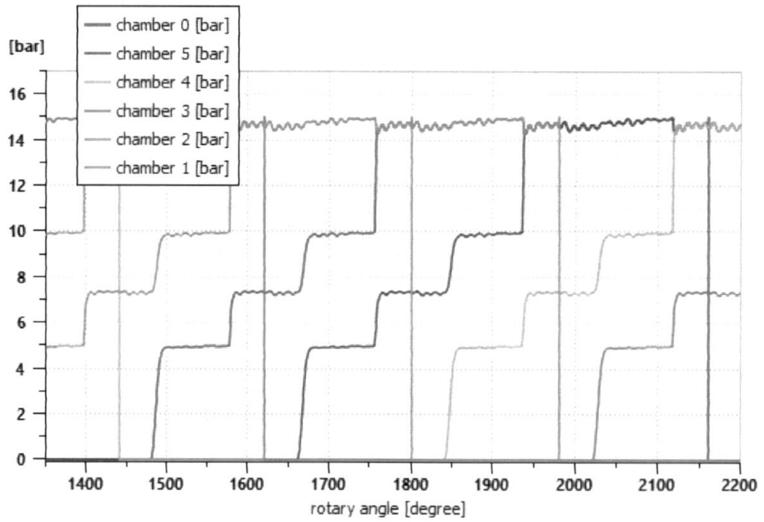

Figure 10. Chamber pressure buildup over the rotary angle at 1500 rpm and 15 bar for the *Leistritz Flexcore 45/70.*

Since there is only one closed chamber at a time for the size 45/90, the chamber pressure curve for the size 45/70 is shown in Figure 10. Because of its lower pitch, this type has a total of 2 simultaneously closed chambers from 1480° to 1570° and from 1660° to 1750°. During this sections the internal pressure reaches 1/3 (5 bar) respectively 2/3 (10 bar) of the discharge pressure. Between these segments, as in size 45/90, chamber 0 is the only one that is completely closed off and contains a pressure of 7.5 bar. This model shows that the distribution of the pressure levels varies with the number of chambers. The section with the incorrectly displayed constant pressure can be seen at about 2160°, but lasts only for a few degrees of spindle rotation.

The results of the simulation agree with Wincek's statement that the pressure increases along the spindle length in linear steps across all chambers [2], as can be seen in Figure 11. It shows a simplified spindle with 4 closed chambers in a fixed position. The pressure is plotted against the length of the spindle, whereas the simulation can only show the pressure of the different chambers over time or rotary angle. To compare the diagrams, the pressure curves of the model must be analyzed in a certain point in time. The difference between two successive chambers is equivalent to a step in Wincek's diagram. Whether the pressure curve along the spindle gaps is linear cannot be determined with the *Simcenter Amesim* model since the leakages have been modeled using orifice type submodels.

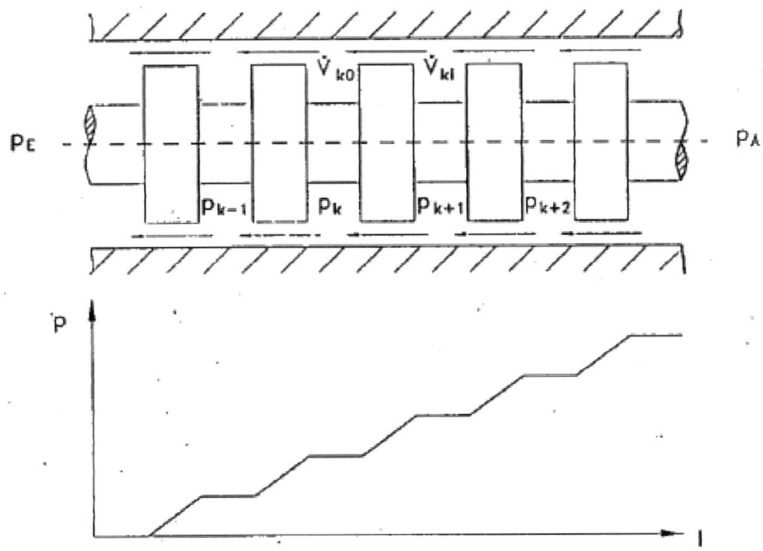

Figure 11. Pressure development along the spindle length by Wincek [2].

5. Conclusion and outlook

The model simulates the pump behavior sufficiently for various investigations. The variations of the volumetric flow and the shaft torque are mostly below 5%. Pressure pulsations show a very good accuracy. The model can be optimized in the investigation of pump behavior and pump design by adding frictional components and an electrical power unit, which would likely improve the accuracy of the simulated parameters. A change of the pump size is easy to manage, because of the deposited tables and the model internal calculations. To simplify the model, most of the internal calculations could be replaced by tables containing the required size-specific information. However, this would imply that all this information would have to be determined again for new pump sizes. In addition, the memory size of the model increases with each table that is included. A disadvantage of the current model is the independence of the leakage from the viscosity of the fluid. When changing the viscosity, the according leakage must be determined again. Correspondingly, an improvement could be made by integrating a viscosity dependent leakage behavior that is not defined by external tables.

The model can be used for different studies, such as investigating the influence of high gas contents, high speeds, high pressures and pressure jumps. It can also be integrated into extensive applications to

IOP Conf. Series: Materials Science and Engineering 1267 (2022) 012011 doi:10.1088/1757-899X/1267/1/012011

study the interaction of a screw pump with its higher level system. Especially the mutual influence of the pressure pulsation is of interest for many implementations. The principle of the *Simcenter Amesim* model works also for different screw pump types. The customization is possible with little effort. For example, the design of a two-flute pump only requires some modifications of the model structure, but the main concept remains the same. First attempts for *Leistritz L4*-pumps show promising results.

Nomenclature

C_p	flow coefficient	
f	frequency	[Hz]
h	spindle pitch	[mm]
k	constant signal	
k_n	chamber number	
K	bulk modulus	[bar]
l	spindle length	[mm]
leak_profile	profile leakage	
leak_rs	running spindle leakage	
leak_throttle	throttle leakage	
M_t	torque	[N/m²]
n	rotary speed	[rpm]
n_{total}	total number of chambers	
p	chamber pressure	[bar]
Δp	differential pressure	[bar]
p_{in}	inlet pressure	[bar]
p_{out}	discharge pressure	[bar]
p_p	pressure pulsation	[bar]
Q	volume flow	[m³/h]
T	torque	[Nm]
t_{print}	print interval	[s]
V	offset	[°]
V_{ch}	chamber volume	[m³]
W	rotary angle	[°]
X	extension	[°]
θ	temperature	[°C]
v	absolute viscosity	[cP]
ρ	density	[kg/m³]
φ_{ex}	existence angle	[°]
φ_{period}	period angle	[°]
φ_{phase}	phase shift	[°]

References

[1] Geimer M 1995 Meßtechnische Untersuchung und Erstellung von Berechnungsgrundlagen zur Ermittlung der Einsatzgrenze dreispindeliger Schraubenpumpen *Doctoral Thesis* Rheinisch-Westfälische Technische Hochschule Aachen p 12 f

[2] Wincek M 1992 Zur Berechnung des Förderverhaltens von Schraubenspindelpumpen bei der Förderung von Flüssigkeits/Gas-Gemischen *Doctoral Thesis* Universität Erlangen-Nürnberg p 55 f

Experimental investigations and 3D simulations by the overset grid method on twin-screw machines in both pump and turbine mode

A H Moghaddam[1], B Kutschelis[2], F Holz[3] and R Skoda[1]

[1]Chair of Hydraulic Fluid Machinery, Ruhr University Bochum, Universitätsstr. 150, 44801 Bochum, Germany

[2]Klaus Union GmbH & Co. KG, Blumenfeldstrasse 18, 44795 Bochum, Germany

[3]ITT Bornemann GmbH, Industriestrasse 2, 31683 Obernkirchen, Germany

E-mail: ali.hassannejadmoghaddam@rub.de

Abstract. Twin-screw pumps due to their particular capabilities, offer unique benefits in comparison to other types of pumps. They can also recover wasted energy from the fluid system when run in a reversed way, i.e., in turbine mode. In this study, the possibility of benefiting twin-screw machines instead of conventional control valves in order to recover energy is investigated. Flow physics and energy recovery are analyzed experimentally and by 3D flow simulations in both, pump and turbine mode. Pump and turbine characteristics under different operating points are experimentally measured for low and high viscous fluids, i.e., water and oil, respectively. It is observed that for higher viscosity the dependency of volume flow rate on the pressure difference imposed on the pump is reduced. This observation indicates the significant effect of viscosity on the gap flow. Based on a simulation method by means of overset grid technique, unsteady 3D flow simulations with a high grid quality and a high spatial resolution particularly in gap regions in terms of $y^+ < 1$ are conducted and results are validated against experiments. The profound assessment of gap flow based on the simulation results shows that for highly viscous fluids such as oil, the rotational speed of spindles as well as the direction of rotation have a significant effect on gap flow characteristics and consequently on the pump and turbine performance. From the experiments, it is clarified that with a lower rotational speed of spindles and a higher pressure difference, a higher amount of energy up to about 50% of the fluid energy in the piping system can be recovered instead of being dissipated by a conventional control valve.

1. Introduction

Considering centrifugal pumps running in reverse mode as turbines instead of pressure control valves (PCVs) in piping systems has been extensively investigated and shown that a considerable amount of wasted energy can be recovered e.g., [1, 2]. For fluids with high viscosity such as oil, the usage of centrifugal pumps cannot be beneficial due to their efficiency drop. In contrast to centrifugal pumps, twin-screw pumps are particularly capable of handling highly viscous fluid with relatively high efficiency. Due to the complex kinematics regarding the helical shape of spindles as well as several narrow gaps inside the pump, the understanding of the internal flow field in twin-screw pumps is yet incomplete. Furthermore, no investigation has been performed yet on twin-screw pumps running in turbine mode based on the authors knowledge. Therefore, this study aims at elaborating the internal flow field in a twin-screw pump by conducting experimental and numerical studies. In particular, the possibility of using a twin-screw pump in turbine mode instead of a PCV in a piping system is investigated. By experimental investigations, only the characteristic curves of a twin-screw pump can be achieved, while the complex internal flow field dynamics cannot be assessed due to the limited optical access. Thus, numerical simulations in terms of three-dimensional (3D) computational fluid dynamics (CFD) are performed to evaluate the internal flow field, after a thorough validation on experimentally measured characteristics. The internal flow field and consequently the twin-screw pump performance is highly affected by the gap flow inside the pump which occurs due to the pressure difference between chambers as well as the rotation of spindles. Therefore, a high-fidelity 3D CFD simulation method is required for resolving the gap flow with high accuracy.

Regarding 3D CFD simulations, there have been few studies on twin-screw pumps which are based on either steady simulations e.g., [3, 4] or fully unsteady simulations by considering the rotation of screw spindles e.g., [5, 6, 7, 8, 9]. In [3, 4] by defining the moving wall boundary condition to the static spindle grid, the movement of spindles and therefore the gap kinematics have been simplified which could lead to an inaccuracy of the gap flow assessment and the pump/turbine performance estimation. For the unsteady simulations in [5, 6, 7], moving grids in terms of deforming body-fitted structured grids have been used. In Räbiger and Holz [5], the movement of cell vertex positions in each time-step is prescribed and the algorithm of Kovacevic et al. [10] for the grid movement has been considered in studies by Yan et al. [6, 7]. By using deforming mesh methods, the local grid refinement in critical regions, i.e., gaps, is difficult, since due to the significant deviation of gap width and chamber volume, strong grid distortion may occur. Zhang et al. [8] have performed unsteady simulations by means of an immersed boundary approach on a double helix screw pump. Although with this method the mentioned drawbacks of deforming mesh method are eliminated, the exact shape of spindles cannot be achieved and is only approximated. In our previous study, Moghaddam et al. [9] have proposed employing the overset grid method on twin-screw pumps. The overset grid method subdivides a complex computational domain into several subdomains which overlap on each others [11, 12, 13]. The flow values are then interpolated between subdomain grids in overlap regions. By benefiting the overset grid method also in this study, rigid and non-deforming grids which rotate with the spindles and overlap with each other as well as with the stationary casing grid are created with a high grid quality and resolution within the gaps as well as the chambers.

This paper is organized as follows: In section 2 the geometry of the studied twin-screw pump together with the experimental setup and test rig facilities is presented. The numerical method and simulation setup are presented in section 3 and the results are presented in section 4. The paper is finalized by conclusions in section 5.

2. The pump test case and experimental test rig

2.1. Twin-screw pump

A double-flow twin-screw pump is investigated in this study which is depicted in Fig. 1(a). The outer parts of the low- and high pressure ports are the inlet and outlet of the twin-screw pump, respectively while in the turbine mode it is vice versa. In Fig. 1(b) and by means of symmetry planes, only one flow path is shown which corresponds to the computational domain used in the simulations. In order to visualize the inside components, an exploded view of one flow path is also shown in Fig. 1(c). The left spindle in Fig. 1(c) is the driving spindle which is connected to the motor/generator. A timing gear is used to connect the right spindle as the driven one to the left spindle with the same rotational speed n. The spindles are counter-rotating. Both, right and left spindle are double-start thread and have the same outer diameter $d_S = 2r_S$ and

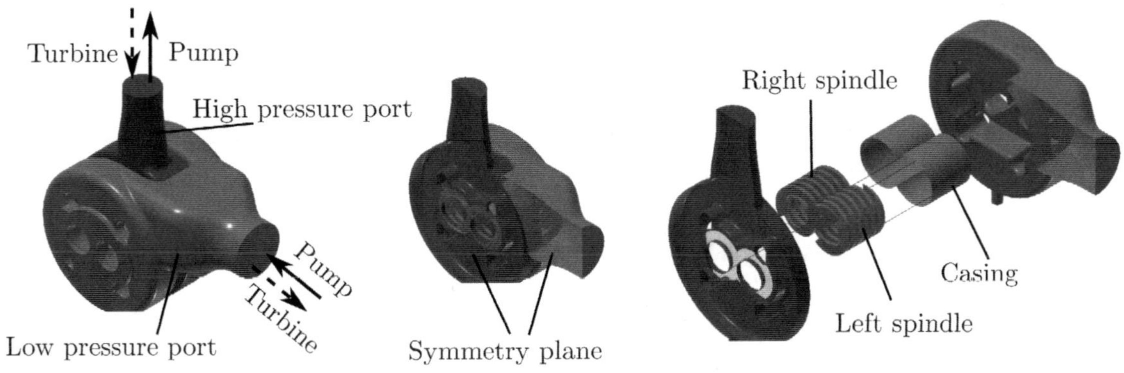

(a) (b) (c)

Figure 1. The geometry illustration of the studied twin-screw pump by (a) the complete double-flow path, (b) one flow path and (c) an exploded view of one flow path.

pitch P which are positioned eccentrically inside the casing with the eccentricity ϵ:

$$\epsilon = \frac{\Delta y}{r_C - r_S} = 0.7 \tag{1}$$

The eccentricity is chosen to allow more bending of spindles due to radial forces. An exaggerated schematic view of the position of spindles is illustrated in Fig. 2(a) where r_C and r_S are the casing and spindle radius, respectively and Δy is the eccentric displacement. Chambers within the spindles and the casing are created repeatedly by rotating intermeshing spindles and delivered to the high pressure port in pump mode or the low pressure port in turbine mode. The fluid is conveyed within the created chambers by the spindle rotation, and the ideal volume flow rate Q_{ideal} is solely governed by the spindle geometry and the angular velocity ω [14]:

$$Q_{\text{ideal}} = 2\,P\,\omega\,A_{\text{fluid}} \tag{2}$$

A_{fluid} is the wetted cross-sectional area of the fluid domain inside the casing and is evaluated by the computer-aided design (CAD) data in our study. In order to avoid any rubbing, spindles do not contact each other nor the casing. Therefore, the created chambers are not completely sealed but connected with each other via different gaps in terms of circumferential, radial and flank gap. The shape of the gaps is discussed in detail by Moghaddam et al. [9] and only briefly summarized here. The circumferential gap is the gap between the spindle tip and the casing which connects each chamber to the immediate adjacent one on the same shaft. The radial gap is located between the tip of one spindle and the shaft of the other one and connects each chamber to the second adjacent one. The position of circumferential and radial gap is schematically shown in Fig. 2(b). Due to the eccentricity, the circumferential gap width varies between $6.2 \times 10^{-4}\,d_S$ and $3.5 \times 10^{-3}\,d_S$ over its length. The radial gap width is independent of eccentricity and has a constant value of $2.2 \times 10^{-3}\,d_S$. The flank gap configuration is more complex than the circumferential and radial gap. It is located between the surface of spindles and connects four different chambers of the two spindles with each other. The position of flank gap and its complex shape are schematically illustrated in Figs. 2(b) and 2(c), respectively. The minimum and maximum flank gap width have the amount of $2.5 \times 10^{-3}\,d_S$ and $12.5 \times 10^{-3}\,d_S$, respectively. In Fig. 2(c), the corresponding two left and right spindle chamber termed L2 and L4 as well as R3 and R5, respectively, are illustrated. As detailed in Moghaddam et al. [9], the pressure relationship $p_{L2} > p_{R3} > p_{L4} > p_{R5}$ holds. As mentioned earlier, due to the pressure difference between chambers and the rotational speed of spindles, a gap flow occurs within the gaps. In pump mode, the gap flow is in the opposite direction of the conveyed fluid and thus corresponds to a backflow. In turbine mode, on the other hand, it is in the same direction as the conveyed fluid and corresponds to an overflow. The total backflow rate $Q_\mathrm{b}^\mathrm{tot}$ or total overflow rate $Q_\mathrm{o}^\mathrm{tot}$ then can be calculated by summation of all gap flow contributions in terms of circumferential, radial and flank gap:

$$Q_\mathrm{b/o}^\mathrm{tot} = Q_\mathrm{b/o}^\mathrm{cir} + Q_\mathrm{b/o}^\mathrm{rad} + Q_\mathrm{b/o}^\mathrm{fla} \tag{3}$$

All gap flow contributions in Eq. 3 are assumed to be positive. Due to the gap flow, the real pump volume flow rate Q_p has a smaller value and the real turbine volume flow rate Q_t has a larger value than the ideal volume flow rate Q_ideal:

$$Q_\mathrm{p} = Q_\mathrm{ideal} - Q_\mathrm{b}^\mathrm{tot} \tag{4}$$

$$Q_\mathrm{t} = Q_\mathrm{ideal} + Q_\mathrm{o}^\mathrm{tot} \tag{5}$$

Therefore, the volumetric efficiencies of pump η_p and turbine η_t are evaluated by:

$$\eta_\mathrm{p}^\mathrm{vol} = \frac{Q_\mathrm{p}}{Q_\mathrm{ideal}} \tag{6}$$

$$\eta_\mathrm{t}^\mathrm{vol} = \frac{Q_\mathrm{ideal}}{Q_\mathrm{t}} \tag{7}$$

2.2. Test rig

In the experiments, pressure difference Δp, volume flow rate Q and the torque M on the shaft between the pump/turbine and electric motor/generator are measured in closed measurement facilities in the lab of the Klaus Union & Co. KG. Q either corresponds to Q_p or Q_t, either if pump or turbine mode is driven. For pump mode, two different liquids, i.e., water and oil ISO-VG-68 have been investigated. The corresponding piping and instrumentation diagram (P& ID) is shown in Fig. 3(a). Q is measured at the high pressure side of the pump with the Micro Motion CMF400 manufactured by Emerson with an accuracy of about ±0.05% of the

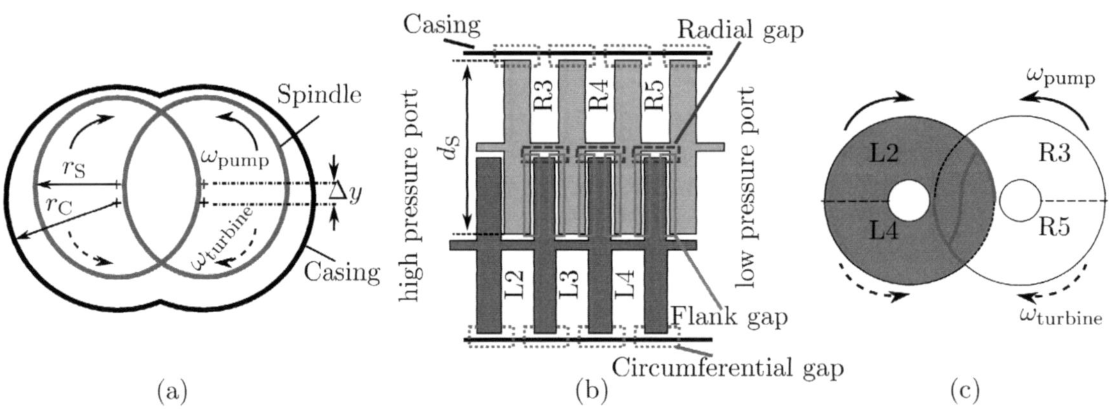

Figure 2. Schematic illustration of (a) eccentricity of spindles inside the casing, (b) positions of circumferential, radial and flank gap and (c) axial view of the flank gap shape and the corresponding chambers connections. For chamber pressure levels in (c), the relationship $p_\mathrm{L2} > p_\mathrm{R3} > p_\mathrm{L4} > p_\mathrm{R5}$ holds.

flow rate. The pressure measurement transducers type 2088 from Emerson with an accuracy of about ±0.1% are located at a distance of 2 times the nominal pipe diameter to the pump flanges at low and high pressure side denoted by p_{low} and p_{high}, respectively. The pressure difference Δp corresponds to $\Delta p = p_{\text{high}} - p_{\text{low}}$.

For turbine mode, only oil is considered whose P& ID is shown in Fig. 3(b). It should be noted that the orientation of the spindle rotation is reversed in turbine mode in comparison to pump mode. For pressurizing the oil before entering the turbine, a booster pump is used. In turbine mode, the driver spindle shaft is coupled with an electric generator and the output power in terms of the shaft torque M is measured. For measuring the torque, DATAFLEX 70/1000 from KTR is used with the accuracy of about ±0.1% of the final value.

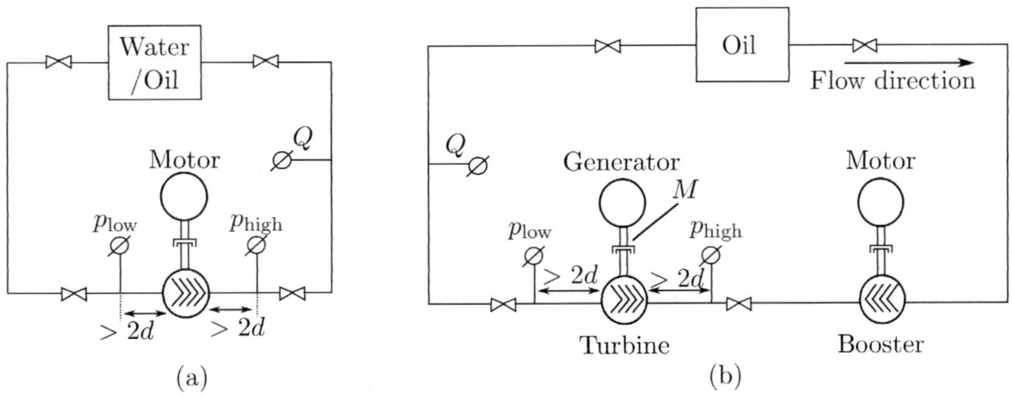

Figure 3. The P&ID of the test rigs for the twin-screw pump in (a) pump mode and (b) turbine mode.

3. Numerical method and simulation setup

3.1. Numerical method

A compressible liquid Navier-Stokes solver with dual time-stepping [15] available in the software CFD++ is used with a statistical turbulence model, i.e., the two equation eddy-viscosity $k - \omega$ shear stress transport (SST) model [16]. The spatial discretization is based on an unstructured finite volume scheme with a second-order multidimensional total variation diminishing interpolation. An approximate Riemann solver is used for the inviscid flow terms and an algebraic multigrid algorithm is benefited for convergence acceleration. No cavitation model is applied. Both water and oil are considered compressible with the constant viscosity of $\mu = 0.001$ Pa · s and $\mu = 0.1$ Pa · s, respectively.

As mentioned earlier, the overset grid technique is used in this study based on Chakravarthy et al. [17, 19, 18]. In the overset grid technique a complex domain is divided into several subdomains that are discretized independently. Each subdomain may overlap with one or more subdomains. A hole-cutting process is utilized to encompass different subdomains into a unique computational domain. The hole-cutting process consists of two general steps. In the first step, the cells of each subdomain which are located out of the computational domain are erased. In the second step, the overlapping cell regions are minimized and zonal interfaces are generated between subdomains. The conservative variables for the nodes located on zonal interfaces are interpolated by a least-square interpolation between subdomains. It should be noted that in unsteady simulations the hole-cutting process has to be performed for each physical time-step since the position of grids varies.

Due to the interpolation error between subdomains, the overset grid technique suffers from poor conservation properties. By a polynomial interpolation [17, 19, 18], mass loss is reduced. In our simulations, the minimum and maximum mass flow deviation between the inlet and outlet is about 1.1% and 3.5%, respectively. The former corresponds to the pump mode simulation with oil and the latter corresponds to the turbine simulation. In our preceding study by Moghaddam et al. [9], details of the overset method and their accuracy assessment have been presented.

3.2. Simulation setup

For all simulations, total pressure at inlet, static pressure at outlet and a constant rotational speed of spindle n are predefined. 1 deg spindle rotation per time-step has been set which has been verified to provide a time-step independent solution. For the domain descretization, the entire computational domain is divided into five subdomains, i.e., high pressure port, low pressure port, casing, right spindle and left spindle, shown in Fig. 1(c). The grid generation has been conducted for each component independently. The high and low pressure ports are meshed with unstructured tetrahedral cells. The casing domain is filled with structured hexahedral cells which are considerably refined toward the wall in order to achieve a high grid resolution near the casing wall and consequently at the outer wall of the circumferential gap. Around the surface of each spindle, a fine layer of hexahedral grid is generated which is significantly refined toward the surfaces of spindles. The grid refinement at casing and spindle walls is performed in a way that a wall resolution in the range of $y^+ \approx 1$ at any wall is obtained. After mesh generation for each component, the hole-cutting process is conducted to encompass the entire computational domain. Thus, firstly, the cells located out of the computational domain have been deleted. Afterwards, the overlapping layer between different subdomains have been minimized and located in the middle of gap walls. More details of the hole-cutting procedure for the twin-screw pump can be found in Moghaddam et al. [9]. The resulting computational domain after the hole-cutting process has a total cell number of about 49 million which can hardly be reduced when maintaining $y^+ \approx 1$. In the normal direction of all gaps about 28 cells are placed which has been verified to produce grid independent results [9]. In addition, in order to eliminate initial transients and assure statistical convergence, 7 complete revolutions of spindles have been considered for each simulation, and the last revolution is evaluated for result presentation in section 4. Each simulation run has been performed on 576 cores at the RWTH Aachen University cluster. For each run, about 0.16 million Core-h are required.

4. Results

4.1. Stationary pump and turbine characteristics

In the experiments, Q_p and Q_t are time-averaged. For the simulations, their average value over the last complete revolution is considered which leads to their time-averaged values in terms of $\langle Q_\mathrm{p} \rangle$ and $\langle Q_\mathrm{t} \rangle$, respectively. $\langle Q_\mathrm{p} \rangle$ and $\langle Q_\mathrm{t} \rangle$ are normalized to the experimentally measured maximum volume flow rate which occurs in pump mode with oil at $n = 3600$ rpm and $\Delta p = 2$ bar yielding $\langle \widetilde{Q}_\mathrm{p} \rangle$ and $\langle \widetilde{Q}_\mathrm{t} \rangle$. While in the experiments, entire characteristic curves in terms of flow rate vs. pressure difference have been measured, in the simulations, only selected operating points have been considered. For pump mode, $n = 3600$ rpm and $\Delta p = 6$ bar are considered for water and oil. For turbine mode with oil, $n = 1000$ rpm and $\Delta p = 6$ bar is considered. Time-averaged normalized volume flow rate in pump mode $\langle \widetilde{Q}_\mathrm{p} \rangle$ and in turbine mode $\langle \widetilde{Q}_\mathrm{t} \rangle$ versus pressure difference Δp are presented in Fig. 4(a) and Fig. 4(b), respectively. The simulation results are in a good agreement with the experiments with a minimum and maximum deviation of about 2.5% and 5.5%, respectively.

According to Figs. 4(a) and 4(b), by rising n at any Δp, the volume flow rate increases due to increasing displacement volume. In pump mode according to Fig. 4(a), by rising Δp at any n, $\langle \widetilde{Q}_\mathrm{p} \rangle$ reduces almost linearly for oil and non-linearly for water. In turbine mode based on Fig.

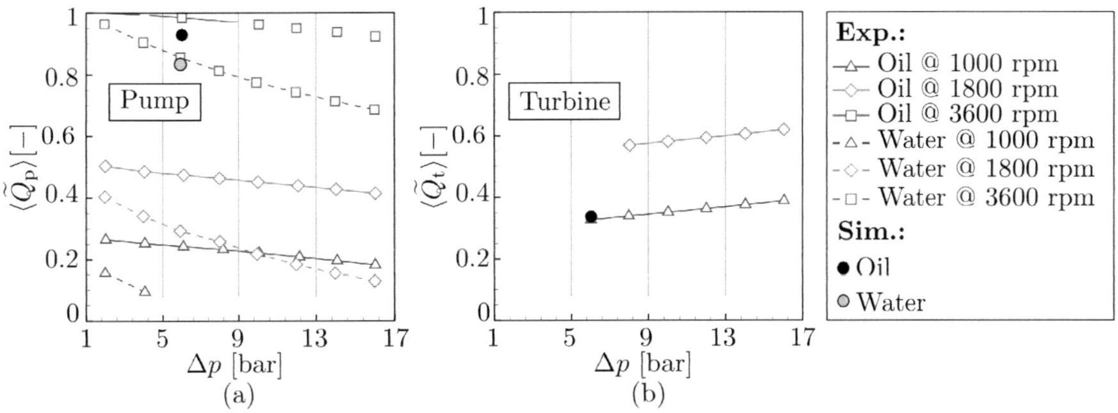

Figure 4. Time-averaged and normalized volume flow rate in (a) pump mode $\langle\widetilde{Q}_\mathrm{p}\rangle$ and (b) turbine mode $\langle\widetilde{Q}_\mathrm{t}\rangle$ at different operating points versus pressure difference Δp. $\langle Q_\mathrm{p}\rangle$ and $\langle Q_\mathrm{t}\rangle$ have been normalized to the experimental result at n = 3600 rpm and Δp = 2 bar in pump mode with oil.

4(b), $\langle\widetilde{Q}_\mathrm{t}\rangle$ increases linearly. The theoretical flow rate Q_ideal linearly correlates with n, according to Eq. 2. Thus, the particular characteristics of either $\langle\widetilde{Q}_\mathrm{p}\rangle$ or $\langle\widetilde{Q}_\mathrm{t}\rangle$ are immediately associated to different gap flow characteristics, according to Eqs. 4 and 5. The gap flow characteristics will be discussed in detail in section 4.2.

4.2. Gap flow characteristics

From measured Q_p and Q_t in the experiments, $Q_\mathrm{b}^\mathrm{tot}$ and $Q_\mathrm{o}^\mathrm{tot}$ are evaluate by Eqs. 4 and 5, respectively. Time-averaged normalized backflow rate $\langle\widetilde{Q}_\mathrm{b}^\mathrm{tot}\rangle$ and overflow rate $\langle\widetilde{Q}_\mathrm{o}^\mathrm{tot}\rangle$ for pump and turbine mode are presented in Figs. 5(a) and 5(b), respectively. $\langle Q_\mathrm{b}^\mathrm{tot}\rangle$ and $\langle Q_\mathrm{o}^\mathrm{tot}\rangle$ have been normalized to the maximum value of backflow rate which is measured at n = 3600 rpm and Δp = 16 bar for water, yielding $\langle\widetilde{Q}_\mathrm{b}^\mathrm{tot}\rangle$ and $\langle\widetilde{Q}_\mathrm{o}^\mathrm{tot}\rangle$, respectively. Different dependency of $\langle\widetilde{Q}_\mathrm{b}^\mathrm{tot}\rangle$

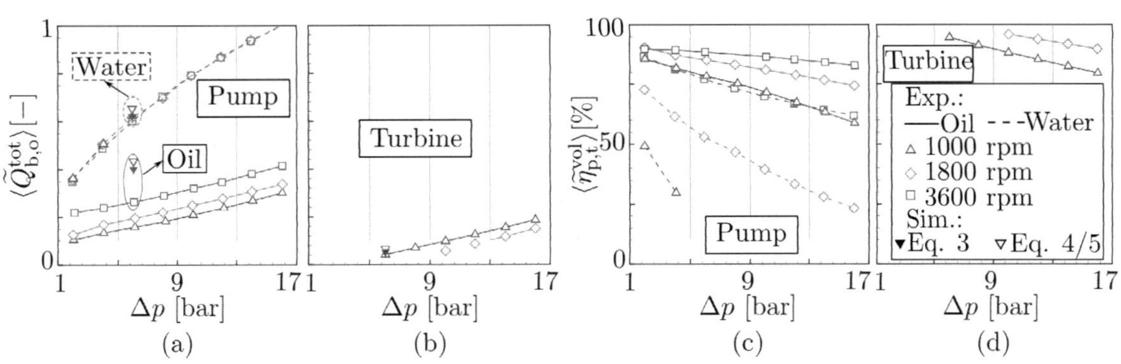

Figure 5. Time-averaged and normalized (a) total volume backflow rate $\langle\widetilde{Q}_\mathrm{b}^\mathrm{tot}\rangle$ in pump mode and (b) total volume overflow rate $\langle\widetilde{Q}_\mathrm{o}^\mathrm{tot}\rangle$ in turbine mode together with the time-averaged volumetric efficiency in (c) pump mode $\langle\widetilde{\eta}_\mathrm{p}^\mathrm{vol}\rangle$ and (d) turbine mode $\langle\widetilde{\eta}_\mathrm{t}^\mathrm{vol}\rangle$ versus pressure difference Δp. $\langle Q_\mathrm{b}^\mathrm{tot}\rangle$ and $\langle Q_\mathrm{o}^\mathrm{tot}\rangle$ have been normalized to the experimental result at n = 3600 rpm and Δp = 16 bar in pump mode with water as the conveyed fluid.

on n and Δp between water and oil can be discerned in Fig. 5(a). For water, $\langle \widetilde{Q}_{\mathrm{b}}^{\mathrm{tot}} \rangle$ increases non-linearly with rising Δp, while it is almost independent of n. This observation indicates that for fluids with low viscosity such as water, the backflow rate is mainly dependent on Δp rather than n, which has also been concluded in [9]. On the other hand, when oil is considered as the conveyed fluid, $\langle \widetilde{Q}_{\mathrm{b}}^{\mathrm{tot}} \rangle$ noticeably rises with increasing n. In turbine mode as illustrated in Fig. 5(b), again, a rising characteristics with Δp is retained as for pump mode, while the n-characteristics is reversed. This contrary dependency on n is traced back to the reverse direction of the spindle rotation in turbine mode compared to pump mode and is associated with the high viscosity of oil, as will be discussed in detail in the following by conducting a detailed gap study by the simulation results. The gap flow rate is lower for turbine mode than for pump mode, as a cross check between Figs. 5(a) and 5(b) shows. Therefore, turbine mode is associated with a higher volumetric efficiency according to Eqs. 6 and 7 and noting that Q_{ideal} is the same for both modes. $\langle \widetilde{\eta}_{\mathrm{p}}^{\mathrm{vol}} \rangle$ and $\langle \widetilde{\eta}_{\mathrm{t}}^{\mathrm{vol}} \rangle$ are depicted in Figs. 5(c) and 5(d), respectively. Regarding pump mode, $\langle \widetilde{\eta}_{\mathrm{p}}^{\mathrm{vol}} \rangle$ is higher for oil than water for any operating point since the amount of $\langle \widetilde{Q}_{\mathrm{b}}^{\mathrm{tot}} \rangle$ is less. These observations show the importance of viscosity on the screw machine performance.

In order to evaluate the gap flow rate in the simulations for the selected operating points, two different methods have been used and compared. The first method is based on Eqs. 4 and 5 which is the same which has been used for the experiments. In the second method, the gap flow portions, i.e., $\widetilde{Q}_{\mathrm{b/o}}^{\mathrm{cir}}$, $\widetilde{Q}_{\mathrm{b/o}}^{\mathrm{rad}}$ and $\widetilde{Q}_{\mathrm{b/o}}^{\mathrm{fla}}$ are calculated individually from the simulation results based on the numerical integration of cell face fluxes, interpolated to a cross-section surface within the gaps [9]. Consequently $\widetilde{Q}_{\mathrm{b}}^{\mathrm{tot}}$ and $\widetilde{Q}_{\mathrm{o}}^{\mathrm{tot}}$ are evaluated by Eq. 3. The resulting $\langle \widetilde{Q}_{\mathrm{b}}^{\mathrm{tot}} \rangle$ and $\langle \widetilde{Q}_{\mathrm{t}}^{\mathrm{tot}} \rangle$ for both methods are also depicted in Figs. 5(a) and 5(b). It can be seen that the deviation of the results obtained by both methods are smaller than the deviations to the experimental results, in particular for oil. This means that the gap flow evaluation by Eq. 3 is feasible. Remaining differences between both methods are traced back to interpolation errors at the cross-section surface within the gaps. The difference to the experimental results is simply another illustration of flow rate characteristics shown in Fig. 4, now evaluated by Eqs. 4 and 5. As already shown in Fig. 4, the mismatch is more pronounced for oil in pump mode which can be due to neglecting the energy equation, since the density and viscosity of oil are highly dependent on the temperature variation. Furthermore, high viscosity promotes heating of the fluid, which is neglecting by omitting the energy equation.

In order to further assess the gap flow characteristics, the separate gap flow rate portions, i.e., $\langle \widetilde{Q}_{\mathrm{b/o}}^{\mathrm{cir}} \rangle$, $\langle \widetilde{Q}_{\mathrm{b/o}}^{\mathrm{rad}} \rangle$ and $\langle \widetilde{Q}_{\mathrm{b/o}}^{\mathrm{fla}} \rangle$ are shown in Fig. 6(a). We focus on the viscosity effect in pump mode, while gap flow rate in turbine mode is much lower as already discussed with Fig. 5. Gap flow rate for turbine mode is illustrated in Fig. 6(a) only for completeness. In pump mode with water as the conveyed fluid, circumferential and flank gap have the maximum contribution to the total backflow rate which is due to their larger cross-sectional area in comparison to the radial gap. It is noticeable that for oil in pump mode, $\langle \widetilde{Q}_{\mathrm{b}}^{\mathrm{cir}} \rangle$ reduces considerably to about 0 which causes a lower amount of $\langle \widetilde{Q}_{\mathrm{b}}^{\mathrm{tot}} \rangle$ in comparison to water. This effect can be traced back to the velocity profile within the circumferential gap whose normal component c_{norm} (normal to the gap cross-section) is plotted in Fig. 6(b) at any circumferential position within the gap. c_{norm} is plotted versus the wall normal coordinate h, normalized to the maximum gap width H. $h/H = 0$ and $h/H = 1$ correspond to the spindle tip and the casing wall, respectively. In Fig. 6(b), it is noticeable that for oil until a certain distance of spindle tip, i.e., $h/H \approx 0.2$, c_{norm} has a positive value which indicates an overflow rather than a backflow. Due to the helical shape of spindles and the direction of rotation, the spindle tip wall has a relative axial movement toward the conveying direction and opposite to the backflow direction, which means in the direction of positive c_{norm}. Due to the no-slip condition and the relative wall movement, a kind of Couette flow occurs in the circumferential gap. Therefore, the fluid is transported with the spindle wall movement, corresponding to $c_{\text{norm}} > 0$ which can be discerned in close-up X in Fig. 6(b). The pressure gradient, on the other hand, counteracts the wall movement effect and causes $c_{\text{norm}} < 0$ in a certain distance from the wall, i.e., $h/H < 0.2$ for oil. Note that this effect is of course also present for water. However, for water due to the low viscosity, the flow reversal region, i.e., the overflow region is very small which can also be discerned in close-up X in Fig. 6(b).

According to Fig. 2(c) spindles are counter rotating toward each other in pump mode and consequently the spindle wall movement is in the same direction as the backflow in flank and radial gap, i.e., from high pressure chambers to low pressure ones (from top chambers in Fig. 2(c) to the bottom ones). This causes a slight rise of $\langle \widetilde{Q}_{\text{b}}^{\text{rad}} \rangle$ and $\langle \widetilde{Q}_{\text{b}}^{\text{fla}} \rangle$ for oil, which can be

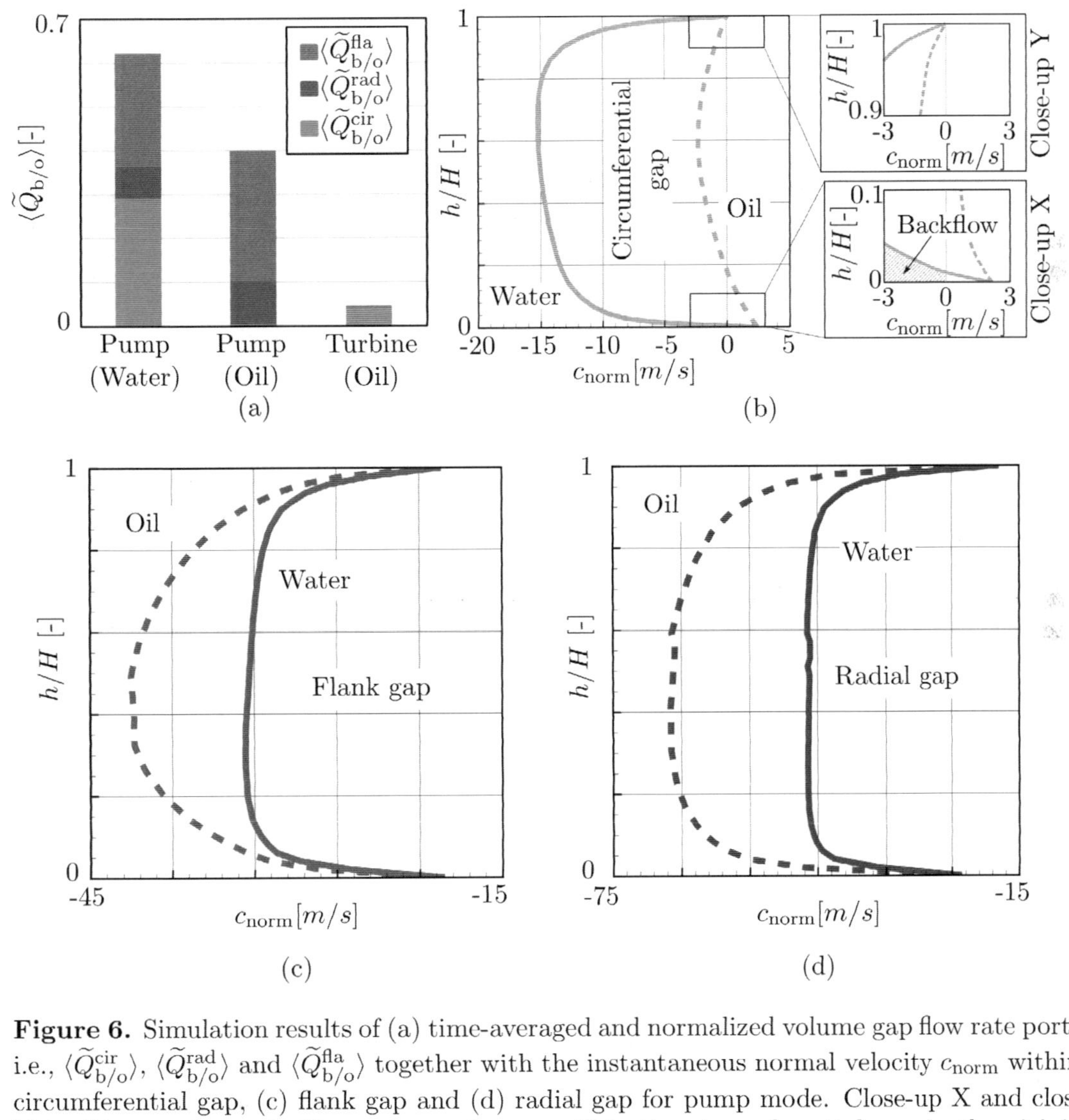

Figure 6. Simulation results of (a) time-averaged and normalized volume gap flow rate portions, i.e., $\langle \widetilde{Q}_{\text{b/o}}^{\text{cir}} \rangle$, $\langle \widetilde{Q}_{\text{b/o}}^{\text{rad}} \rangle$ and $\langle \widetilde{Q}_{\text{b/o}}^{\text{fla}} \rangle$ together with the instantaneous normal velocity c_{norm} within (b) circumferential gap, (c) flank gap and (d) radial gap for pump mode. Close-up X and close-up Y in (b) are schematic illustrations of c_{norm} within the circumferential gap in the vicinity of the spindle tip and the casing wall, respectively. $\langle Q_{\text{b}} \rangle$ and $\langle Q_{\text{o}} \rangle$ have been normalized to the experimental result at $n = 3600$ rpm and $\Delta p = 16$ bar in pump mode with water as the conveyed fluid.

discerned in Fig. 6(a) and is further illustrated by the velocity profiles in Figs. 6(c) and 6(d). In Figs. 6(c) and 6(d) c_{norm} in pump mode is plotted inside the narrowest gap width of the flank and radial gap, respectively. In Fig. 6(c), $h/H = 0$ and $h/H = 1$ correspond to the side surface of the right and the left spindle, respectively. In Fig. 6(d), $h/H = 0$ is associated with the tip of the right spindle and $h/H = 1$ corresponds to the shaft of the left spindles. For both, flank and radial gap, the velocity profile is more developed for oil due to its higher viscosity, resulting in larger $\langle \widetilde{Q}_{\text{b}}^{\text{rad}} \rangle$ and $\langle \widetilde{Q}_{\text{b}}^{\text{fla}} \rangle$ for oil than for water.

4.3. Energy recovery

Finally and in order to demonstrate the benefits of using the twin-screw pump in turbine mode instead of a PCV, the amount of recovered energy in terms of $\langle \widetilde{\eta}_{\text{t}}^{\text{tot}} \rangle = \frac{P_{\text{recovery}}}{P_{\text{fluid}}}$ is measured from the experiments and presented in Fig. 7. P_{fluid} is the total fluid power entering the turbine which is calculated by $Q_{\text{t}} \cdot \Delta p$. P_{recovery} is the amount of power recovered from fluid which is calculated by $M_{\text{t}} \cdot \omega$. It is observed that for lower rotational speed n at any Δp, $\langle \widetilde{\eta}_{\text{t}}^{\text{tot}} \rangle$ has a higher value which is due to less frictional losses between mechanical components e.g., bearings as well as less hydraulic losses inside the fluid domain. In addition, by rising Δp, $\langle \widetilde{\eta}_{\text{t}}^{\text{tot}} \rangle$ also rises. It is observed that at certain operating points, up to about 50% of the fluid power can be recovered by using a twin-screw pump as turbine instead of a PCV.

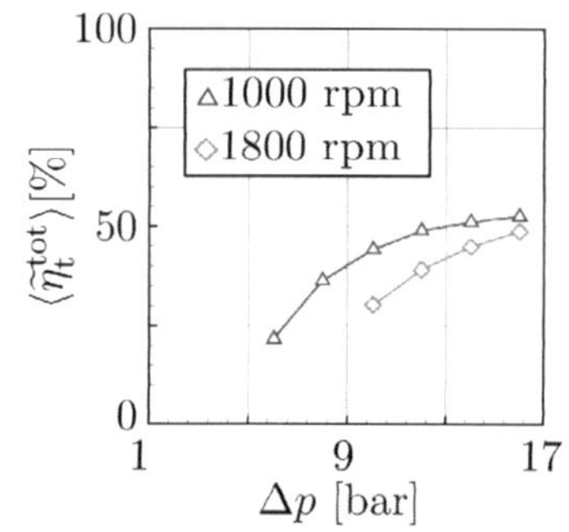

Figure 7. Time-averaged turbine total efficiency $\langle \widetilde{\eta}_{\text{t}}^{\text{tot}} \rangle$ from the experimental results at different operating points versus pressure difference Δp.

5. Conclusions

Experimental investigations and unsteady 3D numerical simulations have been conducted on a twin-screw pump in both pump and turbine mode with two different fluids, i.e., water and oil. By means of simulations and after a careful validation, detailed insights into the internal flow field have been provided, which could not be achieved by experiments. Therefore, they are a valuable complement to the experimental investigations. Once again, the overset grid method has shown its reliability for twin-screw pump simulations, as already demonstrated in [9] for pure pump mode simulations.

Different characteristics of the twin-screw pump, depending on viscosity variation and running in pump or turbine mode have been detected. While rising Δp in pump and turbine mode increases the gap flow rate, the variation of rotational speed has a different effect on the gap flow depending on viscosity. For water, the rotational speed variation has a neglectable effect on the backflow. On the other hand, for oil, while rising rotational speed causes a considerable increase in backflow rate, it reduces the amount of overflow rate. These behaviors could be traced back to different gap flow characteristics.

Our detailed gap study has revealed that due to the no-slip condition and the helical shape of spindles, a Couette flow occurs in gaps which is more pronounced for oil rather than water. For pump mode in the circumferential gap, the Couette flow counteracts the pressure difference effect, leading to a reduction of the gap flow. On the other hand in radial and flank gap, the gap flow increases since the Couette flow is in the same direction as the backflow caused by the pressure difference. These characteristics are found to be reversed in turbine mode.

In addition, by running the twin-screw pump in turbine mode it has been also revealed that up to about 50% of the fluid power in the piping system can be recovered instead of being dissipated by PCVs.

Acknowledgment

The authors gratefully acknowledge the funding by "European Regional Development Fund" (project ID: EFRE-0801026/KEU-2-009B-RekuVent). Simulations were performed with computing resources granted by RWTH Aachen University within the computing project ID: bund0004

References

[1] Lima G M, Luvizotto E, Brentan B M 2017 Selection and location of Pumps as Turbines substituting pressure reducing valves *Renewable Energy* **109** 392-405
[2] Stefanizzi M, Capurso T, Balacco G, Binetti M, Camporeale S M Torresi M 2020 Selection, control and techno-economic feasibility of Pumps as Turbines in Water Distribution Networks *Renewable Energy* **162** 1292-1306
[3] Abhay R P 2013 Performance evaluation and CFD simulation of multiphase twin-screw pumps *PhD Thesis* Texas A & M University
[4] Tang, Qian, Zhang, Yuanxun 2014 Screw optimization for performance enhancement of a twin-screw pump *Proceedings of the Institution of Mechanical Engineers* E **228** 73-84
[5] Räbiger K, Holz F 2012 *2nd International Rotating Equipment Conference* Numerical flow analysis inside two- and three-spindle screw pump Düsseldorf 405-415
[6] Yan, Die, Kovacevic, Ahmed, Tang, Qian, Rane, Sham 2017 Numerical modelling of twin-screw pumps based on computational fluid dynamics *Proceedings of the Institution of Mechanical Engineers* C **231** 4617-4634
[7] Yan, Di, Kovacevic, Ahmed, Tang, Qian, Rane, Sham 2018 Numerical investigation of cavitation twin-screw pumps *Proceedings of the Institution of Mechanical Engineers* C **232** 3733-3750
[8] Zhang, Weibin, Jiang, Qifeng, Bois, Li, Hong 2020 Experimental and Numerical Analysis on Flow Characteristics in a Double Helix Screw Pump *Energies* **12** 3420
[9] Moghaddam, Ali H Kutschelis, Romuald Three-Dimensional 2021 Flow Simulation of a Twin-Screw Pump for the Analysis of Gap Flow Characteristics *Journal of Fluids Engineering* **144**
[10] Kovacevic, Ahmed, Stosic, Nikola, Smith, Ian K 2003 Three Dimensional Numerical Analysis of Screw Compressor Performance *Journal of Computational Methods in Sciences and Engineering* **3** 259-284
[11] Magnus R, Yoshihara H 1970 Inviscid transonic flow over airfoils *AIAA Journal* **8** 2157-2162
[12] Benek J, Buning P, Steger J 1985 *7th Computational Physics Conference* A 3-D chimera grid embedding technique 322-331
[13] Benek J, Steger J, Dougherty F C 1983 *6th Computational Fluid Dynamics Converence Danvers* A flexible grid embedding technique with application to the Euler equations 373-283
[14] Vetter G, Wirth W, Korner H, Pregler S 2000 *Proceedings of the 7th International Pump Users Symposium* Multiphase Pumping With Twin-Screw Pumps - Understand And Model Hydrodynamics And Hydroabrasive Wear Houston, USA 153-169
[15] Weiss J, Smith W 1995 Preconditioning applied to variable and constant density flows *AIAA Journal* **33** 2050-2057
[16] Menter F R 1994 Two-equation eddy-viscosity turbulence models for engineering applications *AIAA Joural* **32** 1598-1605
[17] Chakravarthy, Sukumar, Goldberg U, Peroomian O, Sekar B 1997 *13th Computational Fluid Dynamics Conference*
[18] Chakravarthy, Sukumar 1999 A unified-grid finite volume formulation for computational fluid dynamics *International Journal for Numerical Methods in Fluids* **31** 309-323
[19] Chakravarthy, Sukumar 1998 *34th AIAA/ASME/SAE/ASEE Joint Propulsion Conference and Exhibit*

IOP Conf. Series: Materials Science and Engineering 1267 (2022) 012013 doi:10.1088/1757-899X/1267/1/012013

Rheological modelling of viscoelastic fluid in a generic gap of screw pump

S Mehrnia, L Kerres, M Kuhr and P F Pelz

Chair of Fluid Systems, Technische Universität Darmstadt, Otto-Berndt-Straße 2, 64287 Darmstadt, Germany

E-mail: seyedmajid.Mehrnia@fst.tu-darmstadt.de

Abstract. In this study, the leakage of a non-Newtonian fluid, i.e. silicone oil, in a generic gap was numerically investigated. A CFD tool is used to determine the relationship between leakage flow, gap length and pressure difference. The investigated fluid is viscoelastic and its properties are modelled by a Maxwell equation. The Maxwell model can be used to precisely define the phenomenon of stress relaxation. Moreover, a comparison of the viscosity of measured data with simplified models shows that the Maxwell model is best suited for viscosity prediction. Furthermore, simulation results showed that at low pressures, leakage is reduced by decreasing the gap angle. However, this effect changes with increasing viscosity and relaxation time of the molecule. To determine the pressure drop, the Bagley plot is used. The results confirmed that as the shear rate increases, the elastic pressure drop values increase. In addition, the leakage flow increases with an increasing slenderness ratio.

1. Introduction

The effect of viscoelasticity on lubrication properties has recently become more important due to the effect of lubricant viscosity on energy efficiency. Thus, there are good practical reasons to investigate the general issue of viscoelastic effects in the narrow gaps of screw pumps. Due to their design, leakage occurs inside the pumps, between the rotating and non-rotating parts. Therefore, several experimental, analytical and numerical investigations have been carried out to understand the behaviour of leakage and to describe the flow of Newtonian fluids through these gaps. Rituraj et al. [1-3] modelled and experimentally validated gap losses for a generic gap in positive displacement pumps. Türk et al. [4] experimentally investigated the leakage behaviour of lobe pumps by measuring the volume flow through a blocked pump. Various influencing parameters such as geometry and pressure difference on the leakage were investigated for several pump types by calculating the leakage volume flow [5, 6]. Besides, also analytical and numerical studies have been carried out. It is common to model external gear pumps by using modes ranging from "lumped parameter" (0D) to Computational Fluid Dynamics (3D), cf. Rundo [7]. Those studies aimed to predict the leakage behaviour of the pumps.

It has been shown that there are several investigations on Newtonian fluids, but the literature dealing with non-Newtonian fluids in narrow gaps is scarce. Nevertheless, rotary positive displacement pumps are used daily to convey non-Newtonian viscoelastic fluids like plastic, paints, inks etc. For gear pumps, Rituraj et al. [8] extended an existing modell describing the displacement action of the unit to investigate the behaviour of leakage for non-Newtonian fluids. Rothstein et al. [9] showed that the influence of the inlet geometry on the behaviour of non-Newtonian fluids is crucial. Depending on the type of rotary displacement pump this geometry can strongly vary. Screw pumps, which have complex geometry are hardly researched. Furthermore, it can be shown that the tooth side gap has a major impact on the leakage behaviour, i.e. the efficiency. Hence, this paper aims to systematically analyse the leakage behaviour of non-Newtonian fluid within the generic gap pictured in Figure 1.

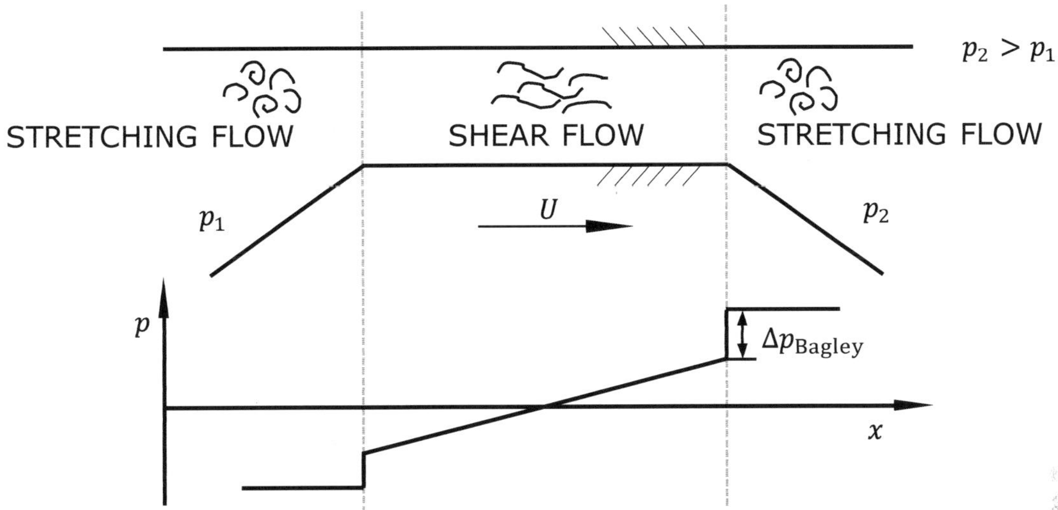

Figure 1. Generic gap inside a positive displacement pump including the stretches exerted on the fluid as well as the pressure in the narrow gap.

Within this paper, the leakage flow in the clearance gap between a stationary and a moving wall is investigated numerically by using Computational Fluid Dynamics (CFD) to find the relationship between leakage flow, gap length, and pressure difference. This work aims to predict the leakage flow rate in a generic gap of a screw pump. Figure 1 shows a schematic view of a generic gap as well as the pressure and the stretches exerted on the viscoelastic fluid. Here, the molecules of the fluid are stretched when entering and relaxed when exiting the gap. The stretching of the molecules leads to an elastic pressure loss in the inflow, i.e. the Bagley pressure loss, which is significant at high molecular mass liquids.

2. Methodology

Focussing on the generic gap, the flow inside is depending on eight parameters, cf. Figure 2.

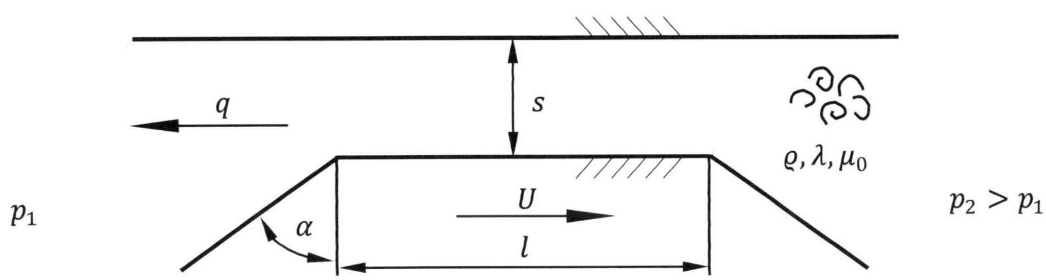

Figure 2. Schematic view of the generic gap and its parameters.

Here, s is the gap clearance, μ_0 is the constant viscosity of a Newtonian fluid, Δp is the pressure difference across the gap $\Delta p = p_2 - p_1$, q is the leakage flow rate (volume flow per width), ϱ is the density of the fluid, λ is the relaxation time of the molecules, U is the velocity of the moving wall, and

T is the characteristic process time. The viscoelastic flow is modelled using the commercial finite element software ANSYS POLYFLOW. The continuity and momentum equations are solved for the incompressible viscous flow field of the continuous phase. The flow is assumed to be stationary and the temperature is assumed to be constant. The fluid under consideration is a non-Newtonian viscoelastic fluid modelled by a Maxwell constitutive equation, cf. section 2.2. Two different molecular masses 10,000 and 20,000 u were simulated. Also, the simulations were carried out at two different gap heights $s = 20$ and 50 µm. The gap length was varied to represent five different slenderness rations s/l ranging from $s/l = 0.02$ to 0.1.

To model the flow in the generic gap a structured grid with approximately 12800 cells was generated by layering cross-sections perpendicular to the axial direction. The results are represented in dimensionless forms by the dimensionless pressure difference Δp^+, the dimensionless leakage flow rate q^+ and the Deborah number De

$$\Delta p^+ := \Delta p \frac{s^2 \varrho}{\mu_0^2}, \quad q^+ := \frac{q\varrho}{\mu_0}, \quad De := \frac{\lambda}{T} = \frac{\lambda U}{s} = \lambda\dot{\gamma}. \tag{1}$$

2.1. Non-Newtonian fluid

Unlike a Newtonian fluid, the viscosity of a non-Newtonian fluid is not constant. When sheared, the viscosity decreases or increases, depending on the type of fluid. This is due to the molecular chain structure of the liquid, which is one of the most characteristic features of non-Newtonian fluids. In some fluids, the molecules tend to orient to surfaces of maximum tension, causing the viscosity to decrease as the velocity gradient increases. These fluids are called shear-thinning fluids, meaning that the fluid becomes thinner as the shear rate increases. In other cases, the fluid is shear-thickening if the viscosity increases with an increasing velocity gradient. Therefore, the Newtonian model, i.e. modelling the liquid with a constant viscosity, is not applicable. The relationship between shear stress and shear rate can be represented as a power law for a viscous fluid:

$$\tau = m\dot{\gamma}^n. \tag{2}$$

Here, τ is the shear stress, $\dot{\gamma}$ is the shear rate, m is the flow consistency index, and n is the power-law index:

$n > 1$ fluid is shear-thickening,
$n = 1$ fluid is Newtonian,
$n < 1$ fluid is shear-thinning.

The shear-thinning behaviour of the fluid is described in terms of a relaxation time λ, defined by the ratio μ_0/G_0 of zero viscosity μ_0 and shear modulus G_0. The shear modulus is defined by the ratio between shear stress τ and shear strain γ, i.e. $G = \tau/\gamma$. Figure 3 shows the shear stress behaviour of a shear-thinning fluid in the function of shear rate. In fact, the relaxation time λ is equal to the inverse of the critical shear rate $1/\dot{\gamma}_0$.

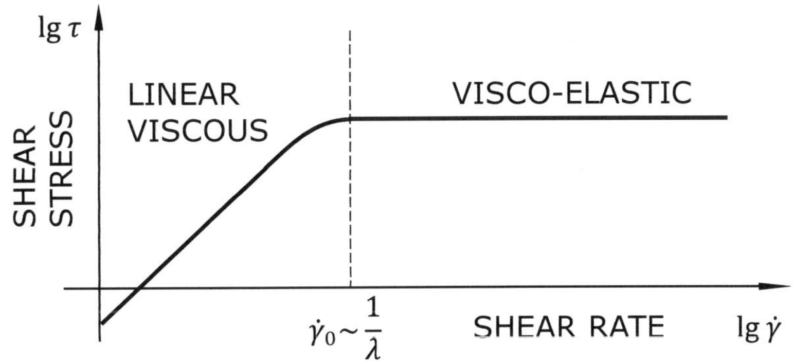

Figure 3. Shear stress behaviour of a shear-thinning fluid in the function of shear rate

In addition, polymeric liquids are viscoelastic, which means that the stress carried by a liquid component depends on the deformation history of this element. This effect is called the memory effect of polymeric fluids. These two nonlinearity and memory effects are responsible for considerable industry-related flow phenomena that must be predicted and controlled using a combination of mathematical and physical models.

We can divide non-Newtonian simulation models into two distinct groups. Some common models used in simulations are more of a fitting method, such as power-law and Carreau, which are called simplified models. Another class of viscoelastic fluid models with whole-order derivatives that have the locality property includes the Maxwell, Oldroyd-B, Kelvin-Voigt, Jeffreys, and Burgers models, which are called differential viscoelastic models. There are several internal variables or inputs for these models. The Maxwell equation is simple and shows less instability at the beginning of the simulation than other differential viscoelastic models. Table 1 has shown the differences between various viscoelastic models.

Table 1. Viscometric fluid functions.

Model	viscous shear stress	viscoelastic normal stress difference
Newton	$\tau = \mu_0 \, \dot{\gamma}$	no elastic effects
Prandl-Eyring	$\tau = \mu_0 \, \dot{\gamma} \dfrac{\text{arsinh}(\lambda\dot{\gamma})}{\lambda\dot{\gamma}}$	$\sigma_i - \sigma_j = 0$
Power-Law	$\tau = m \, \dot{\gamma}^n$	$\sigma_i - \sigma_j = 0$
Carreau	$\tau = \mu_0 \, \dot{\gamma} \, [1 + (\lambda\dot{\gamma})^2]^{(n-1)/2}$	$\sigma_i - \sigma_j = 0$
Maxwell	$\tau = \mu_0 \, \dot{\gamma} \frac{1}{\sqrt{1+(\lambda\dot{\gamma})^2}}$	$\sigma_i - \sigma_j \approx \dfrac{1}{2} \mu_0 \, \lambda \, \dot{\gamma}^2$

2.2. Maxwell model

One of the simple viscoelastic models is the upper convected Maxwell model, including a constant viscosity and a relaxation time. Maxwell model represents the physical properties of the fluid with a mechanical analogue, in terms of a damper and a spring, which can accurately define the phenomenon of stress relaxation, cf. figure 4. In this arrangement, under applied axial stress, the total stress and strain can be represented as the stress-strain relation by equation (3):

$$\tau + \lambda \, \dot{\tau} = \mu_0 \, \dot{\varepsilon}. \tag{3}$$

where λ is relaxation time, μ_0 is zero shear viscosity, $\dot{\varepsilon}(t)$ is the rate of deformation tensor, and $\dot{\tau}(t)$ is upper convected derivative, cf. Li et al. [10]. Input values of the Maxwell model are the zero viscosity μ_0 and the relaxation time of the molecule λ. On the one hand, the equation of the Maxwell model is simple and shows less instability at the start of the simulation. On the other hand, however, the model is complicated from a computational point of view and is only recommended when little knowledge about the liquid is available.

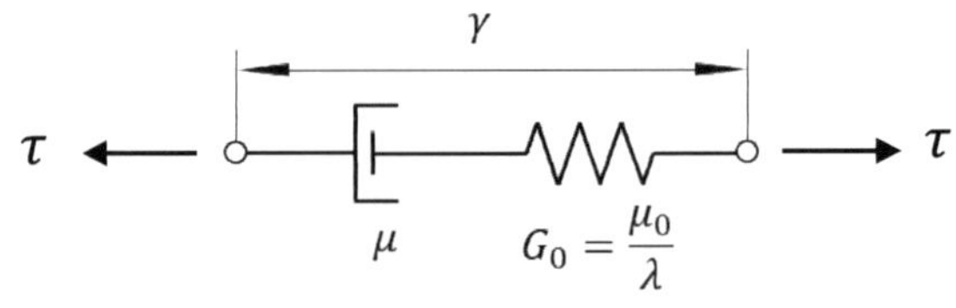

Figure 4. Mechanical analogue of the Maxwell model.

3. Measurement data analysis

The PhD thesis of Andrä [11] was used for the assumption of the measured rheological data and the derivation of the required viscosity from the storage and loss modulus curves G' and G'' given in his work. Andrä [11] has conducted extensive research for silicone oil with a large number of measurements in the selected temperature, molecular mass and shear rate range from a tribological point of view. The authors of this paper have judged the viscosity, storage and loss modulus results with the applied rheological methods in Andrä's work [11] to be reliable, accurate and well studied. The data of the storage and the loss moduli as a function of angular frequency were recorded during the measurements and are shown in figures 5 and 6.

The values of the complex G^* moduli as a function of the angular frequency Ω can be calculated using the storage moduli G' and the loss moduli G'' as a function of angular frequency Ω according to equation (4). Figures 7 and 8 show the master curve of the storage and the loss modulus of silicone oil. The dynamic viscosity μ as a function of shear rate $\dot{\gamma}$ can be derived from the absolute value of the μ^* complex dynamic viscosity data based on the Cox-Merz rule, cf. equation (7) [11]. The Cox–Merz rule is a simple relationship that signifies that complex viscosity and shear viscosity are comparable when the steady shear rate is the same as angular frequency.

$$|\mu^*(\Omega)| = \frac{|G^*|}{\Omega} = \frac{\sqrt{G'^2 + G''^2}}{\Omega} \tag{4}$$

$$G' = \frac{\mu}{\lambda} \frac{(\lambda\Omega)^2}{1 + (\lambda\Omega)^2} \tag{5}$$

$$G'' = \frac{\mu}{\lambda} \frac{\lambda\Omega}{1 + (\lambda\Omega)^2} \tag{6}$$

$$|\mu^*(\Omega)| = \mu\,(\dot{\gamma}) \;\;:\;\; \Omega = \dot{\gamma} \tag{7}$$

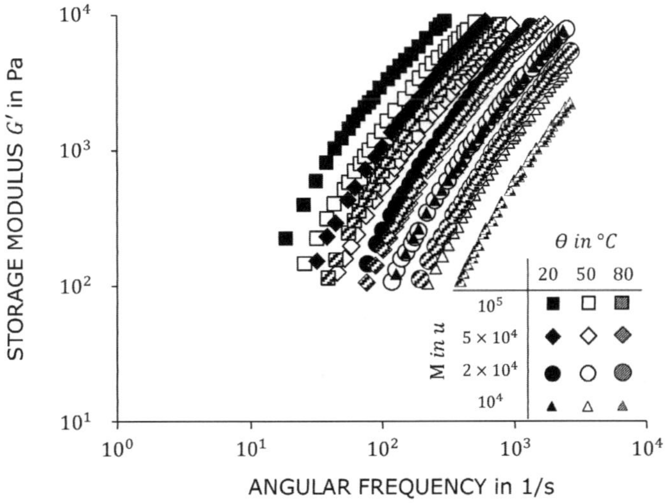

Figure 5. Measured values of storage modulus of silicone oil for different temperatures and molecular masses [11].

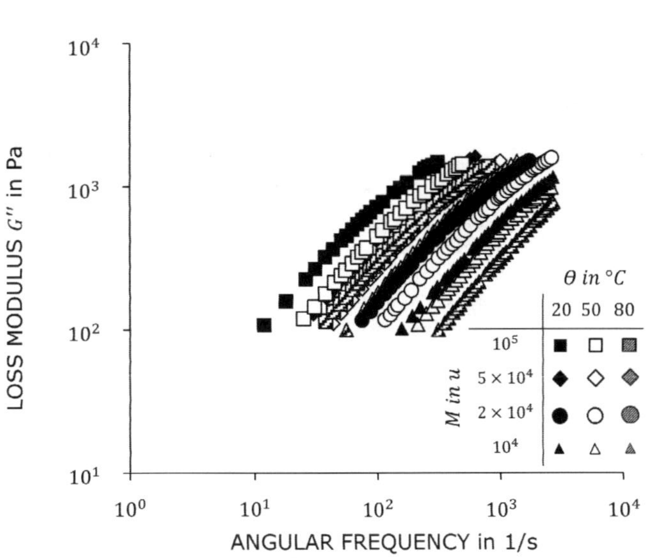

Figure 6. Measured values of loss modulus of silicone oil for different temperatures and molecular masses [11].

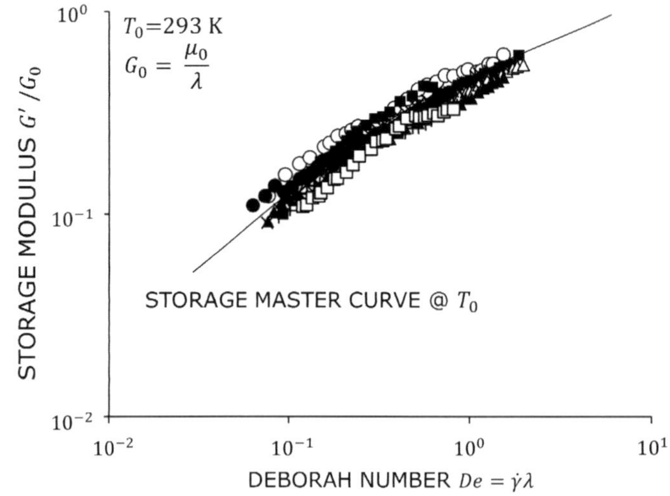

Figure 7. Storage modulus master curve for silicone oil.

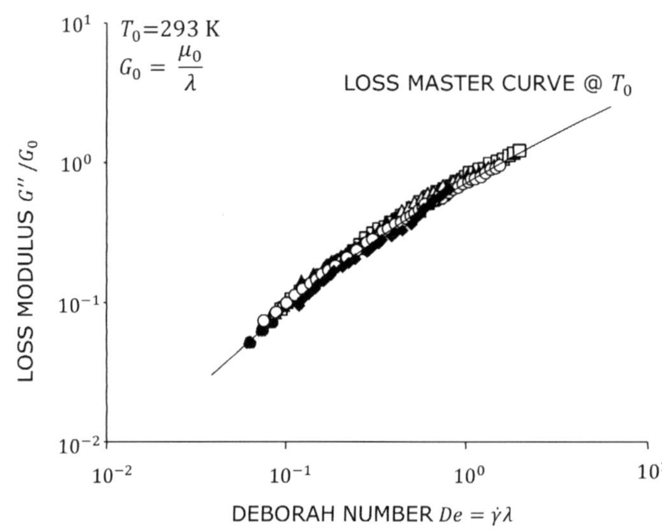

Figure 8. Loss modulus master curve for silicone oil.

Figure 9 shows a comparison of the dimensionless viscosity of measured data with different viscoelastic models. As it was explained, it can be seen that the Maxwell model is best suited for viscosity prediction. It is also evident that the other, simplified models are lacking in the reliable prediction of non-Newtonian behaviour. Here, for the Carreau model, the best power law index to fit the measurement data is $n = 0.3$.

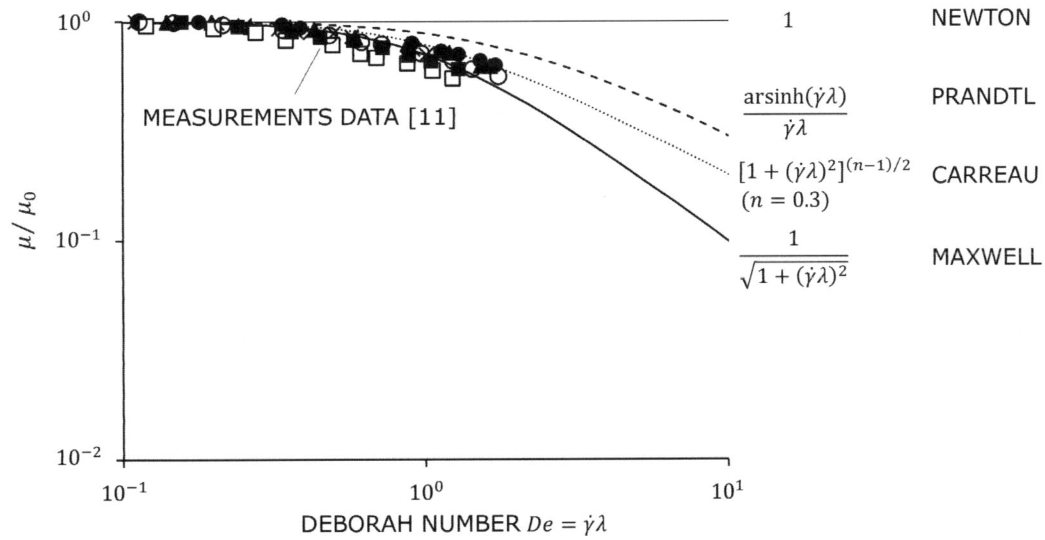

Figure 9. Comparison of dimensionless viscosity of measurement data with different viscoelastic models.

3.1. Relaxation time

The relaxation time of a molecule λ is a function of molecular mass and temperature. In viscoelastic modelling by Maxwell, the zero viscosity μ_0 and relaxation time λ are inputs for the simulation. Hence, it is necessary to calculate the relaxation time of silicone oil from the measurement data. Equation (5) was employed to calculate the relaxation time. Figure 10 has shown Arrhenius plot for the relaxation time of silicone oil molecules in various molecular masses. As it can be seen, the logarithmic ratio of relaxation time is a linear function of the inverse temperature, called the Arrhenius law. However, at high molecular mass, the variations are non-linear.

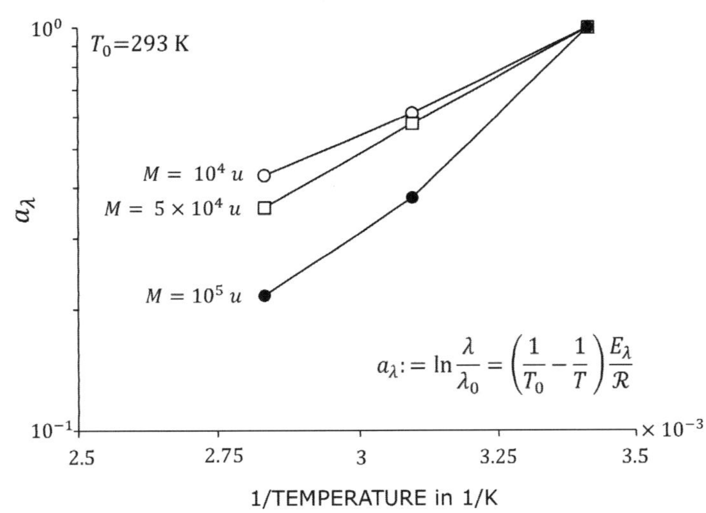

Figure 10. Arrhenius plot for relaxation time of silicone oil molecule in various molecular masses.

4. Simulation

This section investigates the effect of different parameters on the generic gap. Since numerous two-dimensional and three-dimensional simulations have been performed on screw pumps, this work has focused only on the geometry of the generic gap and its parameters. Figure 11 shows the dimensionless ω values in different pressures for the Carreau and Maxwell models. Using measurement data in figure 9, it has been shown that the Carreau model does not have high accuracy in high shear rates. For low molecular masses of 10,000 u, the leakage values are close together for these two models. With increasing molecular mass, the difference in values between the two methods is very large. In figure 11, the difference in the leakage values is shown for the two models. The Carreau model calculated higher leakage values for the higher pressure difference parameter. Here, the power-law index of the Carreau model is equal to 0.3.

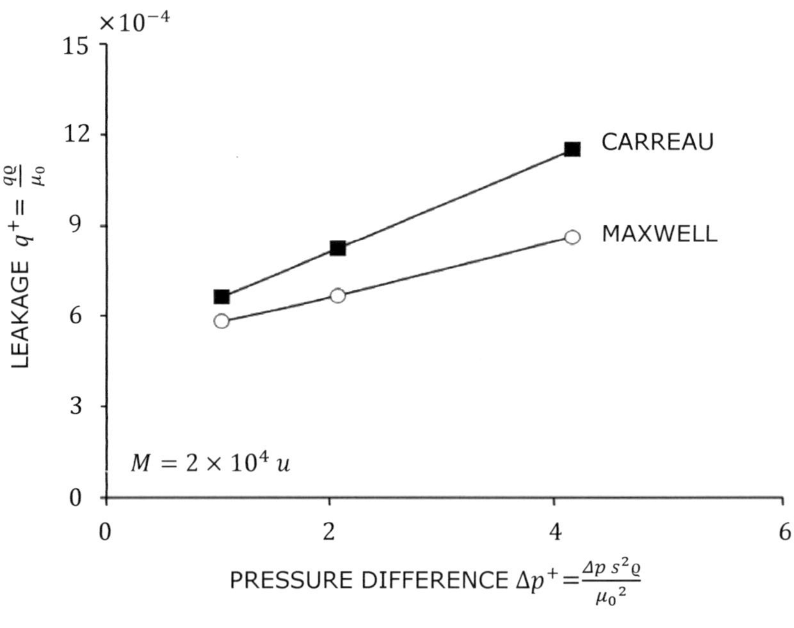

Figure 11. Comparison of dimensionless leakage for Maxwell and Carreau in different pressures

Figure 12 shows velocity values in the generic gap for two different gap angles. The gap angle α was described in figure 2. At low-pressure differences, reducing the gap inlet angle reduces leakage. But this effect changes with increasing viscosity and relaxation time of the molecule. Therefore, a general diagram and conclusion have not been provided here. As can be seen in figure 12, the velocity of the outlet of the gap has decreased as the angle decreases. Hence, we can state that the leakage increased with increasing the gap inlet angle.

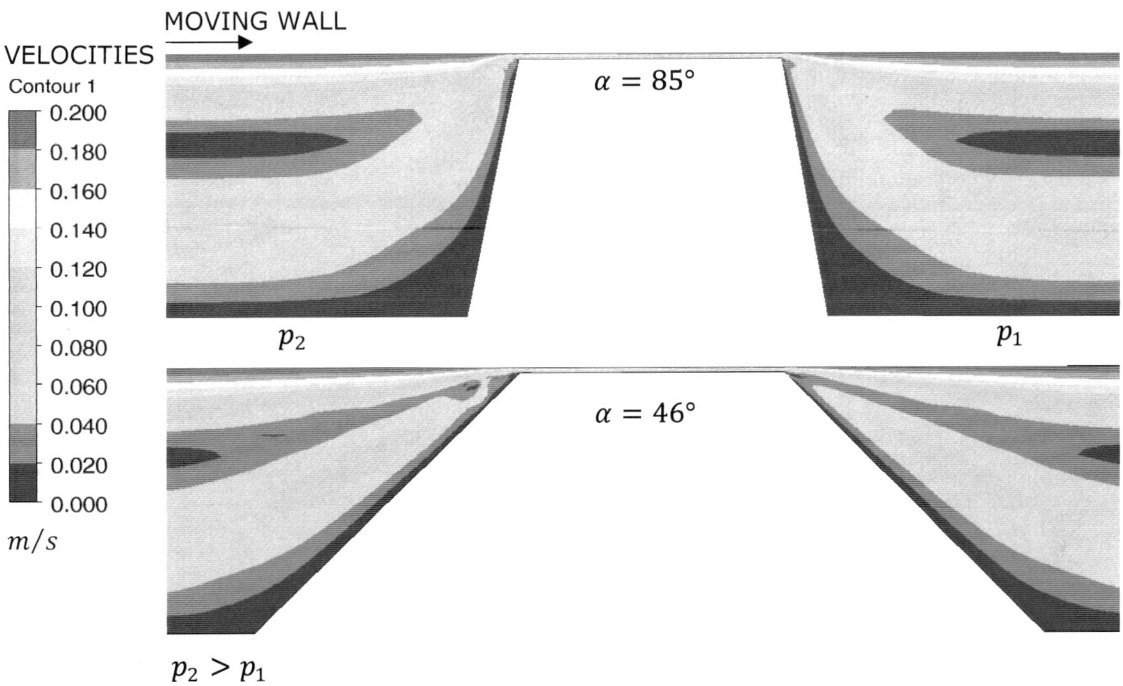

Figure 12. Velocity values in the generic gap for two different gap angles (s/l= 0.02, M= 20,000 $\boldsymbol{u}$, Δp = 20 kPa)

Figure 13 shows dimensionless pressure values across the gap for different Deborah numbers. As it is shown in the figure, with increasing the Deborah number, the pressure drop decreased. In order to explore the reasons, Bagley pressure and elastic loss were studied. Pressure losses versus different gap lengths have shown in figure 14. The Bagley plot supports the separation of the viscose pressure drop and the inlet pressure drop. To determine the inlet pressure drop, the pressure drop is applied with various length ratios of the same gap height but different lengths and extrapolated to zero. Contraction flow at the gap entrance region causes an extra pressure drop due to the stretching of fluid elements. In principle, the significance of the entrance pressure drop compared to the pressure drop across the gap decreases with increasing l/s or relative gap length. The large relative gap length, however, can lead to other errors; the longer the gap is, the greater the pressure effect becomes.

In this graph, the pressure drop is increased with the increasing shear rate. This pressure drop is due to inertia and elastic loss across the gap. For calculating the elastic pressure drop, which depends on the viscosity and relaxation time of fluid, the inertia loss or Carnot loss was calculated and then subtracted from the Bagley pressure drop. For a non-Newtonian fluid, the Carnot loss is negligible compared to the elastic pressure loss. Figure 15 has shown the calculated values of elastic pressure loss at different shear rates. With increasing the shear rate, the values of elastic pressure drop also increased.

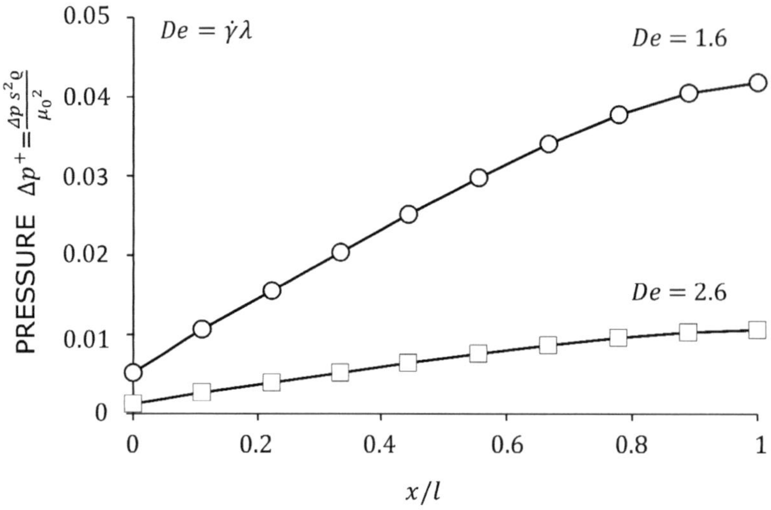

Figure 13. Dimensionless pressure values across the gap for different Deborah numbers.

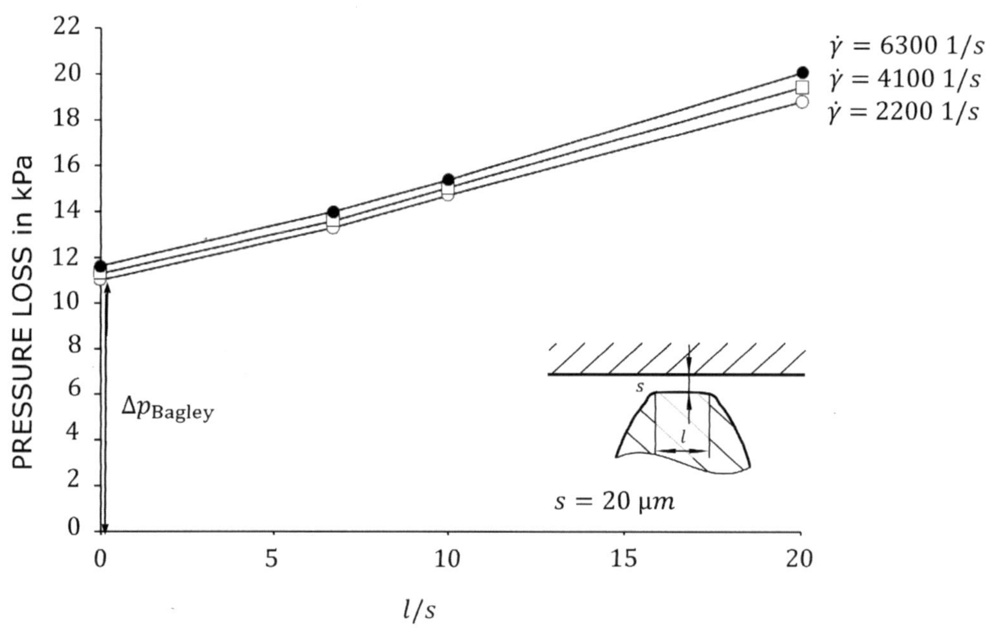

Figure 14. Pressure loss of inflow versus different gap lengths.

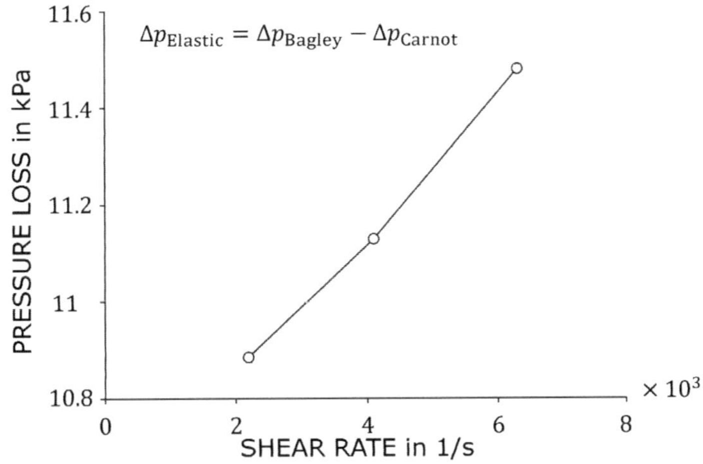

Figure 15. Elastic pressure loss versus shear rate.

Figure 16 shows leakage values versus pressure differences in various slenderness ratios. In this figure, the slenderness ratio s/l is equal to gap height to gap length. As can be seen in this graph, the leakage increases with increasing the slenderness ratio. Also, increasing the pressure increases the leakage.

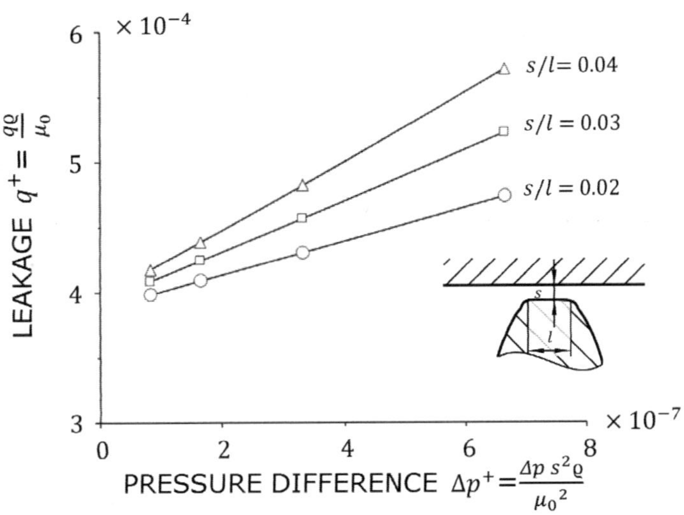

Figure 16. Leakage values versus pressure differences in different slenderness ratios.

5. Conclusion

In the present article, the leakage flow in the generic gap between a stationary wall and a moving wall was investigated numerically by a CFD tool designed for viscoelastic fluid dynamics problems to find the relationship between leakage flow, gap length, and pressure difference. For solving the incompressible, viscous flow field of the continuous phase, the equations for mass and momentum conservation were solved. Also, the flow was assumed to be steady, and the temperature was constant. The non-Newtonian and viscoelastic fluid was modelled by a Maxwell constitutive equation.

Several simplified models used in simulations are the more fitting method, like Carreau. Other viscoelastic fluid models like the Maxwell model, called differential viscoelastic models, have the locality property. Maxwell's model represents the physical characteristics that are equivalent to a damper and a spring in series, which can accurately define the stress relaxation phenomenon. The equation of Maxwell is simple, however, this model is complicated from a computational point of view. A comparison of the viscosity of measurement data with different viscoelastic models shows that Maxwell has the best behaviour for viscosity prediction. It should be noticed, with changing the power-law index, could be had a different fitting for the Carreau model.
The simulation results showed that at low pressures, reducing the gap inlet angle reduces leakage. But this effect was unknown with increasing viscosity and relaxation time of the molecule. The Bagley plot supports the separation of the viscose pressure drop and the inlet pressure drop. To determine the inlet pressure drop, the pressure drop is applied with various length ratios of the same gap height but different lengths and extrapolated to zero. The large relative gap length, however, can lead to other errors; the longer the gap is, the greater the pressure effect becomes. This pressure drop is due to inertia and elastic loss across the gap. For a non-Newtonian fluid, the Carnot loss is negligible compared to the elastic pressure loss. Results showed that with increasing the shear rate, the values of elastic pressure drop also increased. Besides, the leakage increases with increasing the slenderness ratio. We can state with increasing the gap length, the leakage flow rate decreases. In summary, we could conclude:

- The Maxwell model is recommended for use in high molecular masses, however, its computational cost is extremely high.
- By increasing the gap length, the leakage would decrease.
- At low-pressure differences, reducing the gap inlet angle reduces leakage. However, with increasing viscosity, the effect of angle is not significant.
- Leakage decreases with increasing Deborah number.

Acknowledgements
The presented results were obtained within the research project "Verschleißverhalten bei partikelbeladenen Fluiden und Leckage bei nicht-Newton'schen Fluiden in rotierenden Verdrängerpumpen", FKM no. 7051800, funded by Verband Deutscher Maschinen- und Anlagenbau (VDMA), which we carried out for the trade association Pumps + Systems.

6. References

[1] Rituraj F and Vacca A 2019 Modelling and Validation of Tooth Tip Leakages in Gerotor Pumps *ASME-JSME-KSME*10.1115/AJKFluids(2019)**553**.
[2] Rituraj F and Vacca A 2020 An Investigation of Tooth Tip Leakages in Gerotor Pumps: Modeling and Experimental Validation *J. of Verification, Validation and Uncertainty Quantification* 5. 10.1115/1.4046295.
[3] Rituraj F and Vacca A 2021 Investigation of flow through curved constrictions for leakage flow modelling in hydraulic gear pumps, *Mechanical Systems and Signal Processing* (2021)**153**.
[4] Türk M and Verhülsdonk B 2006 Gap leakage behaviour of helical vane rotary lobe pumps *J. World Pumps* (2006) **475**.
[5] Nagamura K, Ikej K, and Tutulan F G 2004 Design and performance of gear pumps with a non-involute tooth profile *Proceedings of the Institution of Mechanical Engineers, Part B: Journal of Engineering Manufacture* **218**(2004)699-711.
[6] Wahab A 2009 Analytical Prediction Technique for Internal Leakage in an External Gear Pump *Proceedings of the ASME Turbo Expo* **5**(2009)85-92.
[7] Rundo M 2017 Models for Flow Rate Simulation in Gear Pumps: A Review *Energies* 10.1261. 10.3390/en10091261.
[8] Rituraj F and Vacca A 2018 External gear pumps operating with non-Newtonian fluids: Modelling and experimental validation *Mechanical Systems and Signal Processing* (2018)**106**.

[9] Rothstein J P McKinley G H 2001 The axisymmetric contraction–expansion: the role of extensional rheology on vortex growth dynamics and the enhanced pressure drop *J. Non-Newtonian Fluid Mech* **98** (2001)33–63.

[10] Li X K, Luo Y, Qi Y and Zhang R 2011 On non-Newtonian lubrication with the upper convected Maxwell model *Applied Mathematical Modelling* **35**(2011)5 2309-2323.

[11] Andrä R 1982 Torsionsschwingungsdämpfer in viskoelastischen Flüssigkeiten, insbesondere in Flüssigkeiten zweiter Ordnung (Book) Ph.D. thesis an der Hochschule Darmstadt.

Conjugate heat transfer analysis of a twin-screw compressor with 4-6 configuration and internal cooling channels

A Saravana[1*], H Liu[1], N Able[2], J Collins[2], E A Groll[1] and D Ziviani[1]

[1] Ray W. Herrick Laboratories, School of Mechanical Engineering, Purdue University West Lafayette, 47907-2099, USA
[2] Ingersoll Rand, Davidson, NC, 28036, USA

E-mail: saravan0@purdue.edu, liu1460@purdue.edu, groll@purdue.edu, dziviani@purdue.edu, nicholas_able@irco.com, james_collins@irco.com

Abstract. Twin-screw compressor technology is widely employed in commercial and industrial HVAC&R applications as well as in air-compression and gas processing industries. Twin-screw compressors are characterized by relatively high efficiencies at part- and full-loads, and high reliability. However, innovative designs and higher efficiency are needed to further reduce power consumption, improve the compression process, and minimize internal losses. This study focuses on developing a simulation model of a twin-screw compressor with internal cooling channels to achieve a near isothermal compression process. A complex coupling between compression process and heat transfer occurs during the operation of the compressor. To analyze the compressor and the interaction between solid (i.e., rotors) and fluid phases (i.e., air and coolant), a CFD model with conjugate heat transfer has been developed and validated. The CFD model is used to predict compressor performance parameters such as isentropic efficiency, heat transfer rate, work input and pressure forces acting on the rotors. The performance of the twin-screw compressor with internal cooling channels is compared with a conventional twin-screw compressor.

1. Introduction

Positive displacement machines such as twin-screw compressors are used extensively in various gas compression applications [1]. Twin-screw compressors involve working chambers created between the gearing of male and female screw rotors that decreases in volume from suction to discharge, leading to the effect of compression [2].

Recently, computational fluid dynamics (CFD) has been gaining increasing popularity in modelling positive displacement machines such as scroll [3], rotary rolling-piston [4], and twin-screw compressors [5]. CFD approaches provide deep insights into the compression process and flow regimes, but also bring major challenges due to the need for deforming grids to describe the volume changes of the working chambers. Regarding twin-screw compressors, Kovacevic [6-7] proposed a method to generate algebraic, adaptive block-structured grid using analytical rack generation methods which led to the development of the software Screw Compressor Rotor Grid Generator (SCORG) [8]. Although large body of work exists for single-phase simulations, multiphase CFD simulations still yield to poor convergence, very long simulation time, and unsatisfactory mass/energy conservations. As an example, Ding and Jiang [9] presented a full 3D transient CFD model of a twin-screw compressor with oil injection. The study employed a volume of fluid (VOF) approach through the software Simerics MP+ [10].

Despite the significant progress on modelling the internal behaviour of compressors, temperature variation during the compression process and the interactions with the solid parts are still challenging to simulate. To accurately predict the thermal behaviour of a compressor, heat transfer between metal components and the working fluid needs to be properly evaluated in CFD simulation. The complexity with this type of model is that the solid side simulation takes longer to converge at a steady-state solution than the fluid side, making it computationally expensive. Ding et al. [11] addressed this concern by modelling conjugate heat transfer (CHT) between the rotor walls and gas chamber using the mixed timescale coupling approach. According to the approach, CHT simulations solves fluid and solid in a sequence of separated models.

The compression process in a regular twin-screw compressor leads to discharge temperatures much higher than that of suction. This makes it diverge away from ideal isothermal behaviour. In this study, a twin-screw compressor with internal cooling channels is modelled to enable isothermal compression and the internal cooling process is studied using the CHT modelling approach. The coolant flowing through the rotors removes the compression heat transfer loads and reduces the surface temperature, which reduces gas discharge temperatures. CHT is used to model heat transfer between solid, liquid and gas phases of the internally cooled screw compressor. The purpose of this research is to conduct computational fluid dynamics to study the gas compression process and internal coolant flow as well as CHT to study the heat transfer process between gas to solid to liquid coolant and analyse how much the efficiency can be improved compared to a solid case and dry running simulation. A full conjugate heat transfer with CFD flow analysis is presented and discussed in detail. To resolve for the varied timescales among fluid, gas and solid CHT, an innovative method of *Super-Cycling* is used and discussed in major detail.

2. Twin screw compressor model

The twin-screw compressor considered in this work has a 4-6 configuration and both male and female rotors feature internal cooling channels to flow coolant through. The compressor working fluid is wet air at 60% relative humidity and specific heat of $1.021\ \mathrm{kJ/kg-K}$ modelled as an ideal gas mixture with an isentropic coefficient of 1.4 and the coolant is water.

The male and female rotors were both additively manufactured with Inconel. For a more conservative approach, the material for both rotors used in the simulation is carbon steel, since carbon steel is a weaker alloy than a nickel-chromium based alloy such as Inconel. Three models needed to be developed to describe the compressor: 1) the geometry model describes the modified geometry of rotors to accommodate the internal fluid domain; 2) theory behind the conjugated heat transfer model in CFD; 3) a lumped heat transfer model describing the heat transfer process using a 1D approach.

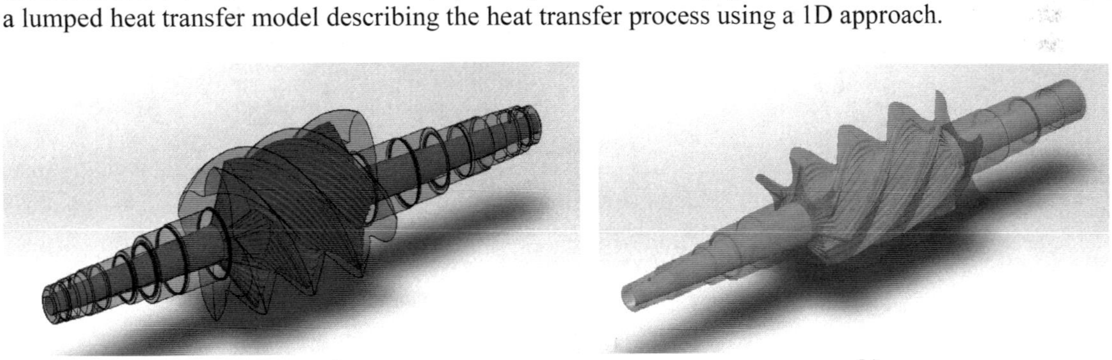

a) b)

Figure 1. CAD model of **a)** male and **b)** female rotors with internal fluid domain projections.

2.1 Geometry model

The solid rotors of the baseline compressor have been modified to include internal cooling channels (i.e. hollow rotors). The interior surface of the rotors follows a wavy pattern to maximize convective heat transfer to the coolant. The internal cooling channels remove the compression heat loads to approach an isothermal compression. The wavy geometry of the channels has been described by Saravana et al. [12] in greater details. The final rotor geometries are displayed in Figure 1.

2.2 Heat transfer modeling

The hollow screw compressor has three phases interacting simultaneously at any point in time: solid rotors and case, liquid coolant and gas being compressed. To simulate the various interactions, a conjugated heat transfer problem needs to be formulated and implemented within a CFD framework. Energy conservation equations are employed for the fluid (Equation 1) and solid (Equation 2) domains.

$$\frac{\partial \rho e}{\partial t}+\frac{\partial u_j \rho e}{\partial x_j}=\underbrace{-P\frac{\partial u_j}{\partial x_j}}_{\text{Pressure work}}+\underbrace{\sigma_{ij}\frac{\partial u_j}{\partial x_j}}_{\text{Viscous dissipation}}+\underbrace{\frac{\partial}{\partial x_j}\left(K\frac{\partial T}{\partial x_j}\right)}_{\text{conduction}}+\underbrace{\frac{\partial}{\partial x_j}\left(\rho D\sum_m h_m\frac{\partial Y_m}{\partial x_j}\right)}_{\text{Species diffusion}}+S \quad (1)$$

$$\frac{\partial \rho e}{\partial t}=\underbrace{\frac{\partial}{\partial x_j}\left(K\frac{\partial T}{\partial x_j}\right)}_{\text{conduction}}+\underbrace{S}_{\text{Heat source}} \quad (2)$$

The heat transfer to and from the interface boundary is balanced and hence the solid and liquid energy conservation equations are coupled using interface boundaries. The heat conduction on the liquid side of the interface at any point in the domain is equated with the convective flux and hence the convective coefficient is calculated. Other aspects of the simulation are outlined below:

- Turbulence model: Reynolds Averaged Navier Stokes (RANS) two equation $k-\varepsilon$ [13].
- Wall model: Law of wall

2.3. Computation of efficiencies

Indicated power in CFD can be calculated using the average pressure and viscous torques for male and female rotors respectively. The total indicated power is computed from the total male and female rotor torques using the Equation 5.

$$\dot{W}_{ind}=\frac{2\pi N_1\left(\tau_1+\frac{z_1}{z_2}\tau_2\right)}{60} \quad (3)$$

where $\dot{W}_{ind}$ is indicated power, N_1 is the male rotational speed in RPM, T_1 and T_2 are total torques for male and female rotors, Z_1 and Z_2 are the number of lobes of male and female rotors respectively. The isothermal and isentropic power and their respective efficiencies are calculated using the following equations,

$$\dot{W}_{isoth}=\dot{m}_{suc}\gamma T_{suc}\cdot ln\left(\frac{P_{dis}}{P_{suc}}\right) \quad (4)$$

$$\dot{W}_{isen\,(ad)}=\dot{m}_{suc}\gamma\left(\frac{k}{k-1}\right)T_{suc}\left[\left(\frac{P_{dis}}{P_{suc}}\right)^{\frac{k-1}{k}}-1\right] \quad (5)$$

$$\eta_{isoth}=\frac{\dot{W}_{isoth}}{\dot{W}_{ind}} \quad (6)$$

$$\eta_{isen(ad)}=\frac{\dot{W}_{isen\,(ad)}}{\dot{W}_{ind}} \quad (7)$$

2.4 Lumped heat transfer model

The heat transfer can be analysed using a lumped heat transfer one dimensional model as described in the Figure 2. The lumped heat transfer model is useful to predict the thermal conductivities of the male and female rotors. Thermal conductivities obtained from the 1D lumped heat transfer model can be compared with external resources for carbon steel. The heat transfer between the interfaces is modelled using the thermal resistance circuit where the convective, conductive resistances and heat transfer rate are given by the following equations.

$$R_{conv} = \frac{1}{hA} \tag{8}$$

$$R_{cond} = \frac{L}{kA} \tag{9}$$

$$\dot{Q} = \frac{\Delta T}{R} \tag{10}$$

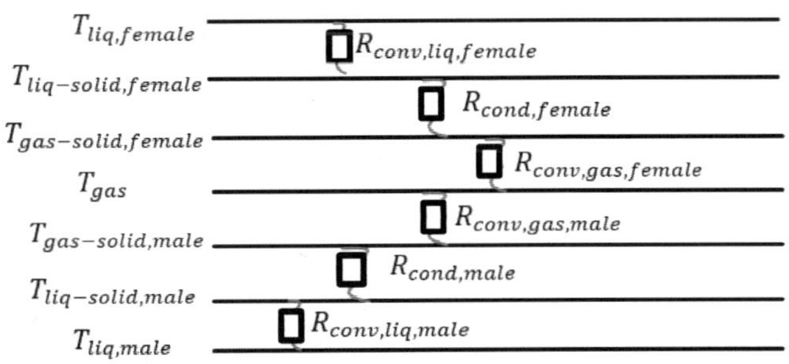

Figure 2. Schematic representation of the lumped capacitance model.

To develop a 1D approach heat transfer model, several assumptions have to be made: 1D heat transfer through the domain; heat transfer rate from gas to solid is same as solid to liquid; heat fluxes, convective coefficients and interface temperatures are averaged across the whole rotor and at steady state. The heat transfer and temperature values are derived from CFD whereas the convective resistance can be calculated using the convective coefficient derived from CFD and known surface areas. Having known the wall thickness of the rotors, using heat transfer balances as in Equations 11-12, the thermal conductivities of the solid rotors can be calculated.

These thermal conductivities can be validated using external resources, thus validating the heat transfer model. The results of the 1D heat transfer approach will be discussed later in the study.

$$\dot{Q}_{gas,male} = \frac{T_{gas,male} - T_{liq,male}}{R_{conv,gas,male} + R_{cond,male} + R_{conv,liq,male}} \tag{11}$$

$$\dot{Q}_{gas,female} = \frac{T_{gas,female} - T_{liq,female}}{R_{conv,gas,female} + R_{cond,female} + R_{conv,liq,female}} \tag{12}$$

3. Simulation setup

The CFD simulation used in this study is performed in CONVERGE 3.0 [14]. The simulation setup in CONVERGE involved several details. To verify the simulation results, it was important to make sure that the mesh was completely resolved. This was concluded by observing the Y^+ numbers at the fluid interfaces. The heat transfer computation in the software is done through a super-cycling process which will also be discussed along with the necessary boundary conditions for setting up the simulation.

3.1 Boundary conditions and inputs

Figure 3 describes the assembly with the twin-screw compressor along with housing and walls for the solid (dry running) and hollow rotor (internally cooled) cases. The heat transfer process is briefly described using the lumped heat transfer model. This heat transfer process is modelled in CFD using various boundary conditions. Table 1a-b summarizes all the boundary conditions employed. Rotation of the screw rotors causes positive displacement which drives the gas flow and progressive reduction of volume leads to gas compression. In this study, the internal coolant is chosen to be water, but this can be easily modified in future applications.

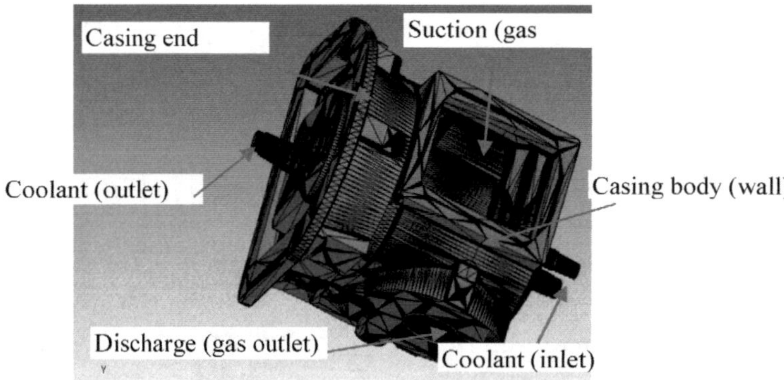

Figure 3. Twin-screw compressor assembly with boundary conditions.

Table 1. Heat transfer and flow boundary conditions for **a)** hollow rotor and **b)** solid rotor.

Table 1a)	**Gas inlet**	**Gas outlet**	**Liquid inlet**	**Liquid outlet**	**Casing (wall)**	**Shaft (wall)**
Temperature (K)	293		278		Convection $h = 10\frac{W}{m^2K}$ $T_\infty = 300$ K	Insulated
Pressure (Pa)	101 325	329 200		101 325		
Velocity (m/s)	Zero normal gradient	Zero normal gradient		Zero normal gradient	Stationary	Rotating
Mass flow rate (kg/s)			0.8			

Table 1b)	**Gas inlet**	**Gas outlet**	**Casing (wall)**	**Shaft (wall)**
Temperature (K)	293		Convection $h = 10\frac{W}{m^2K}$ T = 300 K	Insulated
Pressure (Pa)	101 325	329 200		
Velocity (m/s)	Zero normal gradient	Zero normal gradient	Stationary	Rotating
Mass flow rate (kg/s)		0.8		

3.2 Meshing

The CONVERGE software employs a meshing algorithm that is different from most CFD codes. CONVERGE maintains the original surface definition such that the volumes and areas are correct regardless of the mesh resolution and which grid control strategies have been used. The software automatically generates a body-fitted volume mesh during run time which exactly fits the surface geometry, using a patented cut-cell technique. Figure 4(a) represents how the meshing is done in the software. Interior cells are orthogonal hexahedra which remain stationary to avoid numerical diffusion associated with a moving mesh. Cells intersected by the geometry surface are cut by the surface to form cut polyhedra of any size.

The simulation included a variety of meshing methods to resolve the grid with a total of around 1,300,000 cells. Some methods employed include:

- Base grid: original size of grid which is the largest grid size available in the simulation. The base grid size used in this simulation is around 8 mm × 8 mm × 8 mm.
- Embedded layer: refine a specified portion of the grid at a specified time. Used up to 3 layers near the fluid boundaries to refine Y^+ numbers.
- Adaptive meshing: automatically refine the grid based on local flow conditions such as temperature or velocity.

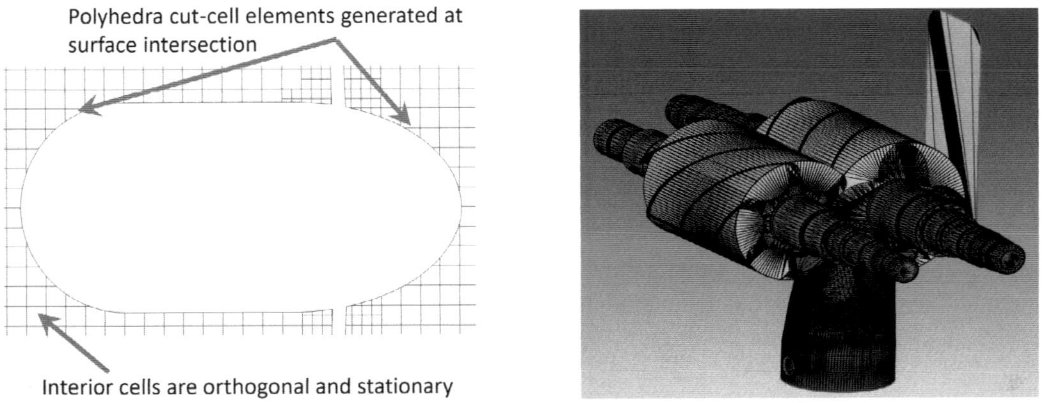

Figure 4 (a). Example of grid generation in CONVERGE; **(b)** Mesh employed for the hollow-rotor twin-screw compressor geometry (adapted from [14]).

3.3 Super-cycling between solid, liquid and gas

Conjugate heat transfer between solid, liquid and gas can be quite computationally expensive as the solid side takes much longer to arrive at a steady state than the fluid interfaces. Super-cycling within CONVERGE is a feature that solves complicated conjugate heat transfer simulations to achieve steady-state within fewer cycles by iterating between transient and steady-state solvers. It uses the transient solver to solve the solid and fluid side; temporarily freezes the transient solver to solve for the solid side temperature; uses this temperature to restart the transient solver to solve for the solid and fluid side and the cycle is repeated. Figure 5 illustrates this numerical approach.

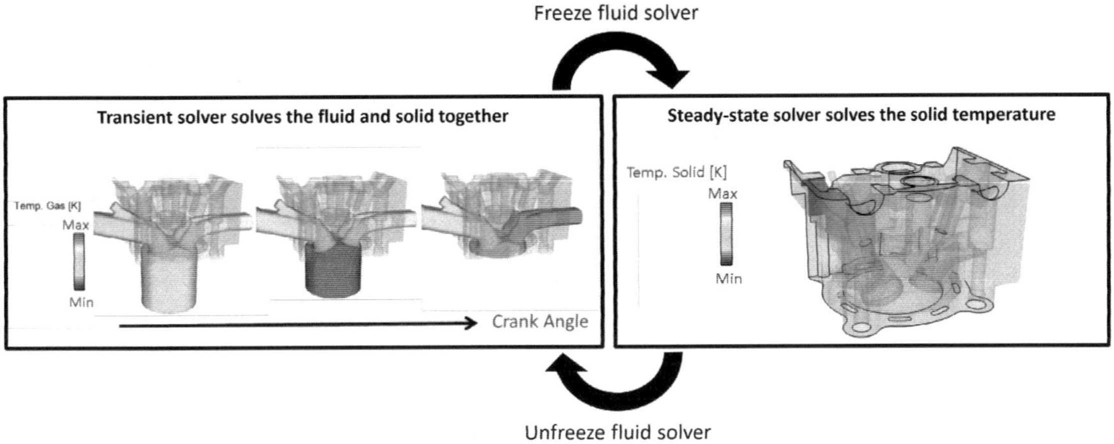

Figure 5. Function ideology behind super-cycling simulation (adapted from [14]).

4. Results and discussions

The detailed model verification, performance comparison, as well as insights of the flow regime of both rotor geometries are reported and discussed.

IOP Conf. Series: Materials Science and Engineering 1267 (2022) 012014 doi:10.1088/1757-899X/1267/1/012014

4.1 Mesh verification

As the mesh involved more than 1 million cells and the compression process is complicated, it can lead to major turbulence flows in the compression chamber. It is important to verify that the simulation resolves turbulent flows effectively. This can be done by analysing the Y^+ numbers at the gas and liquid boundaries. Table 2 displays average Y^+ numbers for liquid side and gas side for male and female hollow rotors as well as that on the gas side for solid male and female rotors. The numbers are within a reasonable range with a minimum of 120 and maximum of around 420, concluding that the turbulence effects are well resolved with the chosen mesh.

Table 2. Summary of Y+ numbers for hollow and solid rotors

	Hollow Rotor				**Solid Rotor**	
	Female rotor		*Male rotor*		*Female rotor*	*Male rotor*
Y+ numbers	*Liquid*	*Gas*	*Liquid*	*Gas*	*Gas*	*Gas*
Maximum	231.6	135.0	475.0	162.0	127.2	153.9
Minimum	225.8	129.0	457.0	156.0	120.5	144.1

Table 3. Lumped heat transfer analyses.

	Female rotor				**Male rotor**			
Quantity	*Liquid*	*Solid-liquid interface*	*Solid-gas interface*	*Female rotor Gas*	*Male rotor Gas*	*Solid-gas interface*	*Solid-liquid interface*	*Liquid*
Area (m^2)	0.053	0.053	0.070	0.070	0.071	0.071	0.063	0.063
Cycle averaged Temperature (K)	282.1	282.1	290.0	290.5	290.2	288.8	284.8	284.8
Cycle averaged Heat Transfer Rate (W)	1650		1350		1600		2500	
Resistance ($\frac{K}{W}$)	0.00189	0.00585		0.00075	0.00074	0.00250		0.00081
Thermal Conductivity ($\frac{W}{mK}$)		14.56				36.72		
Thickness (m)		0.006				0.007		

4.2 Lumped heat transfer

The 1D thermal resistance circuit is analysed to back calculate the thermal conductivities of the male and female rotors using the known temperatures, heat transfer rate and convective resistances. Thermal conductivities of the rotors calculated from the 1D lumped heat transfer model are later used to validate the 3D Conjugate heat transfer model in CONVERGE. Table 3 summarizes the results of the analyses. Note that the heat transfer rate from the solid-liquid interfaces is larger than the solid-gas interfaces due to the additional heat loss to the walls and casing of the compressor.

The thermal conductivities of female and male rotors are computed to be 14.56 $\frac{W}{mK}$ and 36.72 $\frac{W}{mK}$ respectively. These numbers are reasonable and can be compared to 45 $\frac{W}{mK}$, which is the standard thermal conductivity of carbon steel [15]. Due to the material being additively manufactured with thin wall thicknesses, the conductivities could be compromised. This accounts for the difference of about 67% and 18.4% for female and male rotors respectively. Substituting the material properties of Inconel with that of a weaker alloy like carbon steel, the model is conservative towards predicting realistic experimental quantities for Inconel.

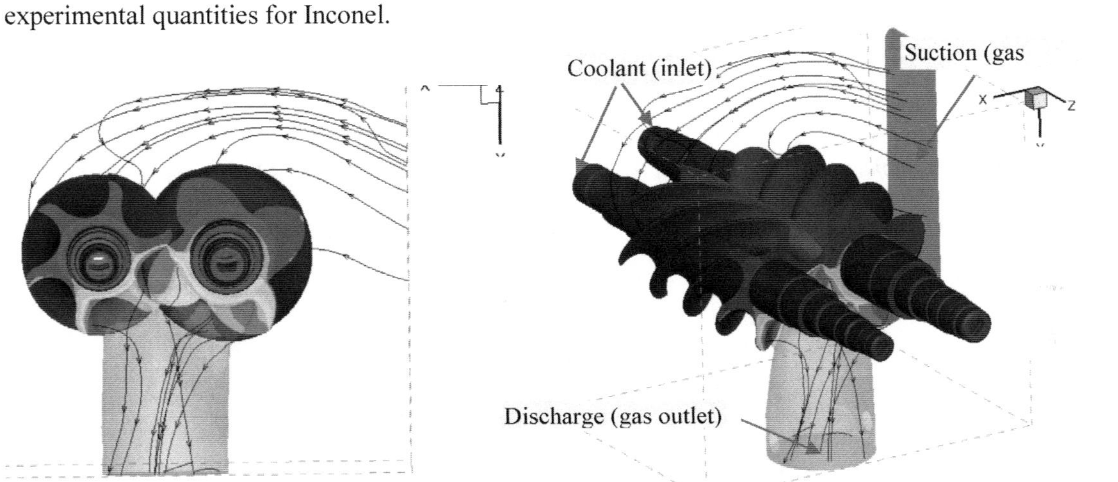

Figure 6. Pressure contour plots for solid rotor configuration with gas velocity streamlines. (Actual numbers have been omitted due to confidentiality)

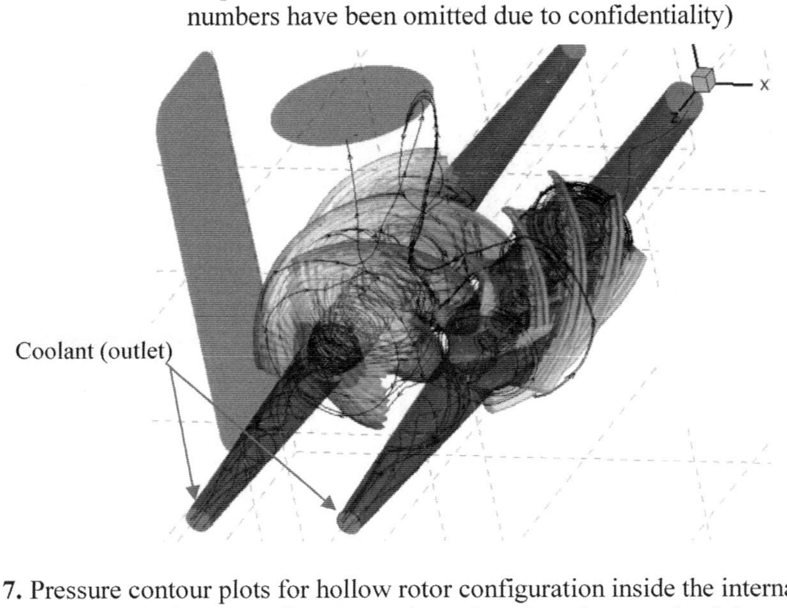

Figure 7. Pressure contour plots for hollow rotor configuration inside the internal fluid domain with liquid coolant velocity streamline. (Actual numbers have been omitted due to confidentiality)

4.3 Internal flow distribution

This section provides an insight into the internal fluid flow of the hollow rotors as well as standard gas compression streamlines with a pressure contour plot as in the case of the solid rotor. The streamlines in Figure 6 indicate a positive displacement of gas flow from inlet or suction port to the outlet or discharge port, compressed to the discharge pressure. The pressure at the discharge end is always kept around the discharge pressure as per the boundary condition. Figure 7 provides an insight into the internal flow domain of the hollow rotor. The rotation forces the liquid to swirl and traverse as much internal surface area of the fluid domain as possible, increasing convective heat transfer from the rotors. The constant input mass flow rate forces the liquid to exit at a steady outlet mass flow rate as well, thereby creating a steady flow of liquid coolant that constantly removes heat from the rotors.

4.4 Model verification

For simulation results, it is important to verify the model, which includes the following steps: verifying steady state by analysing gas mass flow rate projections of both cases; comparing rotor temperatures of solid rotor and hollow rotor simulations; comparison of simulation characteristics such as mass flow rate, suction/discharge temperature and pressure ratios. Steady state is said to be achieved when the gas and liquid outlet mass flow rates become steady-periodic with cycle-to-cycle variations below 0.1%.

Figure 8 displays the gas mass flow rate projections of both simulations. As can be seen, the gas outlet mass flow rate arrives at steady state much quicker for the solid rotor (6000 rotation angle degrees) as compared to the hollow rotor simulation (9000 rotation angle degrees). This is because of the additional internal flow simulation that makes the simulation time consuming. However, since the simulations have arrived at steady state, they can now be trusted.

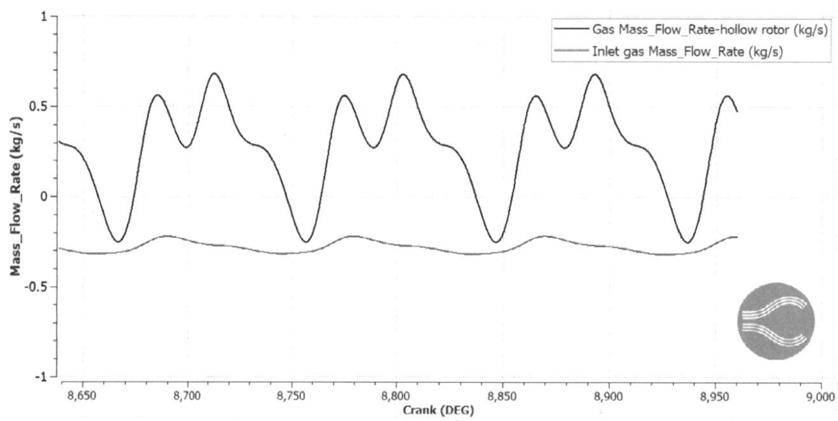

a) Hollow rotor (Average outlet flow rate = 0.247 $\frac{kg}{s}$, average inlet flow rate = -0.278 $\frac{kg}{s}$).

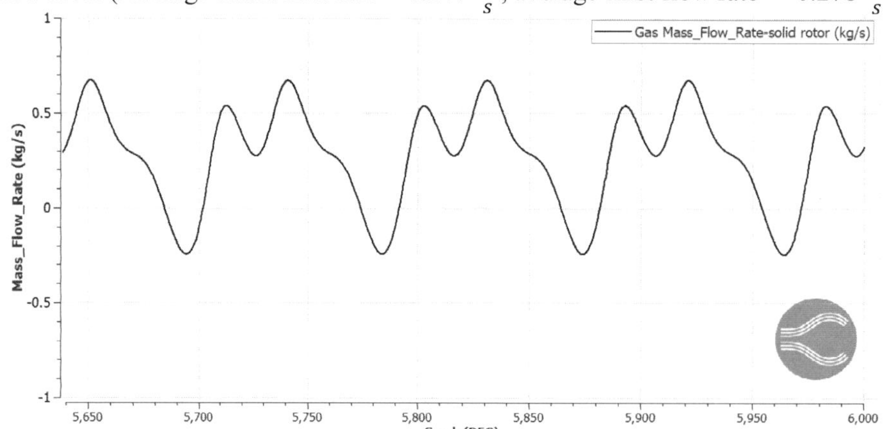

b) Solid rotor (Average outlet flow rate = 0.26 $\frac{kg}{s}$, average inlet flow rate = -0.278 $\frac{kg}{s}$).

Figure 8. Gas outlet and inlet mass flow rates for **a)** hollow and **b)** solid rotor configuration.

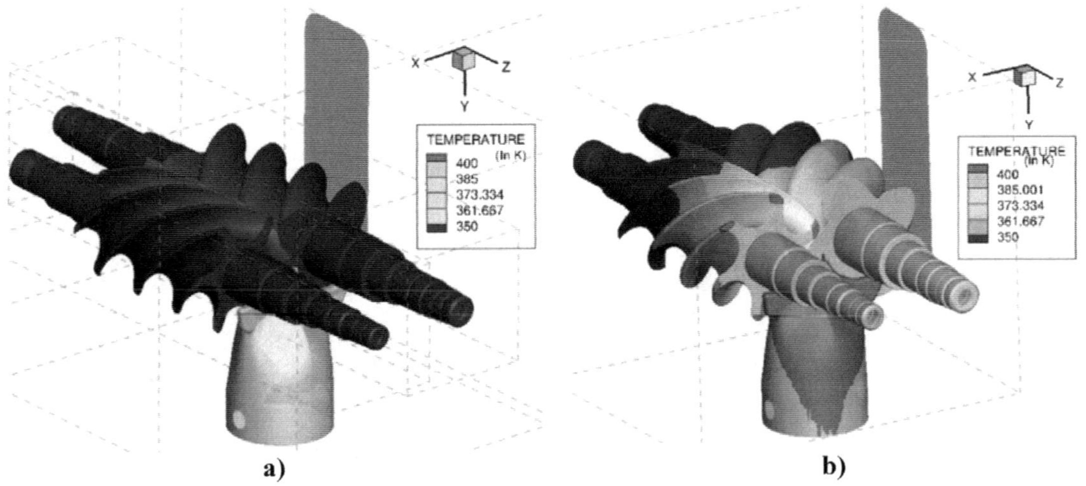

Figure 9. Rotor temperature contour plots for **a)** hollow and **b)** solid rotor configurations at 6000 rotation angle degrees. (Actual numbers have been omitted due to confidentiality).

Table 4. Compression conditions for hollow and solid rotor configurations.

Data	**Outlet cycle average gas Mass flow rate** $(\frac{kg}{hr})$	**Inlet cycle average gas mass flow rate** $(\frac{kg}{hr})$	**Suction pressure (bar)**	**Discharge pressure (bar)**	**Suction temperature (K)**	**Discharge temperature(K)**
CONVERGE – solid	950.00	-1000.00	1.013	3.33	293	440.0
CONVERGE – hollow	889.45	-1000.71	1.013	3.33	293	422.6

Due to the internal cooling, the hollow rotor configuration has a lower rotor temperature than the solid rotor (dry running) configuration. This can be concluded from the rotor temperature contour plots in Figure 9. The hollow rotor maximum temperature is around 325 K compared to 400 K for the solid rotor. The internal fluid cools down the rotors effectively and this cooling effect translates onto lower gas discharge temperatures as well, as will be shown later. Thus, the hollow rotor model can be verified by comparing rotor temperatures and the heat transfer occurs as expected. However, despite the modification in the rotor geometries and boundary conditions, it is also important to ensure that the compression conditions remain the same for both cases. Therefore, it is important to compare pressure ratios, mass flow rates and gas inlet/outlet temperatures. These results are summarized in Table 4. which clearly shows that the discharge temperature is lower as the rotors undergo internal cooling. This contributes to a lower indicated work and higher isothermal and isentropic efficiencies, which will be shown later. It is also important to note that the compression conditions like pressure ratios and mass flow rates do not change as much, verifying the idea that the internal cooled rotors perform better than the solid rotors under similar compression conditions such as pressure ratios and mass flow rates. The discharge pressure from CFD fluctuates periodically and Table 4 shows an average pressure result, which is the reason for a pressure difference of 1.7% from the boundary conditions. The mass flow rates at the outlet are different from that of the inlet with an error of 5% and 10.1% for the solid and hollow rotors due to leakages and backflow caused by adverse pressure gradients occurring within the compression chamber. All of these are accounted by the fluid dynamics process in CFD. Although the mass flow rate in the internally cooled case should be higher than the solid rotor simulation because of higher density at low temperatures, this is not the case. The inlet conditions are the same in both cases and therefore doesn't change the suction density, which ultimately decided the mass flow rate.

4.5 Performance evaluation

The indicated power for each case is calculated using total torque values obtained from CFD. These torque projections can be viewed in Figure 10. Table 5 displays efficiencies using their respective gas inlet/outlet temperatures and equations as described earlier. The internally cooled model reduces outlet temperature causing it to approach isothermal power and thereby reducing indicated power. This idea is corroborated through the results in Table 5. The isentropic and isothermal efficiencies are increased by about 2% for the hollow rotors with internal cooling. The torque numbers are used to compute indicated power and compare with the solid rotor, therefore validating the torque results.

Table 5. Efficiency and power calculations for hollow and solid rotor configurations.

Data	Indicated power (kW)	Isothermal efficiency (%)	Isentropic efficiency (%)
Solid	44.1	63.6	75.4
Hollow	42.8	65.1	77.1

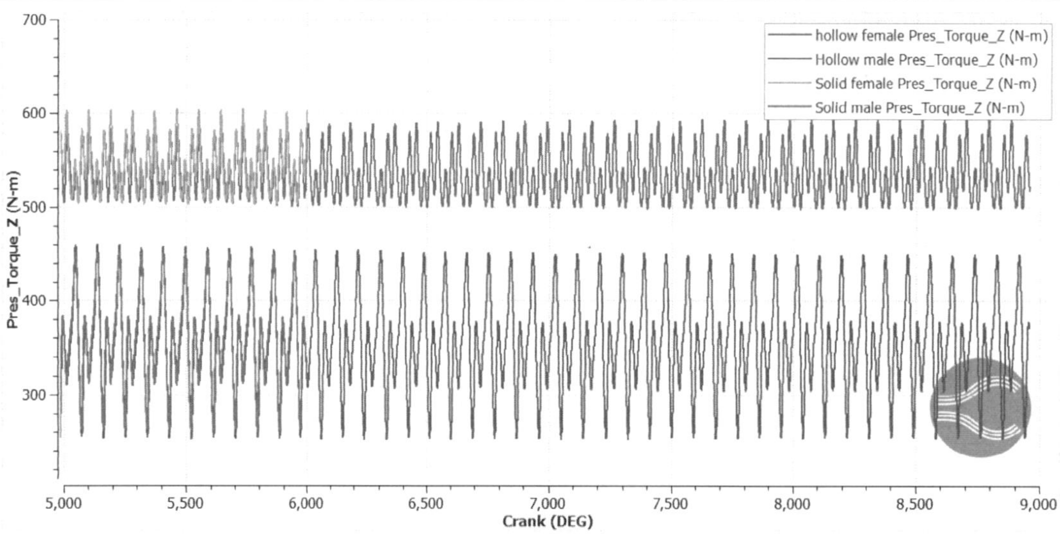

Figure 10. Male and female rotor pressure torques for solid and hollow rotor configurations.

5. Conclusions

In this paper, a CFD model with conjugate heat transfer for a 4-6 configuration screw compressor with internal cooling is presented and verified. The convectional rotor geometry is modified with hollow internal domains that allow fluid to pass through. The internal fluid cools down the rotors and improves efficiency. The model has been validated by comparing thermal conductivities of rotors obtained through a 1D lumped heat transfer model and the goal behind internal cooling channels is demonstrated by comparing results of the hollow rotor with the solid rotor configuration at the same compression conditions. The hollow rotor with internal cooling has an increase in efficiency of 2%, which can be improved further by varying internal flow conditions and replacing the liquid with a refrigerant. Mesh resolution has also been verified by comparing Y+ numbers and by using various mesh resolution strategies as outlined. A more accurate indicated power analysis could be achieved by setting up indicated diagrams for both cases and comparing the indicated powers calculated through the area of the curves. A future study could include a deeper analysis into thermal expansion effects on the internal clearances and how that affects performance.

6. References

[1] Patel H H and Lakhera V J 2019 A critical review of the experimental studies related to twin screw compressors. Proceedings of the Institution of Mechanical Engineers, Part E: *Journal of Process Mechanical Engineering* **234(1)** 157170 https://doi.org/10.1177/0954408919869534

[2] Stosic N, Smith I and Kovacevic A 2005 Screw compressors: mathematical modelling and performance calculation *Springer*

[3] Zhao R, Li W and Zhuge W 2020 Unsteady characteristic and flow mechanism of a scroll compressor with novel discharge port for Electric Vehicle Air Conditioning *International Journal of Refrigeration* **118** 403414 https://doi.org/10.1016/j.ijrefrig.2020.06.022

[4] Sun W, He G, Ning Q, Pang Q and Cai D 2021 Performance investigation and design optimization of a vapor injection rotary compressor without check valve in injection path *Applied Thermal Engineering* **197** 117372 https://doi.org/10.1016/j.applthermaleng.2021.117372

[5] Yang S, Ouyang H, Wu Y, Wang L, Mei L and Wang H 2021 CFD simulation for the internal pressure characteristics of an oil-injected twin-screw refrigeration compressor *International Journal of Refrigeration* **126** 143154 https://doi.org/10.1016/j.ijrefrig.2021.01.020

[6] Kovacevic A 2002 Three-dimensional numerical analysis for flow prediction in positive displacement screw machines. Doctoral Thesis.

[7] Kovacevic A 2005 Boundary adaptation in grid generation for CFD analysis of screw compressors. *International Journal for Numerical Methods in Engineering* **64(3)** 401426

[8] Analysis PDM 2022 SCORG Retrieved from https://pdmanalysis.co.uk/scorg/

[9] Ding H and Jiang Y 2017 CFD simulation of a screw compressor with oil injection. *IOP Conference Series: Materials Science and Engineering* **232** 012020. doi:10.1088/1757-899X/232/1/012020

[10] Inc S 2022 Simerics MP+ Retrieved from https://www.simerics.com/

[11] Ding H, Jiang Y and Dhar S 2019 CFD modelling of coupled heat transfer between solid and fluid in a twin screw compressor. *IOP Conference Series: Materials Science and Engineering*, **604(1)** 012005 https://doi.org/10.1088/1757-899x/604/1/012005

[12] Saravana A, Liu H, Able N, James C, Groll E and Ziviani D Rotordynamics and fatigue analyses of a twin-screw compressor with 4-6 configuration and internal cooling channels 2022 *International Compressor Engineering Conference* Paper **1566**

[13] Launder B E and Spalding D B 1974 The numerical computation of turbulent flows Comput. Methods Appl. Mech. Eng., 3, pp. 269-289

[14] Richards K J, Senecal P K, and Pomraning E CONVERGE 3.0* Convergent Science Madison WI 2022

[15] Franssen J M and Real P V Thermal data for carbon steel and stainless-steel sections 2012 *European Convention for Constructional Steelwork* Retrieved from https://onlinelibrary.wiley.com/doi/pdf/10.1002/9783433601570.app1

Nomenclature

t	time	s
P	theoretical pressure	Pa
u	velocity	m/s
e	internal energy	kJ
S	heat source	$-$
x	distance	$-$
σ	viscosity	$-$
ρ	density	N/m^3
$\dot{W}$	power	W
N_1	male rotor rotational speed	RPM
z_1	number of male rotor lobes	$-$
τ_1	male rotor torque	$N-m$
τ_2	female rotor torque	$N-m$
z_2	number of female rotor lobes	$-$
$\dot{m}$	mass flow rate	kg/s
T	temperature	K
γ	gas constant for air	$N-m/kg-K$
k	polytropic coefficient	$-$
η	efficiency	$-$
R	thermal resistance	K/kW
$\dot{Q}$	heat transfer rate	kW
Δ	change in	$-$
L	length	m
h	convective heat transfer coefficient	$kW/m\text{^}2 - K$
k	thermal conductivity	$kW/m-K$
A	surface area	$m\text{^}2$

Subscript

suc	suction	conv	convection
dis	discharge	cond	conduction
isoth	isothermal	liq	liquid
isen	isentropic		
ind	Indicated	ad	adiabatic

Acknowledgement

The authors would like to thank Ingersoll Rand for their financial and technical support during this project.

Thermodynamic simulation and experimental validation of an oil-free twin-screw expander

M Grieb and A Brümmer

Chair of Fluidics, TU Dortmund University, Leonhard-Euler-Str. 5, 44227 Dortmund, Germany

E-mail: manuel.grieb@tu-dortmund.de, andreas.bruemmer@tu-dortmund.de

Abstract. The chamber model method is still a powerful tool for the thermodynamic simulation of screw machines and is widely used in the industry and in science. When choosing adequate submodels for the compressible flows through machine clearances and the machine ports high simulation accuracy can be reached with minimal computation resources. However, detailed modelling is required when the complex geometry of twin-screw machines is abstracted to a time-dependent zero-dimensional chamber model. In this paper a two-chamber model simulation tool is used for the thermodynamic simulation of an oil-free twin-screw expander operating with dry air as working fluid. Insights into the modelling process are presented and the importance of precise modelling of gap flows between adjacent capacities is discussed. The simulation results are finally validated against experimental data. Integral values obtained from the experiments, such as expander mass flow rate, indicated power, and effective power, are in good agreement with the numerical calculations. To gain a deeper understanding of the thermodynamics of the expander working cycle, additional investigations are carried out at zero speed. The mass flow rates measured for a variation of stationary rotor positions are compared to values calculated with the two-chamber model method.

1. Introduction

Although computational fluid dynamics (CFD) simulations of screw machines advance quickly, the chamber model method is still an important tool for the investigation and design of screw machines. In comparison to CFD significantly fewer computational resources are used in chamber model simulations. In addition, the complex process of grid generation for the dynamically deformed fluid domains becomes obsolete. The quality of chamber model simulations is improved by purposeful employment of more complex submodels, the embedding of experimental databases, and the usage of co-simulations. This approach focusses simulation efforts on the crucial mechanisms, for instance on leakage flows, while still benefiting from the advantages of the chamber model method.

2. Chamber model simulation

The chamber model method has proven itself to be a reliable tool for the design and optimisation of positive displacement machines. It is widely used for the thermodynamic simulation of twin-screw machines and is validated with good agreement for various applications [1, 2]. The complexity of models documented in the literature varies from models where the working chambers are observed as one volume and leakage paths are greatly simplified to models where a distinction between main rotor and female rotor working chambers is made and leakage paths are modelled with higher complexity. Furthermore, the accuracy of the calculation is limited by the choice of submodels, e. g. for mass flows and heat transfers. A review of exemplary submodels for mechanistic chamber models is given in [3]. Furthermore, the chamber model method can be used as a framework for theoretical investigations to gain insights into physical processes that cannot be observed in experimental investigations such as, for instance, the formation of nuclei and growth of droplets during the condensation of vaporous working fluids in screw expanders, as shown by the authors [4, 5].

2.1. Basic thermodynamic concept

The basic presumptions of the chamber model method applied to screw machines are as follows:

- working chambers are formed in the helical grooves of the interlocking rotors
- volumes formed in the intersection of the rotor bores are part of the gate rotor working chamber (arbitrary definition)
- working chambers might contain multiple control volumes (e. g. gaseous and liquid domain)
- assumption that mass flows entering or leaving the control volumes act according to stream filament theory (fluid properties uniform in flow cross-section)
- quasi stationary mass flows through the control volume surface

The basic equations for the chamber model simulation are the mass conservation equation and the energy conservation equation respectively. In this study the equations are derived for a deformable control volume. The integral mass conservation equation for any control volume within the screw machine is:

$$\frac{\mathrm{d}M(t)}{\mathrm{d}t} = \sum_{j} \dot{m}_{rj} \tag{1}$$

The total change of the mass M enclosed in the control volume with respect to the time t is given by the sum of mass flows relative to the control volume surface $\dot{m}_{rj}$, where j is the index of connections to adjacent volumes. The mass flow connections to the control volume are typically gaps between machine parts causing internal leakage, openings to the inlet and outlet ports, chamber connections between main and gate rotor control volumes, and injection bores. The calculation of mass flow rates is carried out in subroutines for any specific case. In this study mass flows are calculated on the basis of the flow coefficient α which is defined as the ratio of the actual mass flow $\dot{m}$ to the theoretical mass flow $\dot{m}_{th}$. Hereby, the model for the theoretical mass flow rate is arbitrary as long as it is consistent throughout the simulation and modelling process.

$$\dot{m} = \alpha\, \dot{m}_{th} = \alpha\, A\, \rho_{2s}\, \sqrt{2\,(h_1 - h_{2s})} \tag{2}$$

The flow coefficient is usually obtained from detailed co-simulations or from experimental data. A is the cross-sectional area of the connection, ρ is the fluid density, and h is the specific enthalpy. The index 1 describes the fluid states within in the connected control volume at higher pressure, and the index $2s$ marks the downstream fluid states in the smallest cross-section of the connection. The density ρ_{2s} and the specific enthalpy h_{2s} are calculated from the equation of state at constant specific entropy s. It is assumed that pressure in the smallest cross-section equals the pressure in the downstream control volume, which corresponds to complete dissipation of the kinetic energy (no pressure recovery). Choking of the flow is considered by limiting the mass flow rate to the critical value. The choked flow state in the smallest cross section of the stream filament is computed by finding the maximal mass flux. The simulation routine allows for various fluid flow models. Flow models for liquids, multiphase flow, and phase transition are available (e. g. condensation).

The energy conversation equation for the control volume is

$$\frac{\mathrm{d}E(t)}{\mathrm{d}t} = \sum_j h_{tj}\, \dot{m}_{rj} - \bar{p}\, \frac{\mathrm{d}V}{\mathrm{d}t} + \sum_k \dot{Q}_k - \dot{\Phi} \tag{3}$$

where E is the total energy within the control volume V. $h_t = h + \frac{c^2}{2}$ is the specific total enthalpy with c being the flow velocity, and the first sum describes the changes in total energy due to enthalpy flows through the control volume surface. The second expression gives the pressure volume power, where $\bar{p}$ is the average pressure acting on the control volume surface. $\dot{Q}_k$ are external heat flows, where k is the index of heat transfer connections. For a diabatic simulation appropriate heat exchange models need to be implemented. Finally, $\dot{\Phi}$ is power induced by tangential shear stresses acting on the moving control volume boundaries. This is often neglected in chamber model simulation tools.

2.2. Implementation

In this work a two-chamber model is used where both the main rotor and the gate rotor working chambers are observed during the complete cycle. The distinction between main rotor and gate rotor working chambers is required since different pairs of teeth are intermeshing during the inlet phase and the discharge phase. During the working cycle of the screw machine the two working chambers might even be disconnected from each other depending on the rotor geometry. The two-chamber model brings the advantage of clarity and a simpler modelling process compared to multi-chamber models, where all working chambers formed in the helical grooves are observed at once. However, when fluid properties of a leading adjacent working chamber are required during run time of the simulation (e. g. calculation of a leakage flow), access to values of the last iteration $i-1$ is required when using a two-chamber model. Flows to leading control volumes calculated at time t are again fed to the observed control volume after a defined time offset t_{offset} due to the periodic working principle of screw machines. Time offset is depending on machine speed and rotor configuration. This periodicity of flows is implemented by saving flow data to leading control volumes in a list during runtime for every connection. The flow data is then continuously accessed during simulation process. Consequently, for periodic flows only connections to leading control volumes are taken into account during the modelling process. The general principle is depicted in figure 1.

Further periodic flows occur during the filling and discharge process of screw machines. Despite only observing one chamber on each rotor during the simulation process, the continuous periodic flow through machine ports is virtually assembled for the whole cycle in order to calculate fluid states in the inlet and outlet ports of the screw machines. This allows for a rapid convergence of discharge temperatures.

Numerical challenges occur in particular when calculating mass flows between control volumes. When the cross section of a connection is large with respect to the control volume size, the forward Euler approach might lead to instabilities, e. g. when adding or removing mass from small chambers that emerge or disappear on the high and low pressure planes of the machine. This is overcome by checking if the mass exchange causes a change of sign in pressure difference and, if so, by iteratively calculating a pressure balance between the control volumes.

The chamber model theory is implemented in a self-developed C++ program named *KAMEL*. The overall objective of developing this two-chamber model simulation tool was creating a flexible solution for screw machines operating with multiphase fluids which was accomplished by creating flexible functionality for multiple fluid regimes per working chamber and general support of fluid-fluid and fluid-solid interaction. This study, however, aims to proof general functionality. Hence, simulation and experimental investigation is carried out with dry air as an ideal gas.

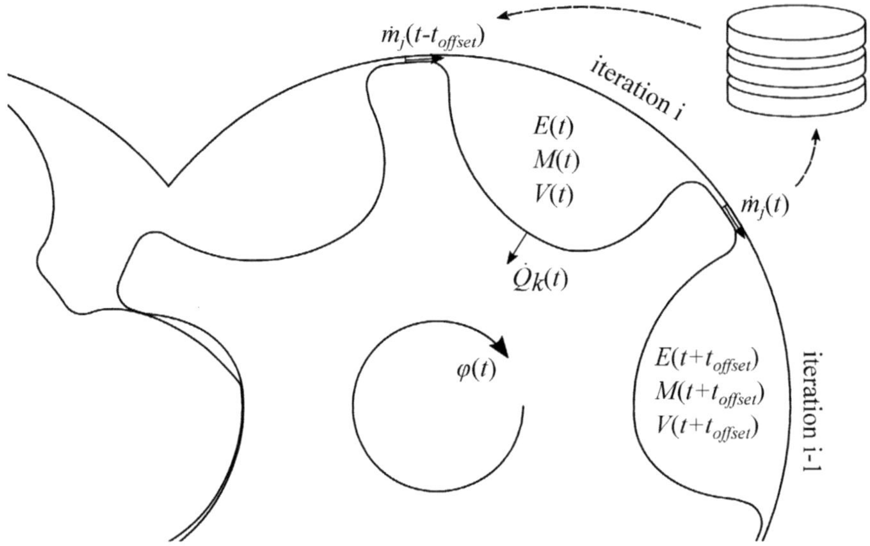

Figure 1. Logical implementation of peridodic flows between adjacent control volumes.

The conservation equations for mass and energy are linearised and solved with the forward Euler method for every time step. In this study main rotor angular step size is $\Delta\varphi_m = 0.03°$ for the numerical calculation which provides adequate simulation results and independence of integral results from the corresponding time step. Convergence is typically reached after 5 to 10 iterations with relative deviations in integral values in the range of $\delta_{\dot{m}} \approx 1 \cdot 10^{-5}$ for mass flow rate and $\delta_P \approx 1 \cdot 10^{-4}$ for indicated power between iterations. Relative mass flow imbalances between inlet and outlet boundaries for the converged simulations are very small and range from $1 \cdot 10^{-5}$ to $1 \cdot 10^{-4}$. Fast convergence and negligible imbalances in the integral conservation of mass and energy are further advantages of the chamber model method.

2.3. Preprocessing

Preprocessing in chamber model simulation includes discretisation of the machine geometry and setting the simulation parameters, which includes numerical parameters and boundary conditions.

Several solutions for automatic discretisation of the machine geometry are introduced in the literature. The algorithms used in this study are based on the analysis of two-dimensional rotor profile and are presented in [6] and [7]. The geometric parameters of the screw expander examined in this paper are listed in table 1. In this context it is noted that the overall size of the prototype is fairly small. It can thus be expected that machine performance is strongly affected by leakage flows through the unavoidable characteristic gaps being relatively large with respect to the control volumes. This aspect and its specific importance in chamber model simulation is further examined in later sections of this study. The combination of a fairly large built-in volume ratio and moderate wrap angles leads to a fairly small inlet and a short filling period due to the rapid volume expansion. The small wrap angle, however, results in a shorter and thus beneficial gap width (transversely to the flow direction) of the housing gaps.

During the discretisation process the machine geometry is analysed and a zero-dimensional model is derived. Geometric quantities (e. g. volume, gap height, gap width) are noted with

Table 1. Screw expander geometry.

Rotor profile	asymmetric SRM; 4+6 configuration
Rotor centre distance	46.2 mm
Rotor diameter	58.9 mm
Rotor length	58.9 mm
Main rotor wrap angle	200°
Built-in volume ratio	4.0

regard to main rotor angular position. Corresponding control volumes need to be allocated and the transport model is to be initialised for all connections. Discretisation of screw machine geometry is a clear process for the majority of connections and capacities, such as for the machine openings or the working chambers illustrated in figure 2.

However, the discretisation of connections becomes challenging in regions where more than two control volumes meet and wherever a clear distinction between control volume and leakage path vanishes, e. g. when modelling the interlobe clearance where the low pressure chambers form pointed regions between the intermeshing lobes. A disadvantage of chamber model simulation is that the assumption of stream filament theory for leakage flows cannot fully capture the complex flow occurring in the gaps of screw machines. In chamber model theory leakages through the interlobe clearance are typically regarded as losses directed towards chambers connected with the low pressure side of the machine [8]. As a consequence of this modelling approach gains in leading working chambers are limited exclusively to flows through the front and housing gaps as well as through the blow hole area. When this was initially implemented in this study the pressure volume plot showed extremely large pressure gradients during the first part of the expansion phase and non-conclusive power calculations. Further geometric analysis of the machine showed that the trailing housing gaps and the trailing blow hole of working chambers just entering the expansion phase are not yet developed. This finding led to the assumption that

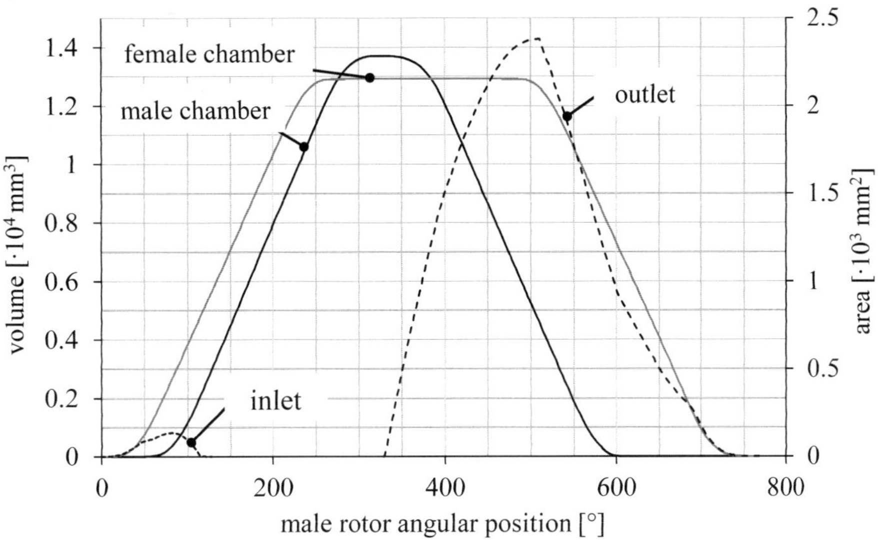

Figure 2. Volume curve and port cross sectional area of screw expander prototype.

gap flows through the interlobe clearance are not exclusively directed towards the low pressure side of the screw expander but partially follow the leakage flow through the blow hole area.

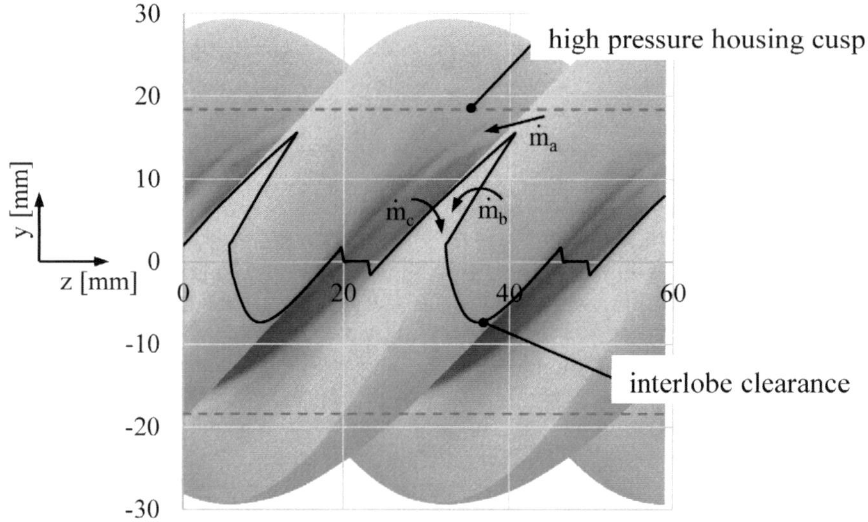

Figure 3. Side view of main rotor with projection of interlobe clearance and simplified illustration of leakage flows in the blow hole area. Blow hole leakage flow $\dot{m}_a$, leakage flow through leading interlobe sealing line $\dot{m}_b$, and leakage flow through trailing interlobe sealing line $\dot{m}_c$.

This becomes more apparent when analysing the working chamber geometry close to the interlobe clearance shown in figure 3. Where the interlobe sealing lines merge into the blow hole area a significant part of the pointed volume of the adjacent low pressure working chamber is characterised by distances between the rotor surfaces being well comparable to typical gap heights in screw machines. This region can thus be considered as a narrow channel connecting three working chambers (two on the high pressure side and one on the low pressure side). It is thus assumed that the leakage flow through the upper part of the leading interlobe sealing line $\dot{m}_b$ follows the flow through the blow hole area $\dot{m}_a$ acting as mass gain in leading working chambers. This is accounted for by modelling 30 % of the leading interlobe clearance width as connection to the leading working chamber.

Furthermore, the modelling of the front gaps between the rotor front side and the high pressure housing plane is a key factor for producing adequate simulation results, especially for small scale machines, like in this investigation, that are strongly affected by leakage flows. In proximity to boundaries of the inlet opening it might be expedient to consider direct gap connections between the inlet plenum and working chambers. Hereby the actual machine geometry is to considered, which eventually complicates the process compared to those modelling approaches based solely on the rotor profile. In this investigation the projected width of the lower inlet contour is considered as gap connection between the inlet plenum and the discharging chambers on the low pressure side of the screw expander. Front gaps in the intermesh region are, however, modelled as connections between the working chambers.

Flow coefficients for the different gap types are chosen in accordance with values given in the literature and based on results obtained in experimental investigations [9].

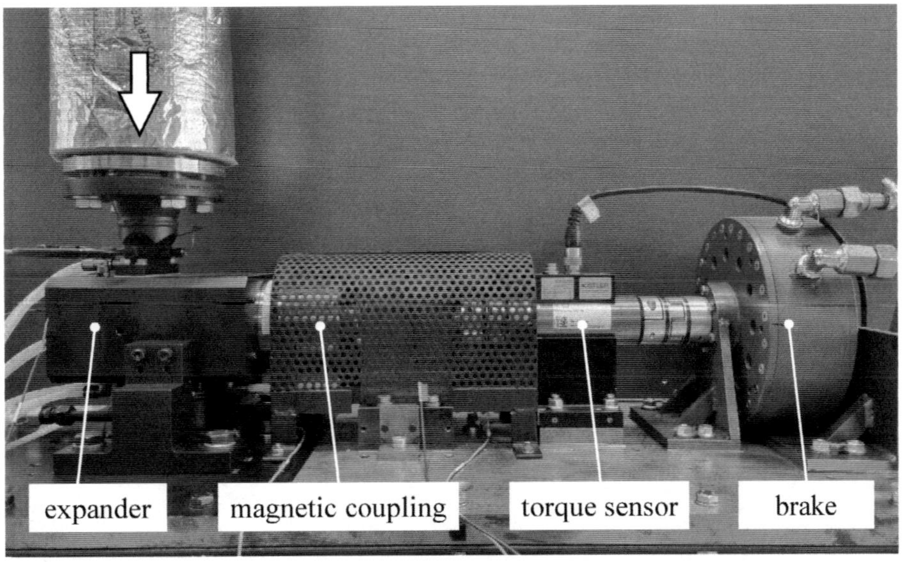

Figure 4. Side view of the expander and drive train mounted on the test bench.

3. Exerimental setup

The prototype expander is an oil-free twin-screw expander with 4+6 configuration. It was initially designed for heat recovery cycles operating with steam or vaporous working fluids, preferably in two-phase state. Rotor surfaces are case-hardened and coated with a diamond-like carbon layer in order to allow for operation without timing gear. The machine is mounted on the expander test bench of the steam cycle at the Chair of Fluidics as shown in figure 4. However, for this study the working fluid is dry air provided by a compressor package and the steam generator is not operating. The screw expander is connected to the drive train via magnetic coupling providing a hermetic seal, albeit requiring an additional shaft for support of the outside magnet in cantilever position, which in turn creates additional frictional loss. A torque sensor is used for precise torque and speed measurement. For the experimental investigation an eddy current brake is used to control expander speed. Torque between the components is transmitted via elastomer couplings. Fluid enters the expander on the top side and is directed through the axial inlet opening. On the low pressure side the machine is equipped with a radial and axial discharge opening.

Figure 5 shows the experimental setup in the in-house test rig. Compressed dry air is provided by an oil-free commercial screw compressor package and passes an electric heater with temperature control. The expander bypass allows for soft startup and the bypass valve is used to control expander inlet pressure. Various thermocouples and precision piezo-resistive pressure sensors provide measurement data during operation. The expander mass flow rate is captured with a coriolis flow meter placed in the upstream piping. All measurement raw data is processed in a digital measurement system and monitored on a computer. The piping located downstream of the expander is open to the ambient.

Experiments are conducted in two steps. Initially, integral performance measurements of the screw expander are carried out where the expander speed is varied at constant inlet and outlet conditions. In consideration of the intended use of the machine and in order to reduce mechanical stress when operating without liquid phase, rotational speed is kept moderate. In a second step the screw expander is investigated at zero speed to gain detailed insights into leakage flow and to further validate the two-chamber model. This is achieved by mounting a

IOP Conf. Series: Materials Science and Engineering 1267 (2022) 012015 doi:10.1088/1757-899X/1267/1/012015

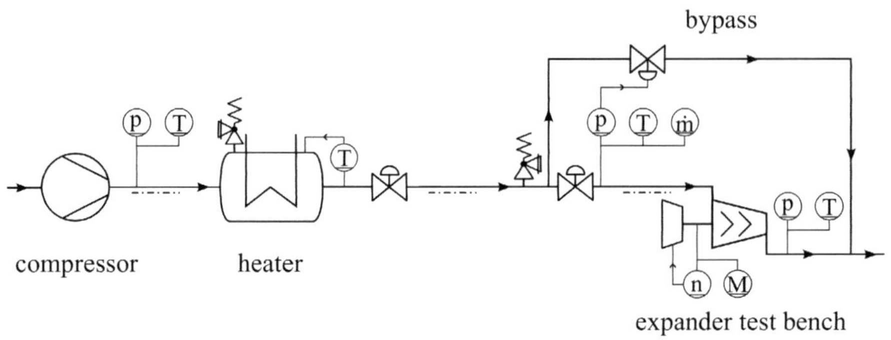

Figure 5. Experimental setup: measurement and piping diagram.

clamp device on the elastic coupling between the torque meter and the eddy current brake that prevents the drive train from rotating. Steady-state flow measurements are then carried out at constant inlet conditions for a variation of rotor positions.

4. Discussion

In this section the data obtained in the experimental investigation is compared against the results from the two-chamber model simulation tool *KAMEL*. Figure 6 shows the mass flow rate for an expander inlet pressure of $p_{in} = 4 \cdot 10^5$ Pa and an inlet temperature of $T_{in} = 333.15$ K. The expander speed varies in the range from $n = 4000$ to $10000\,\text{min}^{-1}$. There is a slight underprediction of the calculated mass flow rate observed in the speed range in question. However, for the most part simulation results are within the error margin of measurement. Furthermore, the gradients show good consistency. In this respect it is noted that the fairly

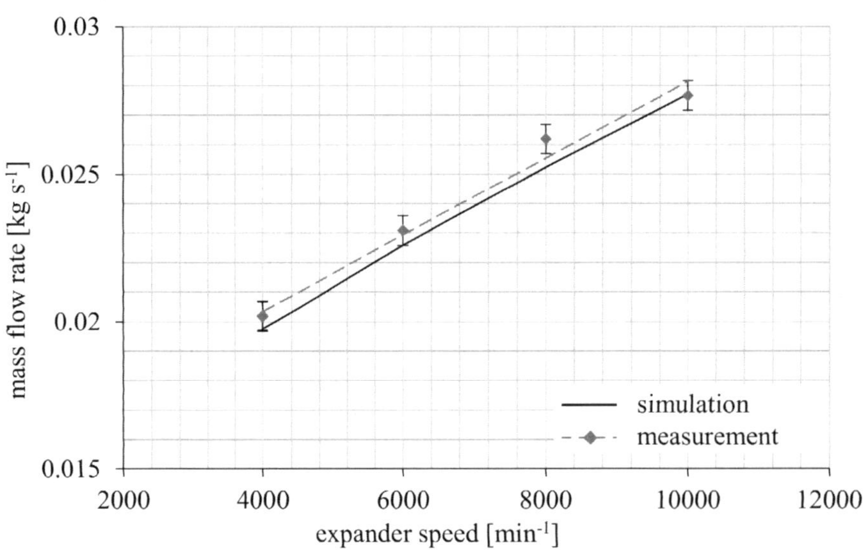

Figure 6. Mass flow rate plotted as a function of expander speed. Expander inlet pressure $p_{in} = 4 \cdot 10^5$ Pa, inlet temperature $T_{in} = 333.15$ K.

small screw expander prototype is set up with gap heights in the region of 0.1 mm, which is to provide safe high temperature operation and compromises on efficiency ratings and volumetric efficiency. The dominant impact of leakage flows on expander operation is challenging for any simulation tool. From this point of view, the overall model accuracy is highly satisfactory.

Figure 7 shows the expander power output for the same parameter range. Without a specific model for mechanical losses the chamber model method provides the indicated power P_i by multiplying the pressure volume work per control volume with the chamber frequency.

$$P_i = n z_m \sum_j \oint p \, \mathrm{d}V \tag{4}$$

Here, z_m is the number of lobes of the main rotor and j is the index of control volumes. An approximate value for the indicated power is derived from the integral equation of energy conservation and calculated from the experimentally obtained outlet temperature measured directly in the outlet port.

$$P_i \approx \dot{m} \left[h_{in}(T_{in}, p_{in}) - h_{out}(T_{out}, p_{out}) \right] \tag{5}$$

This is based on the assumption that the process is adiabatic and that friction-induced heat is mainly conducted through the solids to the environment. The error margin for this value is essentially given by the measurement accuracy of the mass flow rate. The effective power of the screw expander is, however, calculated from the speed n and torque M measurements, where ω is the rotational velocity.

$$P_e = \omega M = 2\pi n M \tag{6}$$

The comparison between simulation results and experimental data shows an appropriate level of accordance. The inner powers calculated from chamber model simulations match the

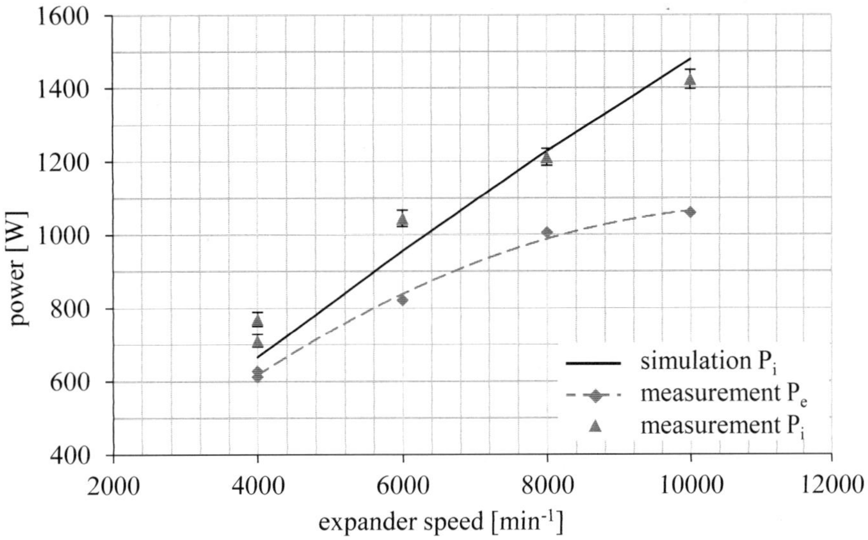

Figure 7. Effective power P_e (measurement) and inner power P_i (measurement & simulation) plotted as a function of expander speed. Expander inlet pressure $p_{in} = 4 \cdot 10^5$ Pa, inlet temperature $T_{in} = 333.15$ K.

experimental values rather good considering the simplification in deriving the data from pressure and temperature measurements. Mechanical losses causing the differences between inner power and effective power are relatively large and increase with expander speed. The nonlinear increase in losses is conclusive. Eddy current loss in the magnetic coupling is to be neglected since the containment shell is made from glass. The majority of mechanical losses is expected to occur in the bearings, especially in those holding the external rotor of the magnetic coupling in cantilever position.

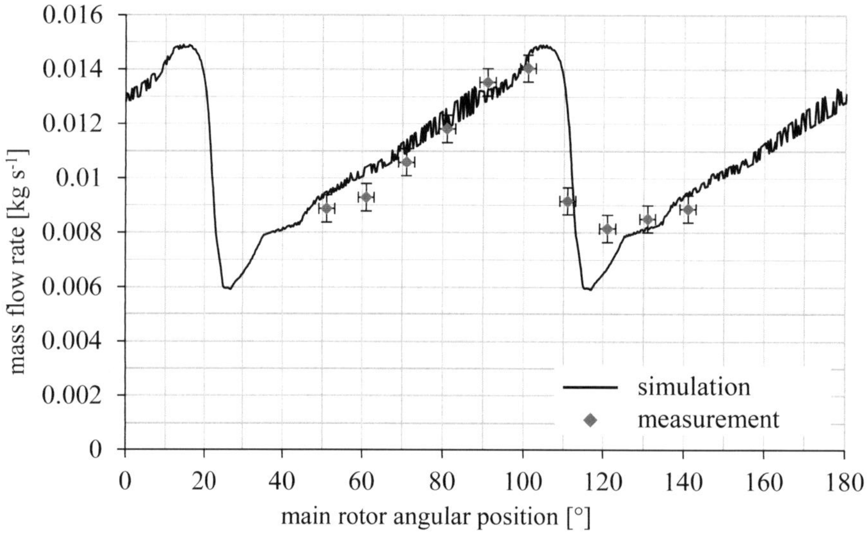

Figure 8. Stationary mass flow rate for a variation of rotor positions (zero speed). Expander inlet pressure $p_{in} = 3 \cdot 10^5\,\mathrm{Pa}$, inlet temperature $T_{in} = 313.15\,\mathrm{K}$. Simulation results calculated for expander speed $n = 1\,\mathrm{s}^{-1}$.

Finally, the expander mass flow rate at zero speed is discussed. The experimental procedure basically corresponds to a cumulative measurement of gap flows since the inlet port opening is large in comparison to the cross sectional area of sealing lines for all rotor positions. Due to the finite torsional stiffness of the magnetic coupling and the missing access to the expander rotor shafts during the measurement it is assumed to reach an error margin for angular positioning of $\pm 2°$. Positioning control is reached by attaching a scale on the static flange of the magnetic coupling and mounting a hand on the external rotor of the coupling. The measurement was carried out for an angular range greater than the expander periodicity and repeatability in the range of $\pm 2\,\%$ was reached. For the two-chamber model simulations an expander speed of $n = 1\,\mathrm{s}^{-1}$ was set to ensure program execution. The angular step size was reduced to $\Delta\varphi_m = 0.003°$. For such small machine speed it can be assumed that the mass flow rate is almost exclusively affected by leakage flows and not by displacement.

Figure 8 shows the comparison of simulation results and the gap flow measurement at zero speed. The measurements are essentially in good agreement with the simulation results. The largest deviation is observed close to the transition between the inlet phase and the expansion phase ($\varphi_m = 113°$). Here, the measurement shows a more steady course and it can be considered to refine the model with respect to clearance connections in this region. When the main and gate rotor chambers reach inlet pressure during the simulation, small fluctuations in mass flow rate are observed. This is due to the numerically challenging calculation of pressure balance between the control volumes (high pressure plenum and chambers). The overall agreement between the two-chamber model simulations and various experimental results is good. In addition to the ability to calculate integral values, such as power or mass flow, the chamber model method proves itself to predict even more detailed aspects of the thermodynamics of screw machines.

5. Conclusion

In this paper the chamber model method is used for the thermodynamic simulation of a small-scale twin-screw expander prototype. The working fluid is dry air and simulations are carried out with a self-developed software tool which applies the chamber model method to screw machines where a distinction is made between the working chambers on the main rotor and on the gate rotor side. An introduction into the basics of the chamber model method is presented and specific modelling aspects concerning the flow directions through the machine clearances are discussed. In particular, modelling part of the interlobe clearance beneath the high pressure blow hole area as connection between high pressure working chambers showed reasonable results. Simulation results were finally validated against experimental data where overall good agreement was reached for integral values such as mass flow rates and power ratings. Furthermore, cumulative clearance flow measurements were carried out at zero speed and the mass flow rates obtained from the experimental investigation were compared to simulation results. In this case good agreement between the simulations and the experiments was reached, too. This indicates the two-chamber model method is an appropriate measure for the thermodynamic calulation of twin-screw machines including even more detailed aspects of the energy conversion within those machines.

References

[1] Janicki M 2007 *Modellierung und Simulation von Rotationsverdrängermaschinen* (Dortmund: Universität Dortmund)
[2] Buckney D, Kovacevic A and Stosic N 2013 Experimental validation of a geometry model for twin screw machines *8th International Conference on Compressors and their Systems*
[3] Tanveer M M, Bradshaw C R, Ding X and Ziviani D 2022 Mechanistic chamber models: A review of geometry, mass flow, valve, and heat transfer sub-models *International Journal of Refrigeration* **142** 111–126
[4] Grieb M and Brümmer A 2017 Effects of spontaneous condensation in steam-driven screw expanders *IOP Conf. Ser.: Mater. Sci. Eng.* **232**
[5] Grieb M and Brümmer A 2019 Simulation-based investigation of spontaneous condensation in steam-driven screw expanders *Part E: Journal of Process Mechanical Engineering*
[6] Nadler K 2017 *Modellierung und Analyse von Schraubenvakuumpumpen im Blower-Betrieb* (Dortmund: Technische Universität Dortmund)
[7] Utri M 2021 *Potenzial von nicht-konstanter Rotorsteigung für Schraubenkompressoren* (Dortmund: Technische Universität Dortmund)
[8] Stosic N, Smith I and Kovacevic A 2005 *Screw Compressors, Mathematical Modelling and Performance Calculation* (Berlin, Heidelberg: Springer)
[9] Utri M, Höckenkamp M and Brümmer A 2018 Fluid flow through male rotor housing clearances of dry running screw machines using dimensionless numbers *IOP Conf. Ser.: Mater. Sci. Eng.* **425**

IOP Conf. Series: Materials Science and Engineering 1267 (2022) 012016 doi:10.1088/1757-899X/1267/1/012016

Transient chamber filling in rotary positive displacement vacuum pumps

H Pleskun[1], A Syring[1] and A Brümmer[1]

[1]TU Dortmund University, Chair of Fluidics, Leonhard-Euler-Str. 5, 44227 Dortmund, Germany

E-mail: heiko.pleskun@tu-dortmund.de

Abstract. Chamber model simulation is a common approach to simulate rotary positive displacement vacuum pumps. Therefore the pump is abstracted into working chambers and connecting clearances, whereby the clearance leakages can be identified as the major loss mechanism in such machines. The clearance mass flow rates are calculated with respect to the thermodynamic states in the adjacent chambers, which are inhomogeneous for rarefied gases due to the movement of the rotors which causes a pressure gradient within the chamber. This effect increases with higher Knudsen numbers, because of the increasingly dominant friction. It is shown that inhomogeneous chamber states cause a non-complete chamber filling. As a result the mass-averaged pressure within the suction chamber is lower than the pressure in the suction port. Due to the non-constant chamber volume over time three-dimensional transient simulations with a Computational Fluid Dynamics (CFD) solver are performed in order to investigate the mass within a geometrically abstracted suction chamber. Based on a dimensionless number, a regression analysis is done to provide a quantitative estimation of this effect by means of analytical calculations. This is implemented in a chamber model simulation software and results of a test machine are compared to measurements and to previous simulations.

1. Introduction

Dry running vacuum pumps have become an important part of vacuum systems due to their oil-free working principle. The need for a clean vacuum, for instance in the semiconductor industry, has pushed the development of dry running positive displacement pumps, such as screw and roots pumps (see reference [1]). Due to contactless operation of the pistons the sealing between the working chambers is realized by small clearances. Regarding figure 1 the effective mass flow rate of a screw machine can be calculated by the delivered mass flow rate $\dot{m}_d$ of a working chamber driven by the rotation of the rotors and reduced by the clearance mass flows $\dot{m}_{clearance}$. Clearances are located between both rotors like the radial clearance, the inter-lobe clearance and the blow hole, and between the rotors and the housing. In order to calculate the thermodynamic operation of these pumps, knowledge of these clearance mass flow rates, which represent the main loss mechanism, is essential. A common approach to calculate the thermodynamic operation of vacuum pumps is chamber model simulation where the pump is divided in a network of chambers that transport gas from low pressure port to high pressure port and connecting clearances which cause a mass flow rate in opposite direction [2][3][4].

Typically in chamber model simulation the thermodynamic state of the gas within the chambers is considered to be homogenous and the clearance boundary conditions are taken

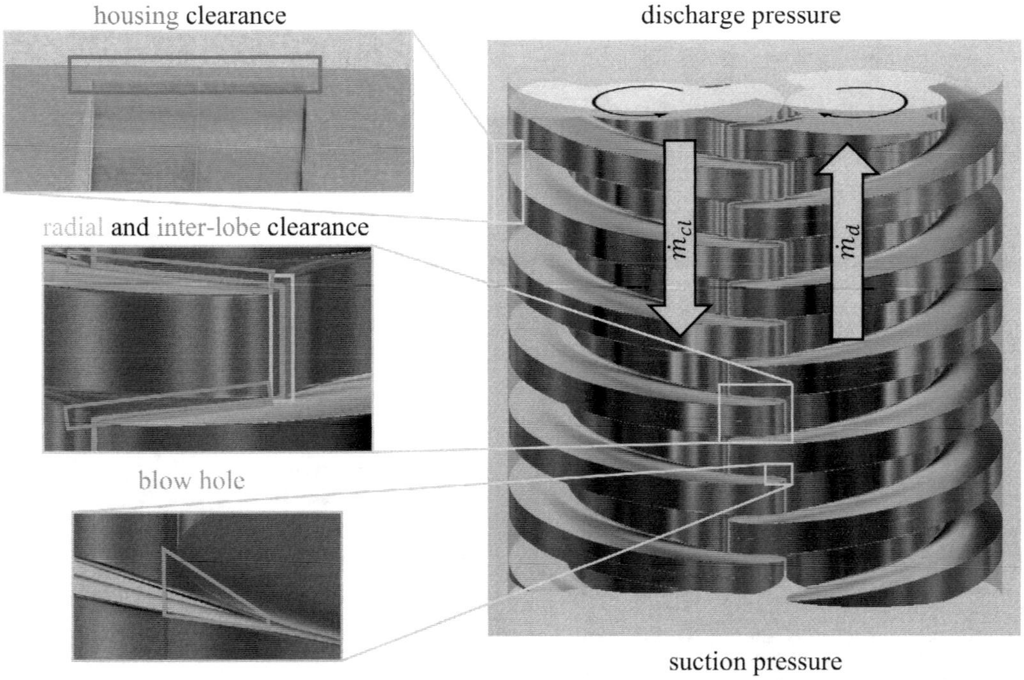

Figure 1. Sketch of an SSVP with 2-lobed cycloid profile.

from mean values within the adjacent chambers. The calculation of clearance flows for rarefied gases including moving boundaries has previously been investigated in references [3],[4] and [5] where a data base for different clearance shapes was created and by [6], [7] and [8] where a 1D model was derived and validated in order to calculate the clearance mass flow rates. The impact of the clearance flow models is shown in reference [9].

In a previous paper it was shown that the chamber states are not homogeneous for rarefied gases but that a pressure gradient occurs in tangential direction of the rotor lead [10]. This is again highlighted in Figure 2 which shows the chamber model abstraction for a screw spindle vacuum pump (SSVP) with the sketched network of chambers and connecting clearances. In this representation pressure p related to the chamber's mean pressure p_{ch} is sketched. The inhomogeneous pressure distribution causes the clearances to not only connect two adjacent chambers, but the high pressure side and a low pressure side of adjacent chambers. Thus the boundary conditions of the clearances have to be corrected for better agreement of simulation results to the pump's operation performance. In a previous work a one-dimensional model was presented to calculate the pressure distribution within the chamber. Although computational fluid dynamics (CFD) solver based on the Navier-Stokes-equation fail if the whole machine is calculated due to the gas rarefaction effects within the clearances, in the chambers the Knudsen numbers are lower due to the characteristic length scales. Therefore CFD simulations including Maxwell slip boundary conditions as presented in reference [11] can be applied within a wide range. In this paper the transient CFD simulations are done in order to investigate the chamber filling process with respect to the non-constant chamber volume over time. Different geometric and thermodynamic boundary conditions as well as the impact of different materials of the rotors on the chamber filling process is investigated. A dimensional analysis is performed and functionals are provided which depict the main impact of the pressure distribution within the suction chamber to reduce the computation time. This is implemented in a chamber model and compared to the results where the chamber filling process is calculated under a quasi-static assumption.

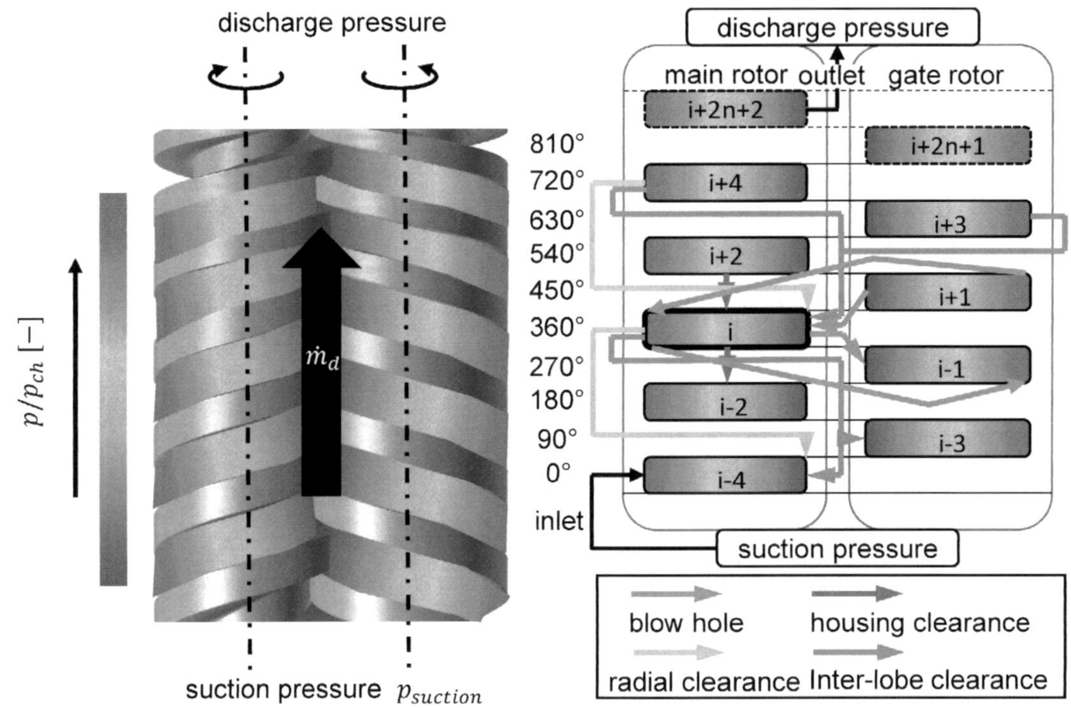

Figure 2. Sketch of an SSVP and the investigated impact of inhomogeneous chamber states on the boundary conditions of adjacent clearances. Clearance connections and models are similar to reference [10].

2. Theoretical approach

In order to show why the pressure gradient occurs within the chambers it is necessary to take the kinematics into account. Then a dimensional analysis is performed and a Strouhal number for the transient flow behaviour is derived.

2.1. Kinematics

Figure 3 shows the sketch of an unwound cylinder cut of one rotor of a two-lobed SSVP, where the chamber and the profile are alternately placed along the rotor axis. The rotor lead and the rotation of the rotors causes a chamber to move from low pressure port to high pressure port along the rotor axis with a velocity

$$U_{ax} = s \cdot f \tag{1}$$

where f is the rotational speed and s is the rotor lead. U_{ax} is the velocity of an encapsulated mass package, that is carried in axial direction. Furthermore each point on the rotor has a movement in circumferential direction

$$U_{circ} = \pi \cdot D \cdot f \tag{2}$$

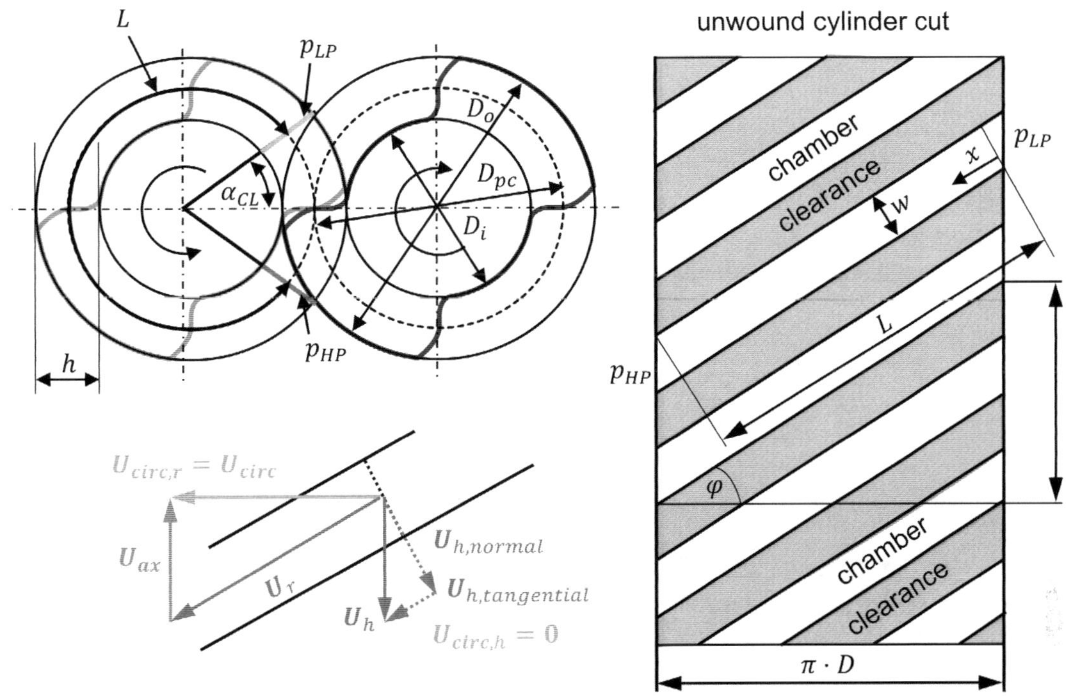

Figure 3. Geometry of a two-lobed cycloid profile and boundary velocities in the relative system of screw vacuum pumps.

which is the absolute velocity, where D is twice the distance from the rotor axis to the considered rotor point. The relation between the absolute velocities of the rotor points due to the rotation and the movement of the mass package in axial direction can be obtained by the angle φ.

Therefore the magnitude of the relative movement U_r of rotor points within the chamber's fixed coordinate system can be obtained by

$$U_r = \frac{\pi D \cdot f}{\cos(\varphi)} \qquad \varphi(D) = \tan^{-1}\left(\frac{s}{\pi D}\right) \qquad D_i \leq D \leq D_o \tag{3}$$

where D_o is the vacuum pump's outer diameter and D_i the inner diameter respectively.

The housing (h) does not move, so the housing's relative velocity $U_h = -U_{ax}$ is negative to the movement of the relative system. This can be divided into a part tangential and normal to the rotor lead:

$$U_{h,tangential} = s \cdot f \cdot \sin(\varphi(D_o)) \tag{4}$$

$$U_{h,normal} = s \cdot f \cdot \cos(\varphi(D_o)) \tag{5}$$

In each cross section of a chamber there are four bounding surfaces causing a Couette flow in the same direction tangential to the rotor lobe and the housing which also causes a flow in direction normal to the rotor lobe. Because of the chamber's encapsulation due to the other rotor and the surrounding housing this flow is dammed up. This leads to a pressure gradient which results in a Poiseuille flow in opposite direction of the Couette flow caused by the moving boundaries.

Geometrically the chamber can be abstracted as a channel with rectangular cross section with height

$$h = \frac{D_o - D_i}{2} \tag{6}$$

and the channel width

$$w = \frac{s}{2z} \cos\left(\varphi(D_{pc})\right) \tag{7}$$

which represents the linearised mean distance between two lobes. D_{pc} is the pitch circle diameter and mean value of inner and outer diameter. The chamber length is defined as

$$L = (\pi - \alpha_{CL}) \cdot D_{pc} \cdot \sqrt{1 + \left(\frac{s}{\pi D_{pc}}\right)^2} \tag{8}$$

with the angle of the cut lens

$$\alpha_{CL} = \tan^{-1}\left(\frac{\sqrt{D_o^2 - D_{pc}^2}}{D_{pc}}\right) \quad \text{(see Figure 3)} \tag{9}$$

This is the length of the helix, at the pitch circle. The end chamber geometry is defined between the intersection lines of rotor axis and the cusps of the cut lens which is the intersection area of the outer diameters of both rotors (see the yellow and red lines in Figure 3). The mean pressure in each of these intersection lines is defined as the chamber's high pressure p_{HP} and low pressure p_{LP} respectively, which is used as inlet and outlet boundary values of adjacent clearances. Thus the chamber is defined with constant cross section area for constant rotor lead s. The impact of a non-constant rotor lead on the pressure distribution within the chamber is neglected in this work.

2.2. *Dimensional analysis*

Assuming isothermal conditions and a constant gas mixture, that can be related to single gas parameters of a perfect gas, the following thirteen parameters influence the pressure distribution within an SSVP's chamber with the above geometric constraints: The specific gas constant R, the mean temperature within the chamber T, the dynamic viscosity μ, the chamber's mass averaged pressure at time t

$$p_{ch}(t) = \frac{m_t(t) \cdot R \cdot T}{V(t)} \tag{10}$$

where $m_t(t)$ is the total mass of the gas within the chamber and $V(t) = A \cdot L(t)$ is the chamber's total volume at time t. Furthermore, a mass flow rate $\dot{m}_{cl,t}$ within the chamber due to adjacent clearances tangential to the lobe is considered. Further parameters are the rotational speed f and the geometric constraints: The rotor lead s, the outer and inner rotor diameter D_o and D_i and the number of lobes z. Two more parameters are the tangential momentum accommodation coefficients (TMAC) of the housing α_h and the rotor α_r. These parameters are dimensionless numbers $0 \leq \alpha \leq 2$ that describe the gas surface interaction in rarefied gases and depend on the wall material and the gas properties and can be obtained by measurements in the literature, see for example reference [12]. If the rotor is made of a different material than the housing (for example due to coating), different TMACs are needed for the certain walls in the model. This also changes if the gas is exchanged for example instead of air, helium or hydrogen is evacuated.

According to Buckingham's theorem [13] this set of fourteen parameters leads to the following set of ten dimensionless parameters:

The pressure ratio

$$\Pi = \frac{p(x,t)}{p_{ch}(t)} \tag{11}$$

for an arbitrary position x within the chamber, the gas rarefaction parameter δ which is inverse to the Knudsen number related to the model of hard sphere molecules [14]

$$\delta = \frac{\min(h, w) \cdot p_{ch}(t)}{\mu \cdot c_m} \tag{12}$$

with $c_m = \sqrt{2RT}$ the molecule's most probable speed, the speed ratio

$$u_0 = \frac{U_{circ,o}}{c_m} = \frac{\pi D_o \cdot f}{\sqrt{2RT}} \tag{13}$$

the dimensionless mass flow rate

$$C_0 = \frac{\dot{m}_{cl,t}}{\dot{m}_{eff}} = \frac{\dot{m}_{cl,t} \cdot \sqrt{\pi} c_m}{p_{ch} \cdot hw} \tag{14}$$

with $\dot{m}_{eff}$ the chamber's effusion mass flow rate and the geometric relations

$$g_1 = \frac{D_o}{D_i} \quad g_2 = \frac{D_o}{s} \quad g_3 = \frac{L(t)}{L_{\max}} \tag{15}$$

where $L(t)$ is the length at time t and $L_{\max}$ is the length calculated with equation 8. The last numbers are the already dimensionless number of lobes z which scales the channel width and the two TMACs α_h and α_r.

2.3. *Strouhal number*

In order to characterize the transient states within the chamber filling process, it is useful to estimate the inertial forces within the chamber:

$$F_{iner} = \frac{\mathrm{d}(m \cdot v)}{\mathrm{d}t} \approx \pi D_{pc} f \left(V \frac{\partial \rho}{\partial t} + \rho \frac{\partial V}{\partial t} \right) = \pi D_{pc} f A \rho \left(\frac{L}{\rho} \frac{\partial \rho}{\partial t} + \frac{\partial L}{\partial t} \right) \tag{16}$$

As the chamber grows with an average constant velocity

$$v = \pi D_{pc} f \tag{17}$$

the derivative $\partial v / \partial t \approx 0$. The mass $m = \rho V$ within the chamber is the mean density times the chamber volume with a constant cross section area $A = hw$. Neglecting the rotor lead, the length of the chamber grows with a linear function and therefore

$$\frac{\partial L}{\partial t} = \pi D_{pc} f = v \tag{18}$$

is a constant and the second term in equation 16 becomes

$$F_{iner,stat} = \rho A v^2 \tag{19}$$

which resembles the estimation of stationary inertia, while the first term is the transient part which can be extended as follows under isothermal conditions:

$$F_{iner,trans} = \rho v A \frac{L}{\rho} \frac{\partial \rho}{\partial t} = \rho v A \frac{L}{p} \frac{\partial p}{\partial t} = \rho v A \frac{L}{p} \left(\frac{\partial p}{\partial x} \frac{\partial x}{\partial t} + \underbrace{\frac{\partial p}{\partial y} \frac{\partial y}{\partial t}}_{\approx 0} + \underbrace{\frac{\partial p}{\partial z} \frac{\partial z}{\partial t}}_{\approx 0} \right) = \rho A v^2 \frac{L}{p} \frac{\partial p}{\partial x} \tag{20}$$

Therefore a parameter resembling a compressible Strouhal number St is obtained through the relation of the first term to the second term

$$St = \frac{F_{iner,trans}}{F_{iner,stat}} = \frac{L}{p}\frac{\partial p}{\partial x} = \frac{L}{\min(h,w)G_P}\left(\frac{2\sum_{i=1}^{4} G_{C,i}U_i}{c_m} - \frac{C_0}{\sqrt{\pi}}\right) \tag{21}$$

which can be evaluated with the mass averaged pressure p_{ch} for any time step. This expression for the dimensionless pressure gradient is derived in Reference [9] and is a result of a superposition of Couette flow and Poiseuille flow within the chamber. The values for the reduced flow rates for the Couette flow G_C and G_P can be taken from References [15] and [16] for any gas rarefaction parameter δ, width to height ratio η and different TMACs for any wall.

3. Three-dimensional transient simulation

In order to investigate the chamber filling in SSVPs, transient Computational Fluid Dynamics simulations (CFD) using ANSYS CFX 2021 R2 are performed where the Navier-Stokes-Fourier equations (NSF) are solved numerically [17].

3.1. Domain and boundary conditions

Therefore the suction chamber is abstracted as shown in Figures 4 and 5. As a first simplification the rotor curvature is neglected. The mean chamber height as well as the chamber length and flank velocities along the pitch circle are then evaluated as shown in section 2.1. This simplification is verified in reference [9], where identical results are obtained using a 1D model with the mentioned geometric abstraction and full 3D simulations including the curvature and linear rising wall velocities along the radius. As a second simplification, the rotor lead and therefore the axial movement of the chamber is neglected.

Figure 4 shows the used boundary conditions of the static domain (Figure 4 a), which is the connection to the machine's low pressure port and the boundary conditions of the moving domain which is the suction chamber (Figure 4 b). The two domains are connected with a conservative interface. The housing is marked in magenta, while the red planes are openings, where the suction pressure is prescribed. The green planes mark the machine's root circle and the black planes are the flanks with the average absolute velocity of the pitch circle. In Figure 4 a, the top plane and the upper front plane is a symmetry plane. The moving domain is just marked with a transparent orange and explained in Figure 4 b where the back plane is the housing while the front plane marks the machine's root circle. The top and bottom plane are the flanks and the yellow plane is the abstracted intersection edge with the cut lens, which is modelled as a no slip wall. For all other walls Maxwell slip boundary conditions as described and validated in reference [11] are used in order to improve the results for the slip regime.

In Figure 5 two different time steps are shown. While the rotor moves clock wise along the y axis, the suction chamber grows according to equation 18. With the assumptions that curvature and rotor lead are neglected, the right hand side of the chamber is fixed and the inlet with the interface moves with constant velocity in negative x direction, in the absolute system. As this would cause problems with the boundary conditions in Ansys CFX because openings are not to be moved in normal direction [17], a change of the relative system is carried out which is the relative system of the rotating inlet area. Thus the inlet domain on the left hand side in Figure 5 is fixed and the chamber elongates in positive x direction. To achieve the same physical state, the relative velocity

$$U_{r,inlet} = -\pi D_{pc} f \tag{22}$$

is subtracted from each boundary condition. Therefore the wall velocities within the CFD simulations are as follows:

$$U_{r,pc} = U_{pc} - U_{r,inlet} = -\pi D_{pc} f + \pi D_{pc} f = 0 \tag{23}$$

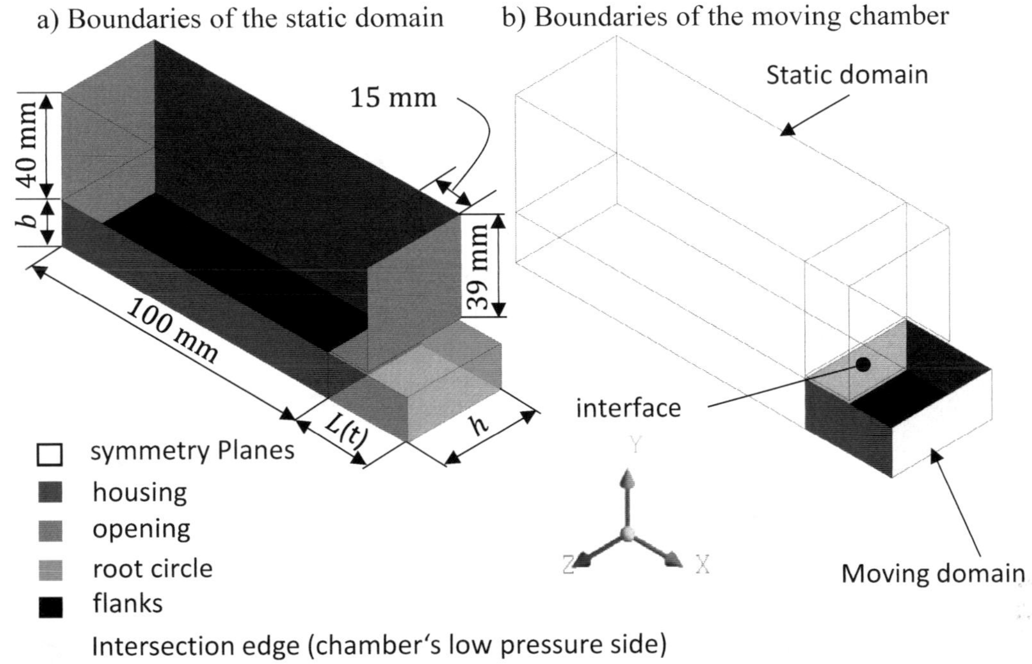

Figure 4. Boundary conditions and 3D domains in Ansys CFX.

$$U_{r,h} = 0 - U_{r,inlet} = \pi D_{pc} f \tag{24}$$

$$U_{r,root} = -\pi D_i f + \pi D_{pc} f = \pi f (D_{pc} - D_i) \tag{25}$$

with $U_{r,pc}$ the relative velocity of the pitch circle, $U_{r,h}$ the relative velocity of the housing and $U_{r,root}$ the relative velocity of the root circle without rotor lead. The simulations are initialised with the suction pressure and the negative relative velocity $-U_{r,inlet}$ with an initial chamber volume of 2.5 % of the whole chamber using a constant node number which is stretched with time. Comparing the pressure and velocity distribution within the two time steps in Figure 5 in the middle of the domain in z direction, a constant pressure is obtained in the static domain, while the pressure within the suction chamber shrinks over x. Regarding the velocity fields a large velocity in x direction is obtained in the static domain with velocity close to zero at the flanks, which cause stagnation points at the chamber's inlet with a large boundary layer due to the low gas rarefaction parameter and Reynolds number. At the earlier time step the velocity at the top is still higher due to the initialization.

As the pressure in the openings p_{HP} with

$$\delta_{HP} = \frac{\min(h, w) \cdot p_{HP}}{\mu \cdot c_m} \tag{26}$$

is prescribed and the mass and volume in the moving domain changes in each time step, the mass averaged pressure as well as the chamber's low pressure p_{LP} are calculated as follows:

$$p_{ch}(t) = \frac{RT}{V(t)} \int \rho(t)\, \mathrm{d}V(t) \tag{27}$$

$$p_{LP}(t) = \frac{1}{A} \int p(t)\, \mathrm{d}S_{LP} \tag{28}$$

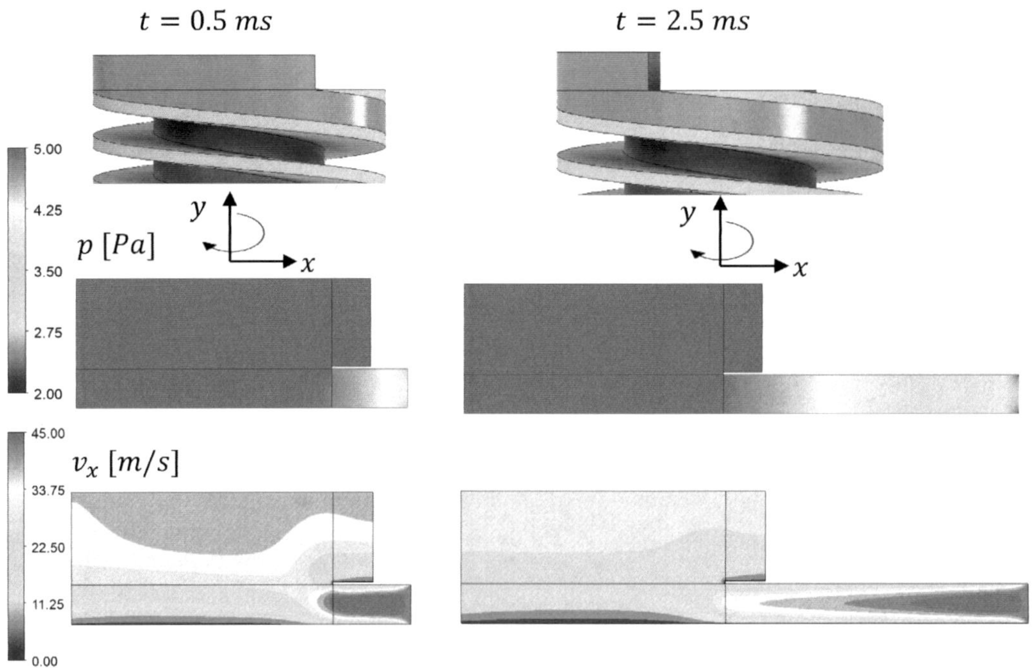

Figure 5. Geometric abstraction of the CFD domain and with pressure and velocity distribution of two time steps.

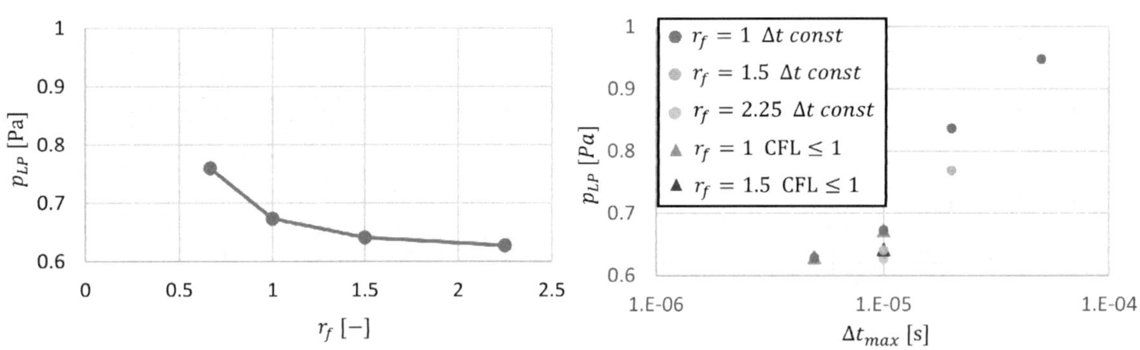

a) Impact of the mesh refinement with $\Delta t = 10^{-5}\ s$ **b)** Impact of the time step Δt

Figure 6. Impact of mesh and time step on the chamber's low pressure p_{LP} at $L = L_{max}$ for the reference case and $\delta = 10$

$V(t)$ is the volume of the moving domain and $\mathrm{d}S_{LP}$ is a surface increment of the chamber's low pressure side.

3.2. Error estimation

All simulations are transient with a 2nd order backward Euler scheme, isothermal, laminar ($Re \ll 1$) with Maxwell slip boundary conditions with diffuse wall scattering and applied high resolution scheme for the convection terms [17]. In order to make an error estimation a mesh study and a time step study is done. Therefore a reference case is created with the machine boundary conditions given in Table 1 with 279500 nodes in the static domain and 255000 nodes in the moving domain. Starting from this reference mesh, different meshes with a constant refinement factor r_f ($r_f = 1$ is the reference mesh) in each direction in the whole domain have been created and the resulting low pressure p_{LP} is evaluated at L_{max} with a constant time step $\Delta t = 10^{-5}$ s. This is shown in Figure 6a, where p_{LP} has an asymptotic course with further refinement. A Richardson extrapolation according to references [18] and [19] provides that the discretization error of the reference mesh is about 7 % regarding p_{LP} and about 2 % regarding the mass within the chamber. It needs to be noted, that the maximum Courant-Friedrichs-Levy (CFL) number (see reference [19]) is about 17 at the start of the simulations and smaller than 0.5 at the end for the reference mesh as the cell size grows with time. Therefore the impact of the time step is investigated in Figure 6b which shows p_{LP} as a function of the maximum time step. Three different meshes are used with a constant time step and two meshes with an adaptive time step with CFL ≤ 1. It is shown, that there is no significant impact whether the CFL number is smaller than one or the time step is set constant, but the choice of the maximum time step is crucial and by reducing the time step the difference between the finest mesh and the reference mesh can be neglected, but the simulation time can be reduced from about a day to about three hours. Therefore the reference mesh is used with a constant time step $\Delta t = 5 \cdot 10^{-6}$ s. The results can be explained by the movement of the mesh. Therefore a smaller time step also provides a better geometric resolution and the change of p with time is smaller.

Table 1. Data of the reference case

D_o [m]	D_i [m]	w [m]	f [Hz]	T [K]	δ_{HP} [−]	R $[J/(kg\ K)]$	μ $[Pa\ s]$
0.17	0.1	0.015	100	298.15	10	287.1	$1.8351e-5$

4. Results

In this section it is investigated how different boundary conditions influence the chamber filling. Therefore starting from the reference case in Table 1, at first different time steps are evaluated. For each time step the mass average pressure and the low pressure within the moving domain are calculated according to equations 27 and 28. Then the Strouhal number related to the mass averaged pressure is evaluated using equation 21. This is done every 0.5 ms until the chamber's full length according to equation 8 is reached. Figure 7 shows the pressure ratio p_{ch}/p_{HP} as a function of the Strouhal number St for different boundary conditions while the impact of a mass flow rate of adjacent clearances is neglected ($C_0 = 0$). The different colours are related to a fixed gas rarefaction parameter which is evaluated at the inlet, while the different symbols show different rotor frequencies which impacts the wall velocities and the chamber speed $\mathrm{d}L/\mathrm{d}t$. The filled symbols are related to the width of the reference case $w = 0.015$ m while the empty symbols have a width $w = 0.03$ m. It is obtained, that all different chamber states are reduced to a single curve which can be approximated with a logarithmic function in equation 29 which is shown as the black dash-dot line (ln is the natural logarithm). As the pressure ratio p_{ch}/p_{HP} is the same as the actual mass within the chamber related to the mass if the chamber was filled homogeneously, it can be deduced, that with higher Strouhal number the chamber gets less and less filled and therefore at $St \approx 3$ only about half of the theoretical mass is sucked into the machine without taking clearances into account. Thus the shown ratio p_{ch}/p_{HP} can be seen as a filling efficiency. Furthermore it is seen, that large Strouhal numbers are obtained by low gas rarefaction parameters, by high rotor frequencies and by small values of the chamber width compared to the height. Now a comparison between the filling efficiency of single-lobed and

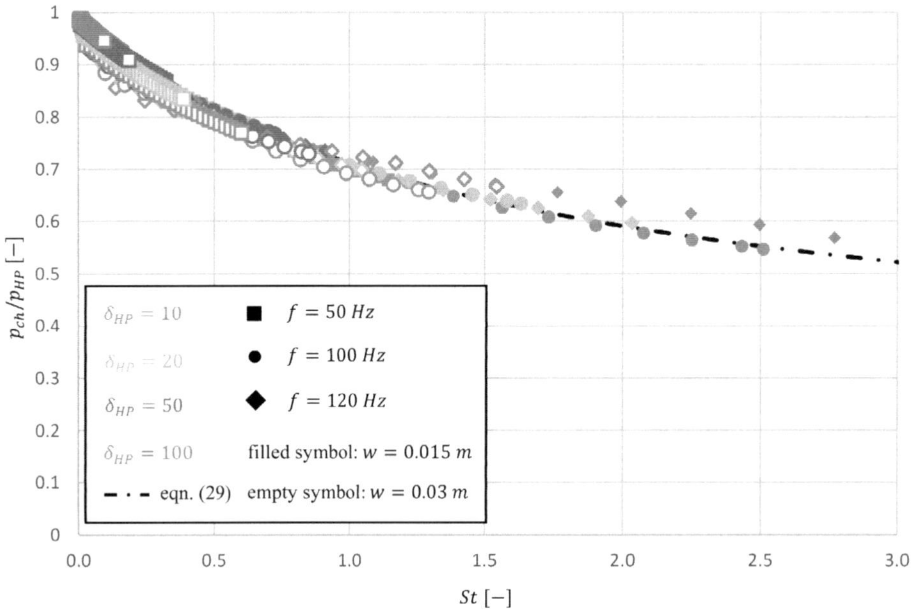

Figure 7. pressure ratio p_{ch}/p_{HP} as a function of the Strouhal number equation 21 for different thermodynamic and geometric boundary conditions and $C_0 = 0$.

two-lobed SSVPs can be done using equation 7 for the chamber width w and equation 12 for the gas rarefaction parameter δ assuming $w \leq h$. It is obtained, that $p_{ch}(t) = \frac{2z\delta\mu c_m}{s\cos(\varphi(D_{pc}))}$. So if the same machine configuration is used and only the number of lobes is set from $z = 2$ to $z = 1$ for same δ the pressure ratio $\frac{p_{ch}(t,z=2)}{p_{ch}(t,z=1)} \approx 2$ and furthermore according to Figure 7 the single-lobed SSVP has a better filling efficiency as the width w is larger. So all in all the filling efficiency of a single-lobed SSVP is better than that of a two-lobed SSVP.

Similar results are obtained when the pressure ratio p_{LP}/p_{ch} is evaluated which is illustrated in Figure 8. This curve can be perfectly fitted by the exponential function in equation 30.

$$\frac{p_{HP}}{p_{ch}} \approx 1 + \ln\left(\frac{St}{2} + 1\right) \tag{29}$$

$$\frac{p_{ch}}{p_{LP}} \approx \exp(0.58 \cdot St) \tag{30}$$

Next the impact of a mass flow rate through adjacent clearances on the transient chamber filling is investigated. Two different cases are studied. The first is a mass flow rate that comes from the direction of a cut lens tangential to the rotor lobe combined by radial clearance, inter-lobe clearance and blow hole. This is modelled by mass sources equally distributed on the chamber's low pressure side (yellow area in Figure 4b). The mass flux per area increment is calculated in a way, that the combined mass flow rate can be normalized with a constant suction pressure which is the chamber's high pressure p_{HP}:

$$C_{0,HP,t} = \frac{\dot{m}_{cl,t} \cdot \sqrt{\pi} c_m}{p_{HP} \cdot hw} \tag{31}$$

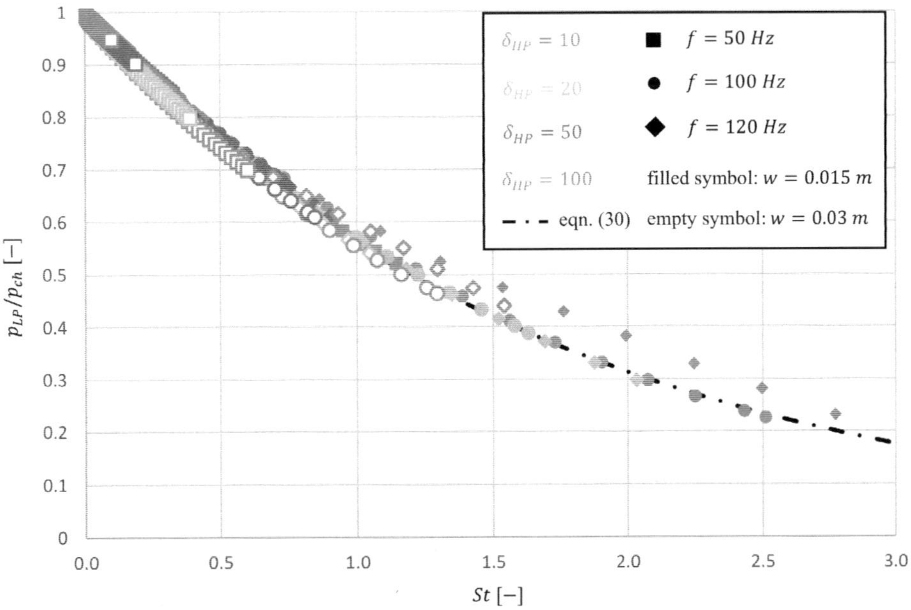

Figure 8. pressure ratio p_{LP}/p_{ch} as a function of the Strouhal number equation 21 for different thermodynamic and geometric boundary conditions and $C_0 = 0$.

The prescribed mass flux per area increment is:

$$\frac{\dot{m}}{hw} = \frac{C_{0,HP,t} \cdot p_{HP}}{\sqrt{\pi} c_m} \tag{32}$$

In order to evaluate the Strouhal number St at the mass averaged pressure according to equation 21, the normalized mass flow rate is calculated as follows:

$$C_{0,t} = C_{0,HP,t} \cdot \frac{p_{HP}}{p_{ch}(t)} \tag{33}$$

The second case is a prescribed mass flux for the housing clearance (purple area in Figure 4b). Again the source term of the resulting inflowing mass flow rate $\dot{m}_h$ of the housing is normalized using the chamber's constant high pressure p_{HP} with the definition:

$$C_{0,HP,h} = \frac{\dot{m}_h \cdot \sqrt{\pi} c_m}{p_{HP} \cdot hw} \tag{34}$$

Due to the different area of the housing clearance, the respective mass flux per area increment is calculated as follows:

$$\frac{\dot{m}_h}{L(t)w} = \frac{C_{0,HP,h} \cdot p_{HP}}{\sqrt{\pi} c_m} \frac{h}{L(t)} \tag{35}$$

As explained in reference [9] for the quasi-static simulations only half of the housing clearance mass flow rate is used to calculated the Strouhal number at the mass averaged pressure, because it is equally distributed along the whole chamber. So when evaluating the pressure ratios p_{ch}/p_{HP} and p_{LP}/p_{ch} for the certain ways from about the middle of the chamber (p_{ch}) to the respective chamber ends only about half of this value impacts the pressure gradient. Therefore

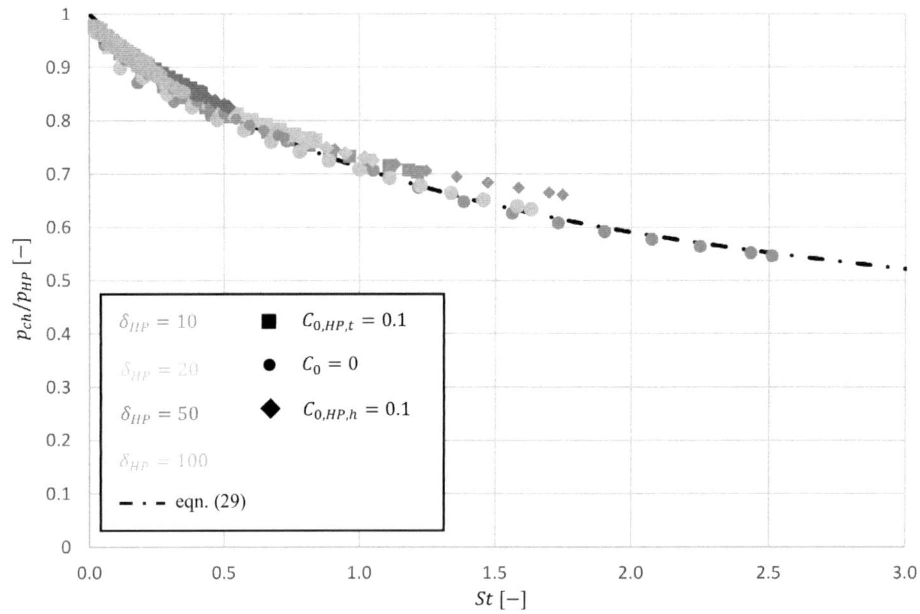

Figure 9. pressure ratio p_{ch}/p_{HP} as a function of the Strouhal number equation 21 for different thermodynamic and geometric boundary conditions with $w = 0.015$ m and $f = 100$ Hz.

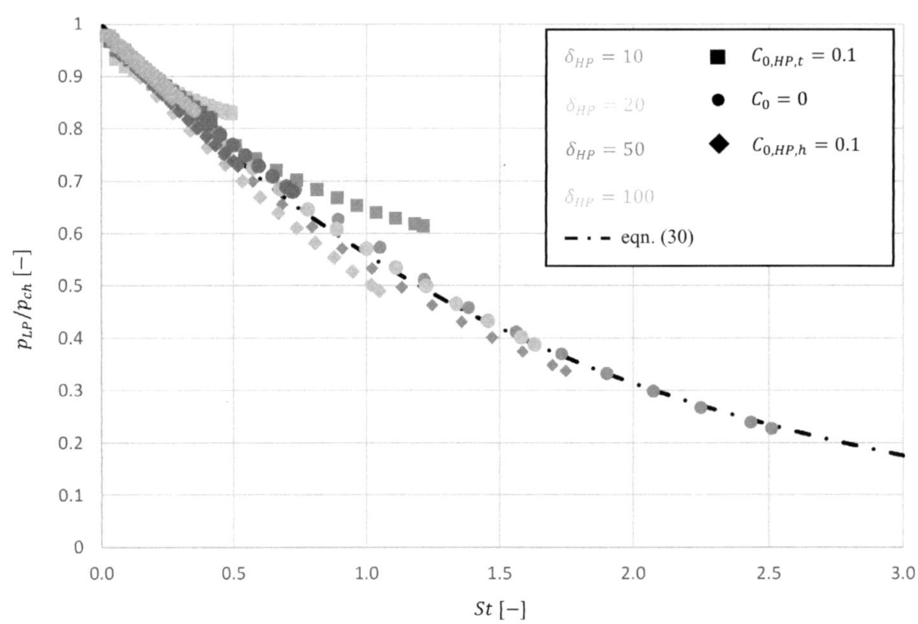

Figure 10. pressure ratio p_{LP}/p_{ch} as a function of the Strouhal number equation 21 for different thermodynamic and geometric boundary conditions with $w = 0.015$ m and $f = 100$ Hz.

the normalized mass flow rate for the housing clearance used to evaluate the Strouhal number at the mass averaged pressure p_{ch} is calculated by

$$C_{0,h} = \frac{C_{0,HP,h}}{2} \cdot \frac{p_{HP}}{p_{ch}(t)} \,. \tag{36}$$

Figure 9 shows the impact oft the adjacent mass flow rates on the pressure ratio p_{ch}/p_{HP} for the reference case with $w = 0.015\,\mathrm{m}$ and $f = 100\,\mathrm{Hz}$. As before the different colours show the different gas rarefaction parameters related to the chambers high pressure p_{HP} and the symbols represent the cases without any adjacent mass flow rates (circles), with a mass flow rate at the chamber's low pressure side (squares) and a mass flow rate through the housing (diamonds). It is seen, that the certain mass flow rates lead to higher pressure ratios and smaller Strouhal numbers, so that equation 29 fits perfectly and the impact of clearance mass flow rates on the filling efficiency is depicted with the above assumptions.

Figure 10 shows the impact of the adjacent clearance mass flow rates on the pressure ratio p_{LP}/p_{ch} for the same boundary conditions. It is obtained, that the housing clearance matches the functional of equation 30 very well. The tangential mass flow rates on the chamber's low pressure side lead to slightly higher pressure ratios which might be due to the fact, that the mass sources are placed on the same plane where the low pressure is evaluated. If a small volume was modelled like the end volumes within the cut lens, this might not be the case. Nevertheless the differences are not too big, so even if there is a modelling error using equation 30 for large values of the tangential mass flow rate, it is neglected in this work. As already stated in reference [9] the chamber's low pressure side has a much less impact on the machine performance, as it only corrects the outlet pressure for the determination of the clearance mass flow rates and this impact is quite low in the low pressure regime.

Now at first the rotor material for the reference case (Table 1) is exchanged to aluminium and therefore the TMAC $\alpha_r = 0.72$ (see reference [12]) while the TMAC of the housing is still $\alpha_h = 1$. Furthermore the geometric data of Table 2 is used with a rotor frequency $f = 83.33\,\mathrm{Hz}$ but diffuse wall scattering. All simulations are shown for $\delta_{HP} = [10, 20, 50, 100]$ in Figures 11 and 12. Again equations 29 and 30 perfectly predict the simulation data. The change of the rotor material can lead to smaller Strouhal numbers with otherwise same boundary conditions and therefore improve the chamber filling process. Nevertheless, the clearance mass flow rates will also be affected by the TMACs and therefore this does not necessarily mean that the overall machine is better if the rotor is made of a different material or is coated. This needs further investigation.

Although the impact of the rotor lead is neglected within the scope of the CFD simulations, it can be considered in order to calculate the Strouhal number St. As the rotor lead slightly effects the chamber length as well as the wall velocities (as shown in section 2.1), the rotor lead slightly increases the resulting Strouhal number and with the help of equations 29 and 30 p_{LP} and p_{HP} can be calculated.

5. Chamber model simulation

In this section the impact of the chamber filling process on the integral machine performance is investigated via chamber model simulation.

5.1. Geometric and thermodynamic model

As sketched in Figure 2 for the chamber model the vacuum pump is divided into chambers that transport mass from low pressure port to high pressure port and connecting clearances. Therefore the geometric properties of the chambers and clearances as a function of the rotation angle are needed. The geometric model used in this paper is the same as in reference [10].

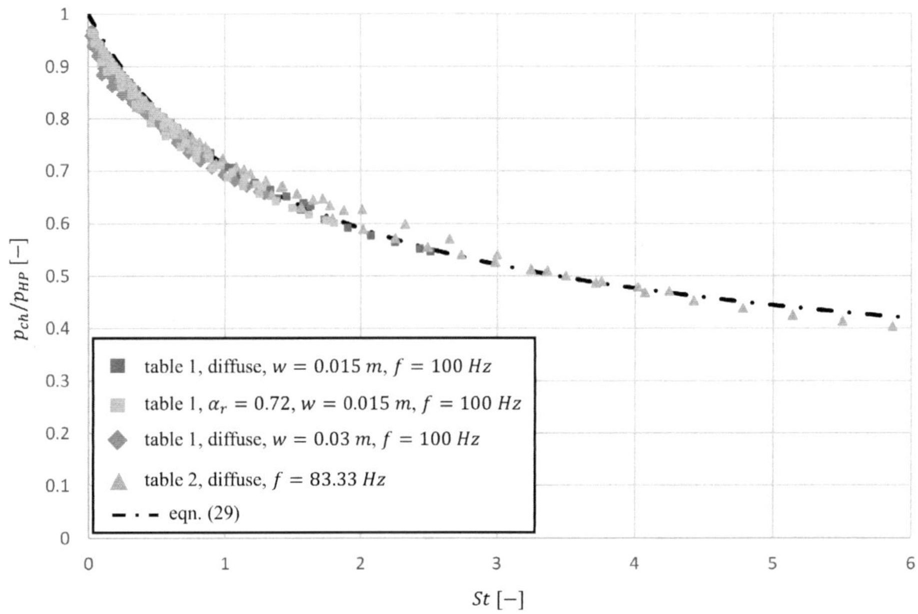

Figure 11. pressure ratio p_{ch}/p_{HP} as a function of the Strouhal number equation 21 for different thermodynamic and geometric boundary conditions.

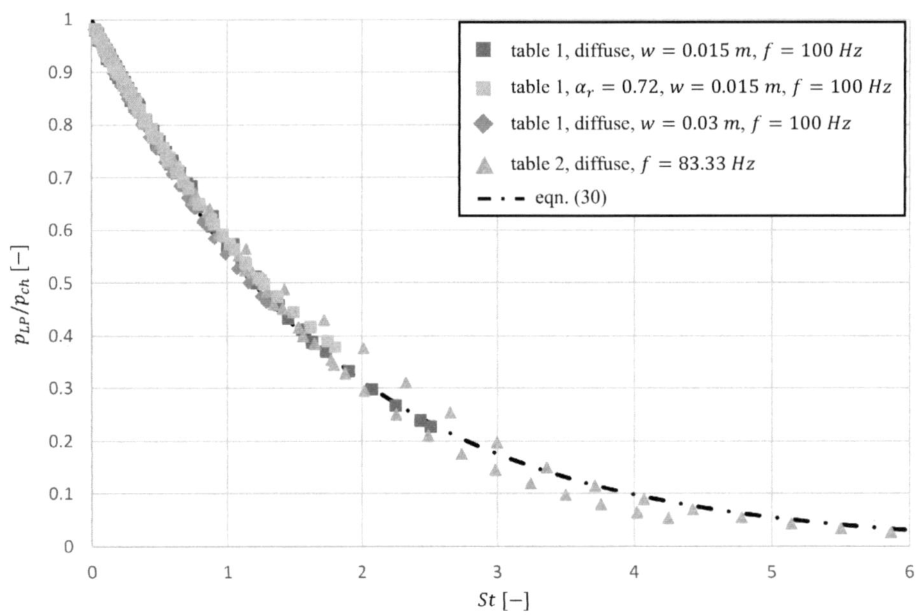

Figure 12. pressure ratio p_{LP}/p_{ch} as a function of the Strouhal number equation 21 for different thermodynamic and geometric boundary conditions.

Therefore the clearances are not simply connected with two chambers, but with a certain chamber position. The blow hole connects the high pressure side of two chambers and the low pressure side of two chambers, while the radial and inter-lobe clearance connect the high pressure side of the chamber at higher rotation angle with the low pressure side of the chamber with lower rotation angle considering the rotation angle to start at the low pressure port. The boundary conditions of the housing clearance are taken from mass averaged mean values of the adjacent chambers. The determination of mass flow rates for given geometric and thermodynamic boundary conditions is based on tabulated data as described in reference [9].

Within the chamber model simulation the total chamber volume with the given chamber mass for each rotation angle is available. But due to the chamber's geometric abstraction for the inhomogeneous model, there is a discrepancy between the model's chamber volume $V_{mod} = L(t)hw$ and the total chamber volume V_t, as the volume within the cut lens is neglected. Thus for chambers that are not connected to the high or low pressure port, the total chamber volume is divided as follows

$$V_t = V_s + V_{mod} + V_e \tag{37}$$

where $V_s = V_e$ are the chamber's start and end volume respectively. This classification is performed as the total mass m_t within the chamber is divided on these three volumes with the linearisation

$$m_s + m_e = \int \rho \, \mathrm{dV_s} + \int \rho \, \mathrm{dV_e} \approx \frac{p_{ch}}{RT} \left(V_s + V_t\right) \tag{38}$$

which assumes a linear pressure distribution within the chamber. As either the start and end volumes are small compared to the total chamber volume or the deviation from the linear pressure distribution is small in case that V_{mod} is small the total error of this assumption is small. Therefore the current chamber length for the model can be calculated by

$$L(t) = \frac{V_t(t) - V_s - V_e}{hw} \tag{39}$$

without changing the mass averaged pressure for the model or violating conservation of mass. Now a case distinction is done whether the chamber is open or closed. As the test machine has a constant rotor lead, the machine is closed when the gradient of the volume curve $\partial V_t/\partial t = 0$. Otherwise the chamber has a connection to the low or high pressure port. Therefore equations 29 and 30 are used to calculate p_{HP} and p_{LP} if $L(t) > 0$ and $\partial V_t/\partial t \neq 0$ and equations 22 and 23 of reference [9] otherwise.

In order to consider the impact of the mass flow rate of adjacent clearances on the pressure distribution within the chamber, the whole mass flow rate $\dot{m}_{cl,t}$ to calculate C_0 in equation 14 is treated explicitly. Therefore the sum of the mass flow rates on the chamber's low pressure side $\sum \dot{m}_{LP}$, the sum of the mass flow rates on the chamber's high pressure side $\sum \dot{m}_{HP}$ and the sum of inflowing and outflowing housing clearance $\sum \dot{m}_h$ of the respective chamber are taken from the last time step. As for closed chambers a stationary approach is used, the mass flow that enters on the low pressure side needs to flow out on the chamber's high pressure side. Which in chamber model simulation typically is not the case. Therefore a simplification

$$\dot{m}_{cl,t} \approx \min\left(\left|\sum \dot{m}_{LP}\right|, \left|\sum \dot{m}_{HP}\right|\right) + \frac{\sum \dot{m}_h}{2} \tag{40}$$

is done in order to improve the model accuracy. For open chambers the min− function is not needed, as only one side has adjacent clearances. The reason why only half of the effective housing clearance mass flow rate is used is explained section 4. Furthermore due to convergence problems the maximum mass flow rate has been capped thus $0 \leq C_0/\pi \leq \sum_i G_{C,i} U_i / c_m$ holds. The new model is implemented in the chamber simulation software *KaSim* developed at the Chair of Fluidics of TU Dortmund University described in reference [20].

5.2. Comparison to measurements

In order to work out the impact of the new investigated model for inhomogeneous chamber states including transient chamber filling, simulations for a test machine are performed and compared to measurements and to homogeneous simulations. This machine is an isochoric two-lobed cycloid profiled SSVP. The dimensionless data is given in Table 2, where L_R is the rotor length, h_h, h_{il} and h_{rc} are clearance heights of the housing clearance, the inter-lobe clearance and the radial clearance respectively and ψ_t is the machine's total wrap angle. As the machine is cooled it is assumed to be fully isothermal at room temperature $T = 293K$. The machine neither has a radial inlet area nor a pressure relief valve.

Table 2. Data of the test machine

D_o/D_i	D_o/s	D_o/L_R	D_o/h_h	D_o/h_{im}	D_o/h_{rc}	ψ_t	z
1.571	0.667	5.641	880	733.33	733.33	3474°	2

Figure 13 shows the ratio of the suction speed

$$\dot{V} = \frac{\dot{m}_{effective} \cdot R \cdot T}{p_{suction}} \tag{41}$$

to the theoretical suction speed

$$\dot{V}_{th} = 2 \cdot V_{ch,max} \cdot f \cdot z \tag{42}$$

as a function of the suction pressure $p_{suction}$. $\dot{m}_{effective}$ is the effective mass flow rate of the machine. The discharge pressure is always 10^5 Pa. Comparing the results of *KaSim* without the new model (red line) to the measurements, a good agreement is obtained for large values of the suction pressure. But with the decrease of the suction pressure those two curves diverge. The measurements show that the machine cannot evacuate the low pressure volume lower than about 6 Pa at the given rotational speed, while the simulated suction speed is still at a high value. Regarding the model of references [10] and [9] (purple line) the curve is not affected in the high pressure regime, but declines at low suction pressures in qualitatively good agreement with the measurements, but it slightly seems to overrate the effect, as the results indicate a worse suction speed than the measurements. In this implementation the tangential mass flow rate in order to calculate the pressure distribution within the chambers is $\dot{m}_{cl,t} = 0$. Furthermore, the open chambers are calculated with the same quasi-static approach as the closed chambers. The simulation results of the orange curve use the same approach, but the impact of the mass flow rate on the pressure distribution is considered ($\dot{m}_{cl,t} \neq 0$). This improves the simulation results and the measured ultimate pressure is almost exactly predicted. The simulation results of the green line include the results of the transient chamber filling using equations 29 and 30. Furthermore, the impact of the mass flow rate on the pressure distribution within the chambers is considered. The overall agreement of the new model compared to measurements is better than that of pure quasi-static approach without taking the mass flow rate into account. The ultimate pressure is calculated to be slightly better than the measurements indicate and therefore is not a conservative estimation any more. Nevertheless, this can also be related to a small leakage at the machine's low pressure side which would highly effect the ultimate pressure while the other operation points would hardly be affected. Another possibility is that the used model to calculate the inter-lobe clearance mass flow rate for given boundary conditions can be improved as stated in reference [9]. Principally the transient model for the chamber filling process seems to be more physically accurate than the quasi-static approach although the deviation to measurements is higher. This contradiction might be a result of the above made statements. Nevertheless the overall impact between the two models is small compared to the homogeneous approach which again is highlighted with the blue curve that shows the suction speed curve in case that a radial inlet would have been provided that ends at the same point as the axial inlet. In this case it is assumed, that the open chamber is homogeneous, but the closed chambers are not. This would highly improve the machine's performance but still would not reach the homogeneous results by far.

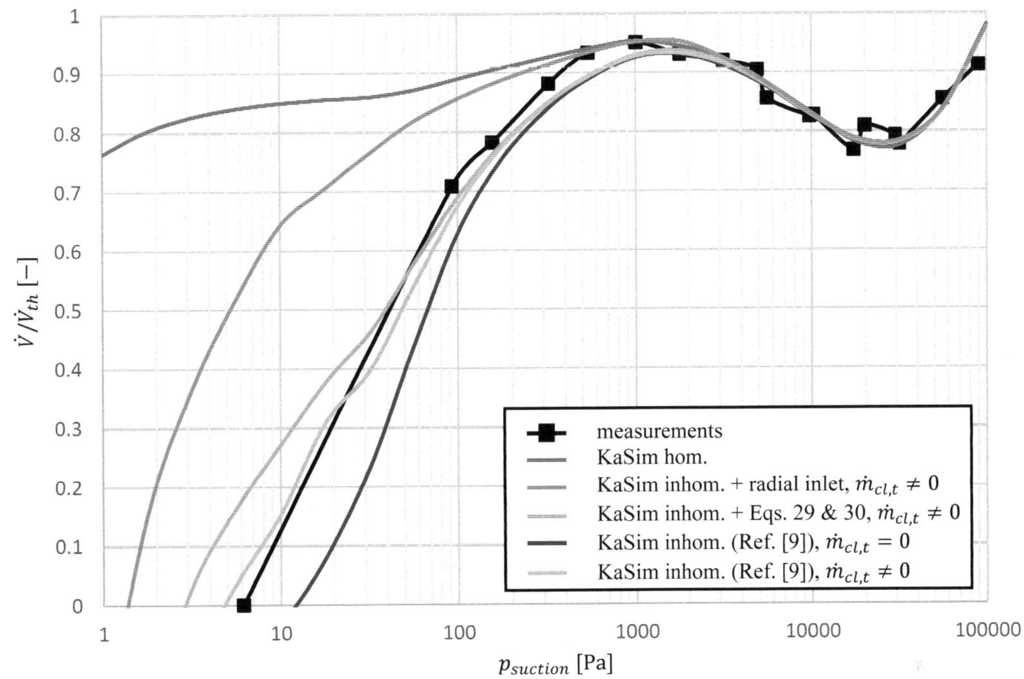

Figure 13. Comparison of the suction speed curves of measurements and chamber model simulation with homogeneous and inhomogeneous pressure distribution within the chambers for a rotational speed $f = 83.33$ Hz at constant temperature $T = 293$ K.

6. Conclusion

In this work the transient chamber filling of screw spindle vacuum pumps is investigated with transient CFD simulations including Maxwell slip boundary conditions. Within these simulations the pressure in the machine's low pressure port is prescribed and the chamber volume grows with time. At different time steps the mass within the chamber as well as the chamber's low pressure is evaluated and compared with the case, that it is filled homogeneously. Due to a dimensional analysis it can be shown, that for each time step and arbitrary geometric and thermodynamic boundary conditions the ratio of the machine's low pressure port to the mass average pressure within the chamber are a function of the Strouhal number, which can be evaluated with the mass averaged pressure within the scope of chamber model simulation. The same holds for the ratio of the mass averaged pressure and the chamber's low pressure. For both pressure ratios a functional is found, that predicts the pressure at both sides of the chamber to correct the boundary conditions for adjacent clearances.

These functionals have been implemented into the chamber model simulation software *KaSim* where a distinction is made between open chambers and closed chambers while for the closed chambers a quasi-static approach described in reference [9] is used. The simulation results of the new model are compared to measurements and to previous simulations, where either a quasi-static approach in order to calculate the chamber filling process is used or the case that the machine would have a radial inlet area and therefore is homogeneous. The results of the new model are in good agreement to measurements. Nevertheless the difference between the transient model and the quasi-static model in order to calculate the chamber filling process is relatively small compared to the case, that the chamber is filled homogeneously.

The new model is easy to implement and can be evaluated explicitly. Therefore the simulation time is hardly affected compared to the homogeneous case. Furthermore, the evaluation of the Strouhal number can help within the machine design. As inhomogeneous chamber states have a negative impact on the machine's operation performance, the evaluation of these numbers can help to change the design. For example it is shown, that the gas surface interactions of the evacuated gas and the surface material of the rotor or housing have an impact on the operation performance. This can be considered by evaluating the Strouhal number. But as this will also affect the adjacent clearances, the impact on the integral machine performance has to be investigated further.

Table 3. List of most important symbols

symbol	explanation	unit	symbol	explanation	unit
A	cross section area	m^2	c_m	most probably speed	m/s
C_0	reduced mass flow rate	–	D	diameter	m
f	rotational speed	Hz	G	reduced flow rate	–
h	chamber/clearance height	m	L	chamber length	m
m	mass	kg	$\dot{m}$	mass flow rate	kg/s
p	pressure	Pa	R	specific gas constant	J/(kg K)
s	rotor lead	m	St	Strouhal number	–
T	temperature	K	u_0	reduced velocity	–
U	velocity	m/s	V	volume	m^3
w	chamber width	m	z	number of lobes	–
δ	gas rarefaction parameter	–	μ	dynamic viscosity	Pa s
ρ	density	kg/m^3	ψ	wrap angle	–

References

[1] Jousten K 2018 *Wutz - Handbuch der Vakuumtechnik* vol 12 (Springer Vieweg)
[2] Huck C, Brümmer A and Sachs R 2014 Thermodynamical operation behaviour of screw vacuum pumps with cycloid- and quimby-toothed rotors *9th Int. Conf. on Screw Machines: VDI Bericht 2228* pp 331–346
[3] Stratmann D 2010 *Kennfeldverhalten zweiwelliger Rotationsverdrängervakuumpumpen* Dissertation Technische Universität Dortmund
[4] Rohe A 2005 *Wärmehaushalt von Schraubenspindel-Vakuumpumpen* Dissertation Technische Universität Dortmund
[5] Wenderott D 2001 *Spaltströmungen im Vakuum* Dissertation Technische Universität Dortmund
[6] Huck C, Pleskun H and Brümmer A 2018 *J. Vac. Sci. Technol. A* **36** 031606
[7] Jünemann T and Brümmer A 2020 *J. Vac. Sci. Technol. B* **38** 044201
[8] Jünemann T and Brümmer A 2022 One-dimensional calculation approach for gaseous clearance flows iOP Conf. Ser.: Mater. Sci. Eng.
[9] Pleskun H, Jünemann T and Brümmer A 2022 Validation of inhomogeneous chamber states in rotary positive displacement vacuum pumps iOP Conf. Ser.: Mater. Sci. Eng
[10] Pleskun H, Jünemann T, Bode T and Brümmer A 2021 *IOP Conf. Ser.: Mater. Sci. Eng* **1180** 012009
[11] Jünemann T, Pleskun H and Brümmer A 2021 *IOP Conf. Ser.: Mater. Sci. Eng* **1180** 012037
[12] Acharya T, Falgoust J, Schoegl I and Martin M J 2019 *J. Thermophy. Heat Trans.* **33** 733
[13] Buckingham E 1914 *Phys. Rev.* **4** 345

[14] Sharipov F 2016 *Rarefied Gas Dynamics - Fundamentals for Research and Practice* (Wiley-VCH)
[15] Pleskun H, Bode T and Brümmer A 2022 *Physics of Fluids* **34** 032004
[16] Pleskun H and Brümmer A 2022 *Physics of Fluids* **34** 082009
[17] CFX 2021 Modeling Guide
[18] Richardson L F and Gaunt J A 1927 *Philosophical Transactions of the Royal Society of London. Series A, Containing Papers of a Mathematical or Physical Character* **226** 299–361
[19] Ferziger J H and Peric M 2002 *Computational Methods for Fluid Dynamics* vol 3 (Springer)
[20] Janicki M 2007 *Modellierung und Simulation von Rotationsverdrängermaschinen* Dissertation Technische Universität Dortmund

IOP Conf. Series: Materials Science and Engineering 1267 (2022) 012017 doi:10.1088/1757-899X/1267/1/012017

Use of CFD to optimize the design of a shunt pulsation trap (SPT) used for noise and vibration mitigation in oil free screw compressors

J Willie[1], **S W Yonkers**[2], **P X Huang**[2] **and R B Ganatra**[1]
[1]CVS Engineering GmbH Grossmattstr. 14, D-79618 Rheinfelden, Germany
[2]Hi-Bar Blowers, Inc. Fayetteville, GA, USA

E-mail: james.willie@cvs-eng.de

Abstract. Noise and vibration are problems that are inherent in screw compressors and other Positive Displacement (PD)Machines. This problem is driven partly by the lobe passing frequencies inside these machines that are generated due to the meshing between the gate and main rotors teeth. Because these compressors are operated over a wide speed range and at various loads they usually run at off-design conditions leading to either over or under-compression. In both scenarios, the pressure within the compressor always changes rapidly to match the discharge or process pressure when the discharge opens thus leading to pulsation and noise. To ensure that the compressor is operated without over and or under compression, which also negatively affects the compressor efficiency Hi- Bar Blowers, Inc developed a Shunt Pulsation Trap (SPT) that will contain and reduce the pulsation inside the compressor cavities thus eliminating the use of a silencer at the inlet and the discharge. During the SPT development the CVS Silo King (SKL 1100) compressor was selected for benchmarking this technology. CFD was used extensively to simulate the SKL 1100 without and with the SPT integrated, which allowed the SPT design to be optimized. This paper will present the results obtained to demonstrate that this technology has the potential to eliminate over and under compression while at the same time leading to energy savings and reducing the compressor footprint and the noise and vibration that are commonplace in screw compressors.

1. Introduction

The issue of noise in gas compression is commonplace and very little has been done to address this issue apart from the use of silencers. Even with the use of these silencers, noise levels of up to 100 dBA and above are possible and this is very uncomfortable and can be damaging to the ear of human beings. To address this issue CVS has teamed up with Hi Bar Blowers to investigate the root-cause of pulsation inside positive displacement (PD) compressors with internal compression. In this case, the pilot project is investigating the Silo King (SKL 1100) screw compressor developed by CVS Engineering. A Shunt Pulsation Trap (SPT) has been designed by Hi-Bar Blowers and its purpose is to trap the pressure wave that is responsible for the pulsation inside the screw housing and allow only the non-pulsating flow to leave the compressor. The pulsation suppression will be done over the load range 0-2.5bar(g) with the following goals:

- Discharge pulsation level (< 0.1bar) and better noise (< 92dBA) than the design with the reactive silencer;

- Smaller footprint. The pulsation trap would be about 25-50 % of the existing reactive silencer ;
- Energy savings. About 5-10% energy savings expected. Torque measurements are expected to be lower due to the lower power consumption by eliminating over and under compression. In addition, the back pressure due to the silencer will be eliminated and for the worst case of maximum speed and maximum discharge pressure this amounts to 132 mbar in the Silo King SKL 1100 compressor

If successful, SPT would address the noise pollution coming from these compressors. It would address not only the loudness but the noise quality as well. It would also reduce the footprint and the weight of these compressors, which is very important for this application. By reducing or eliminating over and under compression, we expect the pressure drop to decrease, thereby reducing the power consumption of the compressor. The elimination of a silencer means that we would also save the pressure drop introduced by it and eliminate the power consumption that goes with this.

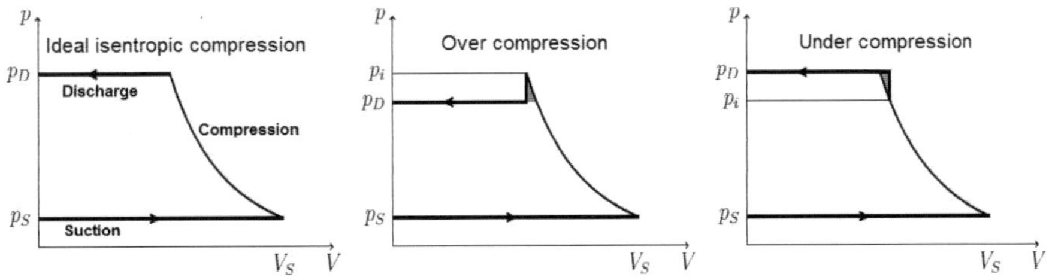

Figure 1. Schematic showing under and over-compression

The authors are of the opinion that the root cause of high pulsations in screw compressors and other PD machines like blowers is not well understood. Based on the work done by Hi Bar Blowers [5, 6, 7, 8] it is believed that apart from the lobe passing frequencies due to the rotors lobes and the noise generated by gears flow induced noise is predominantly due to high pressure waves when the compressor is operated at off design conditions that lead to over and under compression. The screw compressor is designed for a specific internal volume compression or pressure ratio but in application it is seldom operated at this design condition due to process requirements. Operating at off design condition leads to either over or under compression as shown in figure 1. The plot on the left is showing the case when the compressor is operated at the design point. In this case the pressure inside the pump housing at the discharge p_i is equal to the discharge pressure p_D and the process is isentropic. The sketch in the middle is showing the case where we have over-compression and just before the discharge opens the chamber pressure is greater than the discharge pressure. In this case when the discharge opens the chamber pressure drops suddenly from p_i to p_D and this can lead to pulsation or pressure fluctuation and noise generation. The shaded area corresponds to the extra work that must be done because of the over-compression. In the sketch on the right the chamber pressure p_i is less than the discharge pressure before the discharge port is opened and it is referred to as under-compression. When the discharge opens a rapid flow is generated from the discharge tube to the chamber until we have an equalization of pressure. The extra work done as a result of the under-compression is shaded in the p-V diagram.

If a silencer is used, which is the case in practice, an additional loss is introduced due to the back-pressure of the silencer as depicted in the under-compression case shown in figure 2. The main goal of the SPT is to ensure that there is no pressure drop when the discharge opens and so it means using the SPT will lead to additional savings in energy thus optimizing the temperature of the compressor and preventing the use of oil or air cooler, which will lead to additional cost savings.

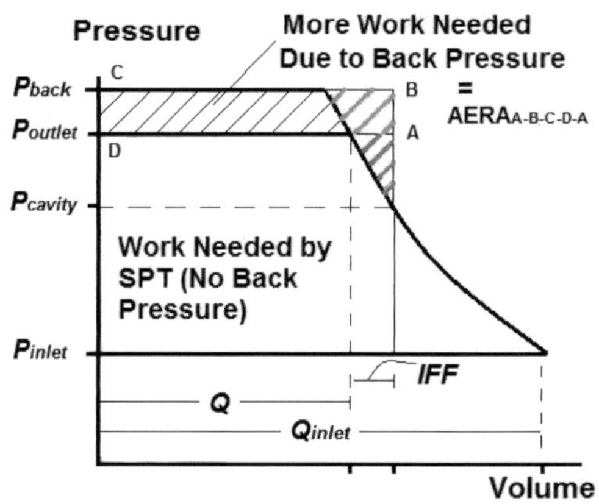

Figure 2. Schematic of the energy saving potential of the SPT technology

In the work by Willie [1] it was shown that for most screw compressors there is gas that is trapped between the rotors and the discharge axial gap just before the discharge opens and this can lead to over-compression and to gas pulsations in screw compressors. In [2, 3, 4], other sources of noise inside screw compressors were presented and discussed. In previous work by Hi Bar Blowers the theory behind the development of the SPT technology has been explained [5, 6, 7, 8]. It has been tested for root blowers, which do not have internal compression, and the results showed upto 10 fold reduction in gas pulsations. However, this technology is yet to be tested for screw compressors, which have internal compression, hence the motivation for this work.

In this work Computational Fluid Dynamics (CFD) is used to simulate the flow inside the Silo King SKL 1100 screw compressor using CONVERGE CFD. The pre-processor of this software (CONVERGE STUDIO) has advantages over others because of the way it handles moving meshes using adaptive and automatic meshing technique including the cut cell technique in order to mesh complicated moving geometries.

The paper begins with the description of the measurement setup and the measurement points used to benchmark the Silo King SKL 1100 compressor without and with a silencer. This is followed by the case setup description of the SPT design used to numerically validate the potential of this technology and after this the CFD setup is presented and this followed by the results and discussions and the conclusions and future work.

2. Measurements

A picture showing the setup of the measurement test stand is presented in figure 3. The stand is used for determining the performance map of the benchmark compressor for testing the SPT technology. The following are measured: Volume flow rate, inlet/discharge temperature and pressure, compressor power, and the silencer pressure drop (Δp). Noise measurement is done using microphone that is located 1 m away from the compressor and similar measurements will be done with the SPT integrated into the compressor for comparison.

The operating point simulated is shown in table 1 with an input speed of 3200rpm. This point is close to the design point that has an internal volume compression of 2.1 and an input

Figure 3. Picture of measurement setup in the laboratory

speed of 3000rpm. The maximum input speed of this compressor is 3600rpm. The discharge pressure p_2 is taken before the silencer directly at the outlet of the compressor and close to the discharge contour.

Table 1. Silo king screw compressor operating point for CFD validation of the SPT technology

Parameter	Measured value
Inlet air pressure (p_1)	1.00 bar(a)
Inlet air temperature (T_1)	28° C
Discharge air pressure (p_2)	2.5bar(a)
Discharge air temperature (T_2)	150.9°C
Volume flow rate ($\dot{V}$)	807.37 m^3/h
Female rotor speed (n_f)	10732.3 rpm
Male rotor speed (n_m)	13415.4 rpm
Compressor power (P)	44 kW
Silener pressure drop (Δp)	132 mbar

3. CFD Case Setup

The setup used in the CFD simulation of the baseline SKL 1100 screw compressor with and without the use of conventional silencer is presented followed by the description of the setup with the SPT included. Due to the need to capture any shocks in the flow a very fine mesh was used and the settings used is similar to the settings used for modeling the NASA supersonic channel benchmark geometry for capturing shock waves [9].

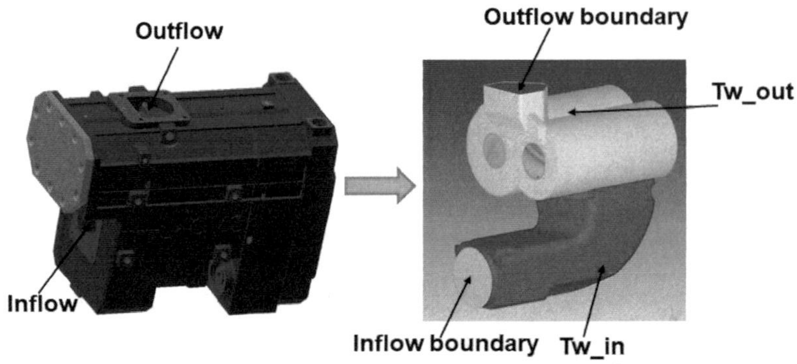

(a) CFD setup of the baseline geometry: Left=3D CAD, Right: Neg. vol.

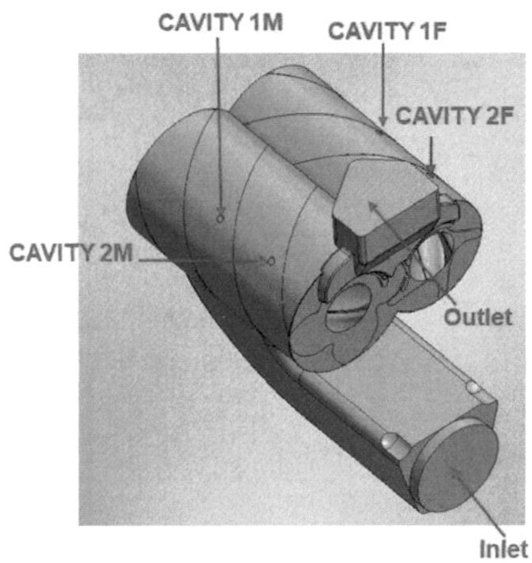

(b) CFD setup: Baseline with monitoring points

Figure 4. CAD geom. + neg. volume of baseline geom. including monitoring points for CFD

3.1. CFD setup of the baseline geometry without silencer

The CAD geometry of the SKL 1100 is first cleaned for CFD simulation and the negative volume is extracted without the discharge silencer as shown in figure 4(a). Included in the geometry are the suction or inlet or inflow, the discharge or outflow and the rotors. The negative volume is exported as an STL file into CONVERGE STUDIO for the definition of the boundary conditions and the solver conditions and the mesh settings. The speed of rotation of the rotors are also defined. The working fluid is air and the inlet and outlet are given pressure boundaries as measured. Using pressure boundaries will allow the solver to compute the mass flow rate from which the volume flow rate can be computed using the computed density. The computed volume flow rate can then be validated with the volume flow measured. Due to the use of cold gaps the leakages are higher than in the reality where due to the thermal expansion of the rotors the hot gaps are smaller. In order to tune the cold gaps to be able to mimic the hot gaps proximity boundaries and porous media are used in CONVERGE [10]. Doing this will allow the solver to accurately predict the volume flow rate inside the machine. Tw_{in} is taken to be the suction temperature and to determine Tw_{out}, the wall temperature at the outlet, the relations

$T_2 = T_1 \left(\frac{p_2}{p_1}\right)^{\frac{\kappa-1}{\kappa}}$, where κ is the polytropic coefficient determined from measurements to be 1.7. This gives Tw_{out}=439 K. For the meshing a base grid of 16mm is used and adaptive mesh refinement (AMR) is used for all regions at level 4 (1mm) on velocity with Sub-Grid-Scale (SGS) 1.0. At the discharge, fixed embedding is used at level 4 (1mm) and finally, grid scaling from -1 to 0 is used at the end of the 3rd cycle. Mesh sensitivity analyses is used to determine that the 16mm base grid will be able to resolve the flow as needed.

To monitor the flow and enable comparison of the various setups and also for validation of the CFD results with measurement data after the modified SPT is tested, four (4) monitoring points are defined as shown in figure 4(b). The pressure and the temperature are monitored at each point for comparison.

3.2. CFD setup of the baseline geometry with the silencer

The setup of the baseline geometry with the discharge silencer included is presented here. The CAD geometry and the negative volume used for the CFD setup are shown in figure 5. The outflow is now located at the discharge of the silencer and the inflow is the same as in the baseline case in figure 4. To determine the pressure at the discharge of the silencer the discharge pressure in table 1, which is measured close to the rotors exit is added to the silencer pressure drop of 132 mbar. When the SPT design is used the silencer pressure drop will be avoided. The settings of the CFD used are similar to those used in the baseline case setup in section 3.1.

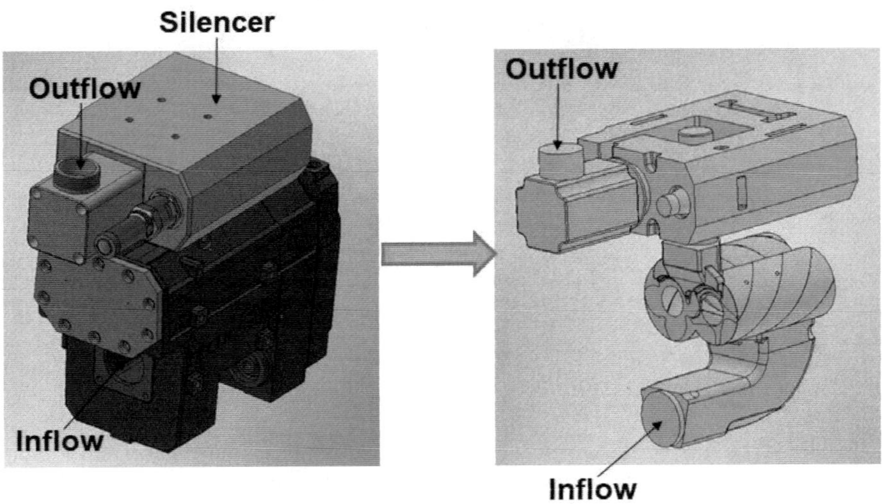

Figure 5. CFD setup of the baseline geom. + silencer: Left=CAD geom., Right=Neg. vol.

3.3. CFD setup of the baseline geometry with the SPT concepts

The setup of the SKL 1100 screw compressor with the SPT Design I and Design II concepts integrated is presented. A two stage SPT design was used in Design I and a one stage design was used in Design II, with each stage having 2 nozzles as shown in figure 6. The purpose of this concept is to ensure that there is no pressure drop at the discharge ensuring that irrespective of the operating point there will be minimum pressure fluctuation or noise and hence optimum operation with no under or over-compression inside the compressor [5, 6, 7, 8, 11, 12]. Based on the results of Design I optimization was done and Design II, which is a single stage design, was proposed. The negative volume used in both cases for the CFD simulations are shown in figure 6(b), with the left plot corresponding to the two stage design and the right corresponding to the one stage design.

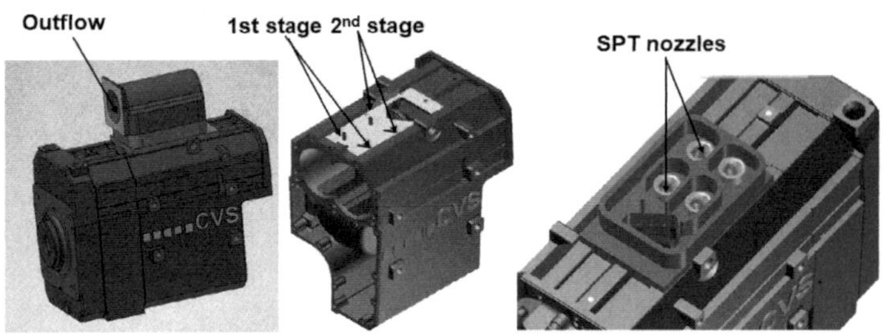

(a) CAD geometry of SKL 1100 with SPT

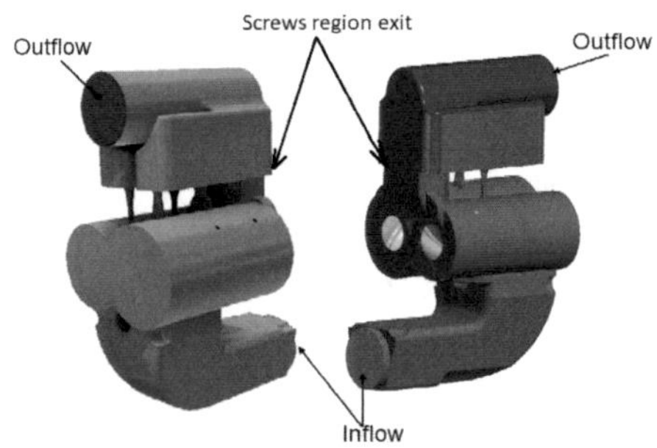

(b) Neg. volume with SPT, left=Design I (two stage STP) & right=Design II

Figure 6. CAD model and negative volume of SKL 1100 with SPT integrated

In total, four setups and mesh strategies are used to ensure that the pressure waves and the flow is being adequately resolved, namely, **A**, **B**, **C** & **D**. The setup associated with each case is summarized in table 2.

A comparison of the averaged results for the 4 setups shows similar results as depicted in table 3. For comparison with the baseline cases, setup **C** is used because it was able to mimic the flow in the screw compressor and capture the pressure fluctuation as expected. Also, grid sensitivity analyses showed that this setup was not numerically expensive. These settings are also used to simulate the SPT Design II for comparison with the SPT Design I and the baseline SKL 1100 compressor.

4. Results and discussions

The CFD simulation results of the baseline SKL 1100 compressor without and with the reactive silencer are first validated with measurement data shown in table 1 to ensure that the results have been able to capture the physics of the flow and are not non-physical due to numerical errors. This is followed by the presentation of the of the SPT results.

The ASCII plots of the average volume flow rate and the compressor power for the baseline case without the silencer are shown in figure 7. The averaging is done considering periods of

Table 2. Averaged results of the various setups of SKL 1100 + SPT Design I

Setup	Base grid	AMR regions	Embedding	Grid scaling	Solver
A	16mm	All, level 4 (1mm) Nozzles level 5 (0.5mm) Vel. SGS 1.0	Fixed at outlet, level 4 (1mm)	1 cycle -2; 2 cycles -1; then 0	Std. RNG k-eps
B	16mm	All, level 4 (1mm) Nozzles level 5 (0.5mm) Vel. SGS 1.0	Fixed at nozzles, level 5 (0.5mm) Fixed at outlet, level 4 (1mm)	1 cycle -2; 2 cycles -1; then 0 (from map file)	Std. RNG k-eps
C	16mm	All, level 4 (1mm) Nozzles level 5 (0.5mm) Vel. SGS 1.0	Fixed at nozzles, level 5 (0.5mm) Fixed at outlet, level 4 (1mm)	1 cycle -2; 2 cycles -1; then 0 (from map file)	CFL=1 nozzles, Total energy Stricter PISO RNG k-eps
D	16mm	All, level 4 (1mm) Nozzles level 5 (0.5mm) Vel. SGS 1.0	Fixed at nozzles, level 5 (0.5mm) Fixed at outlet, level 4 (1mm)	1 cycle -2; 2 cycles -1; then 0 (from map file)	CFL=1 nozzles, Total energy Stricter PISO MUSCL, RNG k-eps

Table 3. Averaged results of the various setups of SKL 1100 + SPT Design I

Setup	Inflow mass(kg/s)	Inflow vel. (m/s)	Outflow mass (kg/s)	Outflow vel. (m/s)	Power(kW)
A	0.207	35.9	0.177	21.1	40.7
B	0.207	35.9	0.176	20.8	40.7
C	0.189	32.9	0.182	21.2	41.0
D	0.190	33.18	0.180	21.5	41.0

rotation of 360°. In the setup used time advancement has been modified and replaced by the number of degrees that the male (main) rotor is rotating. Thus 360° represents a full rotation of the male rotor. So a time average over a period equal to a complete revolution of the male rotor. For statistical convergence at least three revolutions or Crank Angle 1080° are required. The volume flow rate measured is at the inlet of the compressor and the compressor power is computed from CFD using the main (male) and gate (female) rotors torque. For the setup without the use of proximity boundaries (denoted as original in the plot) the inlet mass flow rate at CA 1700deg is 0.234 kg/s and that at the outlet is 0.228 kg/s, giving a percentage error of 2.6%, which is good for mass balance. The average inlet volume flow rate in this case is 733 m^3/h. One reason for the disparity between the measured and the simulated volume flow is due to leakages inside the machine and the fact that the CONVERGE solver used does not resolve the flow inside the gaps but uses proximity boundaries to model the flow inside the gaps of the compressor. The model is tuned by using proximity boundaries (denoted as update in the plot) to mimic the flow inside the gaps. This gives an inlet mass flow rate of 0.257 kg/s and hence a volume flow rate of 805 m^3/hr, which is very close to the measured value.

After the baseline simulations and tuning were completed and the results validated with measurement data, the simulations were extended to include the baseline geometry plus the silencer and the baseline geometry plus the SPT Design I. Based on the SPT Design I results optimization was done, which resulted in the SPT Design II concept that was also simulated and compared with the SPT Design I results.

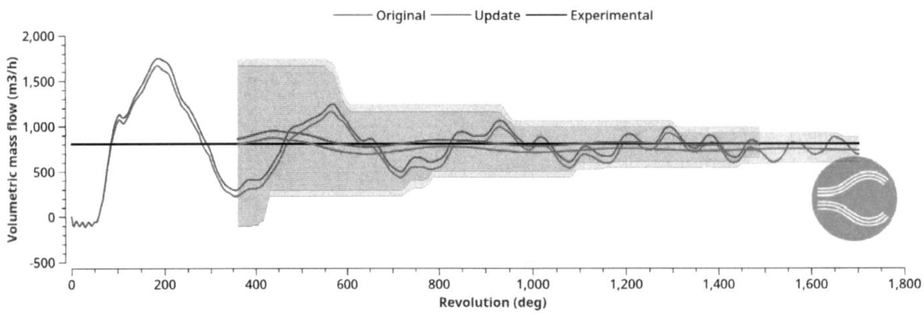

(a) Volume flow rate

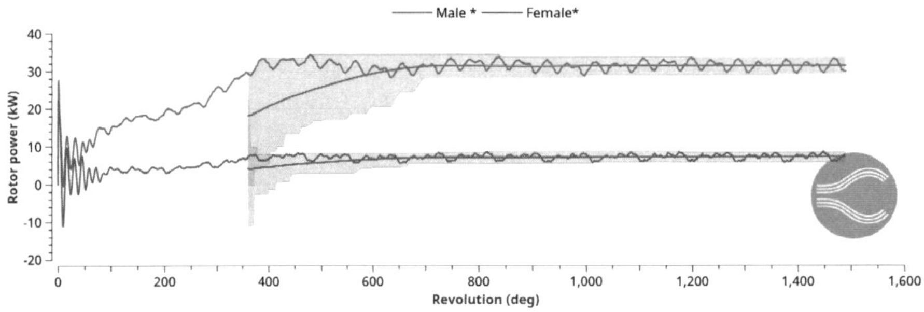

(b) Compressor power

Figure 7. ASCII plots of the volume flow and compressor power predicted, baseline geometry, no silencer

The ASCII results of the average mass flow rate and hence the volume flow rate and the compressor power of the baseline geometry plus the silencer are shown figure 8. The mass flow rate at the inlet is 0.2535 kg/s at CA 1600deg and that at the outlet is 0.222 kg/s, giving a percentage error of 10%. The moving average value was changing slightly at 1600 deg. CA and this could be due to the lager geometry from the addition of the silencer. The average compressor power has increased to 41.1kW at 1600 deg CA due to the addition of the silencer and the back pressure it introduces. This value is closer to that measured in table 1.

Also noticed is the increase in the average discharge velocity from 41.4 m/s in the case without silencer to 53.25 m/s in the case with silencer (6.2%) as shown in figure 9. The increase is due to the fact that the silencer inlet and outlet have smaller diameters when compared to the outlet of the compressor.

The simulation is extended to the baseline geometry with the SPT Design I that has two stage nozzles integrated and after that to the SPT Design II with a single stage nozzle and the results of the mass flow compared to the baseline geometry without and with silencer are shown in figure 10. The average mass flow rate at the inlet of the SPT Design I is reduced compared to the baseline geometry without and with silencer and the baseline geometry plus SPT Design II (See table 4). This could be due to the positioning of the first stage nozzles leading to more leakages inside the compressor. The corresponding power consumption of the compressor with silencer and with the SPT Designs I and II are shown in figure 11. The power consumption is almost identical in the case of the baseline plus silencer compared to the baseline plus SPT Design I and in the comparison between the baseline plus silencer and the baseline plus SPT Design II there a power savings of about 5%.

The pressure contour plot over the rotors in figure 12 shows the distribution of pressure

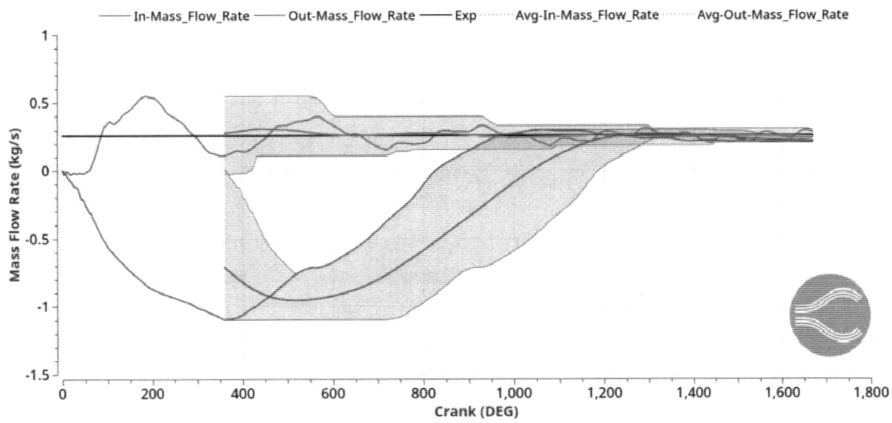

(a) Mass flow rate

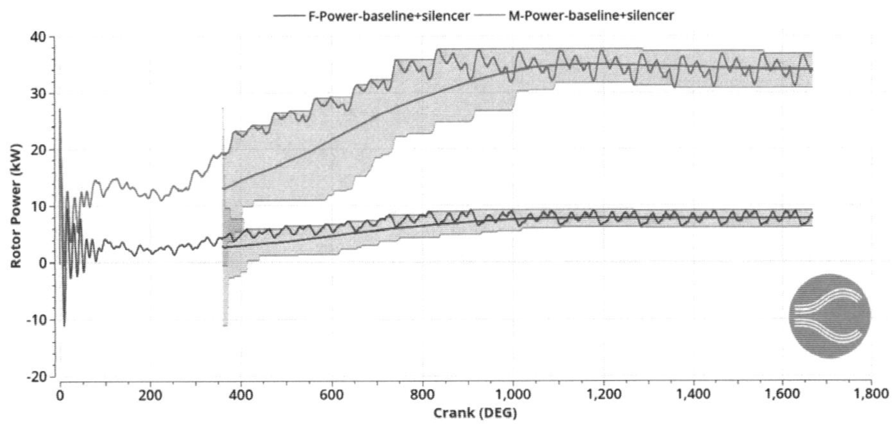

(b) Compressor power

Figure 8. ASCII plots of the mass flow and compressor power predicted, baseline geometry, with silencer

over the rotors in the baseline case without silencer and the baseline with SPT Design II. The pressure distribution is more uniform in the design with the SPT Design II than in the baseline. Another observation is that the timing of the opening of the first stage nozzles are not the same as highlighted in the two rectangular boxes in the figure. This has been corrected in the modified SPT design [12].

An important goal of the SPT design is to size the nozzles such that they are choked (Mach number=1). Doing this would ensure that the Δp at the SPT will match the Δp at the discharge, thus mitigating the effect of over and under compression and hence the pulsation inside the compressor, thus eliminating the need for a silencer. To check this monitoring points are defined at the nozzles to determine the pressure and the temperature as depicted in figure 4(b). Cut planes through the main (male) and gate (female) rotors are used to monitor the Mach number inside the nozzles as shown in figure 13. Based on this result it can be seen that most of the work is being done by the first nozzle with Mach number between 0.6 and 0.8 while the Mach number in the second nozzle is very low. Based on this result and in additional to the high flow loses in the SPT Design I the SPT design was modified by using a single stage nozzle with the nozzle repositioned in order to minimize or eliminate the high flow loses in the SPT Design I.

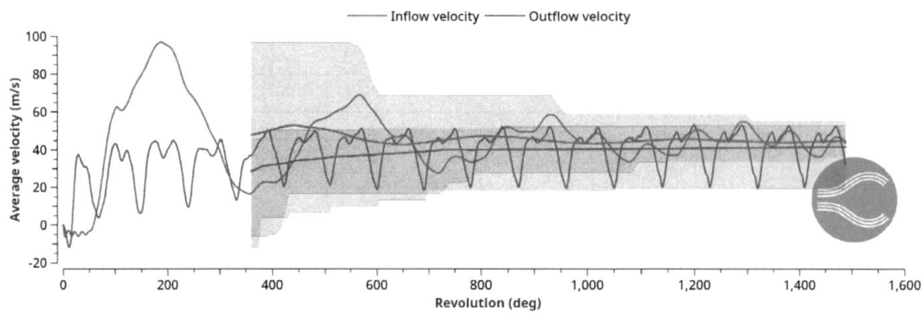

(a) Discharge velocity, no silencer

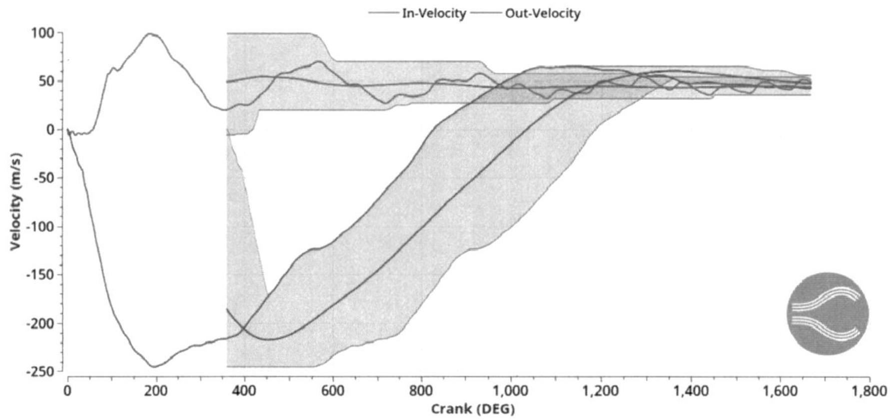

(b) Discharge velocity, with silencer

Figure 9. ASCII plots of the discharge velocity without and with silencer

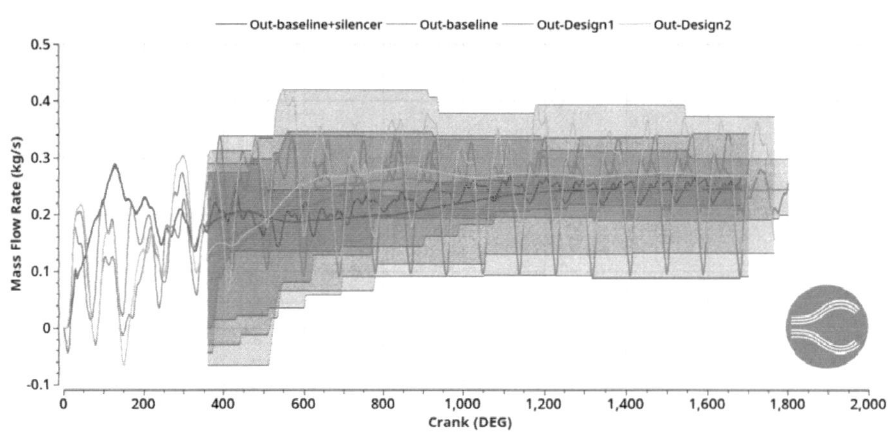

Figure 10. Mass flow rate of the baseline geometry compared to baseline plus silencer and SPT Design I & Design II

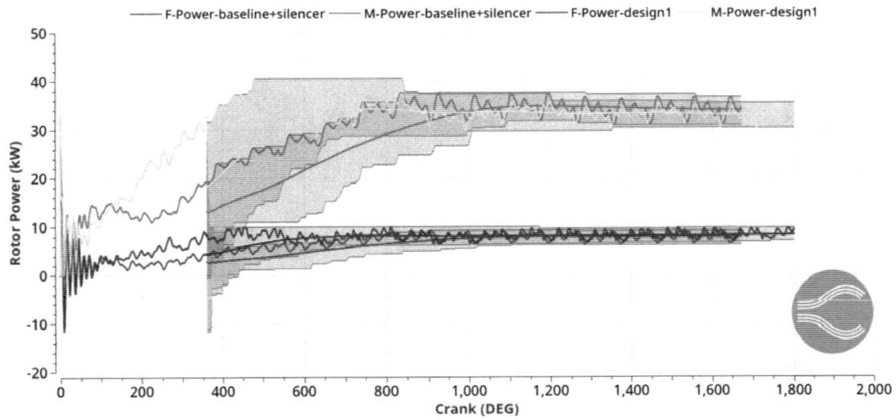

(a) Comp. power, silencer + SPT Design I

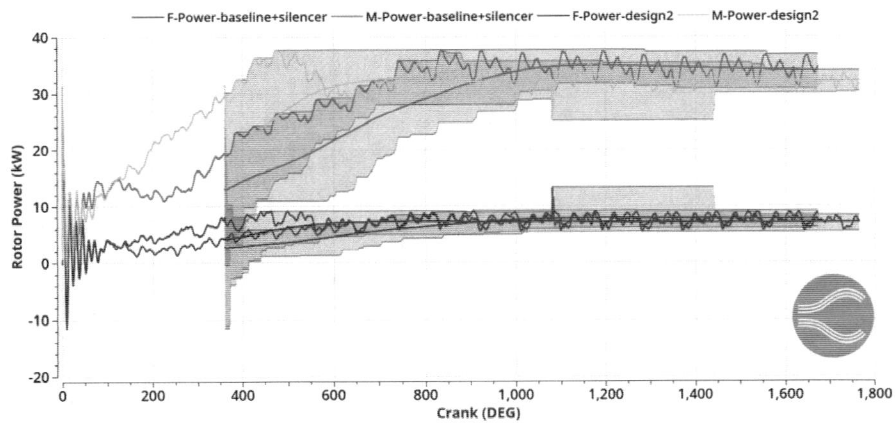

(b) Comp. power, silencer + SPT Design II

Figure 11. ASCII plots of the compressor power compared to silencer plus SPT Design I and silencer plus SPT Design II

This is denoted as SPT Design II (See figure 6(b)) and it is simulated using the same settings used in the other setups and the results compared.

A summary of the average mass flow rate, volume flow rate, compressor power and the total axial loads on the rotors are presented in table 4 for all the cases considered at CA 1600 deg. Using SPT Design II improves the volume flow rate and reduces the compressor power when compared to SPT Design I and the baseline plus silencer cases.

Due to the lack of noise measurement at this stage the improvement in noise or pulsation level is examined by considering the dynamic pressure monitored for all the cases considered in this work as shown in figure 14. There is a direct correlation between the amplitude of the pressure fluctuation and flow induced noise inside the compressor. Using this argument it can be argued that the case with the SPT Design I has the least fluctuation with the lowest pressure amplitude followed by SPT Design II and the baseline without the silencer having the highest fluctuation as expected. It is clear that SPT Design I has better performance in terms of attenuating the noise but the flow losses are higher compared to SPT Design II due to positioning and the timing

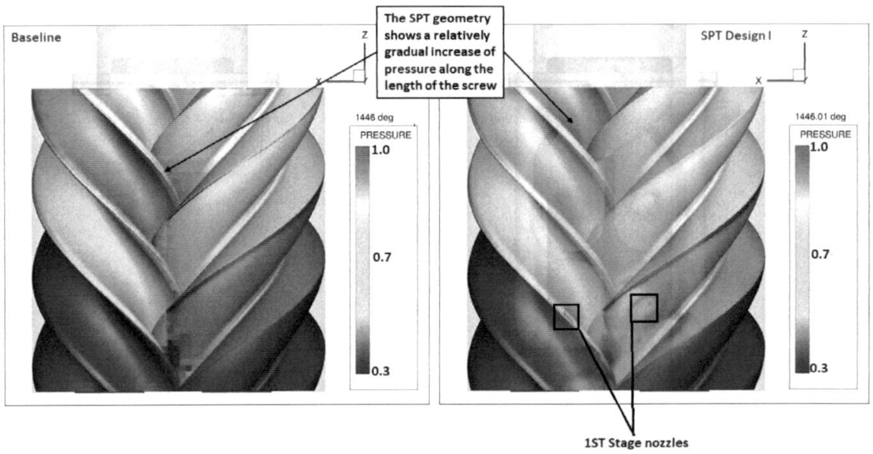

Figure 12. Normalized contour plots of the pressure over the rotors for the baseline geom. without silencer & baseline geom. + SPT Design I

Table 4. Summary of the ASCII plots of average flow and compressor power and axial load on rotors for CFD cases considered

Case	Inflow mass (kg/s)	Volume flow (m^3/h)	Power(kW)	Total axial load (N)
Baseline	0.257	805	38.5	1529
Baseline + silencer	0.2535	794	41.1	1722
SPT Design I	0.229	717	41	1701
SPT Design II	0.2556	800	39.1	1631

of the nozzles. In Design II with the single nozzle the flow losses have reduced but the noise attenuation is slightly reduced compared to Design I but better than the baseline.

5. Conclusions and future work

The flow inside the SKL 1100 oil free screw compressor is successfully simulated using CONVERGE CFD and validated with measurement data. To enable the compressor to be run at other operating points without experiencing increase in pulsation and power consumption due to over and under compression Hi Bar Blowers has developed a Shunt Pulsation Trap (STP) that will trap the pulsations inside the compressor and therefore eliminate the need for a silencer and by so doing reduce power consumption due to the silencer back pressure and reduce the footprint of the compressor. As a first step, CFD was used to simulate the compressor with and without the reactive silencer that is used in the CVS Silo King (SKL 1100) oil free screw compressor followed by the simulation of the Hi-Bar Blowers SPT concepts. The following conclusions can be drawn:

- The SPT concepts are effective in reducing the outflow fluctuations
- The mass flow rate in SPT Design I is about 10% lower than in the baseline without and with silencer whereas for SPT Design II the mass flow is comparable to baseline without and with silencer. The compressor power for the case with SPT Design I and baseline plus silencer are the same whereas SPT Design II has slightly lower power consumption when compared to SPT Design I and baseline plus silencer.

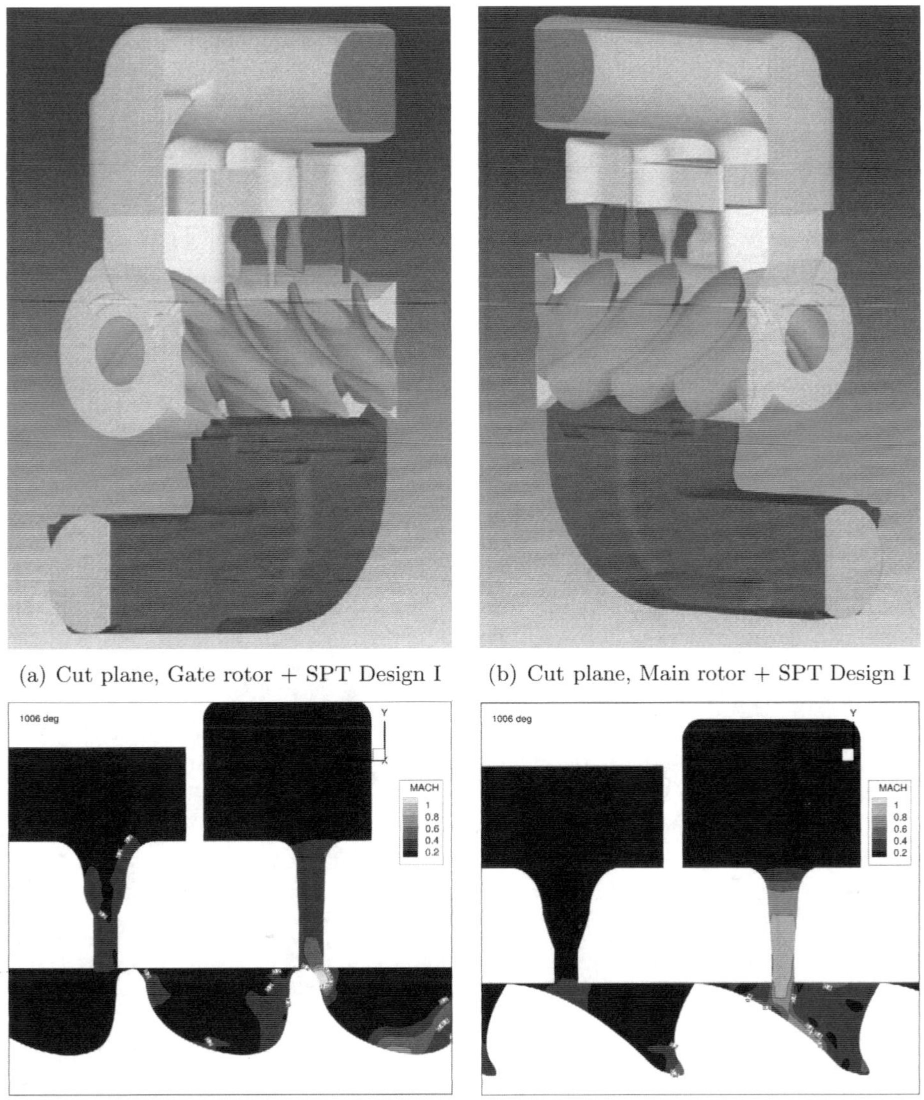

(a) Cut plane, Gate rotor + SPT Design I (b) Cut plane, Main rotor + SPT Design I

(c) Mach number, Gate rotor + SPT Design I (d) Mach number, Main rotor + SPT Design I

Figure 13. Cut planes through and Mach Number showing flow through 1st & 2nd stage nozzles

- The SPT Design I nozzles are not choked with the maximum Mach number of about 0.8 in nozzle 1 and that in nozzle 2 negligible.
- The SPT Design II attenuation of the fluctuations is slightly less effective compared to Design I but it improves the volume flow rate significantly and it has less power requirement relative to Design I
- The SPT Design II nozzles are not choked with maximum Mach number of 0.4 and the female (gate) rotor nozzle contribution is negligible

The results of this work was used by Hi-Bar Blowers in the modification of the SPT design concept leading to the development of the Shunt Enhanced Decompression and Pulsation

Figure 14. Comparison of the pressure fluctuation at the screw region exist for the various designs

Trap (SEDAPT) [11] and the Shunt Enhanced Compression and Pulsation Trap (SECAPT) technologies [12]. In the new concepts modification were made to the SPT to ensure that the opening and closing of the nozzles are correctly timed leading to reduction in the flow losses associated with the noise attenuation.

As future work, the modified SPT concept designed will be fabricated and tested on the CVS Silo King (SKL 1100) oil free screw compressor. For the sake of time one design point was investigated in this work. When the modified SPT concept is tested other operating points, especially minimum speed and maximum load and minimum speed and no load conditions as well as other input speeds and loading conditions will be simulated and validated with measurement data. Dynamic pressure and vibration measurements will be carried out and used to validate the CFD results.

Nomenclature

SPT	Shunt Pulsation Trap	[-]
SEDAPT	Shunt Enhanced Decompression and Pulsation Trap	[-]
SECAPT	Shunt Enhanced Compression and Pulsation Trap	[-]
p_i	Pressure before opening of the outlet	[Pa]
p_D	Discharge pressure	[Pa]
V_S	Suction volume flow rate	[m^3/s]
p_S	Suction pressure	[Pa]
Δp	Pressure drop	[mbar]
p	Pressure	[Pa]
T	Temperature	[K]
$\dot{V}$	Volume flow rate	[m^3/s]
n	Rotor speed	[rpm]
P	Compressor power	[kW]
AMR	Adaptive mesh refinement	[-]
SGS	Sub-grid scale	[-]
PISO	Pressure-implicit with splitting of operators	[-]
MUSCL	Monotonic upstream-centered scheme for conservation laws	[-]
CFL	Courant-Friedrichs-Lewy	[-]
RNG	Re-Normalisation Group	[-]
CA	Crank angle	[degree]

Acknowledgment
The authors would like to thank Pietro Scienza of Convergent Science and Riccardo Gretter for their invaluable support with the CFD setup.

References

[1] Willie J 2017 *Use of CFD to Predict Trapped Gas Excitation as Source of Vibration and Noise in Screw Compressors* (IOP Conf. Ser.: Mater. Sci. Eng.) **232** 012021

[2] Willie J, Asal W and Sachs R 2014 *Analysis of the Noise and Vibration of a Dry Screw Compressor* (In.: 9th Int. Conf. on Screw Machines 2014, VDI-Berichte 2228 (2014), p. 107-123)

[3] Willie J and Sachs R 2015 *Structural and Torsional Vibration Analysis of a Dry Screw Compressor* (IOP Conf. Ser.: Mater. Sci. Eng.) **90** 012005

[4] Willie J and Sachs R 2016 *Structural and Torsional Vibration and Noise Analysis of a Dry Screw Compressor* (Proc IMechE Part E. J Process Mechanical Engineering) **0**, DOI:10.1177/0954408916648989

[5] Huang P X 2012 *Gas Pulsations: A Shock Tube Mechanism.* (International Compressor Engineering Conference). Paper 2092 https://docs.lib.purdue.edu/icec/2092

[6] Huang P X, Yonkers S, Honkey D and Olenick D 2013 *Screw pulsation generation and control: A shock tube mechanism* (8th International Conference oon Compressors and their Systems, 9-10th September 2013, City Universtiy London) p 113

[7] Huang P X, Yonkers S and Hokey D 2014 *Gas Pulsation Control Using a Shunt Pulsation Trap.* (International Compressor Engineering Conference) Paper 2258 http://docs.lib.purdue.edu/icec/2258

[8] Huang P X 2018 *A model for the transient pulsation generation at the discharge of a screw compressor by a shock tube analogy* (IOP Conf. Ser.: Mater. Sci. Eng.) **425** 012022

[9] Hall N 2021 *Oblique shock wave.* https://www.grc.nasa.gov/www/k-12/airplane/oblique.html

[10] Richards J K, Senecal K P and Pomraning E 2021 *CONVERGE 3.0 Manual.* (Convergent Science, Madison, WI)

[11] Huang P X, Yonkers S and Willie J 2022 *A Novel Screw Compressor with a Shunt Enhanced Decomposition and Pulsation Trap (SEDAPT).* (2022 Purdue International Compressor Engineering Conference)

[12] Huang P X, Yonkers S and Willie J 2022 *A Novel Screw Compressor with a Shunt Enhanced Compression and Pulsation Trap (SECAPT).* (Int. Conf. on Screw Machines 2022, TU Dortmund University, Dortmund, Germany, September 7-8 2022)

Pulsating flow velocity profile measurement at an acoustically reflecting and non-reflecting open pipe end using laser doppler anemometry (LDA)

F Nal[1] and A Brümmer[1]

[1]Chair of Fluidics, TU Dortmund University, Leonhard-Euler-Straße 5, 44227 Dortmund, Germany

E-mail: `felix.nal@tu-dortmund.de`

Abstract. Depending on the geometry and operating condition, occasionally strong and destructive acoustic resonances occur in the pipe spool piece between the high-pressure discharge of screw compressors and the connected pulsation damper/silencer inlet. To avoid these resonances, the use of a nozzle with defined cross-section reduction is investigated. By designing the contraction ratio as a function of the nozzle inlet Mach number, a non-reflecting termination/transition can be provided. For a defined operating point, acoustic reflection from the nozzle can be avoided in this way, thus preventing the acoustic resonance in the spool piece.

Within this paper the pulsating flow field directly downstream an acoustically reflecting open pipe end is investigated using laser doppler anemometry (LDA) at the air test rig at the Chair of Fluidics at TU Dortmund University. In the modified experimental setup an operating point-specific designed nozzle is installed at the open pipe end to create a non-reflecting termination. By means of LDA measurements, the velocity profile is determined experimentally at the nozzle outlet for the designed operating condition. The 2D measurement results (radius dependant velocity and phase distribution) extend established 1D decomposition of pressure measurement signals into upstream and downstream travelling waves and allow a deeper understanding of the acoustic behaviour of non-reflecting nozzles.

1. Introduction

Turbulent, unsteady flow occurs in numerous technical systems. Particularly when positive displacement machines are used, the flow is characterized by periodic pressure and velocity fluctuations in the discharge port and the connected piping system due to the machine operating principle. This also applies to rotary displacement machines such as screw compressors. Depending on the geometry and operating condition, strong and destructive resonances occasionally occur in the spool piece between the high-pressure discharge of the compressor and the connected pulsation damper/silencer inlet. For this reason, it is of relevance to study the turbulent, pulsating pipe flow at the acoustically open spool piece outlet in more detail and, if necessary, to find corrective measures to counteract the potentially harmful resonances within the spool piece.

In this context, the pulsating flow directly downstream an acoustically open pipe end is investigated with respect to the radius-dependent flow pattern and reflection behaviour. This is done based on transient velocity profiles recorded by laser doppler anemometry (LDA) directly downstream the open pipe end, and pointed out correlation to plane wave theory. By use of a nozzle with defined contraction ratio, the phase difference between pressure and velocity fluctuations directly downstream the nozzle outlet is significantly reduced. Accordingly, the reflection behaviour at the open pipe end can be substantially influenced by a customized component depending on the operating conditions.

In contrast to already studied pulsating inner pipe flows – laminar [1, 2] or turbulent [3, 4] – the following investigations extend to a solely full turbulent flow ($40\,000 \leq Re \leq 200\,000$) at Womersley numbers of $405 \leq Wo \leq 1070$ directly downstream the open pipe end. The Womersley number relates the transient inertial forces, due to pulsation stimulation, to the viscous forces in the flow. Hence, for the fluctuating flow components, flow behaviour dominated by transient inertial effects is to be expected in this study. This also applies even for low Strouhal numbers in the studied range ($0.82 < Sr < 28.65$), due to high dimensionless frequency ($Wo \geq 405$). Comparison between the measured velocity profile directly downstream the open pipe end shows significant deviation to known inner pipe flow profile at high Sr numbers, attributed to the free jet condition.

Using selected characteristic parameters, the unsteady, radius-dependent flow pattern is classified and evaluated with respect to the phase difference between pressure and velocity fluctuation, whereby reference to plane wave theory is established. Based on analytical considerations for a non-reflecting termination according to [5], a nozzle with defined contraction ratio is designed and installed at the open pipe end. The influence of this nozzle on the radius-dependent phase difference is evaluated accordingly, enabling a predication on the changed reflection behaviour. Presuming small Sr number, the designed nozzle proves to be quite suitable for achieving an almost reflection-free termination for the studied operating point.

2. Experimental setup

The experimental setup used in this study is twofold, summarized in figure 1 and presented in the following. Initial investigations are based on stationary and transient velocity profiles measured at the acoustically open pipe end without nozzle, detail a) in figure 1. For this purpose, the air test rig at the Chair of Fluidics at TU Dortmund University provides the opportunity to generate a sinusoidal pulsating flow with defined frequency and amplitude by means of a pulsation generator, pictured in figure 2. The pulsation frequency is imposed by a slotted rotary disc and thus a function of rotational speed and number of slots. By adjusting the distance to the measuring section inlet using a linear guide unit the pulsation amplitude can be varied as well. The resulting flow profile $c_{\mathrm{LDA}}(r)$ is measured at a distance z downstream the open pipe end by use of a 2D laser/phase doppler anemometer (LDA/PDA) [6] installed normal to the flow direction, figure 1. This arrangement offers the opportunity to record both, main and secondary flow component along the pipe radius.

In addition to the LDA, four piezoelectric pressure sensors ($p_{\mathrm{dyn},1}$ – $p_{\mathrm{dyn},4}$) are installed upstream along the measuring pipe section. These are used to detect the local pressure fluctuations with high temporal resolution (10 kHz). The measured time- and spatial-dependent pressure signals are transformed into frequency domain in post processing. On this basis the spatial-independent down- and upstream travelling wave component inside the pipe are evaluated by means of a decomposition [7]. Hence, pressure and velocity fluctuation at the distance z downstream the pipe end are determined according to plane wave theory. The measuring points 5.0, 5.1 and 5.2 serve as additional monitoring points for the flow condition. Here static pressure sensors and thermocouples meter the corresponding flow variables. Based on the standard volume flow rate (gauged upstream the pulsation generator), the long-time, area-averaged flow velocity is determined based on flow condition measured at 5.2.

As an extension of this setup, a nozzle with defined contraction ratio is attached to the open pipe end, detail b) in figure 1. This is used to influence the reflection of the downstream travelling

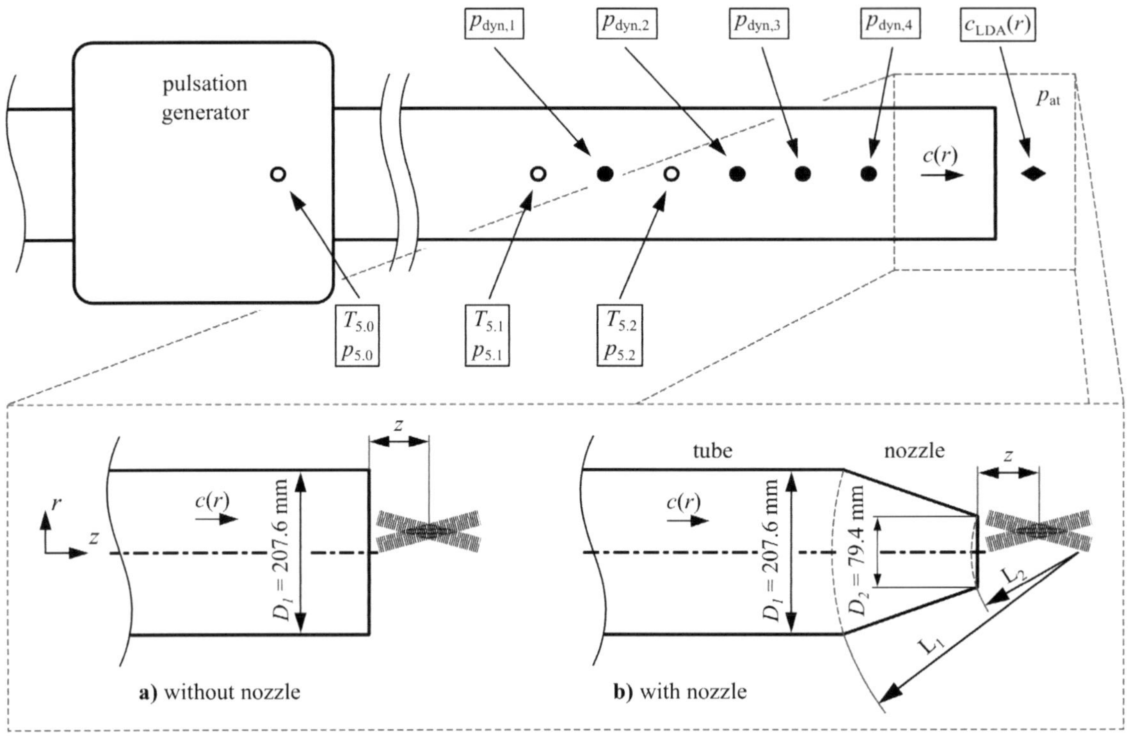

Figure 1. Experimental setup of the measuring section a) without and b) with nozzle at the open pipe end

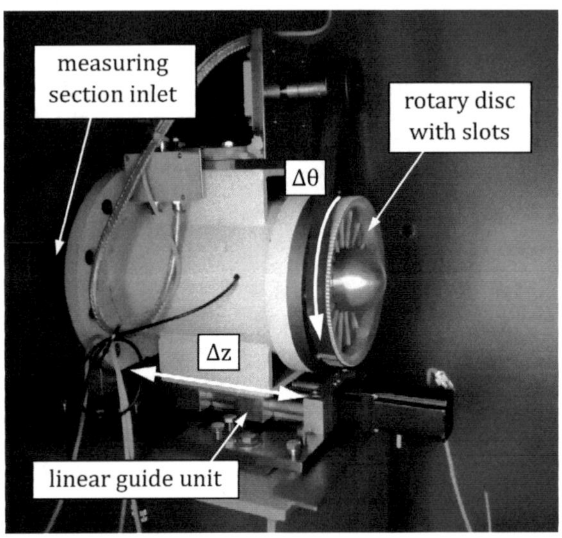

Figure 2. View inside the pulsation generator in front of the measuring section inlet; rotary disc to adjust the frequency and linear guide unit (regulating the bypass) for amplitude variation.

wave at the pipe end with defined impact to create a non-reflecting termination. Likewise in this setup, the velocity profiles are measured at a defined distance z downstream the nozzle opening using LDA. Additionally, the velocity profiles are recorded at different distances z to the modified measuring section outlet. This way the measured velocity profiles provide the ability to understand the radius-dependent reflection behaviour with increasing distance to the pipe outlet. Comparison with the results of plane wave theory provides information about the quality of the one-dimensional modelling.

3. Preliminary theoretical considerations

The theoretical basis for the results presented in this paper is outlined first. Based on the characteristic parameters describing the flow pattern, the experimentally obtained results are evaluated and classified within the studies in known literature. For comparison with plane wave theory the basic equations and definition of phase difference between pressure and velocity are presented. At the section ending the analytical consideration for the non-reflecting termination is presented, which serves as a basis for the nozzle design used at the pipe end.

3.1. Characteristic parameters describing flow pattern

The pulsating flow at the open pipe end presented in section 2 is characterized by eleven independent quantities. Flow velocity

$$c(z,r,t) = \overline{c}(z,r) + \tilde{c}(z,r,t) \tag{1}$$

$$= \overline{c}(z,r) + \sum_{n=1}^{\infty} \hat{c}(z,r)\cos(n\omega t - \varphi_n) \tag{2}$$

consisting of the long-time-averaged velocity $\overline{c}(z,r)$ and the oscillating component $\tilde{c}(z,r,t)$ is defined here as the objective parameter. For analysis the Fourier expansion with $\hat{c}(z,r)$ – amplitude of the oscillating component – is applied to the experimental velocity wave form. In the later evaluation only the fundamental wave (n=1) is considered. The flow guidance is given by pipe inner diameter D_1. With mass flow $\dot{m}$, density ρ and pressure $p(z,t) = \overline{p}(z) + \tilde{p}(z,t)$ the flow condition is specified. Just like the velocity, pressure is divided into a time averaged $\overline{p}(z)$ and a fluctuating part $\tilde{p}(z,t)$. In difference to the velocity, pressure is assumed to be constant over the flow cross section. The fluid, in turn, is considered as an ideal gas, modelled by physical properties dynamic viscosity η, isobaric heat capacity c_p and specific gas constant R_s. In addition, the pulsation is determined by the angular frequency ω. Based on these quantities, the following dimensionless parameters are derived to describe the pulsating flow in the two experimental setups studied.

By means of eleven independent quantities, containing four dimensions, seven known characteristic dimensionless parameter can be derived, which give a complete description of their relationships [8]. The Reynolds number

$$Re = \frac{\overline{c}_m(z)\, D_1\, \rho}{\eta} \tag{3}$$

quantifies the ratio of steady-state inertial forces to steady-state viscous forces – or impulse convection in relation to impulse diffusion – of the flow. $\overline{c}_m(z)$ is defined here as the long-time and area averaged flow velocity. Likewise, the Womersley number

$$Wo = \sqrt{\frac{\omega\, D_1^2\, \rho}{\eta}} \tag{4}$$

relates the transient inertial forces, due to pulsation stimulation, to the steady-state viscous forces. The squared Womersley number is also referred to as the dimensionless frequency in literature. The quotient of Wo^2 and Re is defined as the Strouhal number

$$Sr = \frac{Wo^2}{Re} = \frac{\omega\, D_1}{\overline{c}_m(z)} \tag{5}$$

and thus describes the correlation between transient and stationary inertial forces. The greater the Strouhal number, the greater the influence of transient inertial effects.

The Helmholtz number

$$He = \frac{\omega D_1}{a} = k D_1 \tag{6}$$

can be used to relate the pipe diameter to the circular wave number $k = \frac{\omega}{a}$ as a function of the speed of sound a. Furthermore, the physical fluid properties are described by the isentropic exponent

$$\kappa = \frac{c_\mathrm{p}}{c_\mathrm{p} - R_\mathrm{s}} \,. \tag{7}$$

Up to this point the characteristic parameters relate to the long-time and cross-section averaged flow pattern. In order to provide an insight into the radius-dependent unsteady flow pattern the oscillating momentum equation (no stationary inertial or viscous forces)

$$\underbrace{\rho \frac{\partial \tilde{c}\,(z,r,t)}{\partial t}}_{\text{transient inertia term}} + \underbrace{2 \frac{\partial \tilde{\tau}}{\partial r}\,(z,r,t)}_{\text{transient viscous term}} = \underbrace{\frac{\partial \tilde{p}\,(z)}{\partial z}}_{\text{pressure gradient term}} \tag{8}$$

at an infinitesimal radius increment ∂r is evaluated. Considering only amplitudes of the fluctuating flow components two radius-dependent parameters $\hat{\Phi}_\mathrm{t}\,(z,r)$ and $\hat{\Phi}_\mathrm{z}\,(z,r)$ defined below are introduced [4]:

$$\hat{\Phi}_\mathrm{t}\,(z,r) = \frac{\rho \, \widehat{\frac{\partial \tilde{c}(z,r,t)}{\partial t}}}{\frac{\partial \hat{p}(z)}{\partial z}} = \frac{\rho\,\omega\,\hat{c}\,(z,r)}{\frac{\partial \hat{p}(z)}{\partial z}} \tag{9}$$

$$\hat{\Phi}_\mathrm{z}\,(z,r) = \frac{2\, \widehat{\frac{\partial \tilde{\tau}(z,r,t)}{\partial r}}}{\frac{\partial \hat{p}(z)}{\partial z}} = \frac{2\,\eta\, \widehat{\frac{\partial^2 \tilde{c}(z,r,t)}{\partial r^2}}}{\frac{\partial \hat{p}(z)}{\partial z}} \tag{10}$$

$\hat{\Phi}_\mathrm{t}\,(z,r)$ relates the transient inertial force of the oscillating velocity amplitude $\hat{c}\,(z,r)$ at a certain radius r to the cross-section constant pressure gradient amplitude in flow direction z. Following

$$\tilde{\Phi}_\mathrm{t}\,(z,r,t) = \frac{\rho\,\omega\,\tilde{c}\,(z,r,t)}{\frac{\partial \hat{p}(z)}{\partial z}} \tag{11}$$

is the transient velocity related to the time-independent imposed cross-section constant pressure gradient amplitude in flow direction z. The pressure gradient is derived from the complex sound pressure according to plane wave theory, see equation 15. Analogous $\hat{\Phi}_\mathrm{z}(z,r)$ quantifies the relation between transient viscous forces as a function of the pipe radius and amplitude of the fluctuating pressure gradient, in which shear stress is calculated from oscillating velocity $\tilde{c}\,(z,r,t)$. Due to the two parameters $(\hat{\Phi}_\mathrm{t}, \hat{\Phi}_\mathrm{z})$ the relation between transient inertia and transient viscous forces in a cross section can be quantified, based on the experimentally obtained velocity profiles.

These characteristic parameters allow a quantitative evaluation of the pulsating velocity profile at the open pipe end; as well as the classification within studies in known literature. Accordingly, the following results are presented on the basis of these parameters.

3.2. Plane wave theory

The two-dimensional measured velocity profiles extend the established one-dimensional decomposition of pressure measurement signals into upstream and downstream travelling waves according to the plane wave theory. Assuming an incompressible, isothermal flow, the flow field is influenced only by mean and fluctuating components of pressure $p(z,t) = \overline{p}\,(z) + \tilde{p}\,(z,t)$ and velocity $c_{1d}(z,t) = \overline{c}_{1d}\,(z) + \tilde{c}_{1d}\,(z,t)$. Assuming further small fluctuations, the plane wave equation

$$\frac{\partial^2 \tilde{p}}{\partial z^2} = \frac{1}{a^2}\frac{\partial^2 \tilde{p}}{\partial t^2} \tag{12}$$

can be derived based on the linearized one-dimensional formulations of the continuity equation and momentum conservation. A fundamental solution of this partial differential equation is the complex expression of the sound pressure

$$\underline{\tilde{p}}(z,t) = \underline{\hat{p}}\, e^{i\omega t}\, e^{\pm ikz} \tag{13}$$

as plane wave travelling in one direction in an infinite pipe. The ratio between sound pressure and velocity within this plane wave is described by the characteristic acoustic impedance

$$Z_0 = \frac{\hat{p}}{\hat{c}_{1d}} = \overline{\rho}\,\overline{a} \tag{14}$$

and can be assumed to be constant in a plane wave. [9]

The reflection at an acoustic open pipe end causes an interference in the pipe between the down- and upstream travelling wave p_{e} and p_{r}. Due to the linearity of the wave equation, the complex sound pressure

$$\underline{\tilde{p}}(z,t) = \left(\underline{\hat{p}}_{\mathrm{e}}\, e^{-ikz} + \underline{\hat{p}}_{\mathrm{r}}\, e^{ikz}\right) e^{i\omega t} = \hat{p}\, e^{i\varphi_{\mathrm{p}}}\, e^{i\omega t} \tag{15}$$

results according to the superposition principle as the sum of the down- and upstream travelling wave. Analogously applies to the sound velocity

$$\underline{\tilde{c}}_{1d}(z,t) = \frac{1}{Z_0}\left(\underline{\hat{p}}_{\mathrm{e}}\, e^{-ikz} - \underline{\hat{p}}_{\mathrm{r}}\, e^{ikz}\right) e^{i\omega t} = \hat{c}_{1d}\, e^{i\varphi_{\mathrm{c}}}\, e^{i\omega t} \tag{16}$$

with equation 14. In contrast to the non-reflected passing wave, the phase offset

$$\Delta\varphi = |\varphi_{\mathrm{c}} - \varphi_{\mathrm{p}}| \tag{17}$$

between pressure and velocity occurs in the case of interference in the pipe. This phase offset depends on the reflection behaviour at the pipe end. The complex reflection coefficient

$$\underline{r} = \frac{\underline{\hat{p}}_{\mathrm{r}}\, e^{ikz}}{\underline{\hat{p}}_{\mathrm{e}}\, e^{-ikz}} \tag{18}$$

describes the ratio between incident and reflected sound wave. In theory, a real-valued reflection coefficient of $\underline{r} = -1$ is expected at an acoustically open end. Total reflection corresponds to a phase difference between pressure and velocity of $\Delta\varphi = \frac{\pi}{2}$. [9]

Similar to the phase difference $\Delta\varphi$ of the plane wave theory, the radius-dependent phase difference

$$\Delta\varphi_{\mathrm{LDA}}(r) = |\varphi_{\mathrm{c,LDA}}(r) - \varphi_{\mathrm{p}}| \tag{19}$$

is introduced with $\varphi_{\mathrm{c,LDA}}(r)$ as radius-dependent phase of the transient velocity profile in the measuring plane (z = const). In this way, a comparability between plane wave theory and the measured two-dimensional velocity profile is established by the sound pressure phase φ_{p}. For simplified presentation of the experimental results the normalized phase difference

$$\Delta\varphi_{\mathrm{norm}}(r) = \frac{\Delta\varphi_{\mathrm{LDA}}(r)}{\Delta\varphi} \tag{20}$$

is used, giving a direct relation between the radius-dependent and one-dimensional phase difference.

3.3. Non-reflecting termination

Characteristic for the acoustically open pipe end is a strong reflection of the downstream travelling wave. As mentioned in section 3.2 in theory, the reflection coefficient is given at the acoustically open end by $\underline{r} = -1$. Corresponding to a phase difference between pressure and velocity of $\Delta\varphi = \frac{\pi}{2}$. The installation of a cross-sectional contraction – in the presented case by means of a conical nozzle, detail b) in figure 1 – significantly changes the reflection behaviour at the pipe end.

Under the condition of small Mach number and small nozzle geometry, compared to the acoustic wavelength, an incompressible flow can be assumed [10]. Under this assumption, via the continuity relation, the velocity ratio between nozzle inlet (2) and outlet (1) is directly determined by the contraction ratio α^2

$$\alpha^2 = \frac{D_2^2}{D_1^2} < 1\,. \tag{21}$$

Based on the linearized Bernoulli equation, for a constant nozzle outlet pressure p_1, the normalised impedance at the nozzle inlet

$$\underline{Z}_1 = Ma_2 \left[-i\,\frac{\omega\, L_1}{|\bar{c}_{2,m}|} \left(\frac{1}{\alpha} - 1 \right) + \frac{1}{\alpha^2} \left(1 - \alpha^4 \right) \right] \tag{22}$$

can be determined as a function of the nozzle Strouhal number $Sr_{nozzle} = \frac{\omega\, L_1}{|\bar{c}_{2,m}|}$, the Mach number in the nozzle outlet Ma_2 and the contraction ratio α^2. [5]

A non-reflecting termination is obtained when $\underline{Z}_1 = 1$ is valid for the impedance at the nozzle inlet. Under this constrain a real-valued solution of equation 22 exits for

$$Ma_2 = \alpha^2/(1 - \alpha^4) \tag{23}$$

presuming small nozzle Sr_{nozzle} numbers. Accordingly, for Mach numbers at the nozzle exit that roughly correspond to the contraction ratio, minimal reflections at the nozzle inlet are to be expected [5]. Corresponding to the real and imaginary part of the complex impedance equation 22 the presumption of small Sr_{nozzle} numbers should be valid up to

$$Sr_{nozzle} = \frac{\omega\, L_1}{|\bar{c}_{2,m}|} \leq \frac{1 - \alpha^4}{(1 - \alpha)\,\alpha}\,. \tag{24}$$

4. Velocity profile

The experimentally obtained velocity profiles at the open pipe end are presented for stationary and transient flow conditions. By means of the characteristic parameters $\hat{\Phi}_t$ and $\hat{\Phi}_z$ the transient flow pattern is evaluated as a function of the pipe radius. Based on these examinations the results are classified within the studies in known literature.

4.1. Stationary velocity profile

The measured stationary velocity profiles show good agreement over a wide Re number range compared with the $\frac{1}{7}$ power law according to Nikuradse. Figure 3 shows the comparison for $Re = 40\,000$ up to $Re = 200\,000$ as a function of the dimensionless radius $r^* = \frac{r}{R}$. The analytical solution is determined by the calculated volume flow rate based on the integration of the corresponding measured velocity profile. Consistent with the Re numbers studied, the characteristic turbulent flow profile is visible. The velocity has a plain distribution in a wide range around the pipe centre ($r^* = 0$) and drops rapidly towards the pipe outer radius ($r^* = 1$).

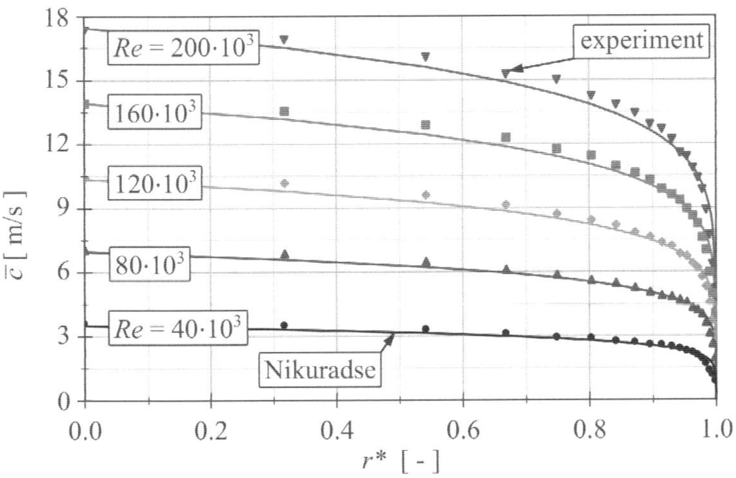

Figure 3. Comparison of the stationary measured velocity profiles with Nikuradse's $\frac{1}{7}$ power law from $Re = 40\,000$ up to $Re = 200\,000$

For all presented Re numbers, two effects are observed. First, in the pipe core region up to $r^* = 0.93$ the velocity profile is slightly underestimated by the analytical solution. On the remaining radius section this relationship is reversed; here the experimental results are overestimated by the analytical solution. Second, in contrast to the calculated profiles – representing the inner pipe flow – the no-slip condition is violated in the measurements, due to the location of the measuring plane $z = 10$ mm downstream the open pipe end. The latter effect is indicated by the velocity offset at $r^* = 1$. Both effects point to the influence of a free-jet boundary condition in the measuring plane. After these preliminary considerations, the measured transient velocity profiles are classified in known flow pattern according to [4] in the following.

4.2. Radius-dependent flow pattern

The transient radius-dependent flow pattern can be classified into three types with respect to Wo number [4]: quasi-steady ($Wo \leq 3$), intermediate ($3 < Wo \leq 100$), and inertia dominant ($Wo > 100$) [4]. Providing constant geometry and fluid property (η) the classification is only dependent on the angular frequency ω. Due to the examination of large Wo numbers in this study ($Wo \geq 405$), a strongly inertia dominated oscillating flow pattern is to be expected compared to the influence of the viscous forces in the flow. Figure 4 shows the comparison between $\hat{\Phi}_t\,(r^*)$ – quantifying the transient inertial forces in the flow – and $\hat{\Phi}_z\,(r^*)$ – determine the transient viscous forces in the flow – as function of the dimensionless radius r^*. The illustrated radius-dependent flow pattern for $Wo = 405$ ($Sr = 0.8$, $Re = 200\,000$) represents the minimal dimensionless frequency experimentally investigated in this study.

Transient inertial forces dominate the flow nearly over the entire radius. Up to approximately $r^* = 0.93$ the transient viscous forces are almost zero. Given this, the oscillating flow component can be regarded as almost inviscid over most part of the radius. On the last seven percent towards the pipe wall radius an increase of the transient viscous forces is visible considering $\hat{\Phi}_z$. At the same time inertial forces decrease in reverse. Although there is a slip due to measuring in the free jet region ($z = 10$ mm downstream the pipe end), transient viscous forces reach their maximum at $r^* = 1$. At the transition between exhausting pipe flow and static ambient fluid, the maximum gradient in the oscillating velocity component occurs here; causing an increasing influence of the transient viscous forces in the near wall radius section. Based on these findings the transient velocity profiles are studied in the following.

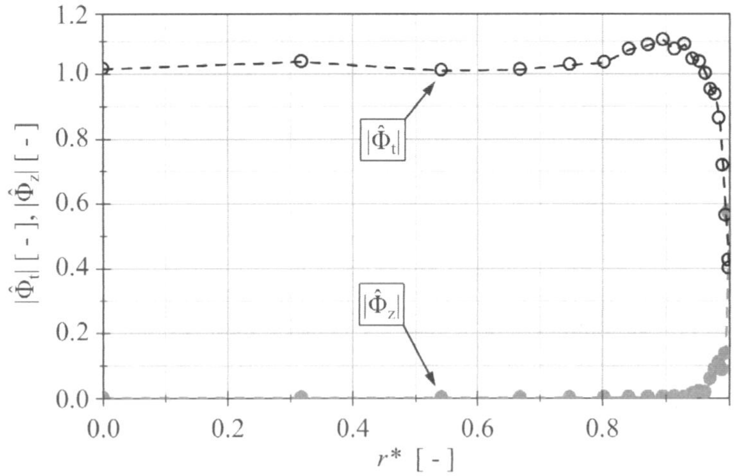

Figure 4. Comparison between transient inertial $\left|\hat{\Phi}_t(r^*)\right|$ and viscous forces $\left|\hat{\Phi}_z(r^*)\right|$ as function of the dimensionless radius r^* for $Wo = 405$ ($Sr = 0.8$, $Re = 200\,000$)

4.3. Pulsating velocity profile

Hereinafter the transient velocity profile is discussed against the background of the classified flow pattern. Figure 5 and figure 6 show the time dependent velocity profiles over one pulsation period considering different Sr numbers. Interpretation of the Sr number as the ratio of transient to stationary inertial forces is illustrated by evolving distortion of the shown velocity profiles. According to the remarks in section 4.2, the oscillating flow component is considered inviscid over most part of the radius. Assuming this the shown Sr number-dependent velocity profiles are mostly influenced by the interaction of steady-state inertial and viscous forces superposed with the imposed transient inertial force in the core flow region. The greater the Sr number, the greater is the influence of transient inertial effects compared to the stationary forces of the flow; resulting in progressive deformation of the time-dependent velocity profiles with increasing Sr number.

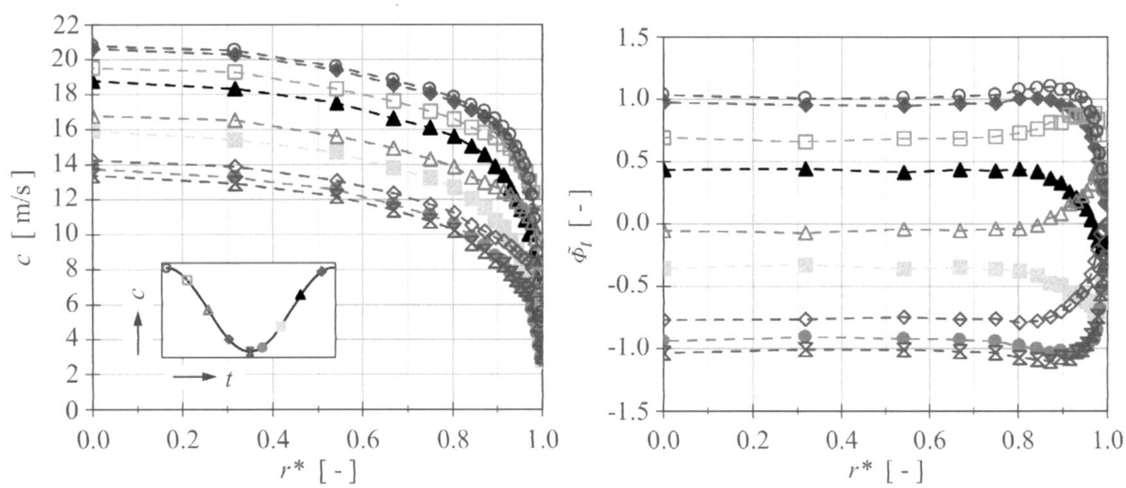

a) Transient velocity profile $Sr = 0.8$ ($Wo = 405$, $Re = 200\,000$)

b) Dimensionless amplitude for $Sr = 0.8$ ($Wo = 405$, $Re = 200\,000$)

Figure 5. Transient velocity profile $c(r^*)$ and corresponding dimensionless amplitude $\tilde{\Phi}_t(r^*)$ over one pulsation period for $Sr = 0.8$ ($Wo = 405$, $Re = 200\,000$)

For small Sr numbers, the known characteristic profiles due to the annular effect[1] are found in the dimensionless velocity fluctuations, figure 5b. As a result of the radius-dependent oscillating momentum equation (equation 8) a difference in phase occurs in the pulsating velocity amplitude toward the outer radius. Here, the influence of the transient shear stress $\hat{\Phi}_z(r^*)$ increases, whereas the influence of the transient inertia $\hat{\Phi}_t(r^*)$ decreases with radius, compare qualitative profiles in figure 4. This is quantified in more detail later in section 5. Compared to the pulsating flow in the pipe center, this phase difference results in a time-dependent higher or lower velocity amplitude in the region near the pipe wall radius.

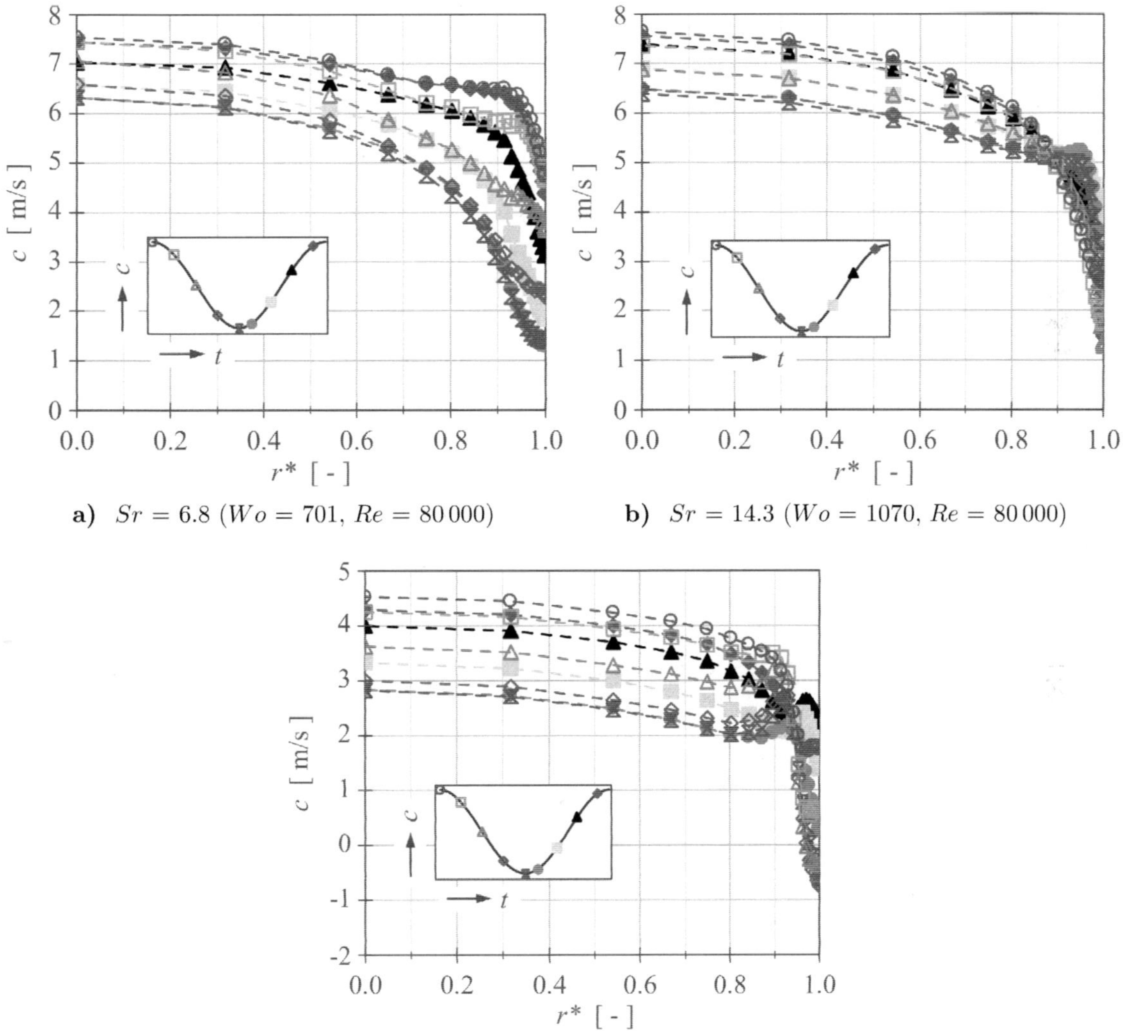

a) $Sr = 6.8$ ($Wo = 701$, $Re = 80\,000$)

b) $Sr = 14.3$ ($Wo = 1070$, $Re = 80\,000$)

c) $Sr = 28.6$ ($Wo = 1070$, $Re = 40\,000$)

Figure 6. Transient velocity profiles $c(r^*)$ over one pulsation period for different Sr numbers

Referred to the velocity profiles of pulsating turbulent inner pipe flow, already known in literature [3], the velocity profiles presented here show a progressively deviating profile in the near-pipe-wall region ($r^* > 0.7$) with increasing Sr number. As already shown in section 4.1,

[1] The annular effect is also known as skin effect in literature.

the presented free jet investigation downstream the open pipe end exhibits a velocity offset at the outer radius $r^* = 1$. Accordingly the no-slip boundary condition $c\,(r^* = 1) = \tilde{c}\,(r^* = 1) = 0$ is violated, in contrast to the inner pipe flow. Thus, the temporal velocity fluctuation at the outermost radius is not constant zero, but follows the pulsating velocity profile time-shifted. With increasing Sr number this offset has an increasing impact on the velocity profile, compare progression figure 5 to figure 6.

Due to this velocity fluctuation at $r^* = 1$ a node is formed in the velocity profile, visible in figure 6b for $Sr = 14.3$ and figure 6c for $Sr = 28.6$, where the velocity amplitude vanishes. Relative to the pipe core flow, the fluctuating velocity component at the outer radius exhibits an increasing phase difference as the Sr number increases. This results in a time-dependent velocity in- or decrease at the pipe wall radius compared to the velocity near the pipe center, see figure 6a. Once a phase skip occurs between the fluctuation in the pipe core and near-pipe-wall region, the mentioned node is formed in the velocity profile, figure 6b. The node formation shifts toward the outer radius as the Sr number increases, accompanied by distortion of the time-averaged velocity profile shown in figure 6c. Accordingly to these considerations, the measured transient velocity profiles in the free jet show significantly different amplitudes and phase differences in the outer radius region than the velocity profiles known in literature, which are solely due to the annular effect including no-slip boundary condition at the pipe wall radius.

Due to the relatively high stationary inertia of the flow, transient free jet effects are of less influence on the velocity profile for small Sr numbers, figure 5. Thereby the annular effect hardly influences the time-dependent velocity profile. This leads to the quasi-stationary appearance of the transient velocity profile for small Sr number.

5. Phase shift

Concerning the observed phase difference in the transient velocity profiles, discussed in section 4.3, following the radius-dependent phase difference is quantified and related to plane wave theory. Further due to this consideration a comparison between the radius-dependent phase difference at the acoustically open pipe end and at the customized nozzle is possible; showing the influence of the nozzle on the reflection behavior.

5.1. Open pipe end – without nozzle

The phase difference along the pipe radius, observed in the transient velocity profiles for different Sr numbers in section 4.3, is quantified using the normalized phase difference $\Delta\varphi_{norm}$ introduced in section 3.2. The definition according to equation 20 establishes a relation between the velocity and pressure phase at the pipe center and near the wall on one hand, and between the radius-dependent transient velocity profile and the plane wave theory on the other hand.

The radius-dependent normalised phase difference, observed in the velocity profiles (see e.g. figure 5 to figure 6c), is shown in figure 7 for different Sr numbers. For all Sr numbers the dimensionless phase difference shows a plain distribution in a wide range around the pipe center up to approximately $r^* = 0.7$. From this radius on the phase difference is strongly dependent on the Sr number, on the last thirty percent towards the pipe wall radius.

For small Sr numbers phase difference between the pressure phase according to the plane wave theory – constant all over the pipe radius – and radius-dependent velocity decreases on the the last thirty percent of the radius compared to the approximately constant phase difference on the seventy percent of the radius around the pipe center. For $5.7 \leq Sr < 14.3$ an increasing phase difference in relation to the pipe center develops between LDA-measured radius-dependent velocity and pressure corresponding to decomposition on the last thirty percent of the pipe radius. Whereas on the last ten percent of the radius a slightly decrease is observed just before reaching the outer radius. The resulting local maximum rises with increasing Sr number. For $Sr \geq 14.3$ the trend continues resulting in a phase skip, visible due to a local unsteadiness.

As indicated beforehand, this unsteadiness in the radius-dependent phase difference coincides with the observed node in the velocity profiles of the corresponding Sr numbers, figure 6c and figure 6b.

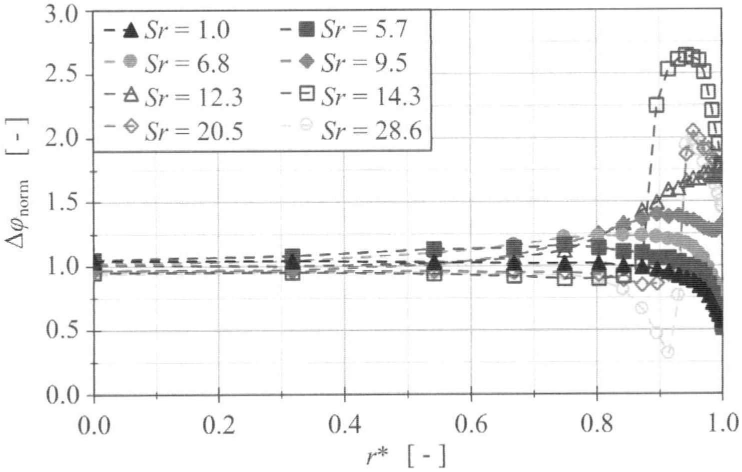

Figure 7. Normalized phase difference $\Delta\varphi_{\mathrm{norm}}(r^*) = \frac{\Delta\varphi_{\mathrm{LDA}}(r^*)}{\Delta\varphi}$ for different Sr numbers from $Sr = 1.0$ up to $Sr = 28.6$

Hence the evaluated radius-dependent phase difference correspondent with the observed radius-dependent interaction between transient and stationary viscous and inertial forces resulting in the shown velocity profiles for different Sr numbers. Moreover for the pulsating flow, measured in free jet directly downstream the open pipe end, one-dimensional wave theory is regarded as a good representation of the pipe core flow.

5.2. Open pipe end – with nozzle (non reflecting)

With the understanding of the radius-dependent flow pattern and the effect on the transient velocity profile concerning different Sr numbers, the influence on flow pattern of the customized nozzle at the open pipe end creating a non reflecting termination is considered. As mentioned in section 3.2 in theory a phase difference of $\Delta\varphi = \frac{\pi}{2}$ is expected at the reflection point at the acoustical open pipe end. The measured radius-dependent phase difference $\Delta\varphi_{LDA}(r^*)$ at the measuring point $z = 10\,\mathrm{mm}$ downstream the pipe end for $Sr = 0.8$ is shown in figure 8. The measured phase difference is $\Delta\varphi_{LDA} = 116°$ over most part of the radius and the already observed qualitative distribution, see section 5.1, is also visible. The circumstance that the phase difference is not exactly $\Delta\varphi = \frac{\pi}{2}$, is most likely due to the fact that the actual reflection doesn't take place in the measuring plane due to the pipe end correction.

In the modified experimental setup a completely different phase difference between measured velocity and pressure phase downstream the nozzle is observed. The contraction ratio of the used nozzle is $\alpha^2 = 0.146$ to create an anechoic termination for Mach number at the nozzle outlet of $Ma_1 = 0.15$. In this setup the phase difference, shown in figure 8, was reached for $Ma_1 = 0.12$ at the nozzle outlet. This results in a phase difference of $\Delta\varphi_{LDA} < 10°$ between measured velocity and sound pressure over most part of the radius for all considered dimensionless distances $\frac{z}{D_1}$. The smallest phase difference $\Delta\varphi_{LDA} = 4°$ is reached for $\frac{z}{D_1} = 0.08$ in the pipe center. Here, the phase difference rises up to $\Delta\varphi_{LDA} = 9°$ for $\frac{z}{D_1} = 1$ with increasing distance to the nozzle outlet.

Considering the region near the nozzle outer radius for $\frac{z}{D_1} = 0.08$ a steep gradient in the radius-dependent phase difference occurs on the last five percent of the radius, resulting in a absolute higher $|\Delta\varphi_{LDA}|$ compared to the the pipe core flow. This is most likely due to the transient flow behavior already discussed on basis of the velocity profiles in section 4.3. With increasing distance to the nozzle outlet, this phase difference is reduced.

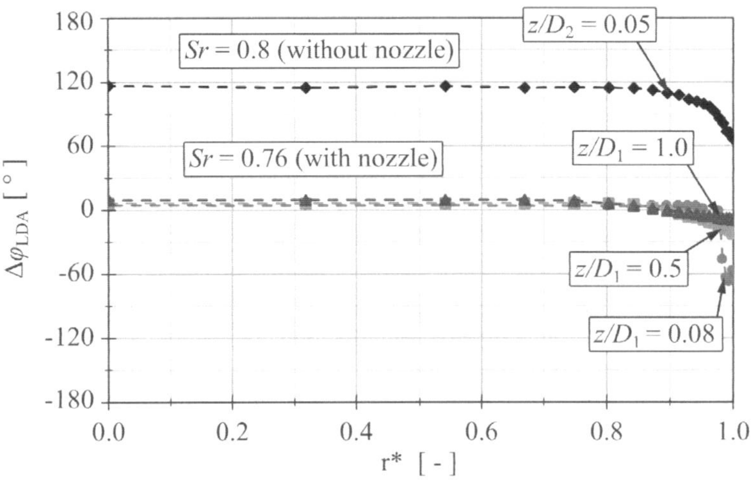

Figure 8. Comparison in phase difference along the pipe between the open pipe end with nozzle $Sr = 0.76$ ($Wo = 405$, $Re = 215\,000$) and without nozzle $Sr = 0.8$ ($Wo = 405$, $Re = 200\,000$); configuration with nozzle is shown in different dimensionless distances $\frac{z}{D_1}$ from the nozzle outlet.

The comparison between the results of the two different experimental setups shows a significant reduction of the observed phase difference just downstream the acoustical open end and thus a significant reduction of the reflection coefficient. In a completely anechoic boundary condition, pressure and velocity would oscillate in phase, thus $\Delta\varphi = 0$ should be also valid for $\Delta\varphi_{LDA}$. Thereby the analytical considerations for the non-reflecting termination introduced in section 3.3 are confirmed in practical application for the considered pulsating pipe flow and used nozzle design.

It has to be outlined that the Mach number at the nozzle outlet (Ma_1) of the adjusted operating point deviates by approximately 20% from the calculated Mach number according to equation 23 for the used nozzle geometry. Since the required low Sr number was maintained, it is assumed that a stationary (or transient) effect occurring in practice is not yet taken into account in the derivation of equation 22. Nevertheless, the high practical use of a nozzle at the acoustically open pipe end remains: By using a cross-section contraction, the reflection at the acoustically open end can be significantly reduced, if not completely eliminated, for a defined operating point. The cross-section contraction could also be realized by a simple orifice [5] and integrated into the existing piping system with minor effort, causing major effect.

6. Conclusion

Summing up in the presented paper pulsation at the open pipe end is evaluated in two modifications. In first instance the pulsating pipe flow just downstream the acoustically open end is examined. In a modified setup the impact of a customized nozzle on the reflection behaviour of the open pipe end is examined.

To gain a deeper understanding of the pulsating flow directly downstream the acoustical open end LDA velocity-profile measurements along the pipe radius combined with dynamic pressure measurements along the upstream measuring section are conducted. Due to this approach, a deeper understanding of the pulsating velocity profile in the free jet area directly downstream the acoustical open end is provided.

By describing the experimental setup the test rig as well as underlying boundary conditions are presented. Preliminary theoretical considerations deal with conducted characteristic parameters describing the flow pattern in an integral (area averaged) and radius-dependent way. For comparison with plane wave theory basic equations and definition of the considered phase difference between pressure and velocity are introduced. The fundamentals conclude with analytical description of the non-reflecting termination, used to customize the examined nozzle to create a reflection free boundary condition at the open pipe end.

Based on this theoretical basis the velocity profile at the open pipe end is examined. First by considering the stationary flow profile in comparison with Nikuradse's $\frac{1}{7}$ power law the plausibility of the LDA measurements over a wider *Re* number range is shown; also indicating the free stream boundary condition at the pipe wall radius. Following the stationary considerations the pulsating radius-dependent flow pattern is examined. By means of the Wo number ($Wo \geq 405$) the integral flow is classified as inertia dominated. This classification is sustained by the introduced radius-dependent characteristic numbers conducted out of the measured flow profiles. Nevertheless evaluation of the radius-dependent momentum conservation shows increasing influence of viscous forces on the flow profile towards the pipe wall radius. This is illustrated in the shown transient velocity profile for small Sr number. With increasing Sr number the influence of the free stream boundary condition at the pipe wall radius is evaluated; resulting in an increasing phase difference towards the pipe wall radius for higher Sr number; following a phase skip at $Sr \geq 14.3$ caused by the increasing influence of transient inertial forces in the flow.

Quantitative evaluation of the resulting phase difference in the velocity profile is done by observing the radius-dependent normalized phase difference for different Sr numbers. Further more the normalized presentation allows a comparison between the phase difference calculated according to plane wave theory and measured two-dimensional velocity profiles. Showing a good agreement between one dimensional wave theory and the measured two-dimensional velocity profile in approximately seventy percent of the flow around the pipe center. The remaining thirty percent of the pipe radius seem strongly effected by transient inertial and viscous effects due to the free stream boundary condition at the pipe wall radius. The occurring phase difference in near-pipe-wall region is of course unrepresentable by plane wave theory.

Based on this background the influence of a customized nozzle at the open pipe end is illustrated by the defined phase difference for small Sr number. By use of the designed nozzle phase difference at the open pipe end was reduced to $\Delta\varphi_{LDA} < 10°$ over 95% percent of the pipe radius for a specific operating point. The remaining phase difference at the last five percent to the pipe outer radius decreases significantly with distance to the nozzle outlet. Showing that the use of the designed nozzle reduces the reflection at the open pipe end significantly over most part of the radius. This provides a simple way to counteract strong and destructive resonances in existing piping networks that occur depending on the operating point, for example in the spool piece between the high-pressure side of a screw compressor and the connected pulsation damper/silencer.

Acknowledgment
The results shown here are based on LDA measurements carried out as part of a master's thesis by Ms. Wiebke Fischer. The authors would like to thank Ms. Fischer for her perseverance and patience in carrying out the measurements.

References
[1] Sexl T 1928 *Annalen der Physik* **392** 570–580
[2] Richardson E G and Tyler E 1929 *The Proceedings of the Physical Society* **42-1** 231
[3] Ohmi M and Usui T 1976 *Bulletin of JSME* **19** 307–313
[4] Ohmi M, Iguchi M, Usui T and Minami H 1980 *Bulletin of JSME* **23** 2013–2020
[5] Bechert D W 1980 *Journal of Sound and Vibration* **70** 389–405
[6] Dantec Dynamics *LDA and PDA: Reference Manual* URL `https://www.dantecdynamics.com`
[7] Åbom M and Bodén H 1988 *The Journal of the Acoustical Society of America* **83** 2429–2438
[8] Buckingham E 1914 *Physical Review* **4** 345–376
[9] Meyer E and Neumann E G 1979 *Physikalische und Technische Akustik: Eine Einführung mit zahlreichen Versuchsbeschreibungen* 3rd ed Schwingungsphysik (Wiesbaden and s.l.: Vieweg+Teubner Verlag)
[10] Bechert D W 1971 *Deutsche Luft- und Raumfahrt – Forschungsberichte* **71-25**

IOP Conf. Series: Materials Science and Engineering 1267 (2022) 012019 doi:10.1088/1757-899X/1267/1/012019

A Novel Screw Compressor with a Shunt Enhanced Compression and Pulsation Trap (SECAPT)

P X Huang[1], S Yonkers[2] and J Willie[3]

[1] Hi-Bar MC Technologies, LLC., Fayetteville, GA, USA
[2] Hi-Bar Blowers, Inc., Fayetteville, GA, USA
[3] CVS Engineering GmbH, Rheinfelden, GERMANY

E-mail: paulxbh@yahoo.com, james.willie@cvs-eng.de

Abstract. Some positive displacement (PD) compressors are equipped with automatic discharge valves such as reed valves that open automatically whenever the cavity pressure is slightly larger than the outlet pressure to deal effectively with varying pressure ratio applications. Screw compressors today do not have such valves, resulting in off-design conditions known as the under-compression or over-compression when the cavity pressure at discharge is deviating from the outlet pressure. Compressor efficiency suffers and pulsation/noise becomes worse under these conditions. Some type of controls are desired such as a discharge pulsation dampener or a variable volume ratio (often called variable Vi) slide valve design to lessen or cure the discharge pressure mismatch problems.

Injecting gas vapor into a screw compressor cavity during internal compression phase has been known to be beneficial in enhancing compressor performance and has been widely used in HVAC&R industry known as the Economizer. However, the underlining principle of the Economizer has thus far not been explored for optimizing compression schemes of a screw compressor in general. This paper introduces a self-sensing and self-correcting compression process that can be "derived or deduced" from the Perfect Gas Law by optimizing for multiple design criteria such as compressor efficiency, pulsation/noise abatement, and cost and footprint reduction for a screw operating over a wide range of pressures. The scheme, called SECAPT (Shunt Enhanced Compression And Pulsation Trap), is then investigated and optimized numerically by a new CFD code for one industrial case: a bulk truck loading application where compressor pressure varies from no pressure rise to maximum load. The numerical simulations illustrate that SECAPT is tentatively capable of achieving multiple targets as theorized.

1. Introduction

1.1 Dichotomy of screw compressor design operating over a wide range of pressures?

Defined narrowly, a screw compressor is a positive displacement device that discretely transports a volume of a gas and at the same time increases its pressure by reducing the volume. An ideal screw compressor converts the shaft energy into gas internal energy by trapping a fixed amount of incoming gas into a series of rotating cavities, then compressing and discharging the gas into an outlet system when the compressor cavity pressure at discharge is equal to the system back pressure, a case called 100% internal compression (IC). But more often, as pointed out by Sangfors [1], Soedel [2] and Stoecker [3], screw compressors operated at off-design conditions where the discharge cavity pressure is either lower or higher than the system back pressure due to the inherent nature of possessing a fixed built-in volume ratio (V1/V2) or pressure ratio. The resulting processes are often called an Under-compression (UC) or an Over-compression (OC) as shown on the P-V diagrams in Figure 1.

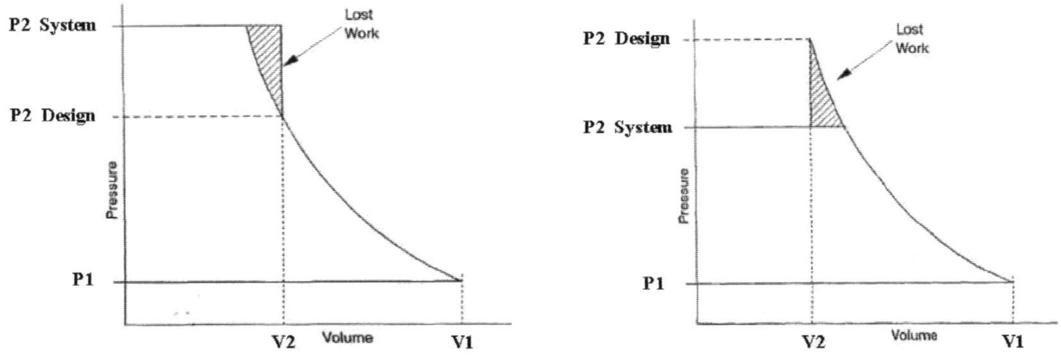

(a) under compression (UC) (b) over compression (OC)

Figure 1. UC and OC of a screw compressor on P-V diagram [3]

All fixed volume ratio screw compressors suffer from UC and/or OC due to the unattainability of trying to fit one fixed design pressure ratio to various operating pressure ratios. Typical applications built-in with variable pressure ratios, as examples, include refrigeration and heat pump systems, and pressurizing or vacuuming a vessel and fuel cell booster. In refrigeration and heat pump, when ambient temperature rises or falls while room temperature stay fixed, the compressor pressure ratios required in the refrigeration cycle have to change accordingly. Often, the range of the pressure ratio variation is significant and the effects of OC and UC are further enhanced by the elevated pressures that refrigerant needs to operate typically around 100 psig (7 bar). In pressurizing a vessel such as bulk loading compressors, the discharge pressure used to push bulk material flow into a vessel continuously increases the pressure ratio as the pressure level gets higher and higher inside the tank. An emerging application for variable pressure compressors is the hydrogen fuel cells used for Electric Vehicles which require oxygen from ambient air to make power. The power density and efficiency of fuel cells are found to be greatly boosted by supercharging the air supply, analogous to supercharging a gasoline car. For these applications, the compressor efficiency and gas pulsation become much worse when operated at off design condition further and UC/OC is most severe. To reduce or eliminate the OC and UC in twin rotor screw compressors for these applications has been a continuing pursuit and challenge since its rebirth at SRM in 1930s. Rotary piston and scroll compressors use automatic discharge valves such as reed valves that can effectively deal with the OC and UC problems as confirmed experimentally by Rohleder & Groll [4], etc. Screw compressors cannot use such valves at discharge due to its intrinsic porting requirement. Instead, tremendous efforts have been devoted to a variable volume ratio scheme, often called variable Vi, since the 1960s as reflected from over dozens of patents filed and many are still pending today.

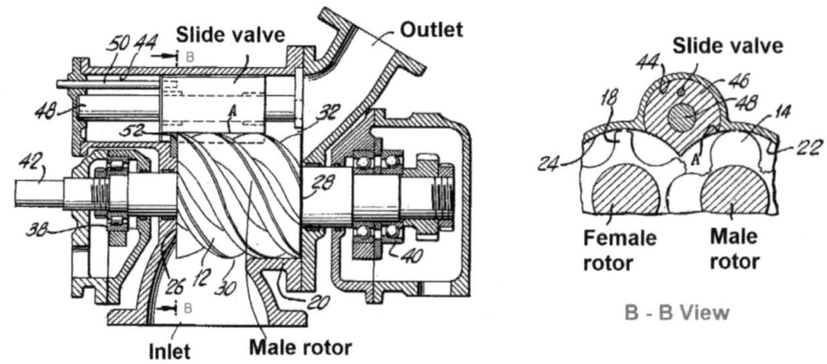

Figure 2. Example of a screw compressor equipped with a slide valve for variable pressure ratio.

An example is taken from SRM's Nilsson patent US3088659 [5] as shown in Figure 2 where a slide valve near the twin screw compressor outlet is used to adjust the gas discharge porting position, hence variable volume ratio, corresponding to the required discharge pressure of the application. This method has been widely used in refrigeration and air-conditioning industry where discharge temperature is relatively low and pressure variation process is slow. However, this scheme fails to work for a very important application – the dry screw compressors where discharge temperature is often high and there is no lubrication available for the slide valve. Variable Vi concept also does not apply to applications where the discharge pressure is changing rapidly.

In the times of global competition and growing environmental concerns, screw compressors are required to achieve multiple goals such as good efficiency, reduced pulsation/noise, and low cost and small footprint not only at one design point, but also over a wide range of operating conditions. No current technology could reliably and economically meet the multiple demands simultaneously. Is it a design dichotomy or a possibility? This paper will search for alternative solutions to the slide valve, and assess them based on how well multiple goals as the above listed can be achieved simultaneously.

1.2 The origin of gas pressure: brief review of kinetic theory of gases

In search for a most desirable compression scheme, we go back to the origin of pressure as disclosed from the kinetic theory of gases that explains the macroscopic (thermodynamic) properties of gases, such as pressure, temperature, by considering their molecular motion at microscopic level. In particular, the pressure is interpreted as the total forces per area exerted by the innumerable flying gas molecules colliding with each other and with the walls of a gas container, as illustrated in Figure 3.

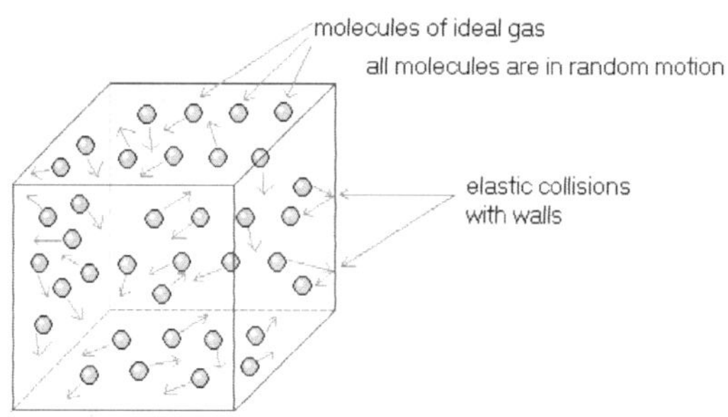

Figure 3. Ideal gas model by the kinetic theory of gases.

Consider a gas of N molecules moving constantly at an average velocity v, each molecule with mass m, as enclosed in a container with volume V, the pressure of the gas inside the container P can be derived from the Newton's 2nd Law of Motion as

$$P = \frac{2}{3}\frac{N\,(mv^2/2)}{V} \tag{1}$$

This is one of the most important results of the kinetic theory because it relates pressure P and volume V, both macroscopic property, to the kinetic energy of the molecules N × [$(mv^2)/2$], which is a microscopic property. In terms of Kelvin temperature T, equation (1) becomes the familiar form of the Ideal Gas Law below, where R is the ideal gas constant.

$$P = \frac{N\,m R\,T}{V} \tag{2}$$

Equations (1) and (2) suggest that gas pressure increase in a closed volume can be achieved by three elemental ways: reducing volume V (Boyle Law), or adding more gas molecules into the volume (Avogadro Law) to increase N, or adding heat to the volume (Amonton Law) to increase the gas temperature T, as exhibited in Figure 4 (a)-(b)-(c) respectively. Moreover, the equations also imply that these methods not only can be used as singles, but also combined in pairs or trios, or arranged in series or parallel, resulting in various schemes of gas compression. It should be pointed out that these options are not favored or discriminated from one another by equations (1) and (2). Rather, the effectiveness of each scheme can only be measured and judged by whether it is capable of achieving simultaneously multiple targets in the context of this paper for screw compressors such as: energy saving, less pulsation/noise and cost/size reduction when operating over a wide range of pressures.

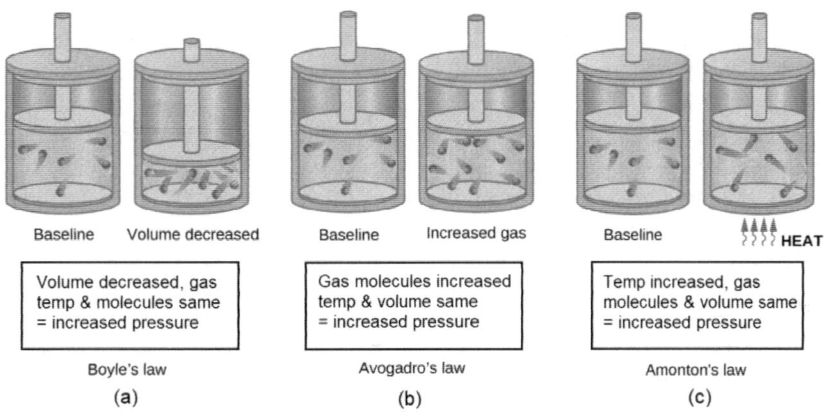

Figure 4. Three elemental ways to increase gas pressure (Internet: BC Open Textbooks)

1.3 Combination of PD compression schemes in singles, pairs or trios

The table below lists various possibilities of PD compression scheme by combining the three fundamental methods illustrated in Figure 4 as singles, pairs or trios:

Table 1. Options of various PD compression schemes.

	Compression Method	Ideal Thermodynamic process	Examples
A	Reduce V only	Isothermal volume reduction	classical Boyle U-tube test
B	Increase N only	Isochoric gas fill-in	Fill-in or inflate a basketball or an auto tire
C	Increase T only	Isochoric heating	Ideal Otto cycle for ICE or constant-volume thermometer

AC	Reduce V adiabatically (add heat internally)	Adiabatic compression also known as Internal Compression (IC) by screw	Adiabatic compression, such as screw operates at design Vi, or other PD with automatic discharge valves
BC	Increase both N and T at end	backflow or under compression (UC)	Roots blower operates under all loads
ABC-1	Reduce V adiabatically and add N at end	IC + UC in series	Traditional Screw with internal compression (IC) and UC
ABC-2	Reduce V adiabatically and subtract N at end	IC + OC in series	Traditional Screw with internal compression (IC) and OC
ABC-3	Reduce V and add N with lower T at start	UC + IC in parallel	Economizer in HVAC&R

Option A is a classical textbook example for isothermal compression but difficult to achieve in reality. Option B is illustrated by the workaday example of a basketball or tire that gets inflated by pumping more air molecules into the ball. Option C is rarely used for PD compressors alone because getting heat transferred into a fast moving cavity is a slow and difficult task. However, the ideal ignition of the fuel-air mixture and the subsequent rapid burning phase of the Otto cycle for an ICE (internal combustion engine) is a good example. Another classic example is a constant-volume thermometer as shown in textbook of physics.

Combining the above three basic ways in pairs or trios results in more options. For examples, option AC is the familiar adiabatic compression method such as a screw compressor operating at design Vi or a screw equipped with a variable Vi such as slide valve or other PDs equipped with automatic discharge valves that open when the cavity pressure is slightly larger than the discharge pressure [2]. Option BC can be exemplified by a Roots compression process or a 100% UC (0% internal compression) where the gas compression is fulfilled by a backflow of hot discharge gas rushing into the cavity. Option ABC could be carried out by either adding more needed gas at end as ABC-1 (For example, IC compensated by UC at discharge as shown in Figure 1a) or subtracting extra gas at end as ABC-2 (IC compensated at discharge by OC as shown in Figure 1b). Finally, a case in point for ABC-3 is the Economizer process, widely used in HVAC&R industry [3], [6], where the internal compression (IC) is boosted in parallel by injecting cooler vapour into the compressor cavity during early stage of IC.

However, the effectiveness of the combo-means BC and ABC as observed in the Roots blower and the Economizer, especially the way (HOW) they are arranged either in series or parallel, have thus far not been explored for optimizing compression scheme of a screw compressor in general. The following will examine and compare two popularly used combo schemes for their respective merits or limits so that the most potentially competitive ones can be evaluated based on whether they can satisfy simultaneously good energy transfer efficiency and low gas pulsation/noise with compact size and low cost.

2. Process analysis of serial vs. parallel compression schemes

2.1 Arrangement of screw compression: IC and UC (OC) in series

Let us take a closer look at the examples ABC-1 and ABC-2 that depict typical problems caused by a traditional serially arranged IC and UC or OC scheme as shown in Figure 1 in a hope to get some insights as to how to devise an improved scheme.

Figure 5a shows a flow chart for the compression cycle of a screw compressor by following a gas cavity for the serial IC and UC scheme. The inlet gas flow Q_{inlet} goes through suction and internal compression (IC) phases, and is then followed by a transient backflow compression due to UC when the cavity is suddenly opened to the outlet, and subsequent discharging and dampening phases. Figure 5b illustrates the same compression cycle in time domain emphasizing the timing of major events especially during the transient backflow phase.

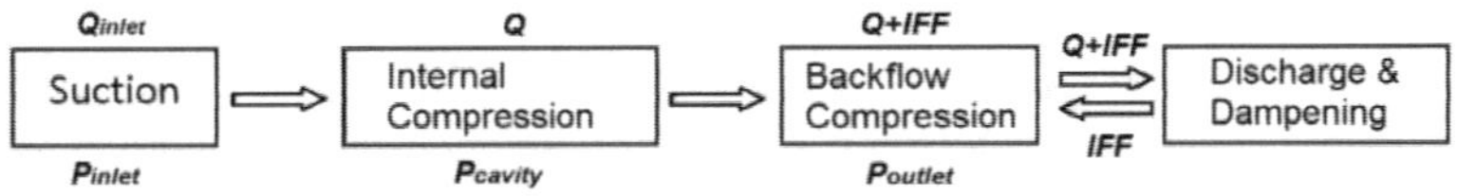

a. Phases of screw compression cycle with serially arranged IC+UC mode

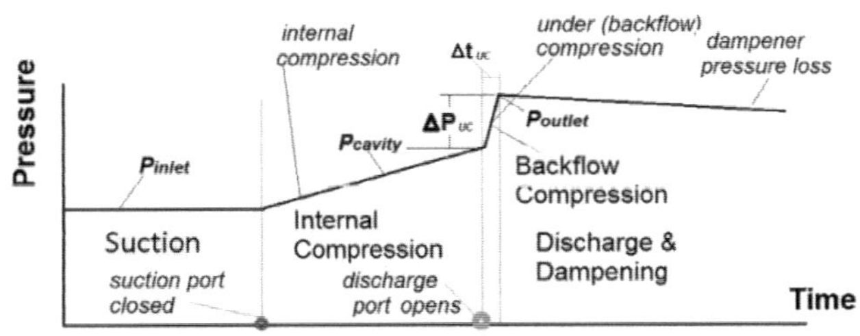

b. Pressure variation during screw compression cycle in time domain

Figure 5. Phases of a screw compression cycle for a serially arranged IC and UC mode.

In a serial configuration, the IC phase stops after the cavity gas finishes the volume reduction as designed and followed by a violent transient UC phase when the discharge port suddenly opens to expose the pressure difference between the design pressure P_{Cavity} in the compressor cavity and the outlet pressure P_{Outlet}, i.e. ΔP_{UC}. According to a transient analysis by Huang [7] using a shock tube analogy (refer to Liepmann & Roshko [8] for shock tube fundamentals), the sudden communication between the ΔP_{UC} would trigger a generation of gas pulsations in the form of CW-IFF-EW (Compression Wave, Induced Fluid Flow, Expansion Wave), which originates at the discharge port opening with the induced fluid velocity over 100 m/s (in contrast, the fluid velocity at design condition is around 20-30 m/s). CW and IFF jet would go into the cavity to further compress the cavity gas Q and impact the rotor to do more work in a negative way as the shaded area shown in Figure 1a, while EW comes towards the outlet pulling IFF in and causing gas pulsation downstream. It should be pointed out that this process is traditionally called backflow compression that neglects the dynamic role played by CW and EW. Since the waves (both CW and EW) travel much faster (5-20 times) than the compressor rotor velocity, the time it takes, Δt_{UC}, to wave compress and induce IFF is much shorter than the time for the IC and discharge phases as shown in Figure 5b. After the transient backflow phase, the combined flow Q+IFF is pushed out into the outlet dampener by mechanical movement of the rotors and there is a dampener loss as depicted by the slight slope of the pressure variation shown in Figure 5b.

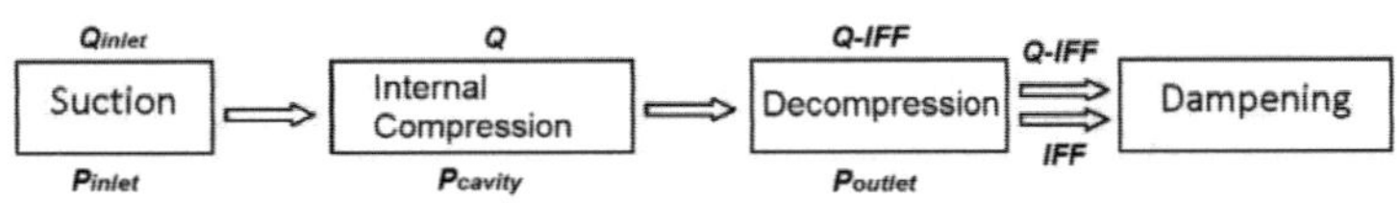

a. Phases of screw compression cycle with serially arranged IC+OC mode

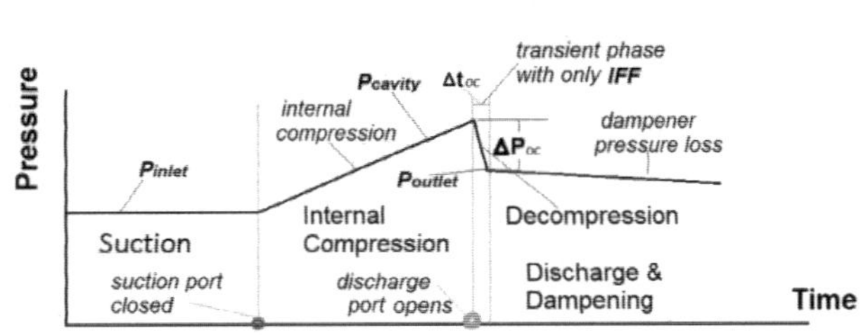

b. Pressure variation during screw compression cycle in time domain

Figure 6. Phases of a screw compression cycle for a serially arranged IC and OC mode.

Just like Figure 5a for the UC mode, Figure 6a shows a compression cycle of a screw cavity for the serial IC and OC scheme in OC mode of Figure 1b and referred to as in example ABC-2. The difference of OC with UC is that, immediately after the IC phase, a violent transient decompression phase due to OC follows when discharge port suddenly opens to expose the pressure difference ΔP_{OC} as shown in Figure 6b. The sudden exposure of ΔP_{OC} between the cavity and the outlet would trigger a generation of gas pulsations in the form of EW-IFF-CW at opening when EW goes into the cavity to decompress the cavity higher pressure gas, a lost work as shown by the shaded area in Figure 1b, by removing extra gas represented as IFF, while CW pushes the extra gas IFF out into the outlet and causing gas pulsation downstream at the same time.

The insights gained from the above transient analyses are twofold. First of all, it shows that compressor discharge system is capable of compensating the deficiency of a fixed Vi IC operating under UC or OC by adding or subtracting extra gas molecules into or out of the compressor cavity at the moment when the discharge port suddenly opens to expose the pressure difference ΔP_{UC} or ΔP_{OC}. Secondly, it is a "blind IC" in the sense that the cavity communicates with the outlet pressure only at the last moment of the port opening that causes the violent transient process when compressor energy loss and pulsation/noise are created. In other words, it is the serial arrangement of UC (or OC) AFTER IC that causes the problem which in turn demands a large size dampener downstream to treat the induced gas pulsations and noises, and causes additional back pressure loss [9]. Naturally, a simple fix should be to arrange IC and UC (or OC) in parallel instead of in series, that is, let the compressor cavity communicate with the outlet pressure way BEFORE the discharge port opens while still taking advantage of the outlet system's ability to "automatically" compensate the cavity pressure by adding or subtracting gas molecules into or out of the cavity. This parallel compression scheme has in fact been illustrated by the Economizer with the difference that it only adds (one way, not subtracts) cooler vapour gas into the cavity during early stage of IC and the vapour source is from condenser outlet instead of the compressor outlet which is critical for minimizing gas pulsation.

2.2 Arrangement of Screw Compression: IC and UC (or OC) in Parallel

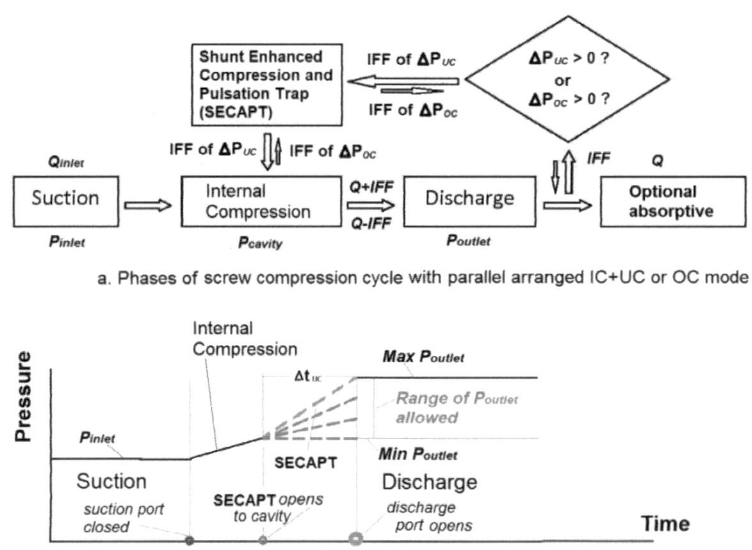

a. Phases of screw compression cycle with parallel arranged IC+UC or OC mode

b. Pressure variation during screw compression cycle in time domain

Figure 7. Phases of a screw compression cycle for a parallel arranged IC and UC (or OC) mode.

Figure 7 shows such a scheme, a partial parallel arrangement called SECAPT (Shunt Enhanced Compression And Pulsation Trap – pending patents of 2020-2022 [10]) that uses a shunt feedback flow loop both to communicate with the outlet and to compensate the IC way before the discharge port opens in such a way that the pressure difference ΔP_{UC} or ΔP_{OC} would be eliminated or reduced close to zero at discharge over a wide range of operating pressures shown as between Min P_{outlet} and Max P_{outlet} in Figure 7b.

As illustrated in Figure 7a for the same example of IC and UC (or OC) as in Figure 5a scheme ABC-1, SECAPT involves modifications to a fixed Vi screw compression cycle from the serial mode, i.e. from backflow compression or decompression AFTER the discharge port opens, to a parallel mode BEFORE the discharge port opens. In this parallel configuration, part of IC and UC (or OC) processes are carried out simultaneously and synergistically during a much longer time interval Δt_{UC} as shown in Figure 7b. Any deviation of the pressure in the compressor cavity from the target outlet pressure, i.e. ΔP (=P_{outlet}-P_{cavity}), would immediately trigger a feedback mechanism in the form of induced fluid flow (IFF) between the cavity and outlet to diminish the pressure difference (ΔP_{UC} or (ΔP_{OC}) just BEFORE the discharge port opens. In other words, SECAPT would automatically eliminate any UC or OC pressure deficit or build-up at compressor discharge port over a wide range of operating pressures; therefore there would be no need to use a mechanical slide valve or a downstream dampener. The following discusses an optimal way to carry out the SECAPT scheme for improved overall compressor efficiency and pulsation reductions.

3. Design optimization of SECAPT scheme and applications

To fulfil the SECAPT idea for a screw compressor, specific design issues have to be addressed such as: where the required compensating mass comes from, where to add it into the cavity and at what amount and rate and geometric form? The following will answer these questions by a specific SECAPT example employing a flow nozzle as a mass exchanger between the outlet and cavity, and then CFD simulations will be applied to two industrial cases.

3.1 Screw compressor equipped with a SECAPT nozzle design

A single dry screw or oil-injected screw compressor can develop a pressure ratio range up to 4.5 or 20 respectively. To achieve an efficient and speedy energy and mass exchange between the cavity and outlet for this pressure range, it is appropriate to choose a flow nozzle (convergent) or a de Laval nozzle (convergent-divergent) capable of accelerating a pressurized gas to a higher sonic or supersonic speed by converting the pressure energy of the flow into kinetic energy.

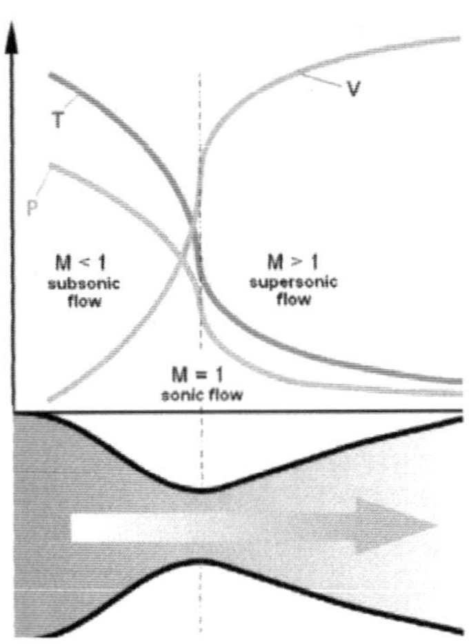

Figure 8a. A supersonic de Laval nozzle with pressure, temperature and velocity distribution [8]

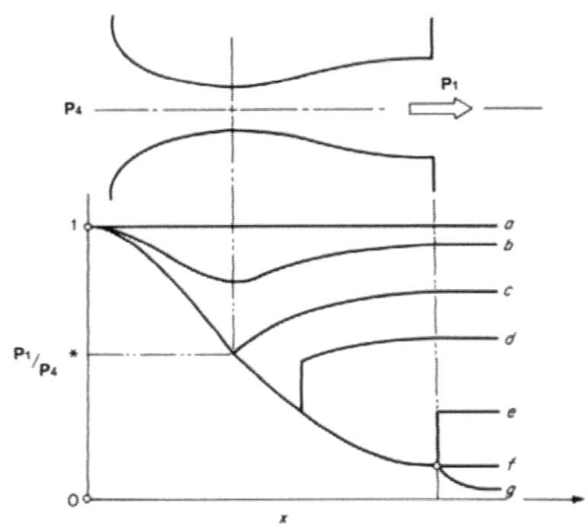

Figure 8b. Pressure distribution for subsonic and supersonic flows under different back pressures [8]

As shown in Figure 8 for a de Laval, both pressure P and temperature T decrease in flow direction as gas with velocity V accelerates from subsonic to supersonic for the ideal isentropic expansion case f in Figure 8b. In general, pressure ratio (p_4/p_1) across the nozzle determines whether it operates at subsonic or supersonic speeds as illustrated in Figure 8b for various pressure distributions under different back pressures ranging from a to g [8]. In a subsonic flow such as case b, some downstream noise, though very low level since jet velocity is low, will propagate upstream through the small nozzle throat. On the other hand for a sonic or supersonic flow when the gas velocity reaches sonic (local Mach number = 1.0) at the "throat" where the cross-sectional area is at its minimum, a condition called **choked flow** as for cases c-g, the jet noise and shock waves generated downstream will not be able to propagate upstream through the throat, forming an effective "trap" isolating jet noise and shock waves inside the cavity. A special case of the de Laval nozzle is ASME flow nozzle when only the convergent section exists as an option for SECAPT to achieve a more compact and economic design. The choked flow is established at the throat as long as the pressure ratio across the nozzle satisfies (p_4/p_1) >= 1.89 (for air) for ASME flow nozzles. It can be seen that the majority range of the maximum application pressure ratio 4-20 for air across the nozzle can effectively form a "noise trap" by the nozzle throat for both forms of nozzles.

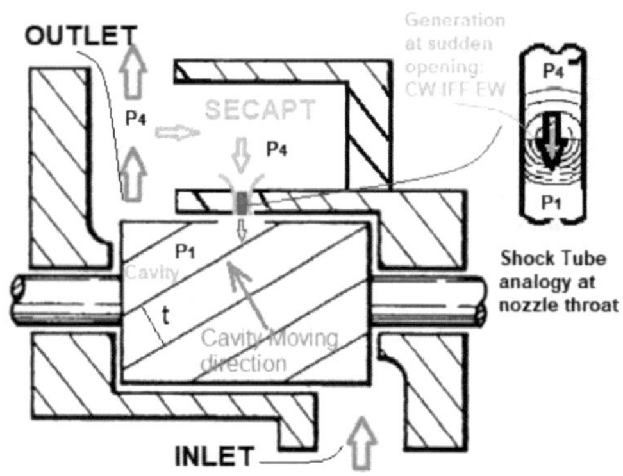

Figure 9a. A schematic of the SECAPT-flow nozzle design for a screw compressor.

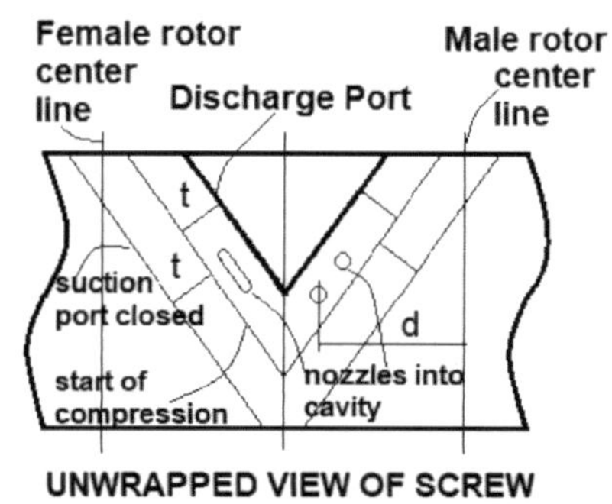

Figure 9b. Unwrapped view of the SECAPT-flow nozzle design.

Figure 9a shows an example of a screw compressor with the SECAPT-flow nozzle schematic while Figure 9b illustrates the unwrapped view of the rotor bore [11] showing the locations for nozzles going into the cavity and the nozzle cross-sectional shape at throat: slot shown on the left and circular on the right. The relative position of the nozzle with respect to inlet and discharge ports of the screw is defined in its thread pitch (t) direction which is perpendicular to the screw lobe thread or cavity orientation.

As a general design rule for SECAPT, the volumetric ratio is designed for the minimum required outlet pressure so that there would be more chances for having shunt flow under UC than OC as shown by the large and small arrows in Figure 7 for optimized nozzle performance. The optimal nozzle (pre-opening) locations and size are designed to be opened to cavity as a compromise: early enough to exchange the required shunt flow but as late as possible to minimize the leakage effect of higher pressure gas injection that reduces the volumetric efficiency. Unlike the Economizer which routes injected gas back from after the condenser by a one-way street, SECAPT communicates and exchange directly in a two-way street with the compressor outlet to minimize the pressure difference ΔP (= $|P_{outlet}-P_{cavity}|$) at discharge that is responsible for gas pulsation generation. Above all, little OC can be developed by SECAPT when operating from the minimum to maximum outlet pressure, but a "sizeable controlled" UC is allowed when operating at outlet pressures above the minimum in order to convert part of the potential energy from ΔP_{UC} to shaft power like a Pelton wheel.

The UC mode of operation is depicted in Figure 9a at the moment when the screw moving cavity with lower pressure P_1 is suddenly opened to the pre-opening of the SECAPT nozzle with quasi-static pressure P_4, triggering a shock tube like reaction as theorized by Huang [7]. This transient event generates, at the nozzle throat where the sudden opening taking place, an instant gas pulsation in the form of CW-IFF-EW with CW and IFF going into the cavity to help "compress" the gas in parallel with IC and impact the rotor to do positive work (jet is going down as shown in Figure 9b to assist rotor rotating) while EW coming out of the nozzle diffused towards the outlet with reduced gas pulsation downstream due to the diverging shape of the flow nozzle. From adding more gas molecules into moving cavity point of view, using an efficient nozzle to transport the required mass into the "starved" or under compressed cavity would minimize the fill-in time and loss while the small nozzle throat minimizes the noise escaping. SECAPT works for high speed dry screw compressors where rotor speed can be over tens of thousands RPM with rotor tip velocity up to 0.3 Mach number, still much lower than the speed of sound of the jet flow. It can be seen from Figures 7a and 9a that the required "assisting mass" comes from the compressor outlet, different from the Economizer, by the shunt feedback flow loop using "a borrow and return strategy" so that both mass and partial energy of the induced flow IFF is recycled back to the outlet, not wasted in the UC mode.

3.2 CFD simulation and optimization of screw-SECAPT application

In this section, the SECAPT scheme is optimized numerically by a 1D CFD code written by Mazzawy [12] for an example dry screw case: a bulk truck loading/unloading application where compressor pressure varies from no pressure rise to maximum load up to 4 Bar absolute.

3.2.1 Code Description

The code is set up to handle multiple PD compressor models of the three types (Screw, Sliding Vane and Roots) with specific geometry templates to enable the necessary analysis of the compression process for each type. In order to optimize the SECAPT design, multiple cases are run for different PD compressor operating points with or without SECAPT nozzles. Options include the selection of the operating gas (Air, Oxygen, Nitrogen or R134A refrigerant). SECAPT nozzle location(s) and geometry are easily varied to determine the best combination of operating efficiency and minimization of UC/OC exit pressure mismatch.

3.2.2 Optimization Procedure

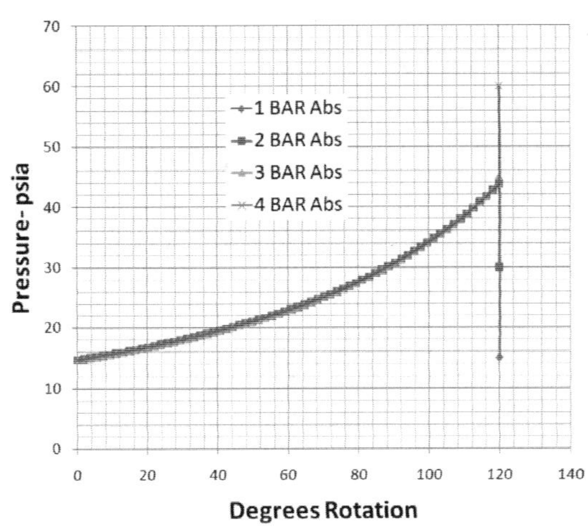

Figure 10a. Base pressure rise with no SECAPT at 4 levels of exit pressure

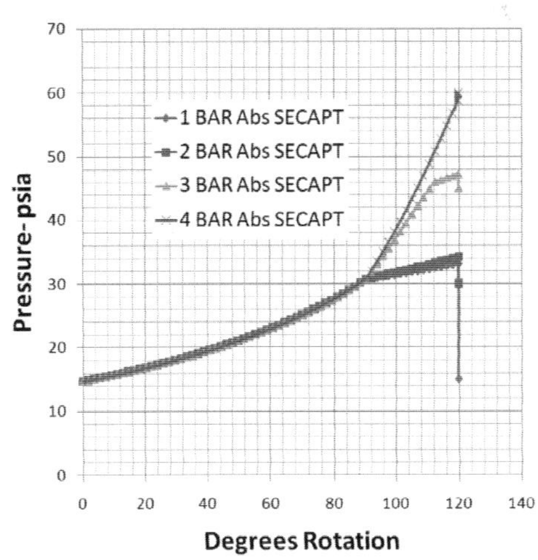

Figure 10b. Pressure rise with SECAPT at 4 levels of exit pressure

In this example case, the pressure level vs degrees of rotation for the screw compressor is shown with no SECAPT and then with a SECAPT nozzle located at 75% of the design level of pressure rise in Figures 10a and 10b, respectively. In the base calculation, there is observed a sudden adjustment to the exit pressure at all pressures except the 3 Bar case which is close to the design level. The addition of SECAPT eliminates the pressure mismatch at 4 Bar, only minorly effects the 3 Bar case and reduces the pressure mismatch at both 2 Bar and 1 Bar. It can be seen that there is an inflection point for the 3 Bar case in Figure 10b, indicating flow at nozzle is changing direction from inflow to outflow to adjust cavity pressure to minimize the mismatch at outlet. Depending upon the primary operating range of interest for the application, one can further optimize the location and size of the SECAPT nozzle. For example, an earlier nozzle opening than 90 degree position would minimize the exit pressure mismatch even more for both 2 Bar and 1 Bar conditions.

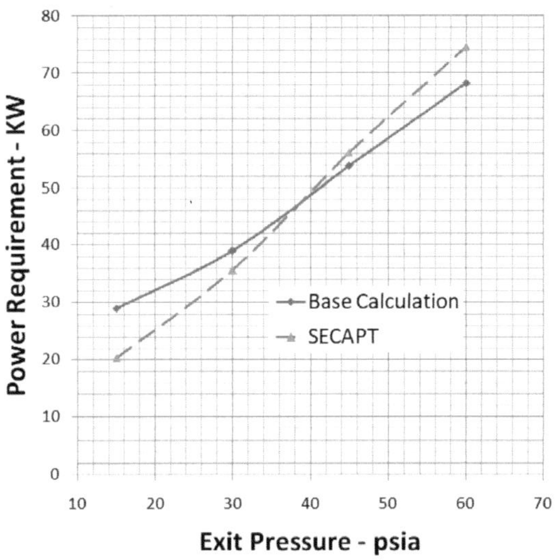

Figure 11. Power Consumption with and without SECAPT

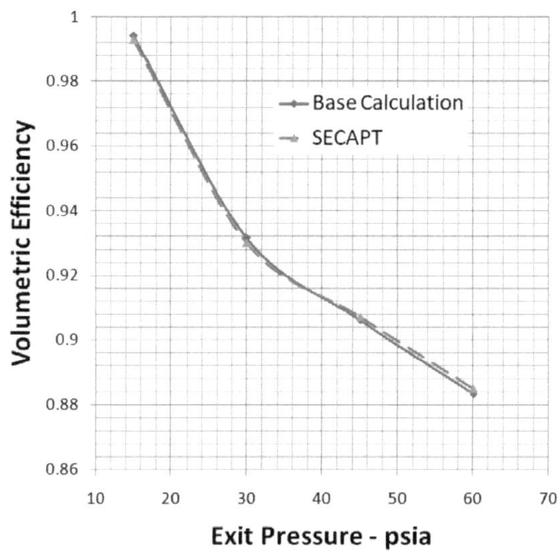

Figure 12. Volumetric Efficiency with and without SECAPT

Power consumption and volumetric efficiency are important considerations for operators and sizing the nozzle area for just reaching the maximum exit pressure of interest has minimized the additional power requirement as shown in Figure 11. The addition of SECAPT requires more power at high exit pressure but less at low exit pressure, consistent with the greater or lesser mass of fluid being compressed downstream of the SECAPT nozzle location. The overall power consumption of SECAPT will be much less than the Base case if the pressure loss of a downstream dampener is included for the Base but not required for SECAPT in general. Volumetric efficiency provided in Figure12 is essentially unchanged by the addition of the SECAPT nozzle

3.3 Comparison of the pros and cons of several widely used compression schemes with SECAPT

Table 2 is a summary of the above analysis intended to assess the effectiveness of various compression schemes in terms of whether they are capable of achieving multiple targets simultaneously, such as suitable applications (wet or dry or both), compression efficiency, pulsation/noise abatement, cost & footprint, and product reliability when operating over a wide range of pressures. The widely used compression schemes of screw compressors equipped with or without a variable Vi slide valve or other PD having automatic discharge valves (like reed valves) will be compared and benchmarked with the SECAPT compression scheme for their respective merit and shortcomings.

Table 2. Benchmarking some popularly used screw compression schemes with SECAPT

	PD with automatic discharge valve	Screw without Variable Vi	Screw with Variable Vi	**Screw with SECAPT**
Screw application	No	For wet & dry screw	Not for dry screw	For wet & dry screw
Energy consumption	Little losses due to OC or UC or serial dampener	High loss due to OC or UC and serial dampener	Little loss due to OC or UC or serial dampener	Little loss due to OC or serial dampener. Slight loss from UC
Pulsation/noise	No discharge dampener needed	Discharge dampener required	No discharge dampener needed	No discharge dampener needed
Cost & footprint	Small footprint and low cost	Large footprint and extra high cost by adding dampener	Large footprint and extra high cost by adding slide valve	Small footprint and medium cost by adding SECAPT
Reliability	Potential fatigue failure of reed valves	Potential failure of serial dampener	Potential failure of slide valve system	Robust, no moving & fatigue parts
Response time	Fast	Fast	Very Slow	Fast

It can be seen from Table 2 that SECAPT is a better or at least an alternative way to achieve the multiple design goals as the popular variable Vi design. It contrasts the latter scheme by exploiting the fluidly gas to compensate variable load conditions rather than by moving the solid mechanical parts (slide valves) that are sensitive to friction, heat, fatigue failure and response frequency. In control terms analogous to the comparison of an unguided rocket with guided missile for the self-propelled weapons, the serial IC+UC (or OC) scheme for screw, ABC-1 and ABC-2 in Table 1, is an unguided and self-propelled fixed pressure-ratio IC process which only "homes in" to target pressure at the last moment at discharge by UC or OC. In contrast, the variable Vi scheme is an externally guided & propelled IC process with electronic-sensing and external power-driven feedback system that will "home in" on its target pressure in a mechanical way using a slide valve. For comparison, SECAPT is a self guided and self-propelled process with an automatic compensating shunt flow loop to "home in" on wide range of target pressures ($\Delta P \rightarrow 0$) by moving fluidly gas instead of solid mechanical parts. The "automatic" sensing and correcting feature of the SECAPT is very much like the PD equipped with automatic discharge valves that are constantly sensing the pressure difference between the cavity and the outlet, then automatically opening as soon as the target pressure is reached. An improved version of SECAPT is taking advantage of the automatic one directional valves and apply them sideways, constituting a new compression scheme called SEDAPT (Shunt Enhanced Decompression And Pulsation Trap) as discussed in a separate paper [13].

4. Conclusions and recommendations for future work

4.1 Conclusions

Automatic elimination of UC and OC for screw compressors operating over a wide range of pressures has been a continuing challenge since its rebirth at SRM in 1930s because they are the sources of gas pulsations and energy losses. Instead of following the traditional methods of employing a variable Vi slide valve, this paper explores a wide range of alternative compression schemes and comes up with an optimized configuration called SECAPT (Shunt Enhanced Compression And Pulsation Trap) that shows most beneficial and promising as follows:

1. Most problems faced by screw compressors today are the result of UC or OC for operating over a wide range of pressures. Even though PD compressors equipped with simple automatic discharge valves have enjoyed a continuing success for their ability to operate under variable loads without suffering from UC or OC, screw compressors have no such option because of possessing a fixed built-in volume ratio defined by porting positions. This is in turn caused by the limit of the traditional design paradigm which employs a "volume reduction" as the only means for raising gas pressure as reflected by the tremendous efforts devoted to the variable Vi concept in the last 60 years.
2. SECAPT breaks away from the "volume reduction only" paradigm by harnessing another effective means to adjust gas pressure: exchange gas molecules between the compressor outlet and cavity directly, synergistically and automatically. Its principle can be "derived or deduced" from the Perfect Gas Law as a combo compression scheme optimized for multiple design criteria such as best compressor efficiency, pulsation/noise abatement, and cost and footprint reduction for a screw to operate over a wide range of pressures. SECAPT is further shown to be superior than the popular variable Vi design in cost and reliability by exploiting the <u>fluidly gas</u> to compensate the variable load conditions rather than by moving the <u>solid mechanical parts</u> that are expensive and sensitive to friction and heat, and response time. More important, SECAPT can be used for both wet and dry, and high speed screw applications.
3. The CFD simulations confirm that SECAPT is <u>tentatively</u> capable of achieving multiple design targets as theorized.

4.2 Future work

The next phase of the project is to validate the SECAPT design for screw compressors by conducting experimental investigation phase, for example, on the Silo King SKL1100 compressor of CVS Engineering GmbH. More CFD modelling will also be carried out by comparing the use of SECAPT with an absorptive type silencer to see which one would result in higher power consumption. In addition, the SECAPT application for the emerging Fuel Cell technology will also be examined. In our continued efforts to understand and improve screw compression mechanism, it is anticipated that better compression schemes can be devised and applied to the design of future generation of screw compressors that would have more energy consumption savings, smaller size and smoother operation.

5. Acknowledgments

This paper is expanded from the concept paper of the 2016 ARPA-E Proposal [14]. The initial SPT design (SPT is the predecessor of SECAPT concept) has been investigated by using 3D CFD on the CVS Silo King SKL1100 screw compressor and valuable insight was gained and used in the development & optimization of the SPT technology [15]. The authors would like to thank Robert Mazzawy, as a "real McCoy from Gas Turbine Industry" (quote Lee Langston), who reviewed both this paper and the concept paper of the 2016 ARPA-E Proposal and provided many rounds of valuable feedbacks and insights. Most importantly, Mazzawy has applied his skills and gas dynamics knowledge to have developed a CFD simulation code to optimize both SECAPT and SEDAPT designs after being inspired by the SECAPT and SEDAPT concepts.

6. Notation

CW	compression wave, pulsation component
d	distance from nozzle to rotor axis
EW	expansion wave, pulsation component
IC	internal compression
IFF	induced fluid flow, pulsation component
m	mass of a molecule
M	Mach number
N	number of molecules
OC	over compression
ρ	gas density
p	gas pulsation pressure
P	absolute gas pressure
PD	positive displacement
Q	volume flow rate
R	ideal gas constant
SECAPT	shunt enhanced compression and pulsation trap
SEDAPT	shunt enhanced decompression and pulsation trap
t	time, or screw thread pitch
T	gas temperature
UC	under compression
v	gas molecule velocity
V	compressor displacement volume
Vi	compressor design volume ratio

Subscripts

1	initial low pressure in shock tube, or inlet pressure of compressor
4	initial high pressure in shock tube or outlet pressure of compressor
cavity	compressor cavity
inlet	compressor inlet
jet	nozzle jet flow
outlet	compressor outlet

IOP Conf. Series: Materials Science and Engineering 1267 (2022) 012019 doi:10.1088/1757-899X/1267/1/012019

References

[1] Sangfors B 1982 *Analytical Modeling of Helical Screw Machine for Analysis and Performance Prediction* The 1982 International Compressor Engineering Conference at Purdue
[2] Soedel W 2007 *Sound and Vibrations of Positive Displacement Compressors* p 16-18 (CRC Press)
[3] Stoecker W 2004 *Industrial Refrigeration Handbook, 2nd Edition* p 125-139 (McGraw-Hill)
[4] Rohleder C and Groll E 2019 *Measurement of Pulsation at the Discharge of Positive Displacement (PD) Compressors* 11th International Conference on Compressors and their Systems City University of London
[5] Nilsson H R 1963 SRM's screw patent with slide valve: US Patent 3088659
[6] Jonsson S 1988 *Performance Simulations of Twin-Screw Compressors with Economizer* The 1988 International Compressor Engineering Conference at Purdue
[7] Huang P 2012 *Gas Pulsations: A Shock Tube Mechanism* The 2012 International Compressor Engineering Conference at Purdue
[8] Liepmann H W and Roshko A 1957 *Elements of Gasdynamics* (Dover Press)
[9] Huang P, Yonkers S and Groll E 2018 *Parallel vs. Serial Noise Suppression Method for Positive Displacement (PD) Compressors & Internal Combustion Engines (ICE)* The 2018 International Compressor Engineering Conference at Purdue
[10] Huang P and Yonkers S SECAPT pending patents:
1. USA, Filed on Sep.8,2020, Patent Application No. 17/014,357;
2. China, Filed on Aug.10, 2021, Patent Application No. 202110911749.1;
3. Europe, Filed on Feb.21, 2022, Patent Application No. 22 157 685.3.
[11] Edstrom S 2000 *Unwrapped Views of Rotor Bores of Twin-Screw Compressors* Design Notes from private communications
[12] Mazzawy R S 2021 *Positive Displacement Compressor Analysis CFD Code Users Manual* Trebor Systems LLC Report Sept 25
[13] Huang P, Yonkers S and Willie J 2022 *A Novel Screw Compressor with a Shunt Enhanced Decompression and Pulsation Trap (SEDAPT)* The 2022 International Compressor Engineering Conference at Purdue
[14] Huang P and Groll E 2016 *Development of a Shunt Pulsation Trap (SPT) - A Parallel Noise Suppression Method for Positive Displacement (PD) Compressors & Internal Combustion Engines (ICE)* ARPA-E Concept Paper Control Number 1428-1904
[15] Willie J, Yonkers S and Huang, P 2022 *Use of CFD to optimize the design of a shunt pulsation trap (SPT) used for noise and vibration mitigation in oil free screw compressors* The 2022 International Conference on Screw Machines at Dortmund

A two-phase approach for simulation of water-flooded twin-screw machines validated for expander applications

A Nikolov, A Brümmer

TU Dortmund University, Faculty of Mechanical Engineering, Chair of Fluidics, Leonhard-Euler-Str. 5, 44227 Dortmund, Germany

E-mail: alexander.nikolov@tu-dortmund.de, andreas.bruemmer@tu-dortmund.de

Abstract. In the lower and medium power range, twin-screw machines offer a high potential for energy conversion in various compressor applications with liquid injection or as expanders with respect to electrical power generation from regenerative and exhaust heat sources in trilateral or wet Rankine cycle systems, for instance. Aiming high efficiencies and reliability, the design of liquid-flooded twin-screw machines as a critical system component presents particular challenges for the engineers. Hence, reliably representative simulations to guide design are mandatory. In this context, this study presents a two-phase approach for simulation of the operational behaviour of water-flooded twin-screw machines. The thermodynamic fluid state is calculated using the humid air model that considers the two-phase mixture in thermal equilibrium. Additionally, dissipative two-phase mass flow rates are predicted regarding a slip-flow model and two-phase discharge coefficients. The proposed two-phase approach including liquid distribution in the working chamber is validated for expander applications considering available experimental data of the test twin-screw expander SE 51.2 in terms of indicator diagrams, indicated power, mass flow rate, and outlet temperature.

1. Introduction

Liquid-flooded twin-screw machines are widely applied as compressors providing benefits in terms of two-phase operation, such as high pressure ratios in a single stage together with low thermal stress at the same time, clearance sealing effects or lubrication of the moving machine parts. While water-flooded process-gas compressors have long been in operation, water-injection in air or hydrogen twin-screw compressors gain more importance in recent years. Recovering exhaust heat from low-grade heat sources, for example, in trilateral or wet organic Rankine cycle systems, twin-screw expanders capable of dealing with large amounts of liquid can be applied as an alternative to turbines avoiding wet expansion. Moreover, injecting an auxiliary liquid during expansion could provide benefits in terms of utilisation of available pressure potentials, such as in (natural) gas reducing stations, where only a small temperature drop during expansion might be allowed. In this context, injected liquid, e.g. water, in combination with an exhaust or other regenerative heat source acts also as heat carrier.

In order to theoretically determine the operational behaviour of water-flooded twin-screw machines with respect to system design, appropriate two-phase models, e.g. for chamber model simulations, considering the thermodynamic fluid properties of the air-water mixture and adequately reflecting the mass flow rates are required. With regard to the thermodynamic modelling of water injection into the working chamber of twin-screw compressors, a study presented in [1] considers a water evaporation model in order to determine the influence of very low amount of liquid injection on the operation of a twin-screw compressor. In terms of two-phase flows, only a few contributions in the literature deal with non-homogeneous two-phase mass flow rate models in rotary displacement machines. Generally, homogeneous approaches are applied as in [2]. In the same manner, in [3] and [4], Sangfors simulates the operation of liquid-flooded twin-screw machines based on homogeneous two-phase clearance flow considerations. In [5], Vasuthevan and Brümmer present thermodynamic modelling of a twin-screw expander in a trilateral flash cycle regarding homogeneous two-phase inlet, outlet, and clearance flows. In contrast, a separated flow model with liquid entrainment as defined by Smith in [6] is applied by Bell in [7] and by Lemort in [8] to determine inlet and outlet flows in scroll expanders. Moreover in [7], Bell implements a constant two-phase discharge coefficient of 0.77 in order to predict the flow losses. In [9], Huagen et al. present a two-phase clearance flow model based on the slip model with liquid entrainment according to [6] and use a discharge flow model as proposed by Lin in [10] to calculate indicator diagrams of an oil-injected twin-screw compressor. In [11], a comprehensive flow path modelling in a water-lubricated air twin-screw compressor is conducted applying different two-phase models to each clearance and opening.

The following study deals with an approach for simulations of water-flooded twin-screw machines. In this context, a two-phase model according to humid air [12] proposed for the calculation of the thermodynamic properties of the working fluid is presented. Moreover for calculation of pressure-driven two-phase mass flow rates according to [13], assumptions as with the modelling of the discharge behaviour of clearances and openings as well as liquid distribution in the working chamber during expander operation are made.

To verify the presented two-phase approach, multi-chamber simulations of the water-flooded twin-screw expander prototype SE 51.2 [14–16] are performed by means of the simulation tool KaSim [17]. The simulation results are compared to experimental data according to [15, 16] by means of expander (two-phase) mass flow rate, internal (simulated) and indicated (measured) power, pressure depending on the working chamber volume, as well as outlet temperature.

2. Thermodynamic properties of air-water two-phase mixtures

The following section addresses the fundamentals for the calculation of the thermodynamic state of the working fluid in water-flooded twin-screw machines. In this context, the humid air concept [12] is proposed considering a mixture of ideal fluids in thermodynamic (thermal) equilibrium. The two-phase mixture consists of the incompressible fluid water in liquid or solid state of matter, its corresponding gaseous phase steam, and dry air. The thermodynamic fluid properties with respect to the humid air model are listed in table A1 (Appendix). Density of the incompressible phase—water or ice—is considered constant. Moreover, within the relevant pressure and temperature range of expander operation, the specific heat capacity of each fluid is assumed constant, too.

The composition of a water-air mixture is specified by water load X—as commonly used in the context of humid air—representing the ratio of water mass m_w to dry-air mass m_{da}:

$$X := \frac{m_w}{m_{da}}. \tag{1}$$

Since in this work a steady-state process is considered, dry air mass m_{da} and water mass m_w in equation (1) can be substituted by the corresponding mass flow rates $\dot{m}_{da}$ and $\dot{m}_w$ with regard to two-phase flows. Hence, water load X can be expressed as follows:

$$X := \frac{\dot{m}_w}{\dot{m}_{da}}. \tag{2}$$

By definition, water mass m_w and mass flow rate $\dot{m}_w$ consider the entire amount of water molecules in liquid, solid, and gaseous state of matter in the two-phase mixture. As schematically illustrated in figure 1, the volume V of the air-water mixture is composed of the sub-volumes of dry air V_{da}, water steam V_{ws}, and the incompressible phase of water V_{liq} (or ice for temperatures below the triple-point temperature of water T_{tr} = 273.16 K):

$$V = V_g + V_{liq} = V_{da} + V_{ws} + V_{liq}. \tag{3}$$

A phase boundary between the volumes of the incompressible and the gaseous phase is present, whereas dry air and steam share the same volume.

Considering the gaseous phase in the humid air mixture consisting of dry air and water steam, the overall static pressure p of the mixture equals the sum of the partial pressures p_j of the single gases in the mixture as expressed by Dalton's law [12]:

$$p = \sum_{j=1}^{n} p_j. \tag{4}$$

Partial pressure p_j of any gaseous component is related to the gaseous volume V_g. Taking the example of humid air, the sum of the partial pressure of dry air p_{da} and water steam p_{ws} provides the overall static pressure p. At the same time, the overall static pressure p imposed on the surface of the incompressible phase represents the pressure of the two-phase mixture. As for ideal gases, the partial pressure of each gaseous fluid can be expressed as follows:

$$p_j := \frac{m_j \cdot R_j \cdot T}{V_g}. \tag{5}$$

A mass-specific gas constant R_j is assigned to each gaseous fluid in the mixture. The maximum mass of water steam $m_{ws,sat}$ in the mixture is related to saturated humid air and can be calculated from the ideal gas law:

$$m_{ws,sat} = \frac{p_{ws,sat} \cdot V_g}{R_{ws} \cdot T}. \tag{6}$$

Here, the saturation pressure of water steam $p_{ws,sat}$ is a function of the temperature and can be calculated, for instance, by means of the Antoine equation [12] or as proposed in [18]. Additionally, the saturation pressure $p_{ws,sat}$ of water steam depends on the overall static pressure p, if solubility of steam molecules in the liquid phase are taken into account [18]. In the following considerations, this effect is neglected.

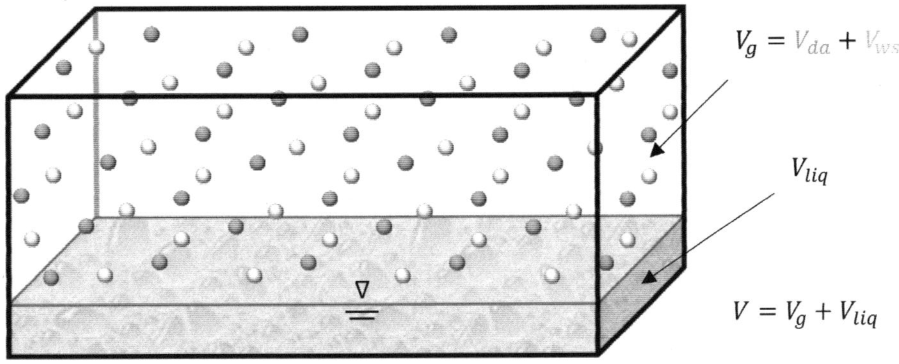

Figure 1. Volume V of the two-phase mixture containing the sub-volumes of dry air V_{da}, water steam V_{ws}, and liquid water V_{liq}.

If less water mass m_w than the maximum mass of water steam $m_{ws,sat}$ at given temperature is available in the humid air mixture, partial pressure p_{ws} is lower than the saturation pressure $p_{ws,sat}$ of water steam. This case can be quantified by means of relative humidity φ dividing the partial pressure p_{ws} by the saturation pressure of water steam $p_{ws,sat}$:

$$\varphi := \frac{p_{ws}}{p_{ws,sat}}. \tag{7}$$

For a relative humidity in the range of $0 < \varphi \leq 1$, all water molecules in the mixture exist as steam. For a relative humidity of 100 % ($\varphi = 1$), saturation water load X_{sat} is calculated as follows:

$$X_{sat} = \frac{m_{ws,sat}}{m_{da}} = \frac{R_{da}}{R_{ws}} \cdot \frac{p_{ws,sat}}{p - p_{ws,sat}}. \tag{8}$$

By analogy with water load X, the composition of the two-phase water-air mixture can be expressed by means of mass dryness fraction x relating the mass m_g of the gaseous humid air phase consisting of dry air and water steam to the mass m of the mixture as follows:

$$x := \frac{m_g}{m} = \frac{m_{da} + m_{ws}}{m_{da} + m_w}. \tag{9}$$

As with two-phase mass flow rate calculations discussed later in section 3.3, mass dryness fraction x is referred to the ratio of the gaseous mass flow rate $\dot{m}_g$ to the mass flow rate $\dot{m}$ representing the sum of the single mixture components dry air and water:

$$x := \frac{\dot{m}_g}{\dot{m}} = \frac{\dot{m}_{da} + \dot{m}_{ws}}{\dot{m}_{da} + \dot{m}_w}. \tag{10}$$

From equation (1), equation (8), and equation (9), mass dryness fraction can be expressed as follows:

$$x = \frac{1 + X_{sat}}{1 + X}. \tag{11}$$

In contrast to the conventional consideration of the triple point of water as a zero energy level, the energy conservation equations regarding (specific) enthalpy and internal energy of humid air are derived with respect to the absolute zero from thermodynamic point of view (0 K). This approach provides throughout positive values for the extensive state variables internal energy and enthalpy as required by the simulation environment in this study. In contrast to internal energy and enthalpy, entropy of the mixture is related to the triple-point temperature of water, since entropy supports the calculations of isentropic change in state as an auxiliary state variable allowing negative values.

For better understanding of the two-phase approach using the absolute zero temperature of 0 K as a zero energy level, specific aspects considering change in aggregate state of water are explained. In figure 2, the correlation between temperature and pressure as well as the different states of matter—steam, liquid, and ice—in terms of specific enthalpy are schematically illustrated. This can be transferred to specific internal energy too.

In the region of two aggregate states—liquid water and steam ($T > T_{tr}$) or ice and steam ($T < T_{tr}$)—pressure is a function of temperature. The boundaries of the two-phase region correspond to a saturated state of the gaseous phase. With regard to the triple point, all three aggregate states are present. Here, the difference of the specific enthalpies (Δh_{me}^{tr}, Δh_{ev}^{tr}, and Δh_{su}^{tr}), in the literature also referred to as latent heats, of two saturated aggregate states define the corresponding specific enthalpy required for a complete change in aggregate state. Outside the two-phase region, only one fluid state of matter exists. The reference condition providing enthalpies for change in state of matter is randomly set in the triple point of water. In general, these energies could be determined at any valid temperature for the transition between two aggregate states.

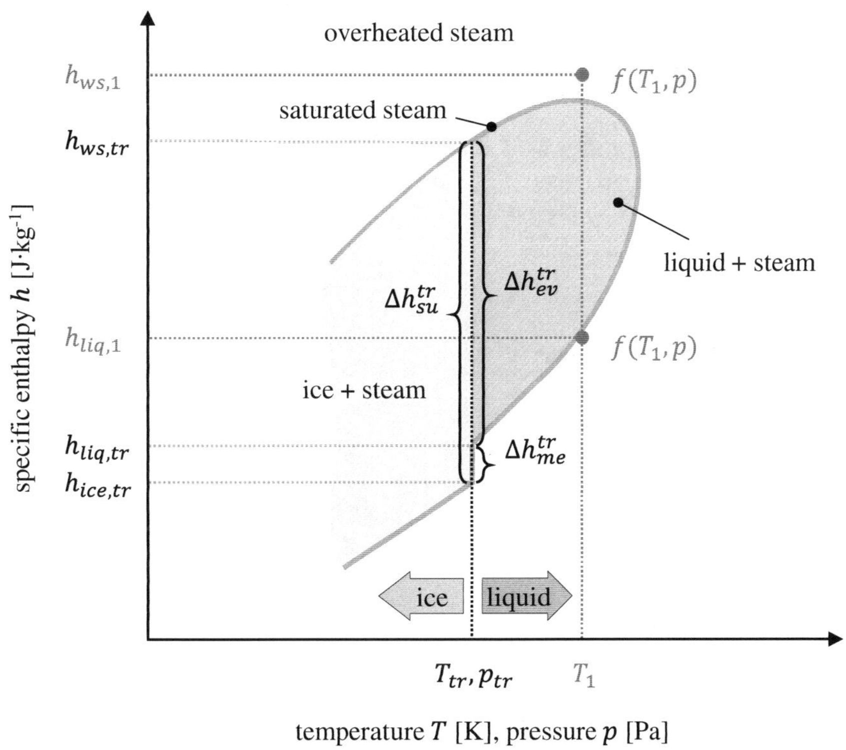

Figure 2. Specific enthalpy h of water as a function of pressure p and temperature T.

As illustrated in figure 2, specific enthalpy can be calculated by addition of different specific enthalpy levels beginning from the absolute zero temperature. For instance, specific enthalpy of liquid water $h_{liq,1}$ equals the sum of specific enthalpy $h_{ice,tr}$ of ice, specific melting enthalpy Δh_{me}^{tr}, and the difference of specific enthalpy of water at temperature T_1 and triple-point temperature T_{tr}. Specific melting enthalpy Δh_{me}^{tr} represents the energy released during freezing of liquid water or required to bring water from solid to liquid aggregate state in the triple point. In the same manner with reference to the thermodynamic absolute zero point, specific enthalpy $h_{ws,1}$ of steam can be calculated considering the specific enthalpy of ice $h_{ice,tr}$ at triple-point temperature T_{tr}, the specific sublimation enthalpy Δh_{su}^{tr}, and the difference of specific enthalpy $h_{ws,1}$ and $h_{ws,tr}$ of steam at temperature T_1 and triple-point temperature T_{tr}.

In the following, the calculation of specific volume, internal energy, enthalpy, and entropy of the two-phase air-water mixture are presented. All fluid specific parameters and enthalpies in terms of change in aggregate state of water can be obtained from table A1 (Appendix).

2.1. Specific volume

Specific volume v_{1+x} of the two-phase mixture is defined as the ratio of the volume V occupied by the two-phase mixture divided by the mass of dry air m_{da}:

$$v_{1+x} := \frac{V}{m_{da}}. \tag{12}$$

The specific volume v referred to the entire mass m of the two-phase mixture equals:

$$v := \frac{V}{m} = \frac{V}{m_{da} + m_w} = v_{1+x} \cdot (1 + X)^{-1}. \tag{13}$$

As illustrated in figure 1, the entire fluid volume V as in equation (3) consists of the partial volumes of the incompressible phase of water as well as the gaseous phase including dry air and steam. Combining equation (3) and equation (12), for saturated humid air ($X > X_{sat}$) and temperatures above the triple point of water ($T > T_{tr}$), specific volume v_{1+x} of the two-phase fluid equals on a pro rata basis the sum of the single specific volumes of each mixture component:

$$v_{1+x} = v_{da} + X_{sat} \cdot v_{ws} + (X - X_{sat}) \cdot v_{liq} = R_{da} \cdot \frac{T}{p} + X_{sat} \cdot R_{ws} \cdot \frac{T}{p} + (X - X_{sat}) \cdot v_{liq}. \tag{14}$$

Here for the incompressible phase, specific volume v_{liq} of liquid water is considered. In contrast, at temperatures below the triple point ($T < T_{tr}$), specific volume v_{1+x} of the two-phase fluid is regarded to ice specific volume v_{ice}:

$$v_{1+x} = R_{da} \cdot \frac{T}{p} + X_{sat} \cdot R_{ws} \cdot \frac{T}{p} + (X - X_{sat}) \cdot v_{ice}. \tag{15}$$

For single-phase humid air ($X \leq X_{sat}$), no incompressible phase is present and the expression for specific volume v_{1+x} reduces to:

$$v_{1+x} = R_{da} \cdot \frac{T}{p} + X \cdot R_{ws} \cdot \frac{T}{p}. \tag{16}$$

2.2. *Internal energy*

Internal energy U of the two-phase fluid is composed of the internal energies of dry air and water:

$$U = U_{da} + U_w = m_{da} \cdot u_{da} + m_w \cdot u_w. \tag{17}$$

Here, u_{da} and u_w are the specific internal energies of dry air and water respectively. On the one hand, referring internal energy U of the mixture to the mass of dry air m_{da} only, specific internal energy u_{1+x}, as commonly used in the literature, is specified:

$$u_{1+x} := \frac{U}{m_{da}} = u_{da} + X \cdot u_w. \tag{18}$$

On the other hand, relating internal energy U to the entire mass m of the mixture, specific internal energy u of the two-phase mixture defined as

$$u := \frac{U}{m} = \frac{U}{m_{da} + m_w} \tag{19}$$

can be calculated from equation (18):

$$u = u_{1+x} \cdot (1 + X)^{-1}. \tag{20}$$

With regard to the thermodynamic absolute zero (T = 0 K), specific internal energy u_{1+x} of saturated and unsaturated humid air ($X \leq X_{sat}$) can be obtained from the specific internal energies of dry air and water steam:

$$u_{1+x} = c_{v,da} \cdot T + X \cdot \{c_{ice} \cdot T_{tr} + \Delta h_{su}^{tr} - p_{tr} \cdot (v_{ws}^{tr} - v_{ice}) + c_{v,ws} \cdot (T - T_{tr})\}. \tag{21}$$

For temperatures above the triple-point temperature of water ($T > T_{tr}$), specific internal energy u_{1+x} of the two-phase fluid can be calculated with reference to saturated humid air and liquid water in the mixture ($X > X_{sat}$):

$$\begin{aligned} u_{1+x} = c_{v,da} \cdot T + X_{sat} \cdot \{c_{ice} \cdot T_{tr} + \Delta h_{su}^{tr} - p_{tr} \cdot (v_{ws}^{tr} - v_{ice}) + c_{v,ws} \cdot (T - T_{tr})\} + \\ (X - X_{sat}) \cdot \{c_{ice} \cdot T_{tr} + \Delta h_{me}^{tr} - p_{tr} \cdot (v_{liq} - v_{ice}) + c_{liq} \cdot (T - T_{tr})\}. \end{aligned} \tag{22}$$

In equation (22), v_{ws}^{tr} is referred to temperature and pressure in the triple point of water. For temperatures below the triple point ($T < T_{tr}$), specific internal energy u_{1+x} is determined as follows:

$$\begin{aligned} u_{1+x} = c_{v,da} \cdot T + X_{sat} \cdot \{c_{ice} \cdot T_{tr} + \Delta h_{su}^{tr} - p_{tr} \cdot (v_{ws}^{tr} - v_{ice}) + c_{v,ws} \cdot (T - T_{tr})\} + \\ (X - X_{sat}) \cdot c_{ice} \cdot \mathrm{T}. \end{aligned} \tag{23}$$

2.3. Enthalpy

Enthalpy H of the two-phase mixture results from internal energy U and the product of pressure p and volume V:

$$H := U + p \cdot V. \tag{24}$$

Relating enthalpy H to the mass of dry air m_{da}, specific enthalpy h_{1+x} of the two-phase mixture can be obtained from equation (24) as follows:

$$h_{1+x} := \frac{H}{m_{da}} = u_{1+x} + p \cdot v_{1+x}. \tag{25}$$

Specific enthalpy h considering the entire mass m of the two-phase mixture can be determined as follows:

$$h := \frac{H}{m} = \frac{H}{m_{da} + m_w} = h_{1+x} \cdot (1 + X)^{-1}. \tag{26}$$

2.4. Entropy

Entropy S of the two-phase fluid is the sum of the entropies of each component in the mixture also including entropy of mixing of the gaseous fluids. Dividing entropy S by the mass m_{da} of dry air, specific entropy s_{1+x} is defined as follows:

$$s_{1+x} := \frac{S}{m_{da}}. \tag{27}$$

Specific entropy s related to the entire mass m of the two-phase mixture is introduced as:

$$s := \frac{S}{m} = \frac{S}{m_{da} + m_w} = s_{1+x} \cdot (1 + X)^{-1}. \tag{28}$$

In contrast to specific internal energy u_{1+x} and specific enthalpy h_{1+x}, specific entropy s_{1+x} is calculated with regard to the triple-point temperature of water T_{tr}. When calculating the specific entropy s_{1+x} of the two-phase fluid, a distinction must again be made between unsaturated and saturated humid air as well as for liquid water or ice as incompressible fluid [12]. Considering saturated and unsaturated humid air ($X \leq X_{sat}$) only, following expression is applicable:

$$\begin{aligned} s_{1+x} = {} & \left(c_{p,da} + X \cdot c_{p,ws}\right) \cdot ln\left(\frac{T}{T_{tr}}\right) - \left(R_{da} + X \cdot R_{ws}\right) \cdot ln\left(\frac{p}{p_{tr}}\right) + \\ & X \cdot \frac{\Delta h_{ev}^{tr}}{T_{tr}} + \Delta^M s(X). \end{aligned} \tag{29}$$

Here, $\Delta^M s(X)$ is referred to as specific entropy of mixing of ideal gases with respect to dry air and steam and can be calculated as follows:

$$\Delta^M s(X) = R_{ws} \cdot \left\{\left(\frac{R_{da}}{R_{ws}} + X\right) \cdot ln\left(\frac{R_{da}}{R_{ws}} + X\right) - X \cdot ln(X) - \frac{R_{da}}{R_{ws}} \cdot ln\left(\frac{R_{da}}{R_{ws}}\right)\right\}. \tag{30}$$

In terms of saturated humid air ($X > X_{sat}$) at temperatures above the triple point ($T > T_{tr}$), specific entropy s_{1+x} of the two-phase fluid can be determined from:

$$\begin{aligned} s_{1+x} = {} & \left(c_{p,da} + X_{sat} \cdot c_{p,ws}\right) \cdot ln\left(\frac{T}{T_{tr}}\right) - \left(R_{da} + X_{sat} \cdot R_{ws}\right) \cdot ln\left(\frac{p}{p_{tr}}\right) + \\ & X_{sat} \cdot \frac{\Delta h_{ev}^{tr}}{T_{tr}} + \Delta^M s(X_{sat}) + (X - X_{sat}) \cdot c_{liq} \cdot ln\left(\frac{T}{T_{tr}}\right). \end{aligned} \tag{31}$$

At temperatures below the triple point ($T < T_{tr}$), specific entropy s_{1+x} is calculated using the expression:

$$\begin{aligned} s_{1+x} = {} & \left(c_{p,da} + X_{sat} \cdot c_{p,ws}\right) \cdot ln\left(\frac{T}{T_{tr}}\right) - \left(R_{da} + X_{sat} \cdot R_{ws}\right) \cdot ln\left(\frac{p}{p_{tr}}\right) + \\ & X_{sat} \cdot \frac{\Delta h_{ev}^{tr}}{T_{tr}} + \Delta^M s(X_{sat}) - (X - X_{sat}) \cdot \left\{\frac{\Delta h_{me}^{tr}}{T_{tr}} - c_{ice} \cdot ln\left(\frac{T}{T_{tr}}\right)\right\}. \end{aligned} \tag{32}$$

3. Two-phase chamber model simulation

Chamber model simulations are commonly applied in the theoretical analysis of positive displacement machines considering one or more cyclically changing working chambers [19]. Including the law of conservation of mass and energy, a numerical solver based on the time-step method—KaSim—solves the thermodynamic and fluid-mechanical equations of the multi-chamber model [17]. For the following theoretical analyses in this work, KaSim is adapted to the requirements of two-phase simulations.

3.1. Two-phase initialisation in KaSim

For purposes of simplicity with respect to chamber model simulations, it is assumed that spatial gradients in the intensive state variables within a fluid capacity are insignificant. The two-phase fluid state is defined by means of the time-dependent extensive state variables internal energy $U(t)$ neglecting kinetic and potential energy, volume $V(t)$ of the fluid capacity, and mass $m(t)$ together with mass dryness fraction $x(t)$, figure 3. Additionally, enthalpy $H(t)$ of the two-phase fluid is specified in order to reduce

IOP Conf. Series: Materials Science and Engineering 1267 (2022) 012020 doi:10.1088/1757-899X/1267/1/012020

the amount of time-consuming iterations during calculation of the intensive state variables pressure p and temperature T. In KaSim, fluid capacities (chambers, ports, pipes etc.) are initialised by means of pressure p, temperature T, capacity volume V, and mass dryness fraction x.

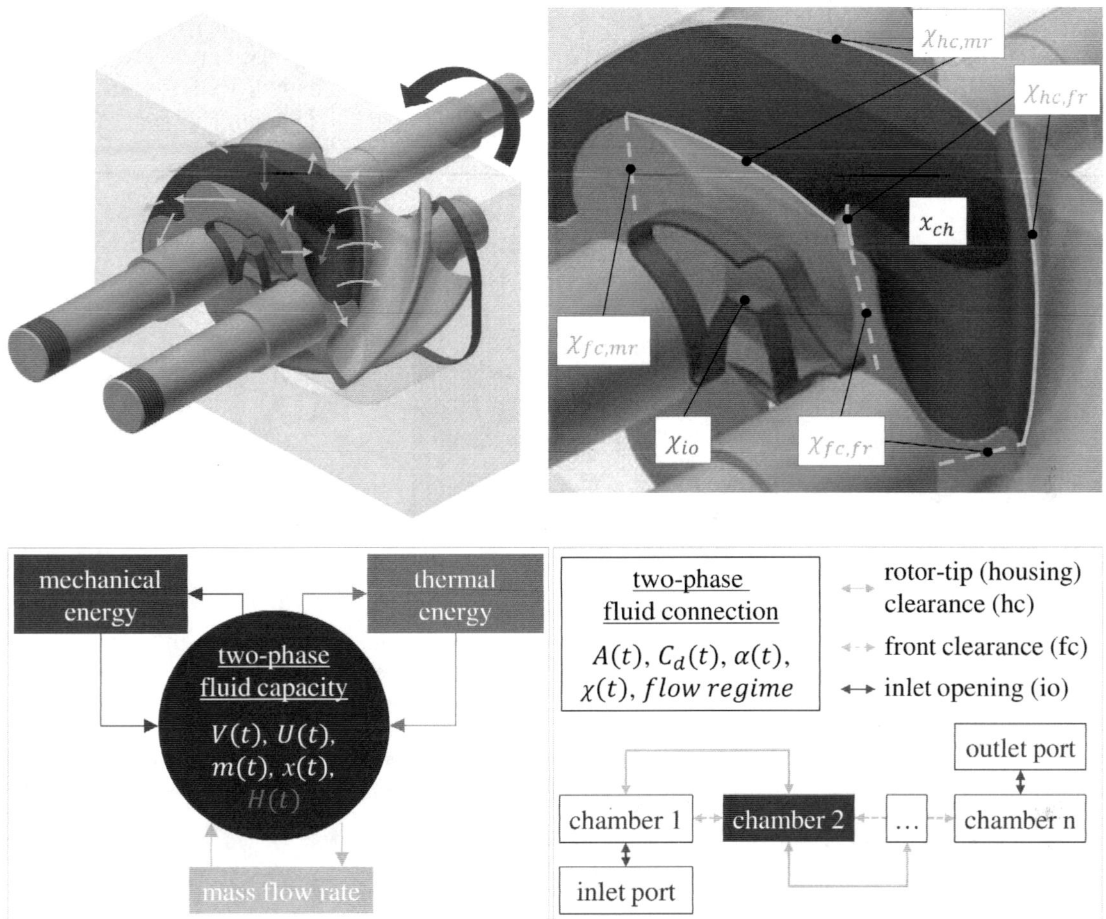

Figure 3. Geometry abstraction of an exemplary twin-screw expander geometry with regard to a chamber model for the simulation tool KaSim [17] including selected two-phase fluid capacities and connections as well as parameters for their specification.

The thermodynamic change in state of the fluid in a capacity results from transfer of, e.g. mechanical or thermal energy and mass. For example, mass exchange is performed using connections representing clearances or openings in the machine. Currently for connections affiliated to moving boundaries, only pressure-driven (Poiseuille) mass flow rates are considered neglecting the effect of Couette flow. Two-phase flow connections (clearances, inlets and outlets, leakage paths etc.) are specified by a cross-sectional area $A(t)$, $orifice$ and $nozzle$ discharge coefficients $C_d(t)$, flow coefficients $\alpha(t)$, a connection-specific mass dryness fraction $\chi(t)$, and a two-phase flow regime ($homogeneous$ or with phase $slip$), see figure 3.

Each connection requires information input with regard to the two-phase flow regime. A $homogeneous$ flow regime implies two-phase flows at equal phase flow velocities. The selection of a $slip$-flow allows for calculation of theoretical mass flow rates as for separated two-phase flows with different phase flow velocities represented by a so-called velocity slip ratio. Considering the discharge behaviour of the connections for two-phase flows, a distinction between $nozzle$ and $orifice$ flows is made. Two-phase flow losses in nozzle or orifice-like connections represented by the discharge coefficient $C_d(t)$ are calculated depending on the fluid state of the connected capacities as proposed in [13]. According to equation (40), the flow coefficient $\alpha(t)$ provides a user-specific constant input reflecting wall friction in addition to the interphase friction loss as calculated by the two-phase nozzle discharge coefficient approach presented in [13]. During simulation, the discharge behaviour of a fluid connection remains fixed and no transition between the nozzle and orifice approach is allowed.

To model liquid distribution in the fluid capacity, a connection-specific mass dryness fraction χ can be explicitly considered in terms of mass flow rate calculation:

$$\chi := x_{con}. \tag{33}$$

Connection mass dryness fraction χ can be specified constant or equal to mass dryness fraction x of the two-phase fluid in the source capacity (capacity of higher pressure considering two connected fluid capacities). With respect to a constant χ, mass dryness fraction x of the source capacity fluid must exhibit values between 0.01 and 0.9. Outside this limits, χ is dynamically initialised during simulation with mass dryness fraction x of the source capacity fluid rather than using the initially specified constant value preventing capacity mass dryness fractions x greater than unity or less than zero.

3.2. Change in state of the two-phase fluid

In this study, adiabatic fluid capacities without heat flows over the control volume boundaries are considered. Hence, change in fluid state results either from change in capacity volume or mass and, correspondingly, enthalpy transfer during a simulation time step. The new internal energy U_1 of the (two-phase) fluid can be calculated from the initial state according to internal energy U_0 as follows:

$$U_1 = U_0 + W + \sum_k H_k. \tag{34}$$

Here, W indicates the work performed in terms of volume change and H_k represents enthalpy of each exchanged fluid element over the control volume boundaries. Assuming isentropic volume change, work is calculated as follows:

$$W = -\int pdV = U_0 - U_{1,s}. \tag{35}$$

At constant initial entropy S_0 and water load X_0 during volume change, internal energy $U_{1,s}$ of humid air is iteratively calculated from the new intensive state variables pressure p_1 and temperature T_1:

$$U_{1,s} = U(V_1, p_1, T_1, X_0, S_0). \tag{36}$$

If the thermodynamic state of water in the two-phase mixture after change in volume corresponds to the triple-point state, internal energy of the two-phase fluid is calculated by a linear interpolation between internal energy according to liquid state of water as in equation (22) and solid state according to equation (23).

For a constant fluid capacity volume, the fluid state alters as a result of mass and corresponding enthalpy transfer. For two-phase flows, enthalpy H_k of each exchanged fluid element is calculated from specific enthalpy h_k according to equation (26) factored by the two-phase mass m_k:

$$H_k = m_k \cdot h_k = m_k \cdot h_{1+x,k} \cdot (1 + X_k)^{-1}. \tag{37}$$

Enthalpy H_k of the fluid elements exchanged between two adjacent fluid capacities is related to the fluid capacity at higher pressure. Further, during balancing internal energy within a specific fluid capacity, elements flowing in are considered positive and such leaving the control volume negative.

3.3. Calculation of two-phase mass flow rate

For the calculation of the two-phase mass flow rate, a two-phase separated flow (slip-flow) method as proposed in [13] is applied. Here, the mass flow rate $\dot{m}$ results from a theoretical two-phase mass flow rate $\dot{m}_{th}$ factored by a discharge coefficient C_d including the entire flow losses:

$$\dot{m} = C_d \cdot \dot{m}_{th}. \quad (38)$$

The theoretical two-phase mass flow rate $\dot{m}_{th}$ through restrictions is derived from the conservation laws of momentum and mass as presented amongst others by Chisholm in [20] and Morris in [21]. The discharge coefficient C_d allows for the entire momentum loss and considers, by definition, a contraction coefficient C_c depending on Mach number effects factored by the so-called coefficient of velocity C_v including friction effects [21–24]:

$$C_d = C_c \cdot C_v. \quad (39)$$

For orifice flows, the discharge coefficient C_d includes momentum losses only due to contraction effects (vena contracta), and the coefficient of velocity C_v equals unity. The orifice approach ($C_d = C_c$) is preferred with respect to inlet and outlet connections in twin-screw expanders. For narrow flow sections or nozzle flows, no flow contraction is considered ($C_c = 1$), and the discharge coefficient primarily allows for momentum losses due to interphase and wall friction ($C_d = C_v$). This method for flow loss calculation is applied to any clearance in the twin-screw-expander or leakage paths where no contraction of the two-phase flow jet is expected. Since the approach regarding two-phase discharge coefficients for nozzle flows presented in [13] reflects only interphase friction effects, in the current study the coefficient of velocity is factored by an empirical flow coefficient a as follows:

$$C_v = C_v' \cdot a. \quad (40)$$

The coefficient of velocity C_v' is addressed to interphase friction as reported in [13] depending mainly on mass dryness fraction while the flow coefficient a reflects wall friction effects.

4. Geometry of SE 51.2

The two-phase approach for the simulation of water-flooded twin-screw expanders proposed in this work is verified by means of existing experimental data of the twin-screw expander SE 51.2 including integral operational parameters as well as indicator diagrams presented in [14–16]. In figure 4, a 3D model of the test twin-screw expander prototype SE 51.2 including positions of the high-resolution pressure indication transmitters used to record indicator diagrams is illustrated.

SE 51.2 is a twin-screw expander prototype without timing gears. The transmission of torque occurs directly via contact between the rotor flanks. The screw rotors are hardened and have a tough tungsten carbide/carbon (WC/C) wear-protection coating, so seizure is avoided in dry-running or water-flooded operation. Both fixed and loose bearing sets are grease lubricated, so no oil supply is necessary. Details about the expander geometry parameters are listed in table 1.

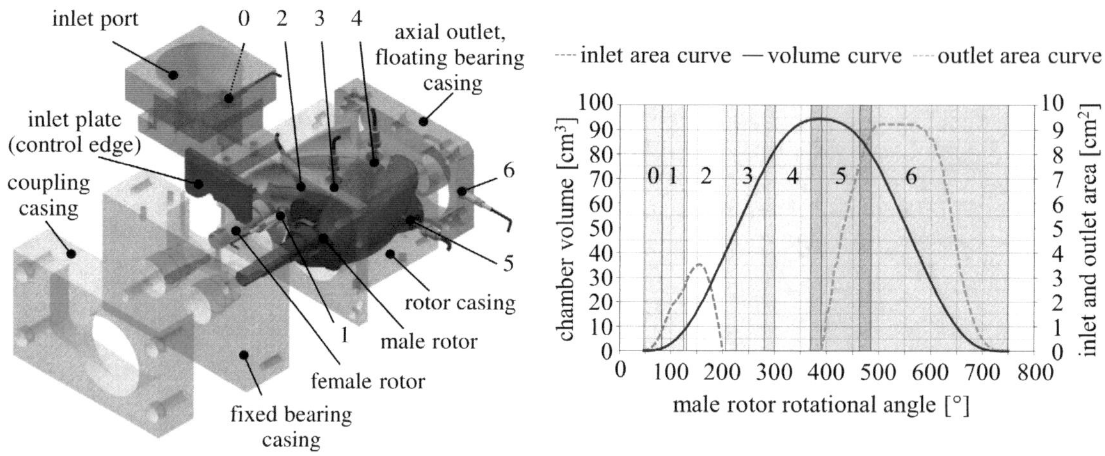

Figure 4. Positions of pressure indication transmitters (0…6), inlet and outlet area, as well as volume curve of the test twin-screw expander SE 51.2 including ranges of male rotor rotational angle corresponding to each pressure transmitter [15, 16].

Table 1. Geometry parameters of the test twin-screw expander SE 51.2.

designation	unit	male rotor (mr)	female rotor (fr)
number of lobes z	[-]	3	5
diameter	[mm]	71.8	67.5
rotor lead	[mm]	181.8	303
wrap angle	[°]	200	-120
rotor length	[mm]	101	
rotor profile	[-]	modified asymmetric SRM	
axis-centre distance	[mm]	51.2	
internal volume ratio	[-]	2.5	
displaced volume per male rotor rotation	[cm³]	286	
front clearance height $h_{fc,hp}$ (high pressure)	[mm]	0.1	
front clearance height $h_{fc,lp}$ (low pressure)	[mm]	0.17	
rotor-tip (housing) clearance height h_{hc}	[mm]	0.08	

5. Simulation setup

In the following section, the multi-chamber model of the twin-screw expander SE 51.2 including specific input with respect to the modelling of liquid distribution and the boundary conditions for the simulations are presented.

5.1. Multi-chamber model of SE 51.2

The multi-chamber model used for the simulation of the twin-screw expander SE 51.2 is illustrated in figure 5. The working chambers are numbered consecutively beginning from the low pressure domain. Each working chamber is divided into two fluid sub-capacities belonging to the male and female rotor. Pressure balance connections provide equal pressures in two corresponding male and female sub-capacities during each calculation time step. Nevertheless, different temperatures and mass dryness fractions of these sub-capacities result after pressure balancing.

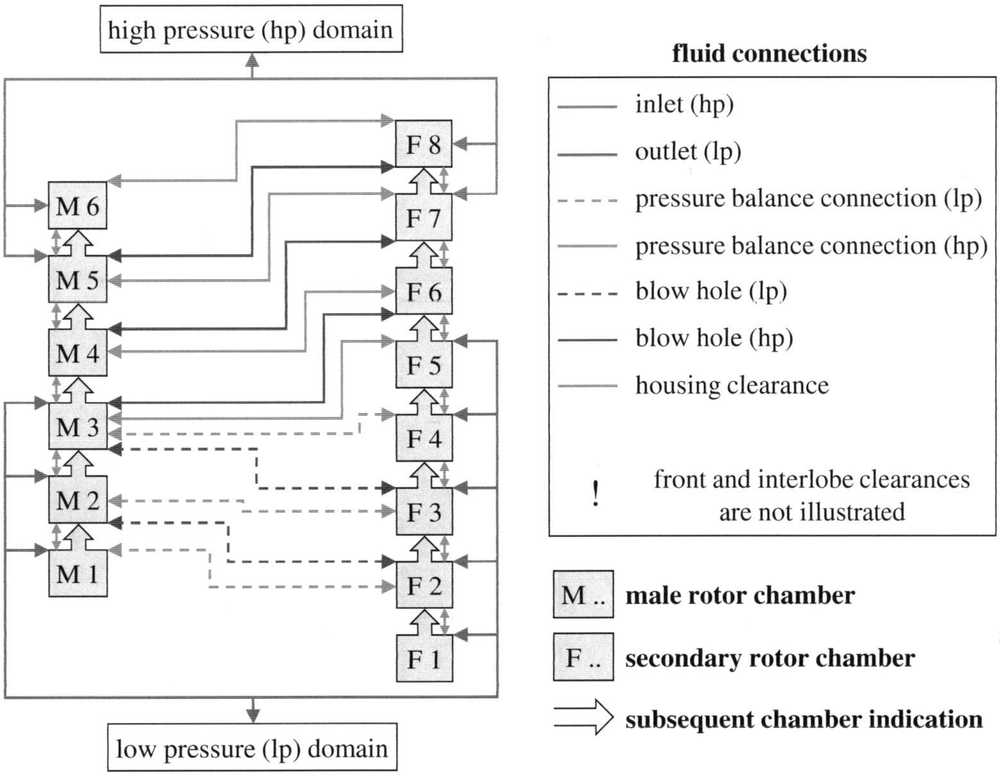

Figure 5. Multi-chamber model (adiabatic, front and interlobe clearances not illustrated) of SE 51.2 containing fluid capacities and connections.

In terms of flow loss calculation, as presented in table 2, all clearances of SE 51.2 are initialised with respect to nozzle discharge coefficient including the coefficient of velocity factored by a user-specified flow coefficient according to equation (40). Inlets and outlets are modelled as orifice flow restrictions according to equation (39) considering only contraction effects.

Table 2. Specific parameters for the initialisation of the chamber model of SE 51.2 with respect to two-phase flow connections.

connection	flow regime	discharge behaviour	dryness fraction χ	flow coefficient a
inlet opening (hp)	slip	orifice	$\chi_{io} = x$	1.0
outlet opening (lp)	slip	orifice	$\chi_{oo} = x$	1.0
rotor-tip clearance (mr)	slip	nozzle	$\chi_{hc,mr} = x$	1.0
rotor-tip clearance (fr)	slip	nozzle	$\chi_{hc,fr} = x$	0.8
blow hole	slip	nozzle	$\chi_{bh} = x$	1.0
interlobe clearance	slip	nozzle	$\chi_{ic} = 0.9$	1.0
front clearance	slip	nozzle	$\chi_{fc} = 0.9$	0.8

Due to the complexity of predicting the liquid distribution in the working chamber and thus at the clearances of twin-screw machines, no holistic models are available in the literature. Therefore, assumptions have to be made. On the one hand, the liquid flows in radial direction as a result of centrifugal forces where it spreads over the housing surface. On the other hand, Couette flows additionally influence the amount of liquid directly at the clearance. Due to centrifugal forces, it could be assumed that all clearances except blow holes and rotor-tip clearances remain relatively dry and are not significantly sealed by the liquid. Therefore, high mass dryness fractions are expected there. For the calculation of the pressure-driven two-phase clearance mass flow rate, rotor-tip clearance and blow hole mass dryness fraction is recommended to be initialised by values equal to those of the chamber fluid or less.

In this work, the liquid distribution in the working chamber is modelled initialising mass dryness fraction of each clearance as proposed in table 2. For example, housing clearances and blow holes are initialised with a mass dryness fraction related to the source chamber value. In contrast, interlobe and front (axial) clearances are considered relatively “dry” according to a dryness faction of 0.9. These assumptions are supported, in particular, by video recordings of the working cycle (not presented here).

5.2. Simulation boundary conditions

The simulations of the operational behaviour of SE 51.2 are performed according to the experimental investigations as presented in [16]. The range of pressure, temperature, inlet and outlet mass dryness fraction, and male rotor tip-speed respectively for simulation and comparison with experimental data are listed in table 3.

Table 3. Boundary conditions for the simulation of SE 51.2.

parameter	**value**
maximum male rotor tip-speed u_{mr}	67.7 $m \cdot s^{-1}$
inlet pressure p_i (absolute)	$3 \cdot 10^5$ Pa, $4 \cdot 10^5$ Pa, $5 \cdot 10^5$ Pa
outlet pressure p_o (absolute)	$1 \cdot 10^5$ Pa
inlet temperature ϑ_i	55 °C ($x_i = 0.5$), 90 °C ($x_i = 1.0$)
inlet/outlet mass dryness fraction x_i/x_o	0.5 (water-flooded), 1.0 (dry-running)

6. Results

In the following section, the two-phase approach for simulation of water-flooded twin-screw machines is validated comparing chamber-model simulation results to experimental data for the expander SE 51.2. In this context, indicator diagrams, expander mass flow rates, experimentally determined indicated and simulated internal powers, as well as outlet temperatures are considered. Indicated power is calculated by means of the experimentally recorded chamber pressure during a working cycle and the corresponding chamber volume as follows:

$$P_{ind} = -n_{mr} \cdot z_{mr} \cdot \oint p \cdot dV. \tag{41}$$

First, calculation and experimental results for dry-running operation ($x_i = 1$) are presented in figure 6 in order to analyse the simulation model regardless of liquid effects. Mass flow rate as well as indicated and internal power are examined as a function of male rotor tip-speed u_{mr} at different inlet pressures p_i. As expected, simulated and experimental mass flow rate and power increase at rising rotor speed and inlet pressure due to increasing number of working cycles and inlet density respectively. With regard to inlet throttling losses at increasing rotor tip-speed, both expander parameters exhibit a degressive progression due to declining chamber pressure during the filling process. Moreover, for a combination of high rotor speeds and low inlet pressures, overexpansion as observed in the indicator diagram for $p_i = 3 \cdot 10^5$ Pa and $u_{mr} = 60$ $m \cdot s^{-1}$ negatively impacts the expander performance.

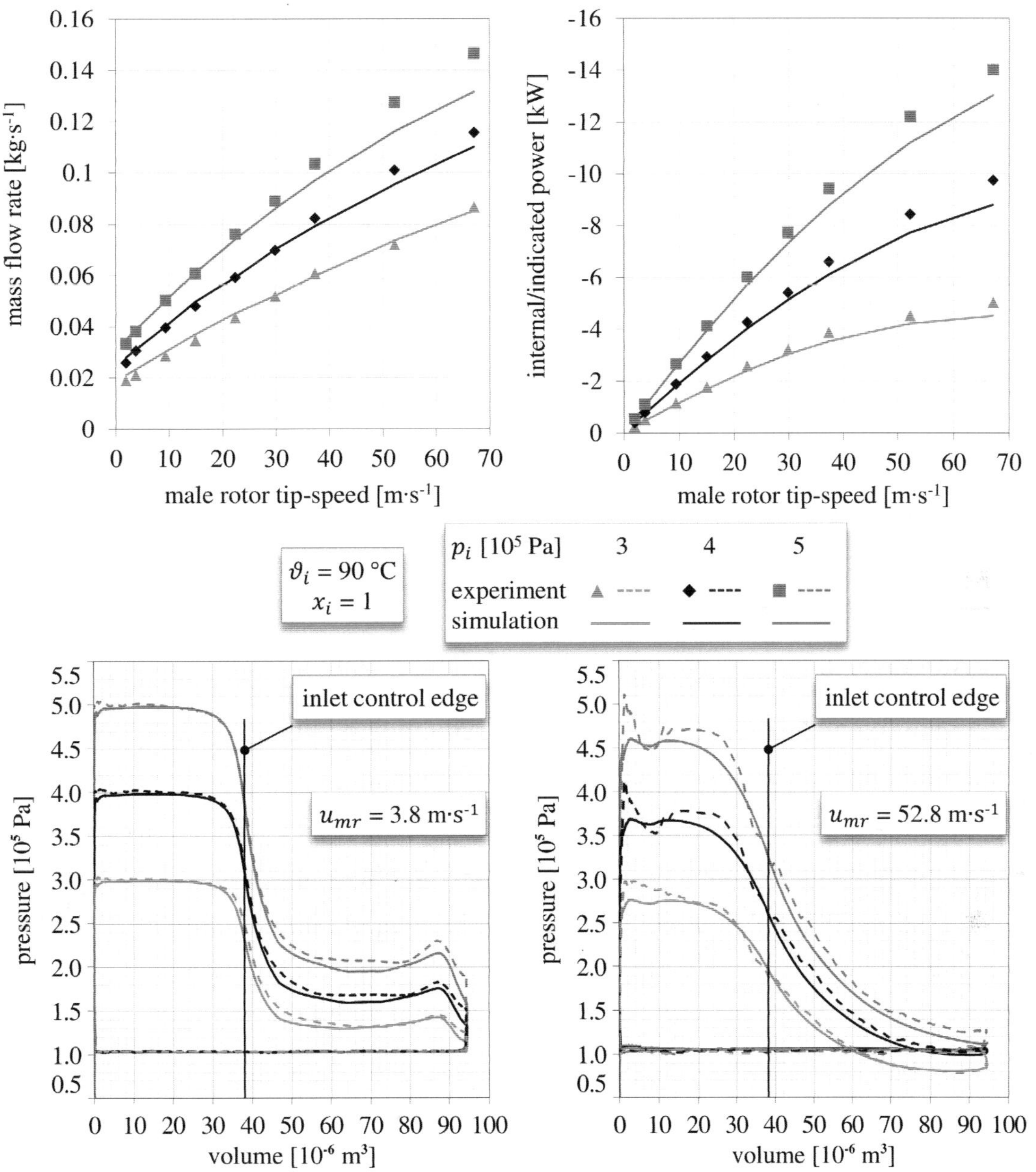

Figure 6. Expander mass flow rate, simulated internal and experimental indicated power, as well as indicator diagrams at different inlet pressures and rotor speeds for dry-running operation (x_i = 1).

Deviations between simulation and experimental results are mainly attributed to either modelling of clearances and openings or dynamic effects during the experiments and uncertainties of the measurement. In figure 6, the greatest disagreement between simulation and experiment is observed for increasing rotor tip-speeds u_{mr} > 30 m·s^{-1} at inlet pressure $p_i = 4{\cdot}10^5$ Pa and $p_i = 5{\cdot}10^5$ Pa. Here, the simulated mass flow rate and internal power are lower than the experimental values. This difference can be traced back to the orifice discharge coefficients applied to the expander inlet during the simulation (see table 2). For single-phase flows, in contrast to nozzles approaching the theoretical mass flow rate, orifices are associated with flow losses due to contraction effects. As reported in [13], the inlet geometry based on the twin-screw expander SE 51.2 exhibits a transition from an orifice to nozzle discharge behaviour as the opening becomes slip-shaped with the end of the chamber filling. Therefore, as with inlet discharge coefficients related to orifices, the flow loss during chamber filling might be overestimated, since, in this study, the inlet discharge behaviour is fixed and cannot be changed during simulation progress. In the same context, the measured indicator diagrams illustrated for u_{mr} = 52.8 m·s^{-1} show higher working chamber pressure during volume increase compared to the simulation results. Additionally, as observed during chamber filling in the indicator diagrams and reported in [15], the pressure oscillation in the inlet port directly influence the chamber pressure. Hence in contrast to the chamber model simulation not considering dynamic fluid inertia effects, the experiments reveal higher fluid density in the working chamber during chamber filling.

With regard to low rotor tip-speed in figure 6, in particular at p_i = 3·10^5 Pa, simulated expander mass flow rate is slightly overpredicted. In the indicator diagrams for u_{mr} = 3.8 m·s^{-1}, the simulated chamber pressure during expansion is lower than the experimentally recorded. Along with a higher simulated than measured mass flow rate, this indicates overestimated internal leakages that could be affiliated with the interlobe clearance directly connecting high-pressure working chambers with chambers discharging into the low-pressure port of the expander. Due to, for example, thermal deformation or bearing clearance, the interlobe clearance could experience significant relative change in height compared to its initial design dimension. The interlobe clearance leakage permanently induces loss flows that reduce the chamber pressure and cannot be utilised at lower pressure in terms of chamber refilling as with housing and front clearances or blow holes.

In figure 7, simulation and experimental results under water-flooded operating conditions (x_i = 0.5) are presented. Generally as a function of rotor speed and inlet pressure, two-phase mass flow rate as well as internal and indicated power exhibit similar characteristics as for dry-running operation in figure 6. The analysis of the water-flooded expander at increasing rotor tip-speed reveals simulated internal power slightly lower than the experimentally determined indicated power. Here again as with dry-running operation, this could be attributed to dynamic effects during the chamber filling or to the inlet area modelling with respect to flow losses. At the same time, for low rotor speeds and increasing inlet pressure, the simulation provides expander two-phase mass flow rates slightly less than the experimentally recorded. The maximum deviation is in the range of 12 % at the lowest rotor tip-speed and highest inlet pressure investigated. In connection with the proposed liquid distribution in this study, this gives a rise to the assumption that a relevant internal leakage is underestimated. As the indicator diagrams show, during the internal expansion, an internal leakage path associated with chamber refilling might be the reason for this difference due to the initially higher and afterwards lower pressure levels compared to the measurements. In this context, the liquid sealing effect with respect to blow holes and the rotor-tip clearances might be overestimated proposing their initialisation as with the chamber mass dryness fraction. As recorded images of the working cycle (not presented here) indicate, liquid distribution at the rotor-tip clearances and blow holes dynamically changes during the expansion progress providing an evidence for the relevance of its appropriate modelling. Hence, the assumption as made within this study has to be considered only as an initial approach. However with the aim of reliable simulations of liquid-flooded twin-screw machines, a high potential for developing appropriate liquid-distribution models in future works exists.

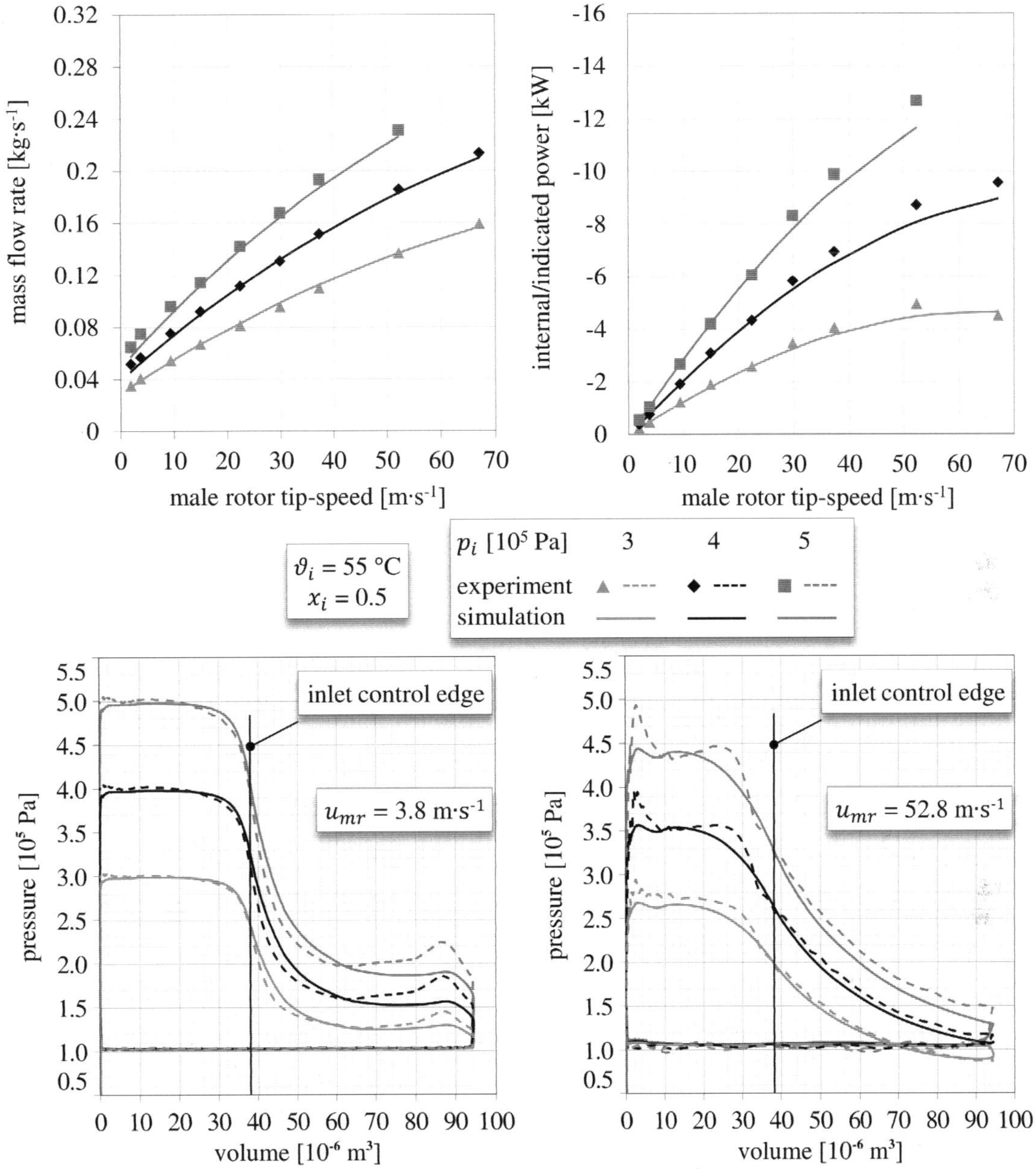

Figure 7. Expander mass flow rate, simulated internal and experimental indicated power, as well as indicator diagrams at different inlet pressures and rotor speeds for dry-running operation (x_i = 0.5).

Figure 8 illustrates the measured and simulated expander outlet temperature depending on rotor speed and inlet pressure. At this point it must be noted that, in contrast to the experimental data, the simulated temperatures do not include any losses as with hydraulic and mechanical friction. Basically, expander outlet temperature decreases due to declining impact of internal leakages at rising rotor speed, on the one hand, and higher utilisable pressure ratio related to increasing inlet pressure, on the other hand. In terms of dry-running operation (x_i = 1), the simulated temperatures deviate from the experimental values by maximum 20 K. As one reason for this result, the greater throttling loss as mentioned above is identified and, thus, lower pressure ratios are available for expansion associated with lower temperature drops. Applying a nozzle discharge coefficients to the inlet opening of SE 51.2, lower expander outlet temperatures are simulated attributed to higher chamber pressures after disconnecting the working chamber from the high-pressure port (not presented here). Hence, the maximum temperature difference declines to 12 K. Additionally, in the simulation, overestimated gaseous flows through the interlobe clearance bypass the working chamber and the relatively high temperatures of the leakage flow resulting from the corresponding isenthalpic change in state contribute to a further temperature increase downstream of the expander. With regard to water-flooded operation ($x_i = 0.5$), expander outlet temperatures are relatively accurately predicted by the simulation. Significant deviations are observed at an inlet pressure of $p_i = 5{\cdot}10^5$ Pa and low rotor speeds $u_{mr} < 30$ m·s^{-1}. For two-phase flows, these can be explained by measurement uncertainties at different phase temperatures unlike the assumptions of thermal equilibrium as with the humid air model.

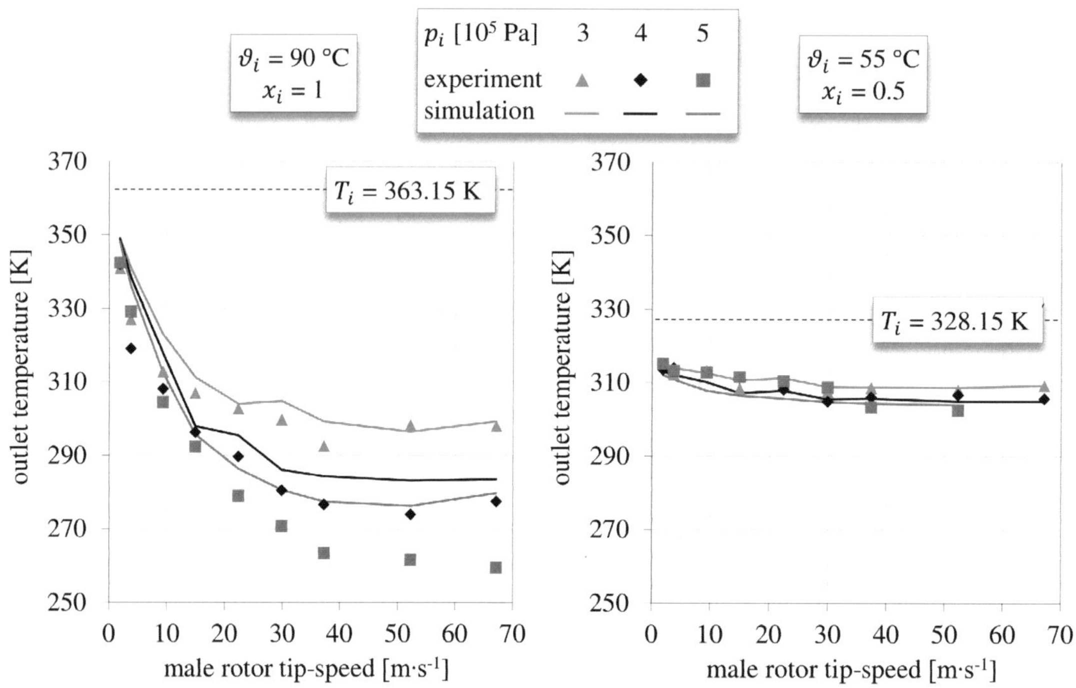

Figure 8. Expander outlet temperature as a function of male rotor tip-speed at different inlet pressures for dry-running (x_i = 1) and water-flooded (x_i = 0.5) operation.

7. Conclusion and outlook

Within the framework of this study, a two-phase approach for simulation of water-flooded twin-screw machines is validated for expander applications. The thermodynamic properties of the two-phase mixture of dry air and water are represented by the humid air model considering both phases in thermal equilibrium. Applying a two-phase slip-flow flow model, mass flow rates are calculated factoring a theoretical mass flow rate by geometry-specific discharge coefficients reflecting flow losses as in nozzle and orifice restrictions. Liquid distribution in the expander working chamber directly influencing the two-phase clearance flows is modelled in a simple way using clearance-specific mass dryness fractions. Due to centrifugal forces, it is assumed that most of the liquid distributes towards the radial clearances, while the axial and interlobe clearances are associated with relatively low amount of liquid. In this context, radial rotor-tip clearances and blow holes are initialised with mass dryness fraction of the source fluid capacity and the remaining clearances are referred to a constant mass dryness fraction of 0.9.

IOP Conf. Series: Materials Science and Engineering 1267 (2022) 012020 doi:10.1088/1757-899X/1267/1/012020

The single- and two-phase simulations reveal a relatively high degree of agreement with experimental data considering expander mass flow rate, internal power, outlet temperature, and chamber pressure depending on volume. At low rotor tip-speeds related to significant clearance flows, for both dry-running and liquid-flooded operation, simulated internal power matches the experimentally determined indicated power within a small range of deviation. Here for dry-running operation, the simulations exhibit a slight overprediction of clearance flows resulting in higher expander mass flow rates. Nevertheless, the comparison of simulated and measured chamber pressure reveals an appropriate prediction of internal leakages contributing to chamber refilling. Regarding the two-phase simulation, the assumed liquid distribution provides greater sealing effect indicated by underestimation of expander mass flow rates and lower chamber pressures during chamber refilling compared to the experimental data. At higher rotor speeds for both dry-running and liquid-flooded operation, simulated internal power and mass flow rate are underpredicted, in particular, due to dynamic effects during chamber filling not reflected by the simulation. Accordingly, higher outlet temperatures are simulated under dry-running conditions. In contrast, for water-flooded operation, deviations between simulated and experimental expander outlet temperatures are mainly traced back to uncertainties linked to two-phase temperature measurements.

In the present work, the liquid distribution in the working chamber is identified as a major challenge for further development of the two-phase simulation approach. In this context, more accurate and detailed models for the initialisation of mass dryness fraction of the clearances of twin-screw machines for expander and compressor applications are required. Since at high rotor speeds the effect of two-phase Couette flow increases, it has to be considered too. Moreover in terms of inlet throttling, flow loss calculation could be improved with respect to transition between an orifice and nozzle discharge coefficient during simulation depending on the shape of the inlet geometry.

Nomenclature

symbol	**unit**	**designation**
A	m^2	area
c	$J \cdot (kg \cdot K)^{-1}$	specific heat capacity of the incompressible phase
C_c	-	(two-phase) contraction coefficient
C_d	-	calculated (two-phase) discharge coefficient
c_p	$J \cdot (kg \cdot K)^{-1}$	isobaric specific heat capacity
c_v	$J \cdot (kg \cdot K)^{-1}$	isochoric specific heat capacity
C_v	-	(two-phase) coefficient of velocity
h	$J \cdot kg^{-1}$	specific enthalpy
H	J	enthalpy
m	kg	mass of the two-phase mixture
m_{da}	kg	mass of dry air
m_{ws}	kg	mass of water steam
m_{liq}	kg	mass of liquid water
$\dot{m}$	$kg \cdot s^{-1}$	mass flow rate of the two-phase mixture
$\dot{m}_{da}$	$kg \cdot s^{-1}$	mass flow rate of dry air
$\dot{m}_{ws}$	$kg \cdot s^{-1}$	mass flow rate of water steam
$\dot{m}_{liq}$	$kg \cdot s^{-1}$	mass flow rate of liquid water
n	s^{-1}	rotational speed
p	Pa	pressure
P	W	power
R	$J \cdot (kg \cdot K)^{-1}$	specific gas constant
s	$J \cdot (kg \cdot K)^{-1}$	specific entropy
S	$J \cdot K^{-1}$	entropy
t	s^{-1}	time

IOP Conf. Series: Materials Science and Engineering 1267 (2022) 012020 doi:10.1088/1757-899X/1267/1/012020

T	K	temperature
u	$m \cdot s^{-1}$	rotor tip-speed
u	$J \cdot kg^{-1}$	specific internal energy
U	J	internal energy
v	$m^3 \cdot kg^{-1}$	specific volume
V	m^3	volume
W	J	work
x	-	mass dryness fraction
X	-	water load
z	-	number of rotor lobes
α	-	flow coefficient
ϑ	$°C$	temperature
φ	-	relative humidity
χ	-	connection mass dryness fraction $\chi \coloneqq x_{con}$

abbreviation/index	**designation**
0	initial fluid state
1	changed fluid state
1+x	related to dry air mass
bh	blow hole
ch	chamber
con	connection
da	dry-air
ev	evaporation
ex	exchanged
fc	front clearance
fr	female rotor
g	gaseous
hc	rotor-tip (housing) clearance
hp	high pressure
i	inlet
ic	interlobe clearance
ice	ice
ind	indicated
io	inlet opening
j	index for a gas component in a gaseous fluid mixture
k	index for exchanged fluid element
liq	liquid (water)
lp	low pressure
me	melting
mr	male rotor
o	outlet
oo	outlet opening
s	isentropic
sat	saturation
su	sublimation
th	theoretical
tr	triple
w	water (liquid and steam)
ws	water steam

Appendix A

Table A1. Constant fluid properties for the thermodynamic state calculation of the two-phase fluid mixture in thermal equilibrium with regard to humid air [12].

designation	unit	value
specific gas constant of dry air R_{da}	[J·kg^{-1}·K^{-1}]	287.05
specific gas constant of water steam R_{ws}	[J·kg^{-1}·K^{-1}]	461.5
isobaric specific heat capacity of dry air $c_{p,da}$	[J·kg^{-1}·K^{-1}]	1004.6
isobaric specific heat capacity of water steam $c_{p,ws}$	[J·kg^{-1}·K^{-1}]	1863
isochoric specific heat capacity of dry air $c_{v,da}$	[J·kg^{-1}·K^{-1}]	717.55
isochoric specific heat capacity of water steam $c_{v,ws}$	[J·kg^{-1}·K^{-1}]	1401.5
liquid water specific heat capacity c_{liq}	[J·kg^{-1}·K^{-1}]	4191
ice specific heat capacity c_{ice}	[J·kg^{-1}·K^{-1}]	2070
specific volume of liquid water v_{liq}	[m^3·kg^{-1}]	0.001002
specific volume of ice v_{ice}	[m^3·kg^{-1}]	0.001087
triple temperature T_{tr}	[K]	273.16
triple pressure p_{tr}	[Pa]	611.657
specific melting enthalpy Δh_{me}^{tr} at T_{tr}	[kJ·kg^{-1}]	333.4
specific vaporisation enthalpy Δh_{ev}^{tr} at T_{tr}	[kJ·kg^{-1}]	2500.9
specific sublimation enthalpy Δh_{su}^{tr} at T_{tr}	[kJ·kg^{-1}]	2834.3

References

[1] Rane S, Kovačević A, Stošić N and Stupple G 2019 On Numerical Investigation of Water Injection to Screw Compressors *Proceedings of the ASME International Mechanical Engineering Congress and Exposition 2018 ASME 2018 International Mechanical Engineering Congress and Exposition (Pittsburgh, Pennsylvania, USA, 09.11.2018 - 15.11.2018)* (New York, N.Y.: The American Society of Mechanical Engineers)

[2] Räbiger K, Maksoud T, Ward J and Hausmann G 2008 Theoretical and experimental analysis of a multiphase screw pump, handling gas–liquid mixtures with very high gas volume fractions *Experimental Thermal and Fluid Science* **32** 1694–701

[3] Sangfors B 1984 Computer Simulation of the Oil Injected Twin Screw Compressor *International Compressor Engineering Conference International Compressor Engineering Conference (Purdue, Indiana) (International Compressor Engineering Conference)* ed Purdue University Libraries (Purdue, Indiana)

[4] Sangfors B 1998 Computer Simulation of Effects From Injection of Different Liquids in Screw Compressors *International Compressor Engineering Conference International Compressor Engineering Conference (Purdue) (International Compressor Engineering Conference)* ed Purdue University Libraries

[5] Vasuthevan H and Brümmer A 2016 Thermodynamic Modeling of Screw Expander in a Trilateral Flash Cycle *23rd International Compressor Engineering Conference at Purdue: West Lafayette, Indiana, USA, 11-14 July 2016 (Purdue)* ed Purdue University Libraries

[6] Smith S L 1969 Void Fractions in Two-Phase Flow: A Correlation Based upon an Equal Velocity Head Model *Proceedings of the Institution of Mechanical Engineers* **184** 647–64

[7] Bell I 2011 Theoretical and Experimental Analysis of Liquid Flooded Compression in Scroll Compressors *PhD thesis* Purdue University

[8] Lemort V December 19th, 2008 *Contribution to the Characterization of Scroll Machines in Compressor and Expander Modes* (Liège: Université de Liège, Belgique)

[9] Huagen W, Ziwen X and Pengcheng S 2004 Theoretical and experimental study on indicator diagram of twin screw refrigeration compressor *International Journal of Refrigeration* **27** 331–8

[10] Lin Z H 1982 Two-phase flow measurements with sharp-edged orifices *International Journal of Multiphase Flow* **8** 683–93
[11] Wang C, Xing Z, Chen W, Sun S and He Z 2019 Analysis of the leakage in a water-lubricated twin-screw air compressor *Applied Thermal Engineering* **155** 217–25
[12] Baehr H D and Kabelac S 2012 *Thermodynamik: Grundlagen und technische Anwendungen (Springer-Lehrbuch)* 15th edn (Berlin u.a.: Springer Vieweg)
[13] Nikolov A and Brümmer A 2022 Two-phase mass flow rate through restrictions in liquid-flooded twin-screw compressors or expanders *International Journal of Refrigeration (submitted)*
[14] Nikolov A and Brümmer A 2014 Influence of water injection on the operating behaviour of screw expanders: Experimental investigation *International Conference on Screw Machines 2014 (TU Dortmund University, Germany, 23rd-24th September) (VDI-Berichte* vol 2228*)* ed VDI (Düsseldorf: VDI-Verl.) pp 43–60
[15] Nikolov A and Brümmer A 2016 Analysis of Indicator Diagrams of a Water Injected Twin-shaft Screw-type Expander *23rd International Compressor Engineering Conference at Purdue: West Lafayette, Indiana, USA, 11-14 July 2016 (Purdue)* ed Purdue University Libraries
[16] Nikolov A and Brümmer A 2018 Impact of different clearance heights on the operation of a water-flooded twin-screw expander—experimental investigations based on indicator diagrams *International Conference on Screw Machines (TU Dortmund University, Germany, 18th-19th September)* ed IoP Conf. Series: Material Science and Engineering
[17] Janicki M 2007 Modellierung und Simulation von Rotationsverdrängermaschinen *PhD thesis* Fakultät Maschinenbau, TU Dortmund University
[18] IAPWS - International Association for the Properties of Water and Steam 2011 Revised Release on the Pressure along the Melting and Sublimation Curves of Ordinary Water Substance *(*IAPWS R14-08(2011)*)* (Plzeň, Czech Republic: IAPWS) http://www.iapws.org/ (accessed 8 Aug 2017)
[19] Janicki M and Kauder K 2003 Adiabatic Modelling and Thermodynamic Simulation of Rotary Displacement Machines *International Conference on Compressors and Their Systems (Cass Business School, City University, London, UK, 7-10 September) (IMechE conference transactions)* ed City University London (Bury St Edmunds: Professional Engineering) pp 511–9
[20] Chisholm D 1983 *Two-phase flow in pipelines and heat exchangers* (London: Godwin)
[21] Morris S D 1991 Compressible gas-liquid flow through pipeline restrictions *Chemical Engineering and Processing: Process Intensification* **30** 39–44
[22] Morris S D 1990 Discharge coefficients for choked gas—liquid flow through nozzles and orifices and applications to safety devices *Journal of Loss Prevention in the Process Industries* **3** 303–10
[23] Trutnovsky K and Komotori K 1981 *Berührungsfreie Dichtungen* (Düsseldorf: VDI-Verlag)
[24] Jobson D A 1955 On the Flow of a Compressible Fluid through Orifices *Proceedings of the Institution of Mechanical Engineers* 767–76

On performance optimisation for oil-injected screw compressors using different evolutionary algorithms

A Kumar[1,2], A Kovacevic[1], S A Ponnusami[1], S Patil[1,2], S Abdan[1,2] and N Asati[2]

[1]Centre for Compressor Technology, City, University of London, London, U.K.
[2]Kirloskar Pneumatic Company Limited, Pune, India

E-mail: abhishek.kumar.2@city.ac.uk

Abstract. Energy consumed for pressurizing air makes a significant proportion of total electrical energy consumption worldwide. To reduce the carbon footprint, it is necessary to have air compressors, which can operate efficiently over a large range of pressures and flow including full load and part load conditions. Several studies have been performed in this area including some which monitor the performance of a large number of compressors to develop strategies for their designs.

This paper focuses on the design optimisation of geometrical and oil parameters of oil-injected screw compressors using different evolutionary algorithms such as genetic algorithm (GA), covariance matrix adaptation evolution strategy (CMA-ES), and so on. A comparison of the performance of these algorithms is presented. SCORG and GT-SUITE (commercial software tools for screw compressor thermodynamic simulations and optimisation) are used in the integrated model producing promising results. Feasibility of the optimum outcomes generated by these algorithms is critically evaluated from machine and system design point of view. Finally, in the context of optimisation presented here, the simplex converges fastest as compared to other algorithms. In the future study, the system design limitations are to be incorporated as constraints for the optimisation along with the objective to improve energy efficiency.

Nomenclature

P	pressure	DIOIL	oil port diameter
Q	flow	FIOIL	oil injection angle
SPC	specific power consumption	TOIL	oil injection temperature
ETAK	adiabatic efficiency	POIL	pressure in oil reservoir
ETAV	volumetric efficiency	VISCOIL	oil viscosity
A	flow area	t	time
γ	adiabatic constant	ρ	density
v	fluid velocity	m	mass
ζ	leakage flow resistance coefficient	$\dot{m}$	mass flow rate
P_{hp}	high pressure	P_{hp}	low pressure
h_{in}	fluid enthalpy (inflow)	h_{out}	fluid enthalpy (outflow)

IOP Conf. Series: Materials Science and Engineering 1267 (2022) 012021 doi:10.1088/1757-899X/1267/1/012021

1. Introduction

A screw compressor is a rotary positive displacement machine preferred for relatively low discharge pressure and a wide range of flow rates than reciprocating, centrifugal, and axial compressors. It is a compact, reliable, and cost-effective machine consisting of a pair of helical rotors with four to six bearings enclosed in a casing. It has been categorized as oil-flooded, oil-free, and water-injected screw compressors. Among all these, the oil-flooded holds the majority percentage of share in the market due to fewer mechanical components, better efficiency and the ability to achieve higher pressure in a single stage itself.

Data given by the US EIA (Energy Information Administration), International Energy Outlook 2021, shows that the projected world energy consumption will increase by 50% from 2019 to 2050 [1]. The majority increase is in the energy produced by burning fossil fuels, with the least in renewable energy. The burning of fossil fuels by the industry leads to the production of carbon dioxide, which contributes nearly about (60-70)% to the greenhouse effect. The emission of greenhouse gasses, in turn, leads to global warming. On the other hand, for a screw compressor, statistical data reveals that in a life-cycle cost, the majority cost is undertaken by the operating cost, which is the power cost. Therefore, from an engineering point of view, the aim should be to reduce the compressor power consumption or increase its efficiency. This is only possible by optimising the screw compressor parameters to their maximum with an aim to improve the performance and reduce power consumption, thereby making them more environment friendly.

Referring to the various kinds of literature in this field, an optimisation of a refrigeration twin screw compressor considering the rotor geometrical parameters such as the wrap angle, relative length (length to diameter ratio) and the slide valve parameters using the computer program has been presented by Fleming et al. (1994) [2]. The optimum rotor geometrical parameters are not only important from the compressor efficiency point of view but also for other factors such as female rotor deflection, bearing loads and the rotor contact force. You et al. (1996) talks about the optimisation of lobe combination, relative length and the wrap angle on the factors mentioned above [3]. Stosic et al. (1997) describes the rotor profile generation. It makes use of the conjugacy conditions when solved explicitly and enables a variety of primary arcs to be defined that allows it to generate a higher efficiency profile [4]. A breakthrough in the optimisation of screw machines using evolution strategy (ES) has been carried out by Berlik et al. (2002) [5]. The limitations of using the ES in optimisation of screw machines has been discussed along with how the unsuitability of the penalty function method led to the extended offspring generation scheme. The problem in optimisation is the number of calculations to be performed in order to reach the optimum and whether this optimum is global optimum or not. Some of the common optimisation methods are listed below:

- Steepest descent: It is often known as gradient descent which is a first order iterative optimisation algorithm for obtaining the local minima of a differentiable function
- Newton's method: It is an iterative method for obtaining the extrema (minima or, maxima) of a differentiable function similar to the steepest descent method.
- Davidon-Fletcher-Powell's method: This method is used to obtain the solution to the secant equation closest to the current estimate satisfying the curvature condition. This method requires both the target function and its gradient.
- Random search: It is a family of numerical optimisation methods that do not require the gradient of the function. It can be used for functions that are not continuous or differentiable. It is often known as direct search, derivative-free or, black box method.

- Grid search: This method is much similar to the random search method that allows to select the best parameters from a range of values thus automating the trial and error method.
- Search along coordinate axes: It is a numerical optimisation method that restricts the search direction to the coordinate axes of the input space alone.
- Powell's method: This is used for obtaining the local minima of a real valued function that need not be differentiable. It passes a set of initial search parameters which are simply the normals aligned to the axis.
- Hooke-Jeeves method: It is also known as the pattern search method which keeps the track of the travel direction as the search propagates from point to point. It doesn't need to start over from the origin at each new search point.

Among all the methods, genetic algorithms are widely used which require only the value of the target function and can handle discontinuous functions as well. The only disadvantage is the slow converging nature due to its small seed size. Apart from this box complex method is also convenient which also requires only the function value and not the gradient. The disadvantage of this method is that it is less suitable for discrete parameters. Stosic et al. (2003) utilizes the box complex method for the optimisation of the compressor shape, size, dimension and operating parameters [6].The application of NURBS (Non-uniform rational B-splines) by Hauser et al. (2008) improves optimisation for the complex geometry machines that include profile variation of screw compressors [8]. On the other hand, Herlemann et al. (2014) highlight the use of a chamber model generator and simulation tool for the optimisation of rotor profiles with the same number of lobe combinations to minimize the energy costs that can be validated experimentally [9]. Last but not least, Zhang et al. (2019) show the structural parameter optimisation of the twin-screw expander and the influence of geometrical parameters on the thermal performance of the machine. Considering a single-objective function and using the sequential quadratic programming algorithm for optimisation the influence of leakage on its functional force is also studied [10].

In light of these considerations, in this work, a SCORG [11] model is developed for the oil-flooded screw compressors, which is validated from the experimental testing data performed at City University. This model has been linked to the optimisation tool, i.e., GT-SUITE from Gamma Technologies [12]. Design space exploration has also been used to find sensitive and important design variables. After identifying significant variables, the formal optimisation using different evolutionary algorithms has been carried out. The post optimality analysis has also been performed to determine the feasible outputs. Finally, a comparative analysis to find out the best-suited method for optimisation producing realistic outputs in a short span of time has been presented.

2. Modelling and optimisation

The machine considered in this work is a 22 kW air screw compressor with built-in volume ratio of 4.6 operated at a pressure ratio of 8.5 and a speed of 2000 rpm, manufactured at Kirloskar Pneumatic Company Limited, Pune (India). Since, being an industrial screw compressor, information about the size and profile cannot be disclosed. However, the normalized performance of the given screw compressor block has been shown in the table-3 for visualization of the improvement after optimisation. Experimental testing has been conducted for this compressor, whose data will be used for modeling the screw compressor in the SCORG (Screw Compressor Rotor Grid Generation) software from the PDM Analysis.

The SCORG is an industry-leading grid generation and performance prediction software for positive displacement screw machines. It is used to design, analyze, and optimise screw machines [11]. All the input parameters as per the compressor specifications have been fed into the SCORG, and then using the multi-chamber model in the background, it evaluates the thermodynamic performance, which is in good agreement with the experimental data. For optimisation, the SCORG offers an option of design exploration, which is basically used for observing the effect of parameters, and then the critical parameters can be used for further optimisation in-direct coupling with the GT-SUITE software from GAMMA Technologies [12].

GT-SUITE is a transient multi-physics tool used to pre-process, build, simulate, and post-process the data using the GT-POST interface. It is uniquely positioned to model the positive displacement pumps and compressors for several applications. This work focuses on the design space exploration and optimisation using the design optimiser available in the GT-ISE build interface.

2.1. Thermodynamic multi-chamber model

SCORG thermodynamic module performs thermodynamic calculation based on the multi-chamber model using equations of conservation of mass and internal energy applied to control volumes. The boundary conditions consist of suction gas pressure and temperature, the discharge gas pressure and oil injection parameters along-with geometry values of the chamber volume, port and leakage areas. Based on the multi-chamber model, the integral parameters such as indicated power, leakage flows, mass flow rate, volumetric and adiabatic efficiencies are computed [11].

Depending on the rotor position, the working chamber formed between the meshing rotor pair and the housing changes the shape and size with time. This chamber is itself connected to the suction and discharge chambers through flow areas which vary with time both in shape and size. The schematic view of a screw machine is shown in figure-1 [11].

According to the mass conservation law, the mass variation in the control volume is given as:

$$\frac{dm}{dt} = \dot{m}_{in} - \dot{m}_{out} \tag{1}$$

The mass inflow into the control volume consists of the suction flow when the chamber is connected to this port, the flow of injected fluid and the leakage flow which enters the control volume.

$$\dot{m}_{in} = \dot{m}_{suc} + \dot{m}_{inj} + \dot{m}_{lg} \tag{2}$$

The mass flow of the leakage flow through the clearance gap is calculated by simplified continuity equation assuming constant temperature which is given as [7]:

$$\dot{m}_{lg} = \rho \dot{V}_{cg} = \rho w_{cg} A_{cg} = A_{cg} \sqrt{\frac{\gamma(p_{hp}^2 - p_{lp}^2)}{a^2(\zeta + \log p_{hp}/p_{lp})}} [kg/s] \tag{3}$$

The mass outflow from the control volume consists of the discharge flow and the leakage flows.

$$\dot{m}_{out} = \dot{m}_{dis} + \dot{m}_{ll} \tag{4}$$

The above mentioned mass flow rates satisfy the continuity equation.

$$\dot{m} = \rho * v * A \tag{5}$$

The conservation of energy within the control volume, neglecting the kinetic energy of the fluid (as the velocity of the fluid within a screw machine is relatively low) is given as:

$$\frac{dU}{dt} = \dot{m}_{in} h_{in} - \dot{m}_{out} h_{out} + \dot{Q} - p\frac{dV}{dt} \tag{6}$$

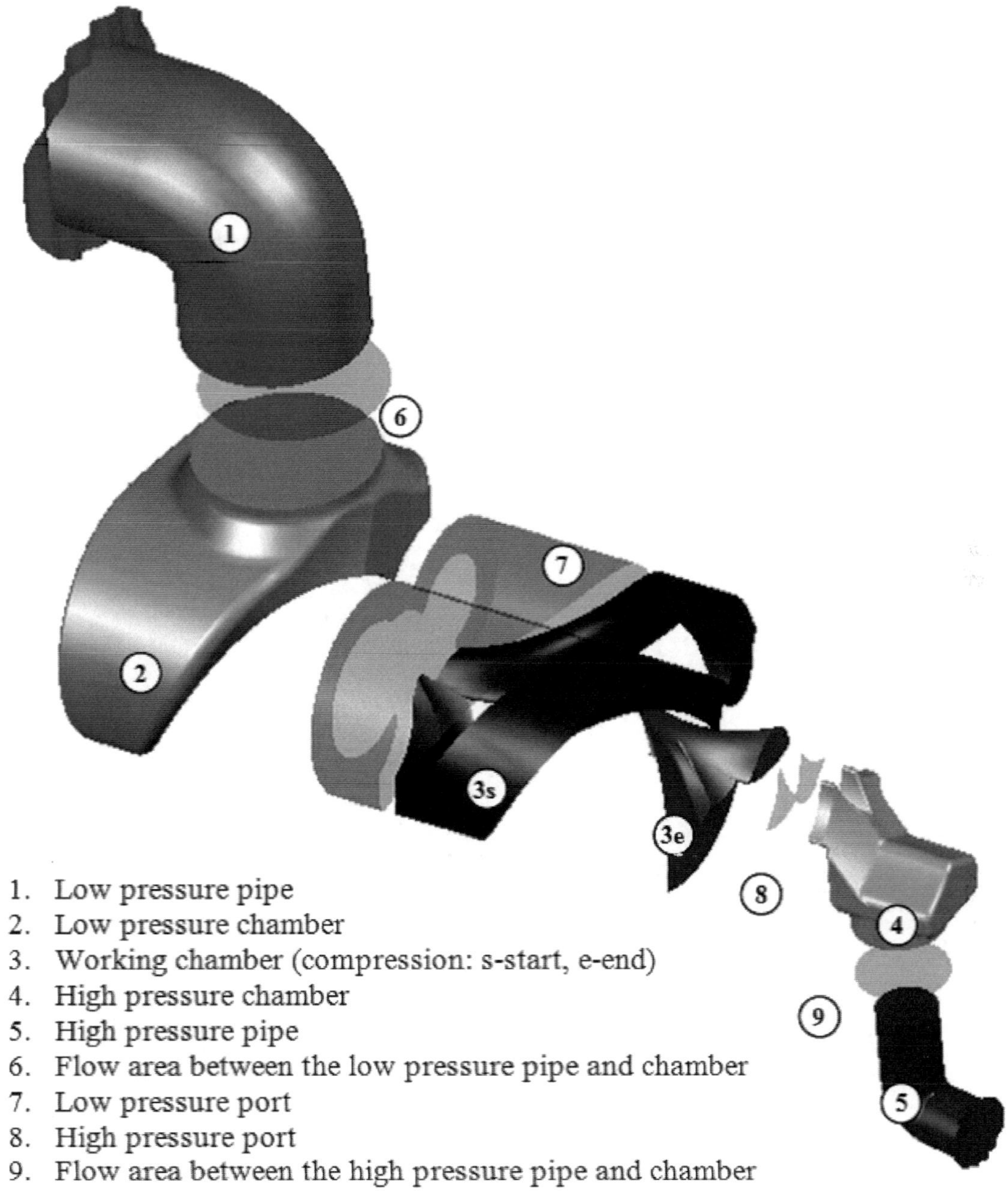

Figure 1. Configuration of a screw machine

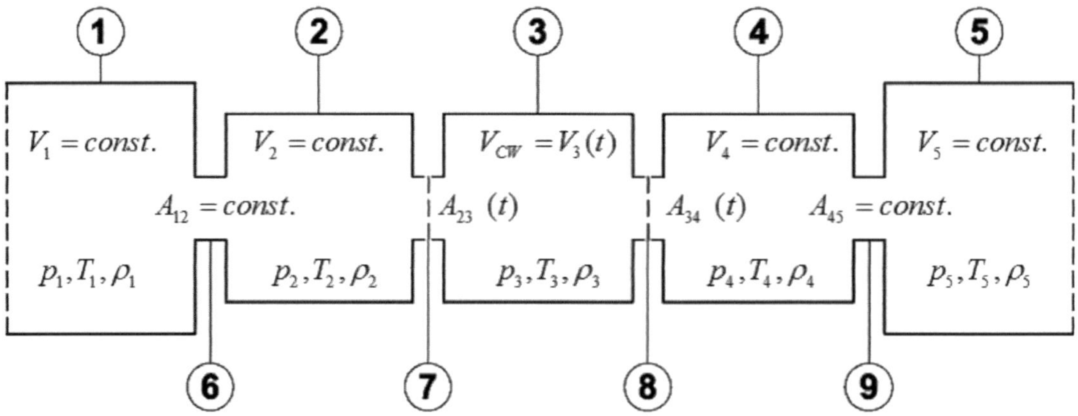

Figure 2. Schematic view of a screw machine chamber configuration

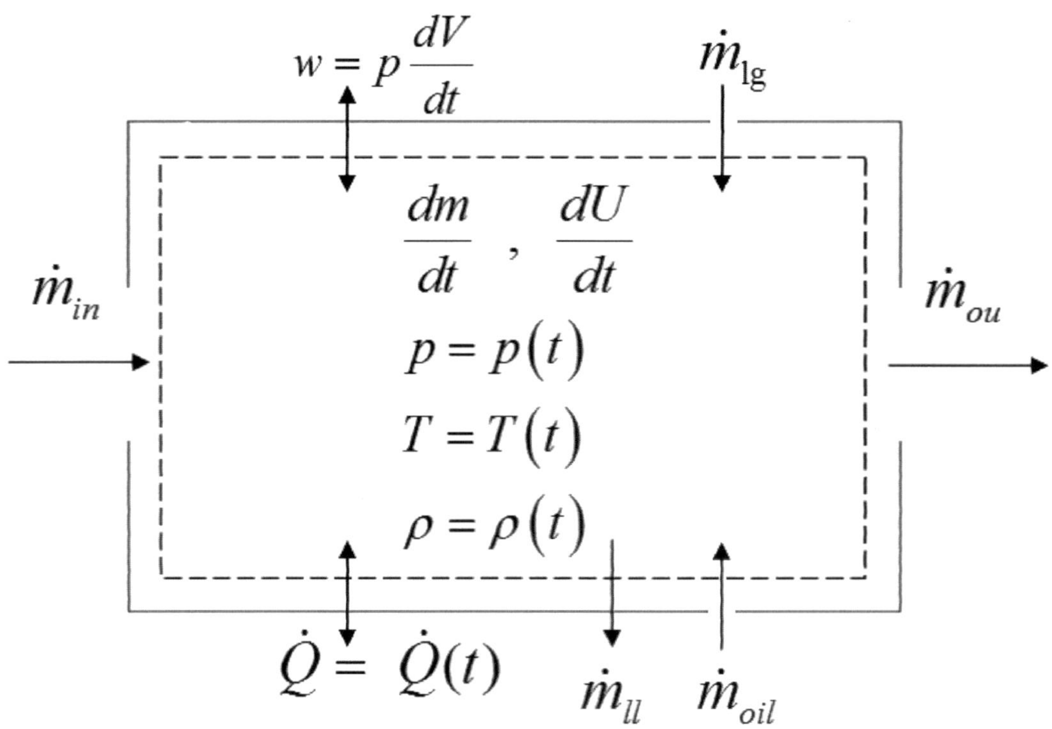

Figure 3. Control volume of the main chamber for thermodynamic analysis

Where,
$\dot{m}_{in}$ = Mass flow rate (inflow)
$\dot{m}_{out}$ = Mass flow rate (outflow)
$\dot{m}_{suc}$ = Mass flow rate (suction)
$\dot{m}_{dis}$ = Mass flow rate (discharge)
$\dot{m}_{inj}$ = Mass flow rate (injected fluid flow)
$\dot{m}_{lg}$ = Mass flow rate (leakage flow entering the control volume)
$\dot{m}_{ll}$ = Mass flow rate (leakage flow leaving from the chamber)
$\dot{V}_{cg}$ = Volume flow rate (clearance gap)
$\dot{Q}$ = Heat transfer through the control volume boundary
$p\frac{dV}{dt}$ = Thermodynamic work
On solving the above differential equation the mass and internal energy within the chamber are obtained. Further using the equation of state for an ideal gas all other thermodynamic properties can be calculated.

2.2. Optimisation algorithms
The GT-ISE build interface offers different types of evolutionary algorithms for optimisation, which are described in the below section:

- Simplex Search: The Nelder-Mead simplex algorithm is a deterministic local search algorithm used to obtain the maxima or minima of the given objective function in a multidimensional space. For example: when working with two influencing variables it simply involves comparing the function values at the three vertices of a triangle. If more variables are taken into account then corresponding geometric shape is considered and thus it is n-dimensional simplex. When using this algorithm, the factor resolution must be 3% or less. This method is only partially parallel capable.
- Genetic Algorithm: An evolutionary global search algorithm is recommended for medium to high complexity problems. The algorithm calculates the fitness value followed by selection, crossover, and mutation.
- Accelerated Genetic Algorithm: It is the modified version of the genetic algorithm with a faster convergence rate. It finds the better solutions in fewer total design iterations.
- Covariance Matrix Adaptation Evolution Strategy (CMA-ES): CMA-ES is also an evolutionary algorithm as the genetic algorithm. In this evolution, strategies are stochastic, derivative-free for numerical optimisation of non-linear or non-convex continuous optimisation problems.

The GT-SUITE design optimiser interface also offers other two optimisation algorithms, such as the Brent and discrete grid methods. A discrete grid is a bisection search algorithm that iteratively reduces the search space by a factor of 2 until convergence is reached. On the other hand, the Brent algorithm is based on the root-finding algorithm. Both are not recommended as the simplex algorithm is expected to perform better for most problems. Therefore, here in this work, for a comparative analysis of optimisation, only the above four listed algorithms are used as they perform better and are recommended by the Gamma Technologies [12].

2.3. Design approach
GT-SUITE design optimiser interface allows two different design approaches for multi-objective optimisation, which are as follows:

- Pareto-design: This approach is used to maximize, minimize or target an output response and study its competing objectives. It gives many optimal designs considered equally optimal and not a single best design.

- Weighted-sum approach: This approach combines the multiple objectives into a single-objective function by assigning a weight function to each output. An example of how this work is shown below:

Let,
P_1= First objective function which is to be maximized
P_2= Second objective function which is to be minimized
The combined single-objective function as per weighted-sum approach is given as:

$$F = \frac{P_1}{P_{norm1}} - \frac{P_2}{P_{norm2}} \tag{7}$$

where, P_{norm1} and P_{norm2} are the normalized value for each output function so that each output function is of the same order of magnitude when combined into a single-objective function. If the values are not normalized, the single-objective function will be dominated by the output function with a larger magnitude of weight function.

In this work, the weighted-sum design approach is preferred for multi-objective optimisation with equal weights so that each output function is given equal importance which is better for comparison of different algorithms.

2.4. Input and output functions

In an air screw compressor, as the pressure rises, the air temperature in the compression chamber also increases. This can be better understood with an example: in an oil-free twin screw air compressor operating at a pressure ratio of 6 and running at 6000 rpm (approx.) will result in an average discharge temperature of more than 320 °C. This temperature can be substantially reduced by injecting oil. That is why oil plays an important role in cooling the compressor chamber. Apart from this, oil also serves in improving the volumetric efficiency by sealing the leakage gaps and providing lubrication to both the rotors as well as the bearings [13].

According to the thumb rule used in any screw compressor industry, the oil contribution by mass to cooling, sealing and lubrication should be in the proportion of 100:10:1, respectively. Injecting oil has many advantages but it should be injected in the right quantity otherwise injecting more or less amount of oil would result in additional power losses. In the figure-4 it shows the oil injection parameters along with the expected effects. The box colour depicts the influence of parameters on the compressor performance. The darker the color intensity the higher is the influence on the performance [13].

The input and output variables need to be specified to optimise the problems. The input parameters considered in this work for both multi-objective and single-objective optimisation along with the effects are listed below:

- Oil port diameter (mm): It determines the flow rate of oil. The heat exchange between gas-oil depends on the mass flow rate of the oil injected. Increased oil mass flow rate can reduce discharge temperature and increase the compressor's volumetric efficiency and adiabatic efficiency. However, excess oil can lead to power losses during injection and transport through compression chambers as its density is several magnitudes greater than the gas.
- Oil injection angle (°): If the oil injection port is positioned so that the oil injection temperature is close to the gas temperature, there will be low convection between gas and oil. If the oil injection port is positioned close to the discharge port, then the residence time for oil will be low, and heat transfer between air and oil will not largely benefit the compression chamber. Therefore, optimal positioning of oil injection will help in controlling the compression chamber temperature.
- Oil injection temperature (°C): A low oil injection temperature is preferred for better heat exchange but cannot be achieved without the high load on the cooling system.

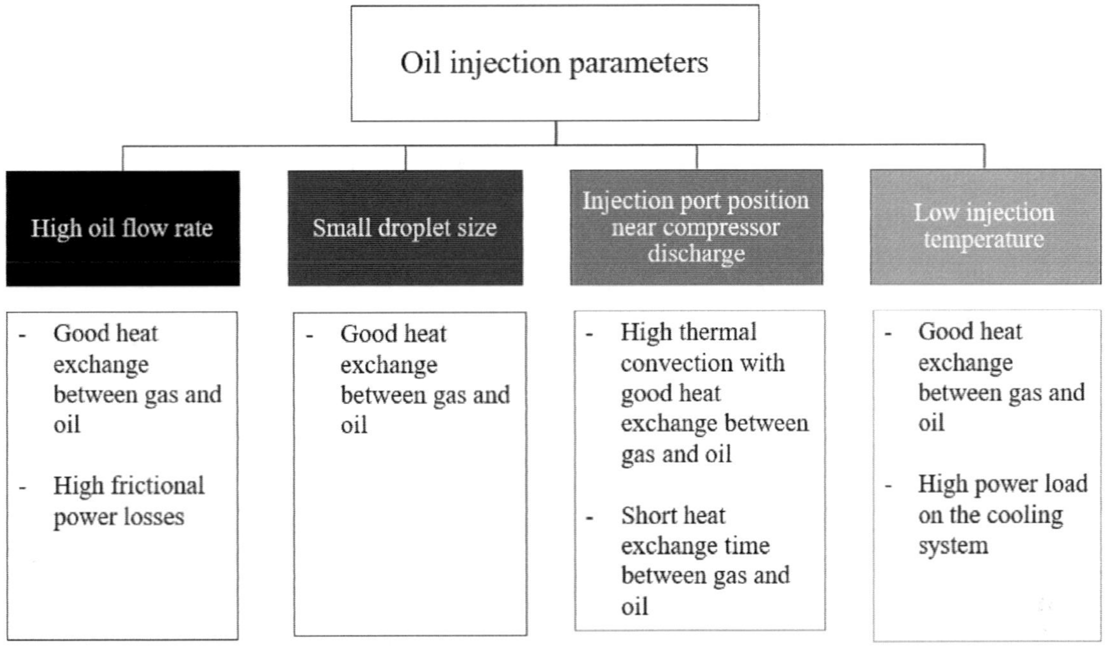

Figure 4. Various oil-injection parameters with their expected effects

- Pressure in the oil reservoir (bar): It affects the oil droplet size. Higher pressure means small oil droplet size which means a high surface area for exchanging heat with the hot gas, ultimately resulting in lower temperature in the compression chamber
- Oil viscosity (m^2/s): If the oil viscosity is low then the compressor experiences metal to metal contact leading to component failures whereas on the other hand if the oil viscosity is too high the compressor will operate less efficiently thus generating excessive heat and more power losses.

The above listed input parameters affect the temperature in the compressor chamber which directly affects the performance of the compressor. Therefore, for this performance optimisation oil parameters are taken into consideration as the temperature in the chamber is directly dependent upon the interaction between gas and oil.
The output function for multi-objective optimisation are as follows:

- Flow (m^3/min) : Mass flow rate is the mass of gas delivered per unit time. It is obtained from the mass of the gas in the working chamber just before opening the discharge port multiplied by the rotational speed and number of cycles per revolution of the male rotor.

$$\dot{m} = mz_1\frac{n}{60}[kg/sec] \quad (8)$$

The volume flow rate is always normalised to the suction conditions.

$$\dot{V} = \frac{60\dot{m}}{\rho_o}[m^3/min] \quad (9)$$

- Specific power consumption ($kW/m^3/min$) : Specific power is power required to deliver a unit quantity of mass of the gas . It is calculated as the ratio of the compressor power and volume flow rate.

$$SPC = \frac{P}{\dot{V}}[kW/m^3/min] \tag{10}$$

- Volumetric efficiency (%) : Volumetric efficiency is the ratio of the volume flow rate and the theoretical capacity of the machine.

$$\eta_v = \frac{\dot{V}}{\dot{V}_t} * 100[\%] \tag{11}$$

- Adiabatic efficiency (%) : Adiabatic efficiency is the comparison of actual indicated power and theoretical adiabatic power for the compression.

$$\eta_{ad} = \frac{P_{ad}}{P} * 100[\%] \tag{12}$$

The output function for single-objective optimisation is specific power consumption of the oil-injected twin screw air compressor.

The range of input parameters as per general operating conditions of a typical screw compressor are listed below in the table-1.

Table 1. Range of input parameters

Input parameters	Range
DIOIL	(1-12) mm
FIOIL	(35-130) °
TOIL	(35-80) °C
POIL	(3.5-8) bar
VISCOIL	(1-8)*E-06 m^2/s

2.5. SCORG_GT-SUITE Integration model

The SCORG model for the 22 kW oil-injected twin screw air compressor is prepared using the geometrical and design input parameters which is validated from the experimental data.

The GT-SUITE model has been set up in the GT-ISE interface using the SCORG batch file for the 22 kW machine. Once the model has been linked to the SCORG thermodynamics, the design optimiser in the GT-SUITE can be activated. The simulation is made to run iteratively in the design optimiser with the parameters values changing each time until the stopping criteria is satisfied. The results can be viewed and post analysis can be carried out in the GT-POST interface. The flow chart showing the complete process of SCORG and GT-SUITE integration is shown in figure-5.

3. Results

3.1. Single-objective optimisation

The simulation results from the GT-SUITE_SCORG model are presented in this section. In the single-objective optimisation using the different evolutionary algorithms, a comparative analysis has been presented. An important thing to note is that all the different algorithms

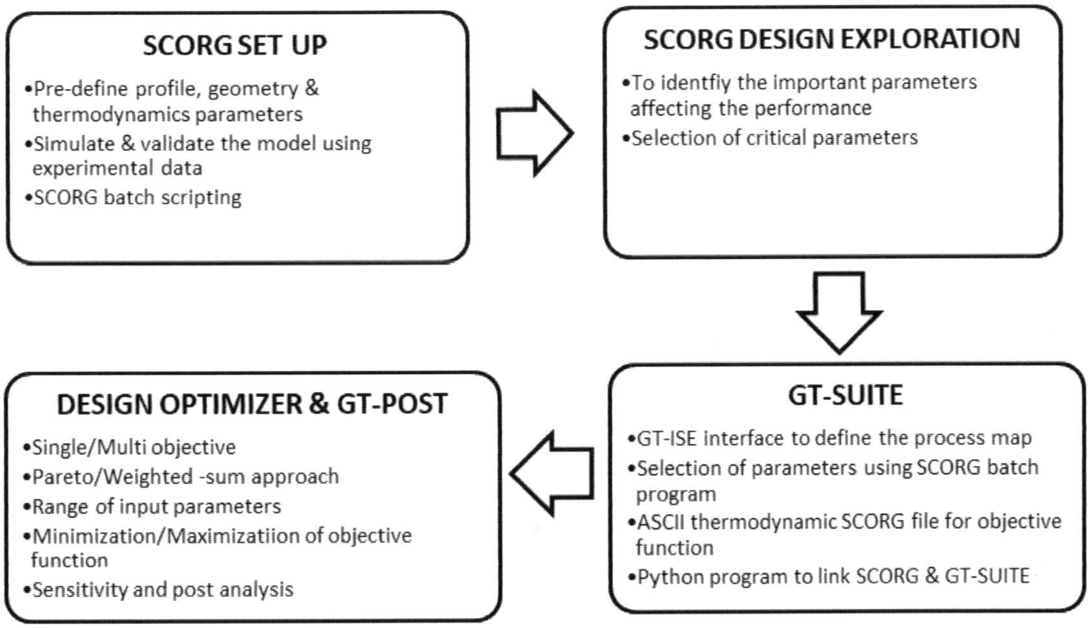

Figure 5. SCORG_GT-SUITE process map

for optimisation are simulated independently with the same input range for variables and the same objective function. For a single objective optimisation the specific power consumption is considered. The input parameters are the same for both the multi-objective and single-objective optimisation so that a proper conclusion can be obtained and the results are validated.

From figure-6 it can be seen that the change in specific power consumption for GA is around 10.8 % , for simplex is 10.5%, for CMA-ES is 9.6% and for accelerated GA is 9.2% respectively.

The time elapsed for each search algorithm is shown in figure-7. Simplex method takes the least time whereas the GA takes the highest time for single objective optimisation.

3.2. Multi-objective optimisation

In this section, a comparative analysis using the different evolutionary algorithms are presented for each objective function. Figure-8 shows the percentage change in specific power consumption for different algorithms with GA showing the highest variation of 10.8% and the accelerated GA with the lowest of 9.2% which is validated from single-objective optimisation.

It is also evident from figure-9 and 10 that the change in flow and the volumetric efficiency is maximum for GA and the least for accelerated GA. Apart from the thermodynamics performance analysis a comparative study of the elapsed time is also presented in figure-11. Here, it can be seen that the simplex search algorithm takes the least time for multi-objective optimisation and the GA takes the highest time for predicting the optimum output. This depends on the step-length and the seed size taken during the optimisation.

The seed size taken here for optimisation using the evolutionary algorithms for both the single-objective and multi-objective are listed below in the table-2.

The fitness function determines how fit an individual is (the ability of an individual to compete with other individuals). It gives a fitness score to each individual. The probability that an

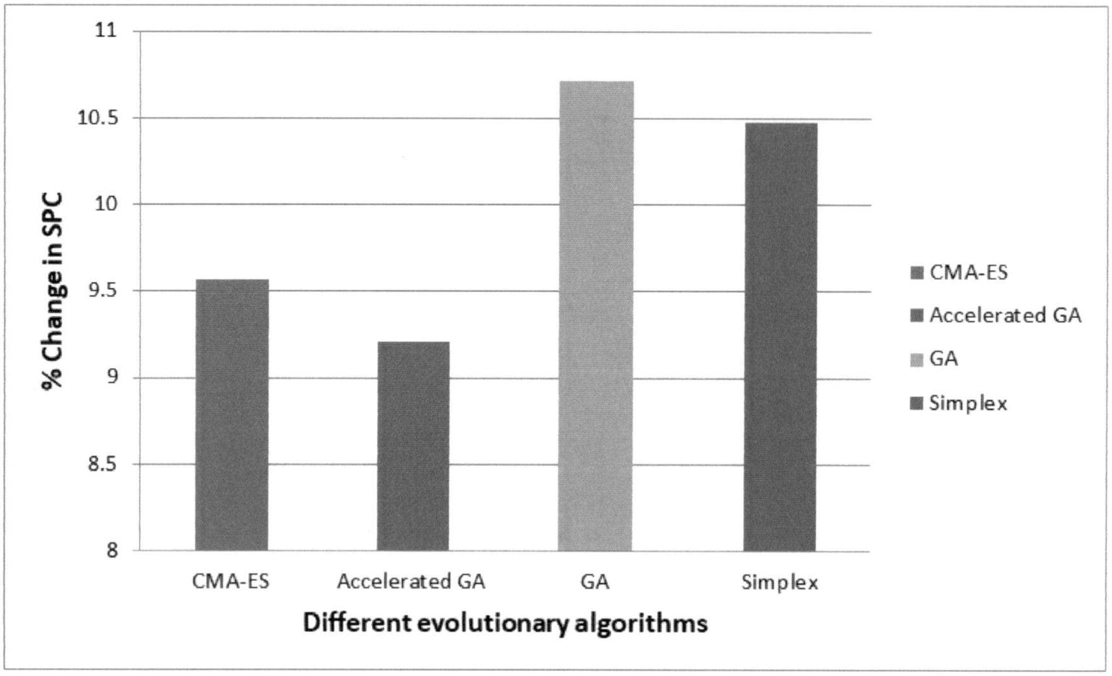

Figure 6. % Change in SPC (Single-objective optimisation)

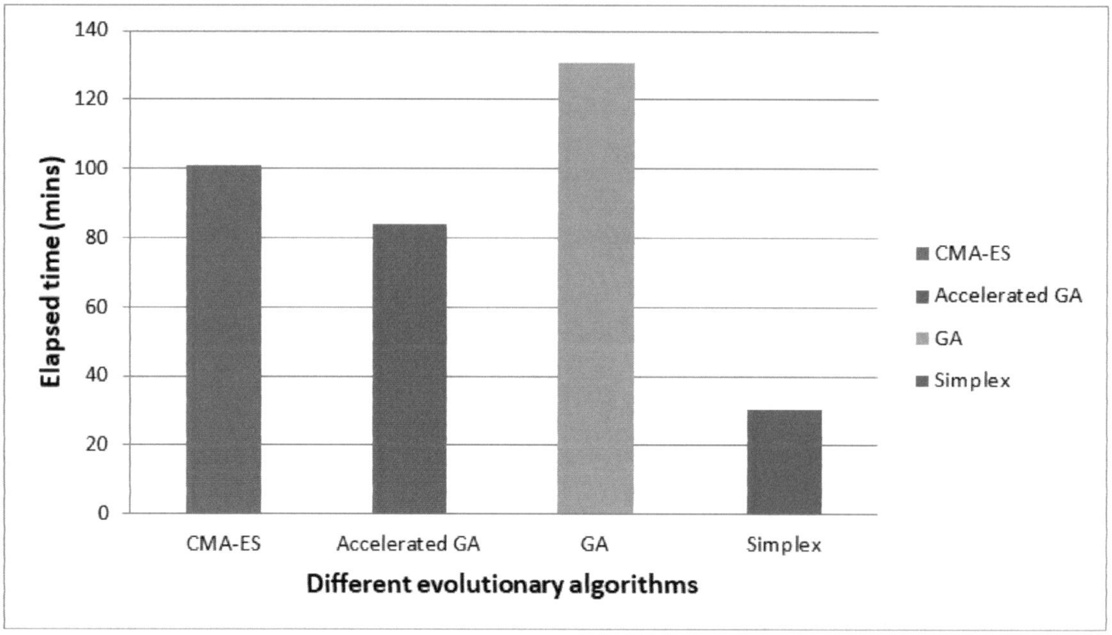

Figure 7. Time elapsed for single-objective optimisation

individual will be selected for reproduction is based on its fitness score. Functions with higher fitness value are recommended. Also, the fitness value is inversely proportional to the seed size.

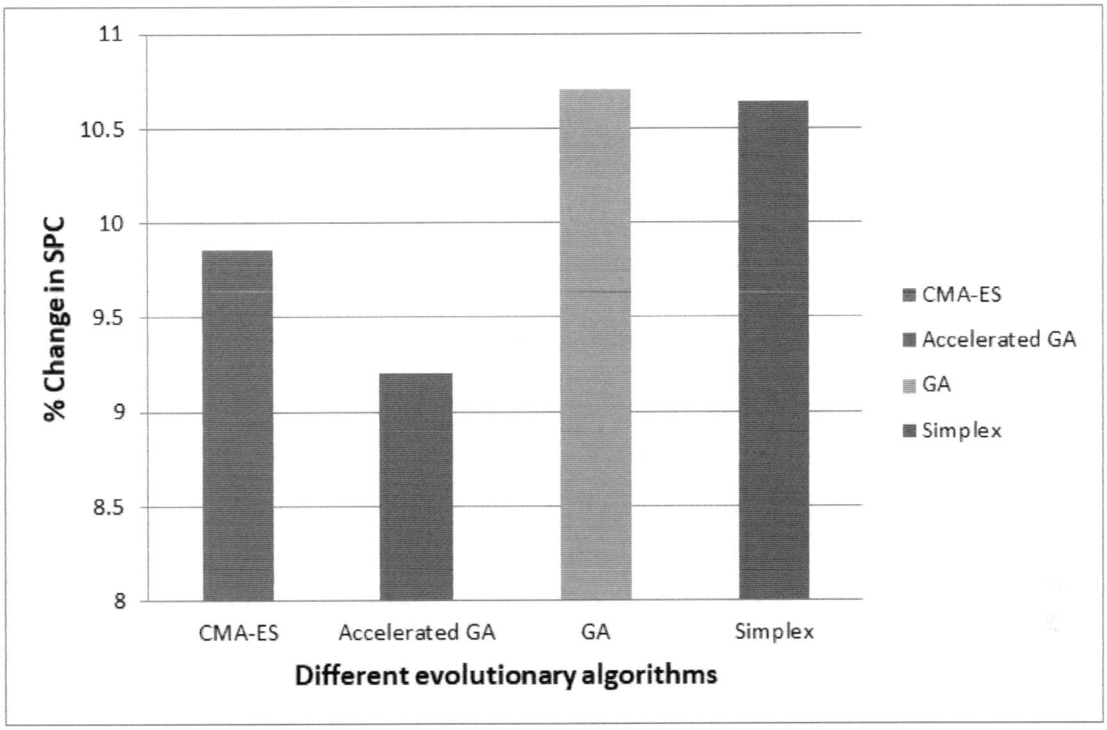

Figure 8. % Change in SPC (Multi-objective optimisation)

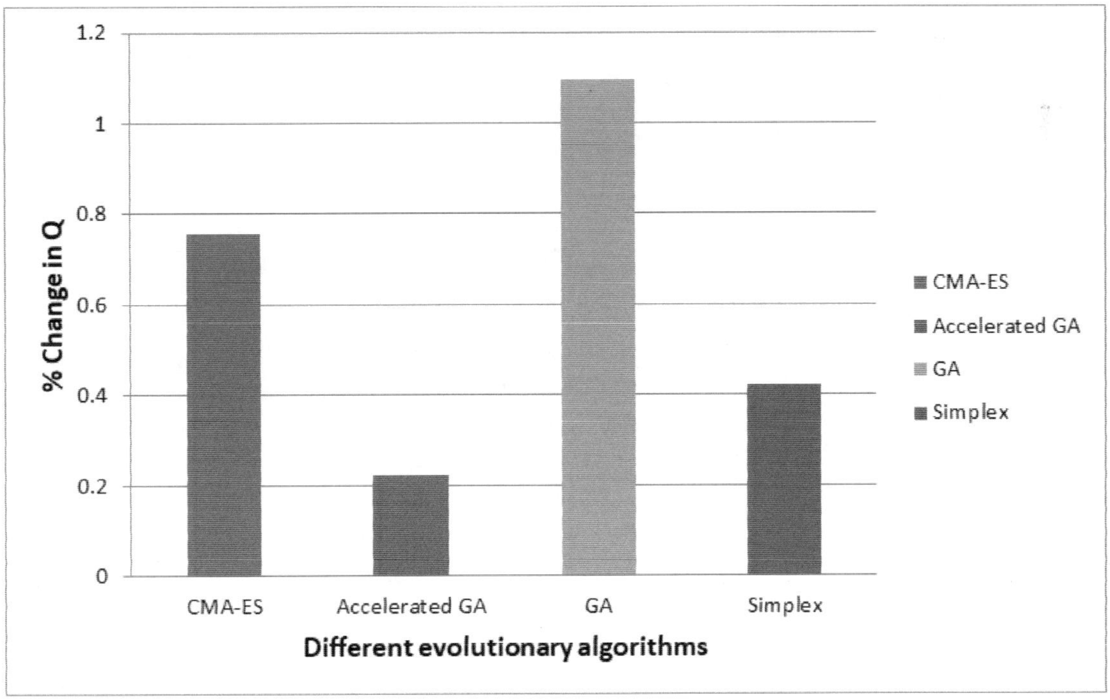

Figure 9. % Change in Q (Multi-objective optimisation)

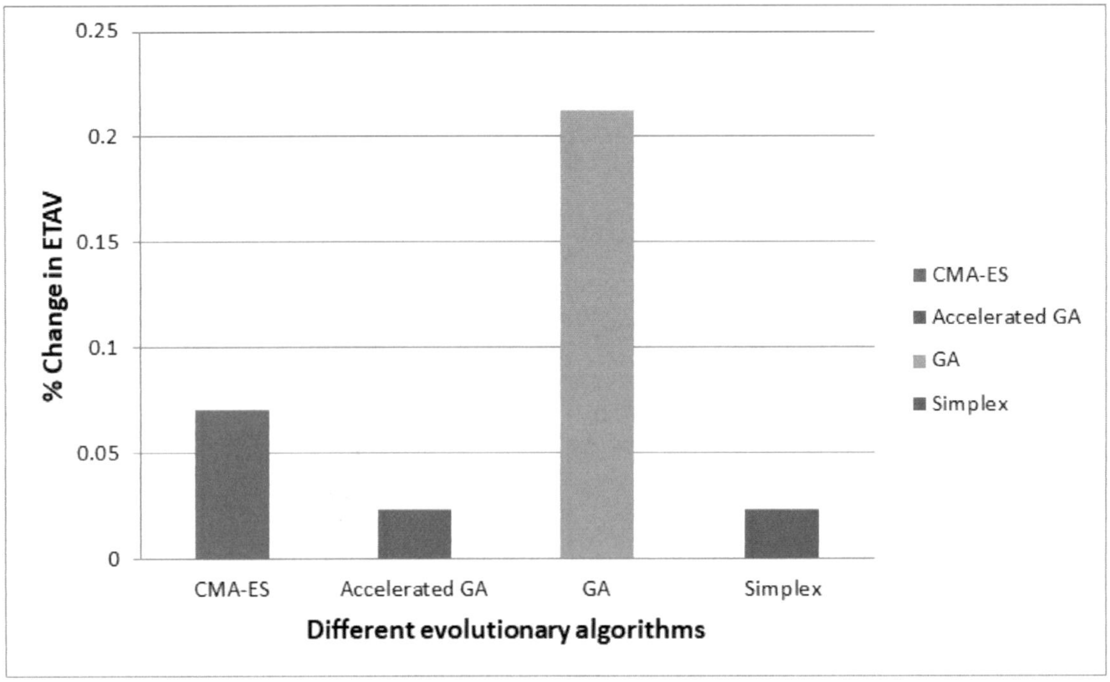

Figure 10. % Change in ETAV (Multi-objective optimisation)

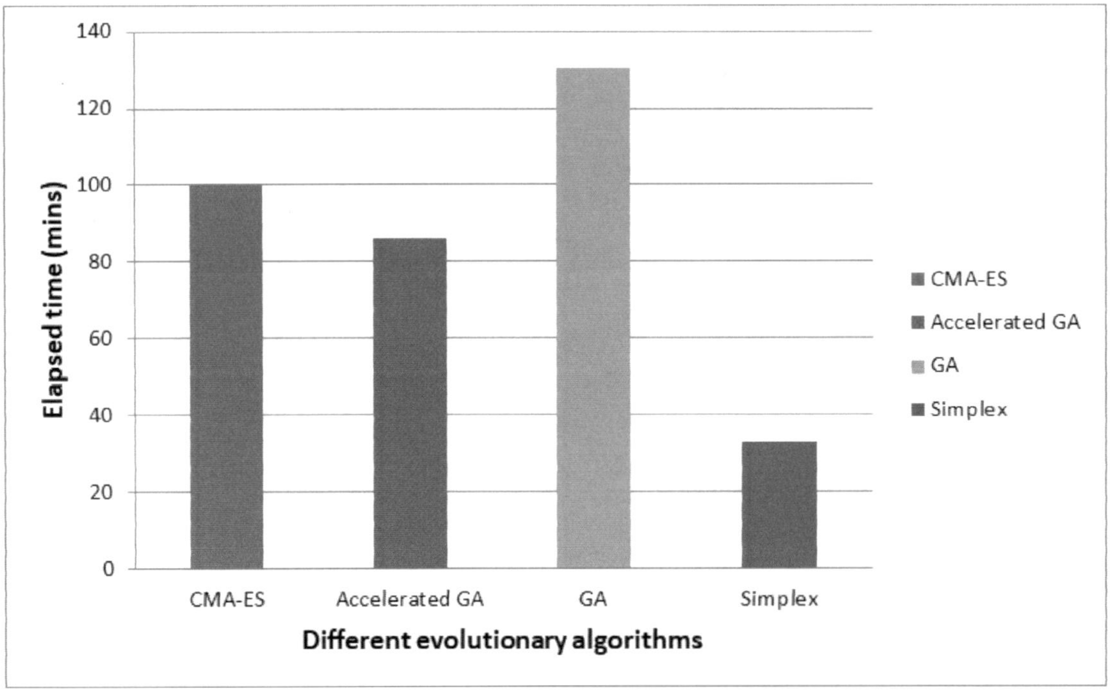

Figure 11. Time elapsed for multi-objective optimisation

Table 2. Seed size for optimisation

Search algorithm	Seed size
GA	0.033005270
CMA-ES	0.10439938
Accelerated GA	0.24159176

Therefore, the one with the smallest seed size will have the highest fitness value. In this case, the GA will have the highest fitness value followed by the CMA-ES and then the accelerated GA.

Table 3. Tabular column showing normalised performance map

MODEL	*SPC*	*Q*	*ETAV*
KPCL MODEL	1.120	0.989	0.998
CMA-ES	1.009	0.997	0.999
ACC. GA	1.017	0.991	0.998
GA	1.000	1.000	1.000
SIMPLEX	1.001	0.993	0.998

The normalized performance map after the optimisation is shown in table-3 in comparison to the original KPCL model of screw compressor. The specific power consumption is normalized with the one having the minimum value and the remaining two i.e flow and volumetric efficiency are normalized with the one having the maximum value.

On account of the optimisation performed here, the optimum parameters obtained from any of the evolutionary algorithms shows that they are almost in the same range such as optimum oil injection angle being in the range of (40-65) °, the oil injection temperature in the range of (30-40) °C, the oil port diameter in the range of (6-7) mm, pressure in the oil reservoir in the range of (5-6.5) bar and the oil viscosity in the range of (1-2)*E-06 m^2/s. This signifies that injecting oil at a lower temperature with lower oil viscosity and injecting earlier will result in improvement of performance by more than 5% as it will lead to near isothermal efficiency which is indeed a good choice of compressor. Thus from the screw compressor block point of view these optimum parameters are the best suited ones.

However, from the system design point of view there are other important conditions which need to be incorporated as constraints to the optimisation algorithms for reliable performance of the compressor.

For better understanding a sensitivity analysis has been shown in figure-12 to understand the effect of parameters on the objective functions. The analysis clearly shows that the major parameters which are contributing to the significant amount of change in the performance of the 22 kW twin screw air compressor are the oil viscosity followed by the oil injection temperature. To achieve an improvement in specific power the oil viscosity needs to be drastically reduced. But, there is a limitation on the oil viscosity to keep the bearings operating smoothly, provide an feasible bearing life and form a film for bearing lubrication.

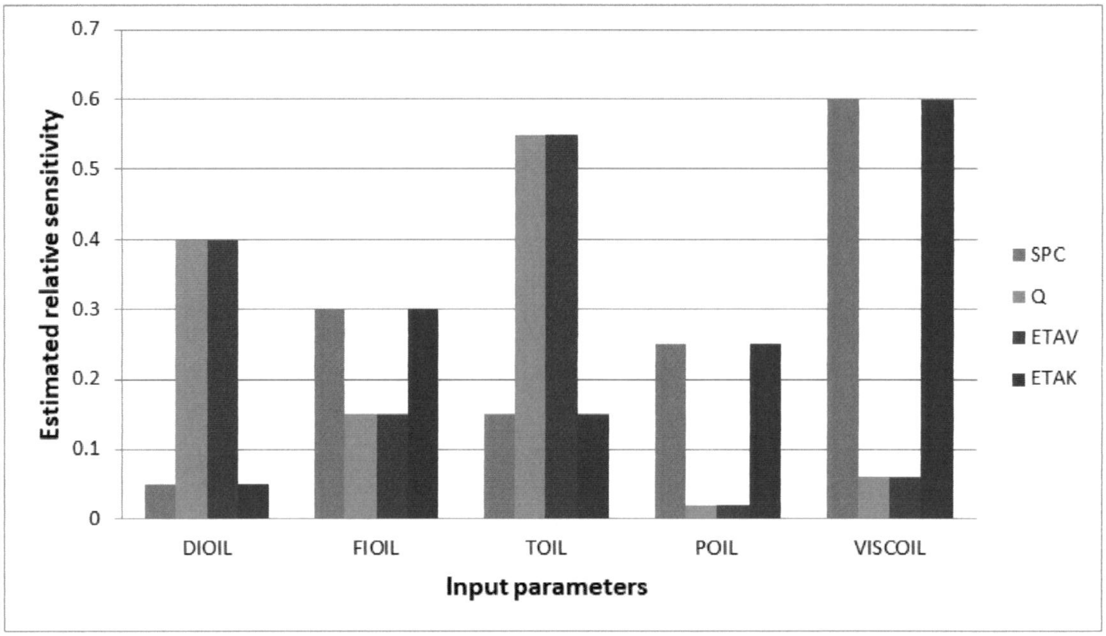

Figure 12. Sensitivity comparison for output vs. input

According to the bearings manufacturer for the bearings to work properly the viscosity ratio i.e., the kappa value must be greater than or equal to one. Kappa is the ratio of the actual oil viscosity to the minimum viscosity required to support a dynamic load within a bearing to provide full film lubrication. The kappa value for the current KPCL model is already at the verge of limit provided by the manufacturer thus using an oil with lower viscosity is not recommended from bearings point of view.

Lastly in this optimisation context a lower oil injection temperature would be desirable from the compressor block point of view but considering the system design point of view a lower oil injecting temperature signifies the temperature difference between the inlet and outlet of the compressor would increase. The desirable discharge temperature from the compressor is in the range of (80-90)°C. Thus, as the temperature difference increases the heat load on the cooling unit also increases which is an important consideration from system point of view. Therefore, the optimum oil injecting temperature should be in the range of (40-60)°C.

4. Conclusions

In the present paper, the optimisation of a 22 kW oil-injected twin screw air compressor using the different evolutionary algorithms has been carried out. Firstly, the SCORG model is set up in good agreement with the experimental results. The batch file of this model is linked to the GT-SUITE model of the screw compressor. Secondly, a GT-Process Map for the model is generated with the GT-ISE interface, which uses the SCORG thermodynamic calculation in the background for performance optimisation. Finally, the design optimiser is executed from the GT-ISE interface by providing the limits for the variation of the input parameters and defining the search algorithm. The input parameters considered in this case are the oil parameters as the oil plays a vital role in the oil-injected twin screw compressor performance. The output functions are the integral parameters used for the performance comparison of these machines such as specific power consumption, adiabatic efficiency, volumetric efficiency, and flow. Different

search algorithms have been made to run one by one independently for optimisation.
A comparative analysis has been carried out for both the single as well as the multi-objective optimisation. After the analysis, it is observed that the simplex algorithm converges the fastest and provides almost the same range of optimum parameters as other evolutionary algorithms. However, GA is an evolutionary global search algorithm recommended for all problems having medium to high complexity. The seed size is the smallest for the GA, thus having the highest fitness value.
In the future scope it is recommended that since the optimum oil parameters obtained in this context are good from the screw compressor block point of view but from the system design point of view other design and manufacturing constraints need to be incorporated to the algorithms for a feasible performance. Therefore, these conditions have to be applied for the best screw compressor optimisation algorithms, and then the simulation needs to be performed, which needs to be validated experimentally.

References

[1] *International Energy Outlook- US Energy Information Administration, 2021* https://www.eia.gov/outlooks/ieo/ (accessed Mar. 15, 2022)
[2] Fleming J S, Tang Y and Anderson H 1994 *Optimisation Techniques Applied to the Design of a Refrigeration TiN Screw Compressor* (United States, Purdue University) In International Compressor Engineering Conference paper 1049
[3] You C X, Tang Y and Fleming J S 1996 *Optimum Rotor Geometrical Parameters in Refrigeration Helical Twin Screw Compressors* (United States: Purdue University) In International Compressor Engineering Conference paper 1074
[4] Stosic N and Hanjalic K 1997 *Development and optimisation of Screw Machines With a Simulation Model - Part I: Profile Generation* (London: City University) ASME Transactions Journal of Fluids Engineering Vol 119 p 659
[5] Berlik S and Helpertz M 2002 *Optimising Screw Machines using an Evolution Strategy with Extended Creation Scheme* (Germany: University of Dortmund) In IEEE International Conference on Systems, Man and Cybernetics Vol 5, pp 6-pp
[6] Stosic N, Smith I K and Kovacevic A 2003 *Optimisation of Screw Compressors* (London: City University) Journal of Applied Thermal Engineering 23 pp 1177-1195
[7] Stosic N, Smith I and Kovacevic A 2005 *Screw compressors: mathematical modelling and performance calculation* (London: City University) Springer Science & Business Media
[8] Hauser J, Bruemmer A and Kauder K 2008 *Rotor profile generation and optimisation of screw machines using NURBS* (United States: Purdue University) In International Compressor Engineering Conference, paper 1861
[9] Herlemann S, Hauser J and Henning N 2014 *Optimisation of rotor profiles for energy efficiency by using chamber-based screw model* (Germany: University of Dortmund) In International conference on Screw machines pp 211-224
[10] Zhang C, Gao J, Fu J, Li Y and Lin J 2019 *Research on structural parameter optimisation and thermal performance of twin-screw expander* (China: Harbin Institute) International Journal of Low-Carbon Technologies 14(3) pp 386-393
[11] PDM Analysis Ltd. 2014 *SCORG - screw compressor rotor grid generation* (London: City University) Software Package for Design and Analysis of Positive Displacement Machines Retrieved from https://pdmanalysis.co.uk/scorg/
[12] Gamma Technologies LLC 2022 *GT-SUITE* (Illinois: USA) Multi-Physics Software for Design Optimisation Component Analysis and Root Cause Investigation Retrieved from https://www.gtisoft.com/
[13] Basha N 2021 *Numerical analysis of oil injection in twin-screw compressors* (Doctoral dissertation, City, University of London).

Acknowledgments

We gratefully thank Kirloskar Pneumatic Company Limited, Pune, India for sponsoring this research. We also want to acknowledge the Centre for Compressor Technology, City, University of London for their continued guidance and support in this research.

IOP Conf. Series: Materials Science and Engineering 1267 (2022) 012022 doi:10.1088/1757-899X/1267/1/012022

Experimental and numerical analyses of the thermodynamic and mechanical performance of an oil-injected and economized 4/6 twin-screw compressor

A Saravana[1*], H Liu[1], E A Groll[1] and D Ziviani[1]
[1] Ray W. Herrick Laboratories, School of Mechanical Engineering, Purdue University West Lafayette, 47907-2099, USA
E-mail: saravan0@purdue.edu, liu1460@purdue.edu, groll@purdue.edu, dziviani@purdue.edu

Abstract. To advance oil-injected twin-screw compressors, it is necessary to understand and model the complex physical phenomena occurring during the compression process (e.g., mass and heat transfer mechanisms) as well as analyze the mechanical behavior of the compressor (e.g., rotordynamics, bearing loads, variation of clearance gaps). To validate these models, the in-chamber compression process as well as mechanical loads need to be accurately measured. This paper presents a comparison between experimental results and numerical modeling of a 4/6 oil-injected twin-screw compressor with slide-valve part-load modulation and economization. The compressor has been equipped with high-frequency pressure sensors, load-cells at the bearings and torque sensor on the main rotor. Bearing forces are analyzed to quantify and validate the experimentally obtained axial force data. The validated model can later be used to analyze the mechanical performance of the compressor, such as bearing loss computation. This study also discusses isentropic and mechanical efficiencies computed for the various data points.

1. Introduction

Oil injected twin-screw compressors are widely used in various industrial applications due to their ability to operate at a wide range of operating conditions (mass flow rate, pressure ratios, volume ratios). However, based on the lobe configurations and operating conditions, the rotors are subjected to radial and axial compression loads that must be compensated by bearings. Therefore, it is important to analyse the compression loads, including the axial and radial forces, torque, and moments with both theoretical and experimental methods.

There have been several attempts to compute compression forces and moments on twin-screw compressors and thereby calculate bearing loads and frictional losses. Fujiwara *et al.* [1] showcased a model to compute rotor loads such as torque and moments using gas pressure loads and sealing lines between the male and female rotors. They also presented a method to compute area and volume curves using 2D profile projection, which serves as the basis for bearing load computations. This idea was explored in more detail in Adams [2], where he used an analytical approach to compute bearing loads and geometrical characteristics. Rotor geometrical characteristics such as wrap angles, lead etc. were used to determine the 3D profile analytically which were later projected onto 2D planes, and the loads were computed. A simplified method was presented in Zhou *et al.* [3] where the pressure loads were assumed to be constant and acting evenly on the rotor. However, since in a realistic scenario gas loads are always varying across the rotor, this assumption sacrificed the accuracy of the results. Additional studies on computation of gas loads on twin-screw compressors were conducted by You [4], where he examined the effect of geometrical characteristics on rotor loads. Wu and Tran [5] implemented a numerical model where the rotor profiles were given as point coordinates and rotor loads were calculated using a 2D surface mapping approach. With respect to experimental investigations, Hou *et al.* [6] presented an experimental method to compute axial loads based on measured in-chamber pressure data as well as load-cells installed at the bearings to capture positive and negative components of the forces. The actual axial loads were estimated by applying a force balance equation to the male rotor.

Despite numerous studies on load computation of twin-screw compressors, in-depth experimental investigations on fluid and mechanical losses are still limited. To this end, a 4/6 oil-injected twin-screw compressors was previously instrumented (Salts *et al.* [7]) to collect experimental data at both full and part-load conditions as well as with and without economization. The twin-screw compressor was equipped with axial force cells, static and dynamic pressure transducers, temperature measurements and torque sensor at the male rotor shaft. Oil injection flow rate, pressure and temperature were also monitored. Such experimental approach allowed to obtain indicated diagrams at various operating conditions and validate the compressor mechanistic model. The methodology used to process the data, align pressure traces, and compute the indicated power was described by Saravana *et al.* [8].

In this work, the mechanical loads obtained from the axial force cells have been analysed and numerically validated. Dynamic pressure data obtained from the experiments can be used to compute rotor loads through force decomposition and 2D surface area projections, which can be used for a deeper analysis on bearing losses. This study presents an experimental method to measure and validate axial forces, numerical method for force decomposition based on pressure data and 2D surface area projections for all the working chambers, and reports estimated mechanical efficiency at different operating conditions (i.e., pressure ratios and volume ratios)

2. Twin Screw Compressor Model

2.1 Geometry model

A twin-screw compressor with a 4/6 configurations is modelled and tested as part of this study. The compressor is oil injected with an economization port, and the working fluid is Nitrogen. Specific geometry parameters of the rotors are reported in Table 1. Additional information regarding the rotor profiles can be found in a previous study conducted by the authors [7,8].

Table 1. Screw rotor main geometric parameters

Parameter	Value
Theoretical displacement	0.025599 m^3/rev (0.90402 ft^3/rev)
Compressor volume ratio	2.0-4.1
Rotor wrap angle, γ	325°
Male rotor lead (right hand helix)	0.752 m (29.6063 in)
Interlobe nominal clearance	0.00005 m (0.002 in)
Axial clearance	0.000165 m (0.0065 in)
Radial clearance	0.000229 m (0.009 in)
Male lobes, z_1	4
Female lobes, z_2	6
Center distance, $A_{O_1O_2}$	0.222 m (8.74 in)
Length to male rotor diameter ratio, L_R/D_m	2.40

2.2 *Test Matrix*

The test matrix consisted of various operating conditions with a wide range of parameters, including suction and discharge pressure, volume ratio, load capacity dictated by the slide valves and slide stops as well as rotational speed. These parameters control the compression process and affect the bearing loads as will be shown later. Table 2 reports Test #11 and #16 as a reference case to showcase the calculations. Although this study presents validation on only two cases, other data points have been summarized to better describe the experimental conditions. The table also summarizes additional testing conditions employed to map the behaviour of the compressor.

Table 2. Experimental operating conditions

Test #	Suction/Discharge Pressure (psi)	RPM	Volume Ratio					Load (%)		
2	25/175	1800			4.1					100
3 4 12 6 5	25/175	3600	2	3.5	3.5	4	4.1			100
9 7 10 8 **11**	**40/125**	**3600**	2	2.3	3	3.5	**4.1**			**100**
13 14 15 **16**	**40/125**	**1800**	2		3	3.5	**4.1**			**100**
17 18	40/125	1800		2.3				50	75	
19	40/175	1800		2.7						100
20 21 22	40/225	1800	2	2.3			4.1			100
26 24 25	60/175	1800		2.5	3.1		4.1			100
1 0	60/225	1800	2						75	100

2.3 *Mechanistic force analysis*

The gas pressure loads act normal to the surfaces of the rotors. The surface of the rotor under pressure load is however constrained by the lobe tip sealing curves or the inter-lobe sealing curves depending on the stage and angle of rotation, as illustrated in Figure 1. The pressure loads that act normal to the sealing boundary formed between the lobe tips or at inter-lobe contact of male and female rotor, need to be decomposed into radial and axial components due to the 3D nature of the rotor surface.

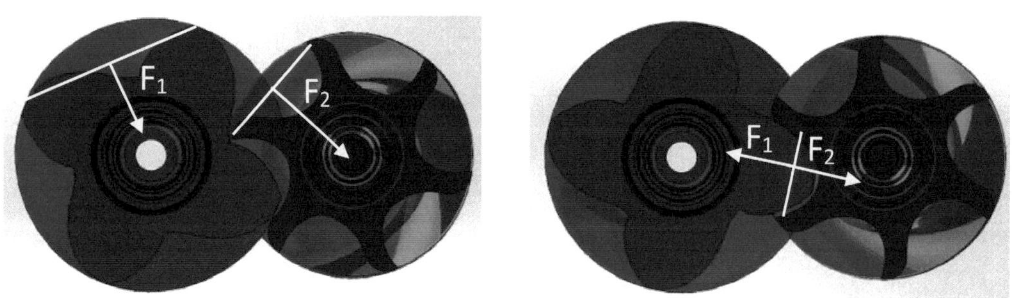

a) b)

Figure 1. Simplified schematic of pressure loads acting normal to the **a)** lobe tips and **b)** inter-lobe seal line (reproduced from [2])

Figure 2 displays the boundary sealing curves for the male rotor for a particular compression chamber on X and Y planes. These are obtained from a 2D projection of 3D sealing curves obtained from profile generation as will be shown in later sections. As it can be seen in Figure 2, the sealing boundary for a single chamber consists of the male rotor lobe tip sealing line and the inter-lobe sealing line between the male and female rotor lobes as the gas approaches the compression pocket between the male and female rotors. This boundary translates as the rotors rotate and therefore is a function of rotation angle.

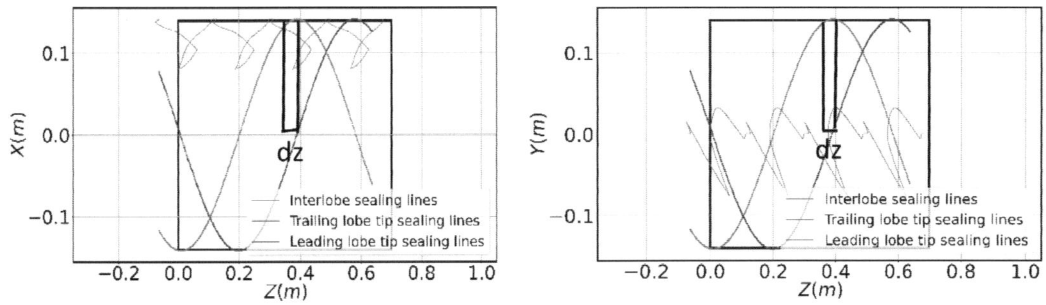

Figure 2. Translation of male rotor sealing line boundary with rotation angle along with cavity surface area projection on X and Y planes.

The surface area projections on X and Y between the sealing line boundary are used to compute area moment of inertias which are then used to compute torque, axial, and radial forces according to the below equations [5]. The following equations summarize a numerical method for computing bearing losses whose results will be demonstrated in a future publication.

$$M_{yz} = \frac{1}{2}\int_0^{L_R} y^2\, dz \tag{1}$$

$$M_{xz} = \frac{1}{2}\int_0^{L_R} x^2\, dz \tag{2}$$

$$T_g = \sum_{i=0}^{z_1}\left(P\left(M_{yz} + M_{xz}\right)_{\Phi=\frac{2\pi i}{z_1}} + P\left(M_{yz} + M_{xz}\right)_{\Phi=2\pi\left(1+\frac{i}{z_1}\right)} \right) \tag{3}$$

where M_{yz} and M_{xz} are area inertial moments on the Z-Y and Z-X planes as shown in Figure 2 respectively and L_R is the length of the rotor in Z coordinate. Equation 3 lists the calculation for the gas-induced total torque T_g based on pressure P and area moment of inertia. As the compression chambers between the rotors are periodical, it is sufficient to perform the computation for one chamber and apply a phase shift as shown, where Φ represents phase angle and z_1 is the number of male rotor lobes. Equation 4 and Equation 5 describe the computation of the axial and radial forces based on the computed torques and area moment of inertia,

$$F_{ga} = \frac{2\pi T_g}{l_d} \tag{4}$$

$$F_{gr} = P\sqrt{\left(\frac{M_{xz}}{S_x}\right)^2 + \left(\frac{M_{yz}}{S_y}\right)^2} \tag{5}$$

where F_{ga}, F_{gr} refer to axial gas and radial gas loads, l_d is the helical lead of the screw rotor (defined as the axial length of helix in one complete turn), S_x and S_y are the centroid distance of the sealing boundary along the X and Y axis, respectively.

Since the sealing boundary is a function of rotation angle, there is a functional dependance on the area moment of inertias and the rotational angle. Pressure data from the transducers as a function of crank angle, for this experimental study, has been previously obtained and analyzed [8]. Therefore, the force and torque loads can be computed as a function of rotation angle using experimentally obtained pressure loads. This study describes the basic equations employed to perform necessary computations. A spline approach is used to perform the above integration whose process is not described in this paper but can be found in Wu and Tran [5].

3. Experimental Setup

There were three different types of sensors employed within the experiment. A set of five pressure transducers to measure pressure values at various points in the compression chamber, one torque sensor and a set of three compression load-cells at the discharge to measure axial load make up the experimental setup [7]. As shown in Figure 3, the load-cells are equally placed around the thrust bearings at 120° apart from each other and are under a spring pre-load such that they can measure forces in both the positive and negative direction. Table 3 lists all the different sensors used within the experiment, their types, and uncertainties.

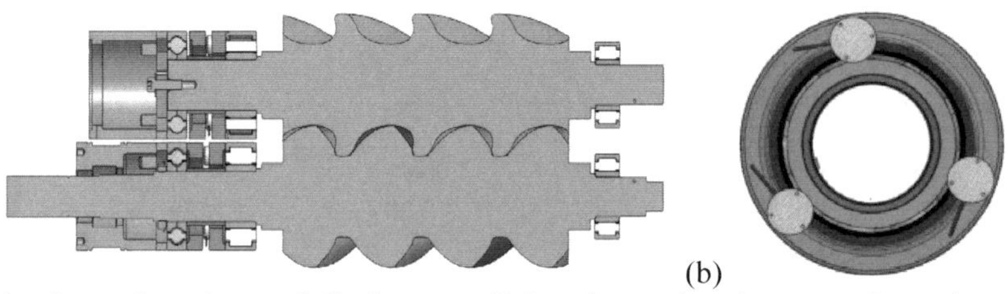
(a) (b)

Figure 3. a) Side view of suction and discharge radial and trust bearings; **b)** close view of axial load-cells on the thrust bearing (obtained from [7])

Table 3. Summary of all sensors used and their properties (obtained from [7])

Measurement	Type	Accuracy
Dynamic Pressure	KULITE XTEL-190SM	±0.5% FSO (Max.)
Axial Force	Tacsis XLCH–005T-TS	±0.25% BFSL
Torque Meter	Lebow 1807-100K	±0.1% FS
Optical Tachometer	SPSR-115/230, 6150-020	

The sampling rate is determined by the compressor RPM and sensor with the shortest angle of rotation between opening and closing angle. For example, if sensor #1 has 50° of rotation between its opening and closing angle and the compressor operates at 6000 RPM, the minimum sampling rate of 1440 Hz is required to ensure that the sample is taken every 25° of rotation completing roughly 2 samples per working chamber. In this study, the axial force data obtained from the axial force sensors is visualized and validated. Trends of mechanical performance parameters such as mechanical efficiency is also plotted as a variation of pressure and volume ratios in subsequent sections. As described earlier, a more in-depth computation of mechanical losses can also be done based on the axial force data and surface area projections, but this will be demonstrated in a future publication. Pressure data visualization, thermodynamic and frictional losses have been previously analyzed by the authors [8].

4. Data processing

The detailed post processing and resolution of experimental force data, comparison of experimental results with numerical, rotor profile reconstruction and surface area plot generation are reported in this section.

4.1 Force resolution

Axial force data is measured from all the 3 sensors for each of the rotors. Force data from these sensors are categorized according to their time periods and sampling rates and each period is overlaid onto each other. Accurate force values can be obtained by averaging force results across all the periods for each sensor as in Figure 4. As the three sensors for each rotor are positioned at 120° apart on the axial force component, they measure the same force acting in the same direction but different positions. Accurate force data acting in the axial direction can be obtained by computing the averages of three of the sensors as in Figure 6.

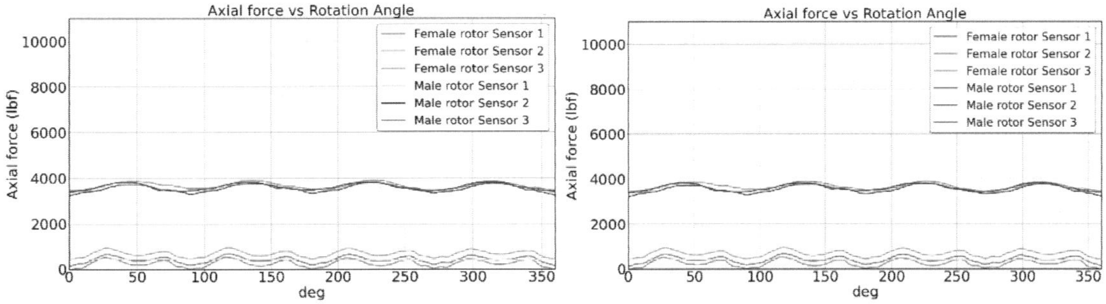

a) b)

Figure 4. Experimental measurements depicting period averaged force data for each sensor on male and female rotors for a) test case #11 b) test case #16

4.2 Indicator diagram processing

Pressure data from the 5 pressure transducers are averaged according to a specific sampling rate to produce mean pressure vs crank angle curve. The averaged pressure data from each of the sensors are further averaged and put together to produce the pressure vs crank angle curve. The pressure vs crank angle curve is compared with the volume vs crank angle curve and the indicator diagram is generated. Figure 5 displays the indicator diagram obtained from the analysis. Saravana et al. [8] describes this process and thermodynamic analysis based on the generated indicator diagrams.

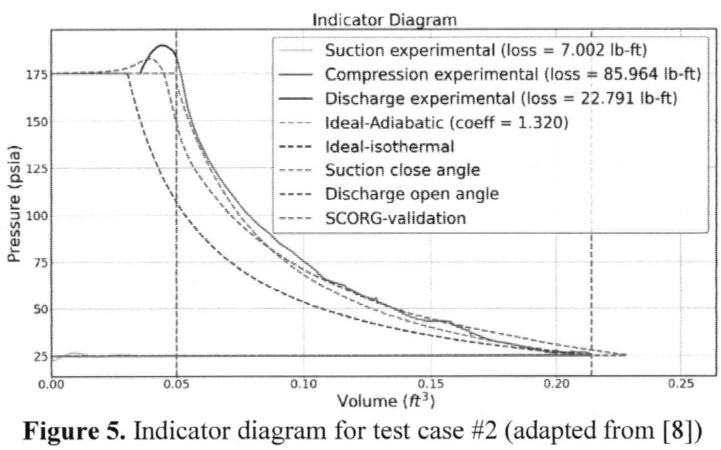

Figure 5. Indicator diagram for test case #2 (adapted from [8])

5. Results and Discussion

5.1 Model Comparison

The resultant axial force obtained from the period averaged force data for each of the sensors are compared with the numerical force results from SCORG [9]. Figure 6 describes the comparison between the numerically obtained forces from SCORG and the experimentally obtained force data. As can be seen, the male rotor forces are well predicted but the female rotor numerical and experimental forces are different. This could be due to missing axial forces on the shaft shoulders which are not accounted for in the numerical model, or due to constant offsets in the experimental data arising from the sensors.

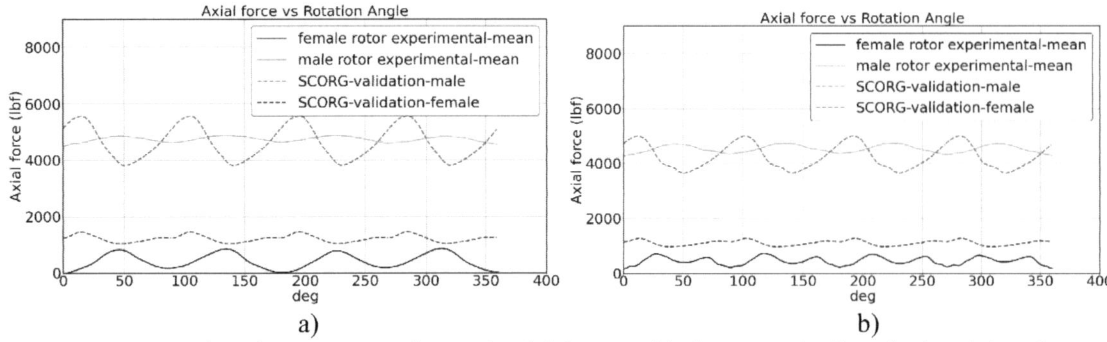

Figure 6. Comparison between experimental axial forces with the numerically calculated data from SCORG for a) test case #11 b) test case #16

5.2 Surface area plot generation

The 2D profile coordinates for one lobe of the male and female rotors are revolved and transformed in the form of a helix to create a 3D projection of one groove profile, as shown in Figure 7. The figure shows the leading and trailing lobe tip for the male and female rotors. The rack curves for both rotors are rotated and translated until the wrap angle and the length of the rotor is reached. The blue and green markers highlight the lobe tip sealing lines for the rotors. The continuous traces are then projected onto the 2D planes to run the surface integrations.

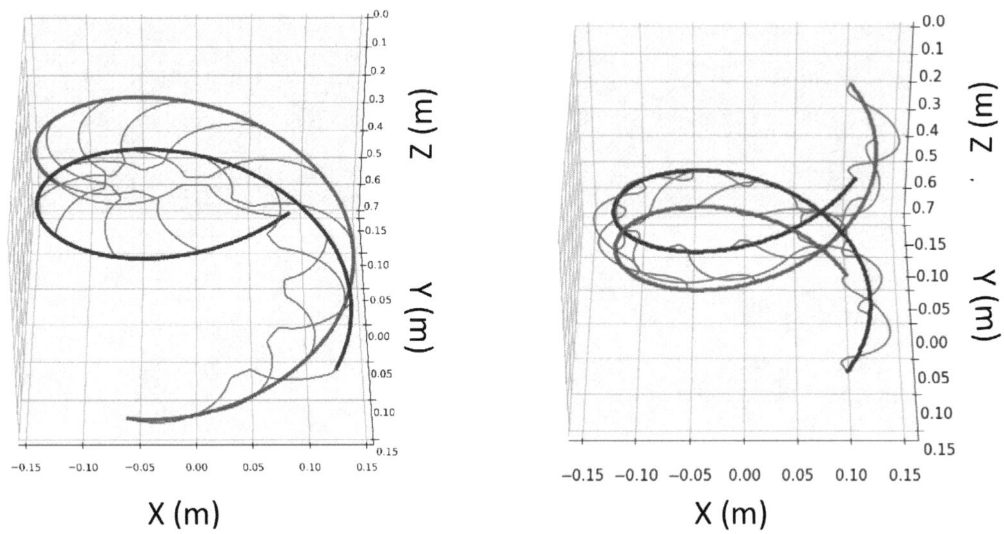

Figure 7. 3D construction of **a)** male rotor profile **b)** female rotor profile

The lobe tip sealing lines from the 3D rotor reconstruction is projected onto the XZ and YZ planes to obtain a 2D projection of the groove profile. Figures 8 and 9 displays the YZ and XZ projection for the male rotor. The rectangle is modeled using the height and width based on the diameter and length of the rotors respectively, and the lobe tip sealing lines propagate through the compression chamber depending on the compressor crank angle. The rectangle depicts the entire rotor whose flow volume is constrained by lobe tip sealing lines and inter-lobe sealing lines. These together form the active compression chamber. The lobe tip sealing lines translate along with the rotation of the screw rotors which is shown in Figure 8 and 9.

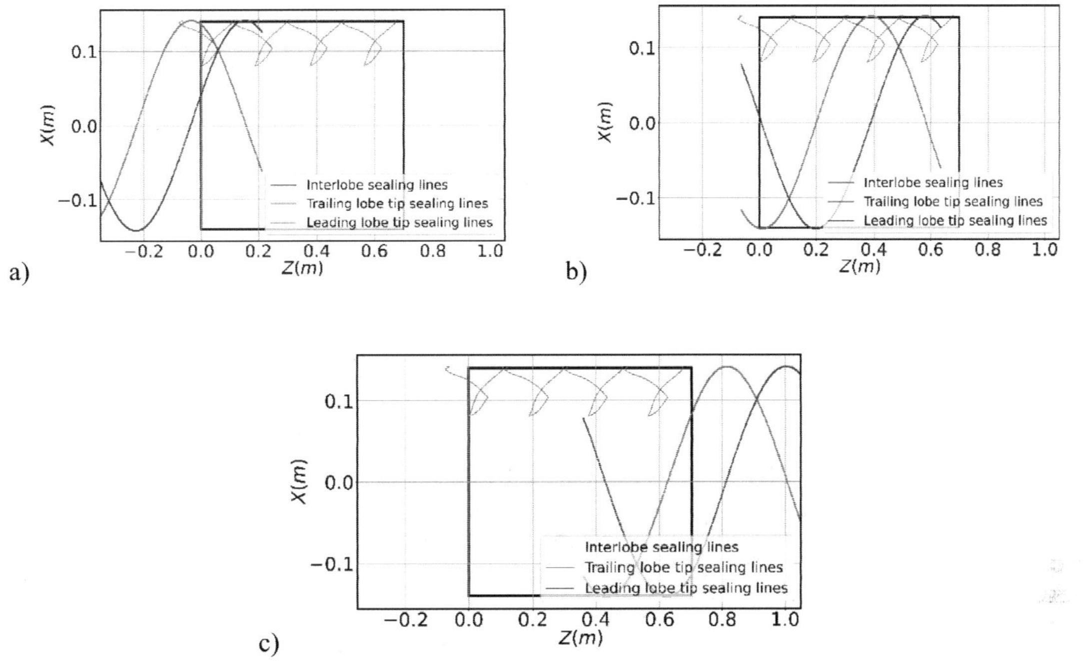

Figure 8. 2D projection of lobe tip sealing lines for male rotor on the YZ plane at **a)** 100 crank angle degrees **b)** 300 crank angle degrees and **c)** 500 crank angle degrees

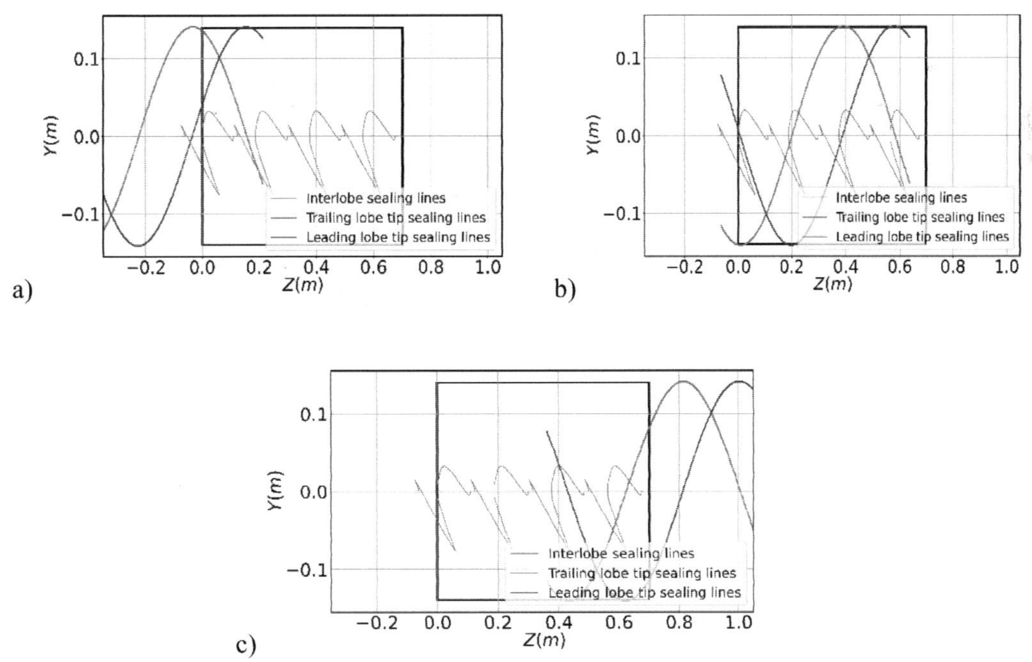

Figure 9. 2D projection of lobe tip sealing lines for male rotor on the XZ plane at **a)** 100 crank angle degrees **b)** 300 crank angle degrees and **c)** 500 crank angle degrees

IOP Conf. Series: Materials Science and Engineering 1267 (2022) 012022 doi:10.1088/1757-899X/1267/1/012022

The goal of 2D projection is to derive the variation of surface areas (XZ and YZ planes) of the groove profile with crank angle degrees. This is done by computing the area between the lobe tip sealing lines and inter-lobe sealing lines by integrating the areas in between the plots for each crank angle.

5.3 Surface area curves

Figure 10 displays a complete evolution of all the grooves over one rotation of the male rotor. The control volume C1 is taken as reference. The groove control volumes C2, C3 and C4 (solid lines) are the remaining grooves phased by β, where β is dependent on number of lobes of the male and female rotors z_1 and z_2 and wrap angle γ, given by equation 6

$$\beta = \frac{z_1}{z_2} . \gamma \tag{6}$$

After the grooves C2, C3 and C4 complete a revolution, they start wrapping from the discharge side of the rotor (dashed lines). This model with phase shifts handles all the grooves at once without any excess complication by using simplification through symmetry. The total surface area plots are generated by taking a sum of all the groove surface areas from C1 to C4 and this is displayed in Figure 11.

The surface area projections on the XZ and YZ planes can be resolved along with the pressure vs crank angle curves to compute radial forces using the equations presented earlier. These surface area plots are numerically modeled separately from SCORG.

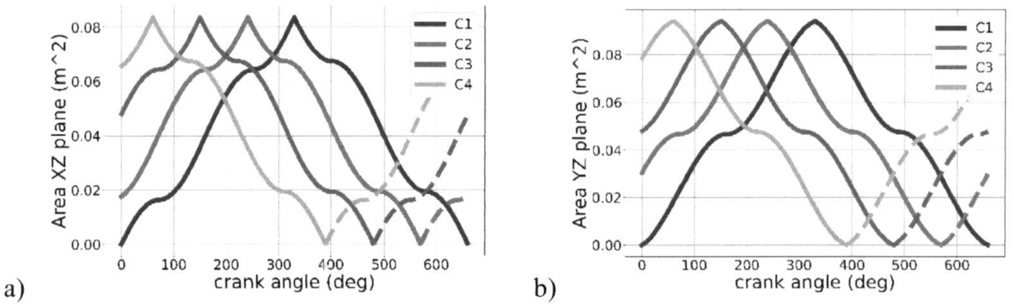

Figure 10. Surface area of all the groove control volumes plotted over each other and projected on the **a)** YZ and **b)** XZ plane

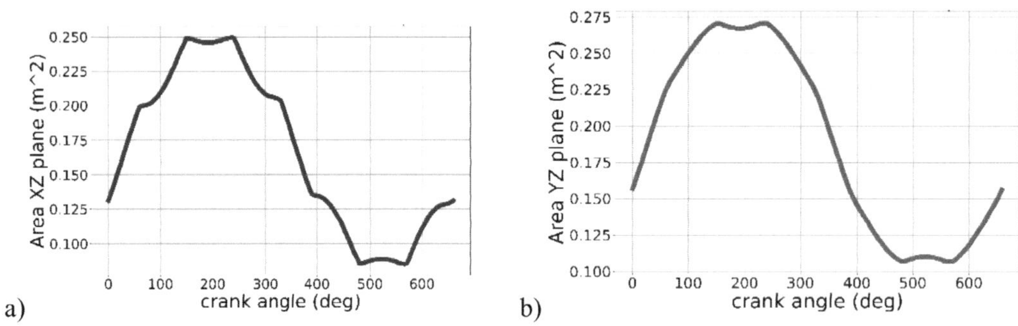

Figure 11. Total surface area of all the groove control volumes projected on the **a)** YZ and **b)** XZ plane

5.4 Mechanical and Isentropic Efficiencies

Mechanical and isentropic efficiencies can be analyzed to study mechanical and thermodynamic performance of the compressor [8]. These parameters are plotted as a variation of pressure and volume ratios as shown in Figure 12.

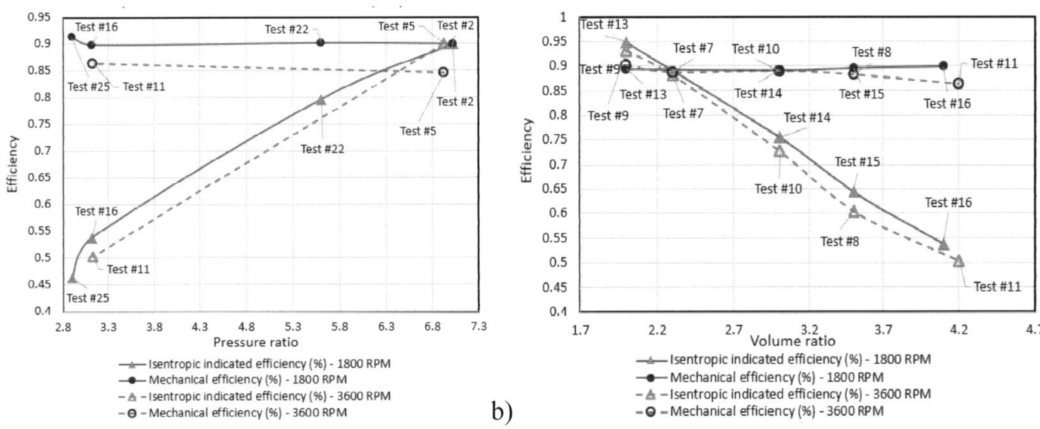

a) b)

Figure 12. Numerical estimations of isentropic and mechanical efficiencies as a function of **a)** pressure ratio with constant volume ratios (4.1) and **b)** volume ratios with constant pressure ratios (3.125)

A clear trend can be observed through the figures. Isentropic efficiency increases with pressure ratio and decreases with volume ratio. Mechanical efficiency, however, is dependent on the geometry and rotor clearances and remains constant. It can be noted that both efficiencies also decrease with increase in rotational speeds. This is because higher rotational speeds are associated with higher thermodynamic and mechanical losses.

6. Conclusion

In this paper, a comprehensive experimental and data processing procedure has been presented and compared with numerical results. Five pressure transducers, six axial force sensors and one torque cell sensor are used in the experiment. A post processing method for the axial force sensors is demonstrated and compared with numerical results from SCORG. The difference was significant for the female rotor. However, the male rotor showed a much better comparison with the numerical results. A method to use the rotor profiles to regenerate the lobe tip sealing lines and to compute surface area projections on the XZ and YZ planes is also presented. These surface area projections could be used to compute axial forces and radial forces using a numerical method, as shown in SCORG, which can be correlated with the experimental data. Although the correlation for only one data point is presented, this process can be repeated for the other experimental data points as well. A future study would include computation of bearing losses using the surface area projections and pressure data obtained from the indicator diagram and mapping the compressor based on the thermodynamic and mechanical losses.

7. References

[1] Fujiwara M, Kasuya K, Matsunaga T and Watanabe M 1984 Computer Modeling for Performance Analysis of Rotary Screw Compressor *International Compressor Engineering Conference* Paper **503** https://docs.lib.purdue.edu/icec/503

[2] Adams G P 1997 Modelling and computer simulation of rotor chatter and oscillating bearing forces in twin screw compressors, *PhD. Dissertation*, Purdue University

[3] Zhou Z, Wang D, Zhu T and Mao J 1990 Analysis of the Applied Forces in Twin-Screw Refrigerant Compressors *International Compressor Engineering Conference* Paper **684** https://docs.lib.purdue.edu/icec/684

[4] You C X 1997 A theoretical study of rotor forces and torques in helical twin screw compressors, *PhD. Dissertation*, University of Strathclyde, Glasgow

[5] Wu Y R and Tran V T 2016 Dynamic response prediction of a twin-screw compressor with gas-induced cyclic loads based on multi-body dynamics *International Journal of Refrigeration* **65** 111128 https://doi.org/10.1016/j.ijrefrig.2015.12.002

[6] Hou F, Zhao Z, Yu Z and Xing Z 2017 Experimental study of the axial force on the rotors in a twin-screw refrigeration compressor *International Journal of refrigeration* **75** 155163 DOI:http://dx.doi.org/10.1016/j.ijrefrig.2017.01.002
[7] Salts N, Ziviani D and Groll E A 2019 Application of a generalized compressor modeling framework for simulating an oil-injected twin-screw compressor *IOP Conference Series: Materials Science and Engineering* **604(1)** 012009 https://doi.org/10.1088/1757 899x/604/1/012009
[8] Saravana A, Liu H, Groll E A and Ziviani D 2022 Experimental quantification of the thermodynamic and mechanical losses of an oil-injected and economized 4/6 twin-screw compressor *International Compressor Engineering Conference* Paper **1564**
[9] Analysis PDM 2022 SCORG Retrieved from https://pdmanalysis.co.uk/scorg/

Nomenclature

P	experimental pressure	psi	
γ	rotor wrap angle	deg	
β	groove phase angle	deg	
Φ	torque phase angle	deg	
$A_{O_1O_2}$	center distance	in	
z_1	number of male rotor lobes	$-$	
z_2	number of female rotor lobes	$-$	
L_R	rotor length	m	
D_m	male rotor diameter	m	
M_{yz}	area inertial moments in the YZ plane	ft^3	
M_{xz}	area inertial moments in the XZ plane	ft^3	$-$
T_g	torque	$lb - ft$	
F_{ga}	axial force	lb	
F_{gr}	radial force	lb	
S_x	centroid distance in x	ft	
S_y	centroid distance in y	ft	

Subscript

1	Male rotor	2	Female rotor
R	rotor	gr	radial
ga	axial		

Acknowledgement

The authors would like to thank Johnson Controls for the financial and technical support of this project.

Experimental research on influence of interstage oil injection in a dual-motor-driven two-stage screw air compressor

X Liao[1], C Wang[1*], H Yuan[1] and Z Xing[1]
[1]School of Energy and Power Engineering, Xi'an Jiaotong University, Xi'an, Shaanxi, PR China

E-mail: chuangwang@xjtu.edu.cn

Abstract. Energy saving for compressed air production is of great importance all over the world due to its great energy cost and the indispensable utility of compressed air in the industrial sector. In recent years, oil-injected two-stage screw air compressors, especially driven by dual motors, are becoming a promising alternative to other compression equipment for their extraordinary efficiency and high reliability in a wide range of applications. However, two-stage screw air compressors have a different and complicated oil injection system compared with single-stage compressors. And an appropriate setting of oil injection parameters might result in a considerable performance improvement of the compressor. In this paper, a dual-motor-driven two-stage oil-injected screw compressor was tested under the actual service conditions and control modes experimentally. The effect of interstage oil injection on the compressor performance was studied in detail by varying the interstage oil injection flow rate and rotating speed. The performance parameters such as specific power, volumetric efficiency and adiabatic efficiency were calculated from the measured data. The results indicate that an excess amount of oil exists in the second stage compressor and reducing the oil moderately could decrease the power consumption of the two-stage air compressor unit.

1. Introduction

Compressed air is a basic energy source and it is widely utilized in various industrial applications, including power tools, conveying systems, process control, energy storage, etc. Approximately 10% of the world's generated electrical power is consumed in the production of compressed air[1]. Thus, even a minor improvement in the energy efficiency of compressed air systems could lead to a substantial reduction in energy consumption. The compressor is the key component in the compressed air system, which is used for air compression and accounts for the majority of energy consumption of the whole system. In recent years, oil-injected screw compressors have become a promising alternative to other compressor types for their extraordinary efficiency and high reliability in a wide range of applications. The great energy cost and market share give strong incentives to designers to further improve the performance of oil-injected screw compressors.

Oil is injected into the working chamber of the compressor for three main purposes: lubricating bearings and rotors, sealing the gas leakage paths and cooling the air during compression. The introduction of oil substantially modifies the thermodynamic process in the working chamber, leading to a considerable improvement in efficiency. Nevertheless, the oil in the chamber also has negative effects. Oil with viscosity and density several orders of magnitude higher than compressed air fills the clearances between rotors and casing. The rapid rotation of rotors leads to the acceleration and friction of oil in the clearances, which contributes to additional mechanical losses. The oil-induced losses in screw compressors have been studied in detail by Deipenwisch and Kauder[2], and it is clear that an excess amount of oil would lead to undesirable mechanical losses and increase the power consumption of screw compressors. Given the opposite effects of oil, oil injection parameters, including the mass flow rate, temperature, viscosity and injection position, are crucial for the performance of screw compressors, and the highest efficiency could be achieved with an optimal choice of oil injection parameters.

In order to get a better understanding of the influence mechanisms of oil injection and optimize the oil injection parameters, many researchers have investigated the oil-injected screw compressor detailly in the past three decades. Stosic et al.[3] developed a mathematical model to simulate the working process of oil-injected screw compressors and analyzed the influence of various oil injection parameters. It was concluded that the oil temperature and oil injection position have a significant effect on the working process of the compressor, but the oil viscosity and oil/gas mass ratio have a surprisingly small effect. And a new modified oil injection system with a change of the nozzle diameter and position was applied and tested on a compressor, which results in a considerable decrease of specific power up to 7%. Fujiwara et al.[4] studied the flow and heat transfer characteristics between air and oil in the screw compressor and determined the heat transfer coefficient required for simulation from the experimental data. Peng et al.[5] investigated some processes related to oil injection in screw compressors experimentally by measuring the *p-V* indicator diagrams and visualizing the atomization effects at various oil injection parameters. It was discovered that an optimal diameter of oil injection ports exists, which could result in perfect oil atomization and enhance the internal cooling in the working chamber. Lee et al.[6] proposed a theoretical model to study the performance of oil-less and oil-injected screw compressors under different design parameters including rotor profile, geometric clearance, oil-injected angle, oil temperature and oil flow rate. It was concluded that lower oil temperature could decrease the compression work but increase the dynamic loss due to higher oil viscosity. Thus, an optimal oil temperature might exist, resulting in the lowest specific power. De Paepe et al.[7] studied the effect of oil atomization in an oil-injected screw compressor. It was found that lowering the oil droplet diameter and raising the oil flow rate could give a better heat transfer effectiveness, but it doesn't definitely give a considerable gain in specific work. And lowering the oil temperature gives larger gains than changing the oil flow rate. Basha et al.[8] studied the performance of oil-injected screw compressors by changing the oil flow rate with various injection port sizes. For lower discharge pressures (6.5 bar and 8.5 bar), the power consumption increases with the increase in oil flow rate. But for higher discharge pressures (10.5 bar and 12.5 bar), the power consumption has an opposite trend related to the oil flow rate. Basha et al.[9] also developed a CFD analysis on the oil distribution in a screw compressor chamber. It was shown that injecting oil through two separate ports, one on each rotor, could ensure a more uniform oil distribution in the working chamber than a single port, which leads to a reduction in discharge temperature by 30~35 °C and in specific power by 1.8%. He et al.[10] investigated the influence of oil flow rate in a variable speed screw compressor experimentally. As the oil flow rate increases, the adiabatic and volumetric efficiency both increase while the discharge temperature and specific power of the compressor decrease, which indicate better performance but a higher crisis of condensation of moisture. It was proposed that the optimal oil flow rate is limited by the discharge temperature at low rotating speed and is a trade-off among two efficiencies and specific power at high rotating speed.

Multi-stage compression with interstage cooling could achieve considerable energy savings compared to single-stage compression in the view of thermodynamic theory. And compressors with this technology have been widely used in some applications demanding a high-pressure ratio[11-14]. In recent years, two-stage compression technology is commonly applied in air compressors to realize better performance. Some researchers have studied the characteristic of two-stage air compressors. Hauser et al.[15] developed a chamber-based model to study the operational characteristics of an oil-free two-stage screw compressor. The simulated results including power consumption and volumetric efficiency of both stages matched well with the measured results. It was concluded that the intermediate pressure is a function of the rotating speed of compressors with a fixed gear ratio between the first and second stage compressors. A reduction in power consumption could be achieved by optimizing the intermediate pressure. Liu et al.[16] investigated the performance of a two-stage oil-free screw compressor under various operating conditions with a mathematical model. The results showed that the volumetric and indicated efficiency of the second stage is more sensitive to the operating condition than that of the first stage. Andres et al.[17] carried out a transient simulation on a gear-driven oil-free screw compressor by CFD, and the calculated flow rate and power consumption had a good match with the experimental measurements. Cipollone et al.[18] developed a comprehensive mathematical model to study the performance of a dual-stage sliding vane rotary compressor, which applied pressure swirled oil atomizers in each stage and an air to air intercooler between two stages. Overall energy-saving potential of up to 9.5% was demonstrated in the proposed dual-stage compressor system. Murgia et al.[19] compared the performance of conventional single-stage sliding vane rotary compressor, two-stage compressor with interstage cooling by an intercooler and by oil injection theoretically. It was proved that the proposed two-stage compressor with interstage cooling by oil injection could achieve an energy saving of up to 10%. An experiment was also carried out to study the interstage cooling effectiveness by oil injection, but only the second stage compressor was applied in the experimental setup.

As evident from the above literature review, the characteristic of oil-injected screw compressors has been investigated comprehensively and deeply in the past few decades. And researches related to two-stage air compressors are mainly aiming at the oil-free screw compressors with a fixed gear ratio and oil-injected sliding vane rotary compressors. To the best knowledge of the authors, no detailed research about the oil injection of dual-motor-driven two-stage screw air compressors, in which the compressor of each stage is driven by a motor independently, has been presented so far. In a two-stage screw air compressor, oil is injected into the intermediate pipe to cool the air/oil mixture flow discharged from the first stage and an efficient cooling could result in the reduction of the compression work and discharge temperature of the second stage. However, all of the oil injected in previous processes would enter the working chamber of the second stage compressor, which generally has a small size compared to the first stage compressor. Hence, an excess amount of oil might exist in the second stage compressor, causing extra mechanical losses and deteriorating the performance. Taking measures to reduce the oil flowing into the second stage might lead to an improvement in the two-stage compressor performance.

In this paper, an experimental test rig was established to investigate the performance of a dual-motor-driven two-stage screw air compressor. In order to acquire data close to reality, the compressor unit was tested under the actual service conditions and control modes. The performance parameters including volumetric efficiency, adiabatic efficiency and specific power were calculated to evaluate the performance of the two-stage compressor unit and each stage compressor. The effect of interstage oil injection on the compressor performance was studied in detail by varying the interstage oil injection flow rate and rotating speed. The results might provide some references for the performance improvement of two-stage screw air compressors in future development.

2. Experimental setup

2.1. Specification of compressor

In order to study the effect of oil injection in the two-stage screw compressors, a dual-motor-driven two-stage oil-injected screw compressor with a rated power of 22 kW was tested experimentally. The two-stage compressor consists of two screw compressors connected in series by an intermediate pipe. The compressor of each stage is driven by a permanent magnet synchronous motor coaxially and independently with a rated power of 11 kW. Main parameters of the tested compressor are listed in Table 1. The application of dual motors makes it possible to discard the conventional transmission gears in two-stage screw air compressors, and the rotating speed of each stage could be controlled respectively without the limitation of a fixed gear ratio between the first and second stage compressors. Thus, the noise and power losses caused by the transmission gears could be removed, and the accurate control of interstage pressure could be achieved by adjusting the rotating speed ratio of compressors. The rated speeds of the first and second stage compressors are 3100 rpm and 3600 rpm respectively, and both of them have a lobe combination of 5:6.

Table 1. Main parameters of the two-stage compressor unit.

Item	Unit	Value
Rated electrical power	kW	22
Rated volumetric flow rate	m^3/min	4.6
Rated discharge pressure	MPa(A)	0.8
Rated speed of first stage compressor	rpm	3100
Rated speed of second stage compressor	rpm	3600
Theoretical displacement per rotation of first stage	cm^3	1520
Theoretical displacement per rotation of second stage	cm^3	500

2.2. Compressor test rig

A compressor test rig was designed and constructed to investigate the performance characteristic of the dual-motor-driven two-stage screw air compressor in accordance with ISO 1217 standard. As shown in Figure 1, the experimental compressor test rig consists of compressors, air circuit, oil injection system and data measurement system. Picture of the compressor test rig is shown in Figure 2.

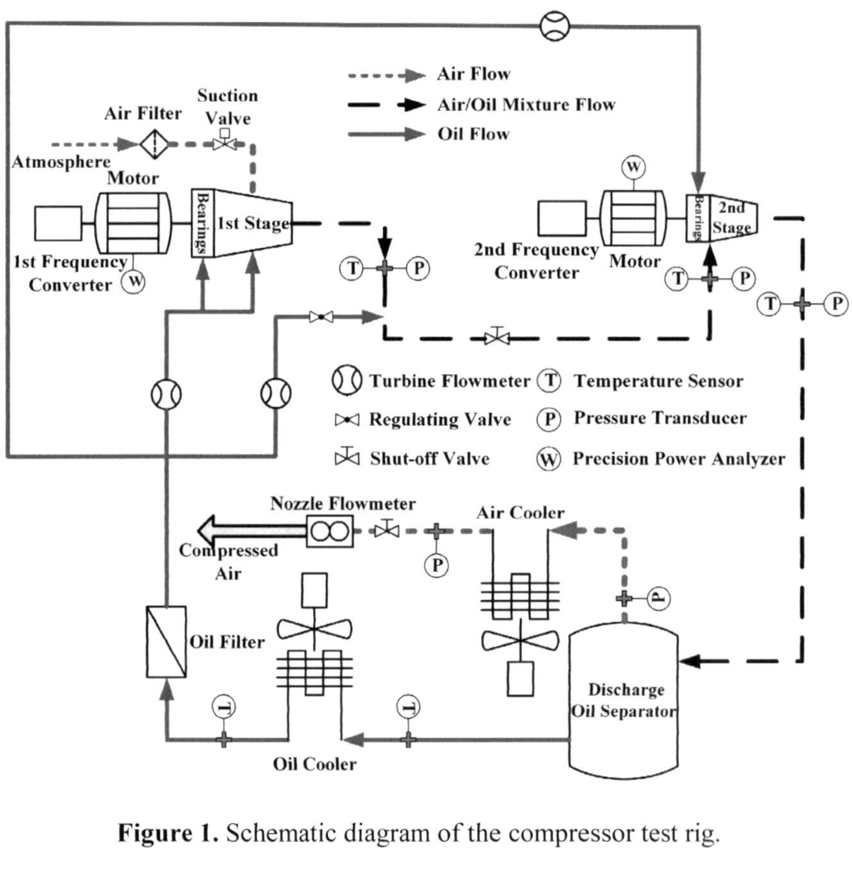

Figure 1. Schematic diagram of the compressor test rig.

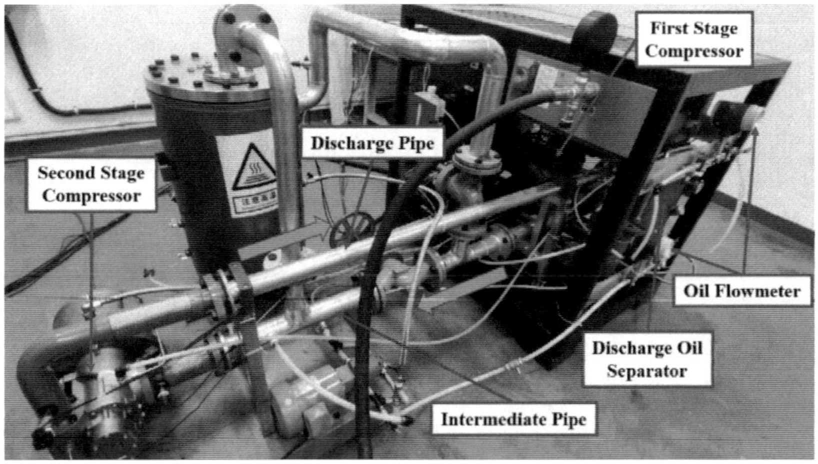

Figure 2. Picture of the compressor test rig.

The oil injection system includes a discharge oil separator, a fan-assisted oil cooler, an oil filter and regulating valves. Oil stored in the discharge oil separator is driven by the pressure differences between the oil separator and injection positions. Oil could be injected into the bearing chamber and working chamber of the first stage, the intermediate pipe and the bearing chamber of the second stage respectively after being cooled by the oil cooler and filtrated by the oil filter. The oil flow rate could be adjusted by varying the openings of regulating valves.

In order to analyze the performance of the compressor, the data measurement system is well instrumented and the measurement points are indicated in Figure 1. The volumetric flow rate of compressed air was measured with a standard ASME nozzle flowmeter as per ASME PTC 9-1970. The oil flow rate could be measured using turbine flowmeters on the oil pipes. The power consumption of the compressors was captured using a precision power analyzer provided by Yokogawa Co., LTD. The rotating speed of compressors was obtained by the frequency converters used to drive the motors. The static temperature at different positions was measured by the platinum resistance thermometer PT100 with an accuracy of class A and the static pressure at different positions was measured by the pressure transducers with a reasonable range. The atmosphere temperature and pressure were obtained by a thermometer with an accuracy of 0.1 °C and a barometer with an accuracy of 0.2 kPa. All the data from the temperature and pressure sensors were recorded and monitored by a data acquisition system. Detailed information about the measurement instruments and their accuracy is presented in Table 2.

Table 2. Measurement instruments adopted in the compressor test rig.

Parameter	Measurement instrument	Full scale	Accuracy
Temperature	PT100(Class A)	-20~+850 °C	±0.35 °C
Pressure	Pressure transducer	0~600 kPa, 0~1000 kPa	±0.5% of full scale
Volumetric flow rate	ASME nozzle flowmeter	1.0~6.7 m^3/min	±1.66% of measured value
Oil flow rate	Turbine flowmeter	150~1500 L/h, 40~400 L/h	±0.5% of full scale
Power	Precision power analyzer	~	±0.5% of measured value

2.3. Performance evaluation and uncertainty analysis

Performance parameters including volumetric efficiency, adiabatic efficiency and specific power could be calculated from the measured data. The volumetric efficiency η_{Vi} of each stage compressor is defined as the ratio of actual volumetric flow rate q_{Vi} to theoretical volumetric flow rate q_{Vti}, which could be expressed as:

$$\eta_{Vi} = \frac{q_{Vi}}{q_{Vti}} = \frac{q_{Vi}}{n_i V_i} \quad i = 1,2 \tag{1}$$

where n_i refers to the rotating speed, V_i is the theoretical displacement per rotation, $i = 1$ and $i = 2$ refer to the parameters of the first stage and second stage compressor, respectively.

The adiabatic efficiency $\eta_{\text{ad}i}$ of the single stage compressor is defined as the ratio of adiabatic power $P_{\text{ad}i}$ to the actual shaft power P_{si}, as shown in the following equations:

$$\eta_{\text{ad}i} = \frac{P_{\text{ad}i}}{P_{si}} = \frac{P_{\text{ad}i}}{P_{ei}\eta_{ei}} \quad i = 1,2 \tag{2}$$

$$P_{\text{ad}i} = \frac{\kappa}{\kappa - 1} p_{si} q_{Vi} \left(\left(\frac{p_{di}}{p_{si}} \right)^{\frac{\kappa-1}{\kappa}} - 1 \right) \quad i = 1,2 \tag{3}$$

where P_{ei} refers to the power consumption of the motor, η_{ei} refers to the motor efficiency with a nearly invariable value of 94% provided by the manufacturer, κ is the specific heat ratio (1.4 for air), p_{si} and p_{di} refer to the suction pressure and discharge pressure of the compressor, respectively.

Especially, the adiabatic efficiency $\eta_{\text{ad,dual}}$ of the two-stage compressor unit could be expressed as:

$$\eta_{\text{ad,dual}} = \frac{P_{\text{ad,dual}}}{P_{s1} + P_{s2}} \tag{4}$$

$$P_{\text{ad,dual}} = \frac{\kappa}{\kappa - 1} p_{s1} q_{V1} \left(\left(\frac{p_{d2}}{p_{s1}} \right)^{\frac{\kappa-1}{\kappa}} - 1 \right) \tag{5}$$

The specific power P_{sp} of the two-stage compressor unit is defined as:

$$P_{\text{sp}} = \frac{P}{q_V} \tag{6}$$

where P refers to the input power consumption of the two-stage air compressor unit, and q_V refers to the actual volumetric flow rate of the compressor unit.

In addition, the measurement uncertainties in relation to performance parameters were obtained by the error propagation analysis with the Kline and McClintock method[20], which could be expressed as:

$$y_R = \sqrt{\left(\frac{\partial R}{\partial x_1} y_1 \right)^2 + \left(\frac{\partial R}{\partial x_2} y_2 \right)^2 + \ldots + \left(\frac{\partial R}{\partial x_n} y_n \right)^2} \times 100\% \tag{7}$$

where y_R is the uncertainty of the result, $x_1, x_2, \ldots, x_n$ are the uncertainties of the independent variables, and R is a given function of the independent variables $x_1, x_2, \ldots, x_n$. By using Equation (7) and the accuracy listed in Table 2, the maximum measurement uncertainties are calculated to be 1.66% for volumetric efficiency, 1.75% for adiabatic efficiency and 1.73% for specific power.

3. Results and discussion

In a two-stage screw compressor, a large amount of oil, generally accounting for 30%~40% of the total oil flow rate, is injected into the intermediate pipe to cool the gas flowing to the second stage. In order to evaluate the influence of interstage oil injection, performance parameters were tested under different interstage oil injection flow rate and rotating speed. The interstage oil injection flow rate varied from 2 L/min to 10 L/min in the interval of 2 L/min. The first stage compressor operated at rotating speeds of 1200 rpm, 2200 rpm and 3100 rpm, and the rotating speed of the second stage compressor was controlled by the frequency converter to achieve a constant interstage pressure. The discharge pressure and interstage pressure of the two-stage compressor unit were maintained at 0.8 MPa and 0.3 MPa, respectively.

In the actual application scenario, the discharge temperature of the air compressor is usually controlled in a reasonable temperature range(70~85 °C) to avoid a low discharge temperature with the condensation of moisture in the air or a high discharge temperature which shortens the life of oil. In order to acquire data close to reality, the two-stage compressor was tested with a control of the discharge temperature. In the experiment, the discharge temperature of the second stage compressor was controlled at 75±1 °C by switching on or off the fan of the oil cooler manually. Additionally, when the compressor operated at a low speed(below 1400 rpm), the discharge temperature could not reach 75 °C and the fan of the oil cooler was kept in a turn-off state. In order to exclude the extra power consumption introduced by the fan, all of the test data was recorded when the fan was in a turn-off state.

3.1. Effect of interstage oil injection on temperature

The interstage temperature drop, which refers to the difference between discharge temperature of first stage and suction temperature of second stage, reflects the cooling effect resulting from the interstage oil injection. Figure 3 shows the variation of interstage temperature drop under different rotating speed and interstage oil injection flow rate. With the decrease of interstage oil injection flow rate from 10 L/min to 2 L/min, the interstage temperature drop under different rotating speed decreases, which indicates a worse cooling performance in the intermediate pipe. But the change of oil injection flow rate does not have an obvious effect on the cooling performance, and the maximum change of interstage temperature drop is lower than 3 °C under different rotating speed. However, with the increase of rotating speed, the interstage temperature drop increases rapidly, achieving an interstage temperature drop of 12 °C at 3100 rpm and 3 °C at 1200 rpm under an oil injection flow rate of 10 L/min. In a two-stage screw air compressor, the oil injection is driven by the pressure difference and the oil injection flow rate is nearly constant under a fixed discharge pressure. With the increase of rotating speed, the air flow rate in the intermediate pipe increases, which leads to a higher velocity and a lower oil/gas ratio. A higher flow velocity would improve the convective heat transfer coefficient and a lower oil/gas ratio would decrease the heat transfer area between oil and gas. Thus, results in Figure 3 indicate that the flow velocity had a comparatively higher effect on the heat transfer of interstage cooling than the oil/gas ratio.

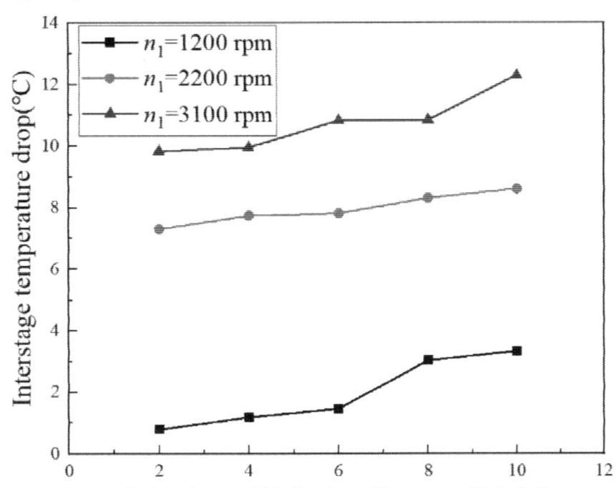

Figure 3. Variation of interstage temperature drop under different rotating speed and interstage oil injection flow rate.

Figure 4 shows the variation of oil injection temperature T_{oil}, discharge temperature of first stage T_{d1} and suction temperature of second stage T_{s2} under different rotating speed and interstage oil injection flow rate. As is mentioned above, the two-stage compressor which operates at a fixed discharge pressure and different rotating speed has the same oil injection flow rate. As shown in Figure 4(a), at a low rotating speed of 1200 rpm, a large amount of oil could achieve a sufficient cooling effect with a discharge temperature lower than 65 °C for the first stage and second stage compressor. Hence, under this condition, the fan was kept in a turn-off state and oil in the oil cooler was cooled by the natural convection. With the decrease of interstage oil injection flow rate, the discharge temperature of first stage is nearly constant due to a stable oil injection temperature at 1200 rpm. As a consequence of the worse cooling performance in the intermediate pipe, the suction temperature of second stage is in an increasing trend, which also leads to a slightly grow of the discharge temperature of second stage. However, as shown in Figure 4(b), at higher rotating speeds of 2200 rpm and 3100 rpm, the discharge temperature of second stage would exceed the limit of 75 °C without the forced cooling and thus the fan of the oil cooler was started intermittently. With the decrease of interstage oil injection flow rate, the amount of oil in the compression process of second stage is reduced at 2200 rpm and 3100 rpm. In order to control the discharge temperature at the same level, the oil injection temperature decreases with the decrease of interstage oil injection flow rate, which also leads to the reduction of the discharge temperature of first stage and the suction temperature of second stage.

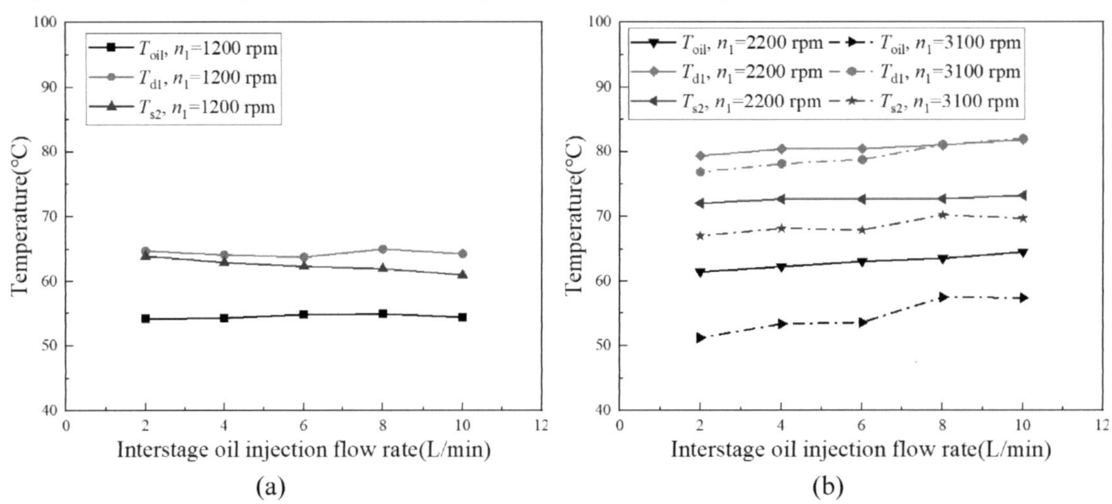

Figure 4. Variation of oil injection temperature, discharge temperature of first stage and suction temperature of second stage under different rotating speed and interstage oil injection flow rate: (a)1200 rpm; (b)2200 rpm and 3100 rpm.

3.2. Effect of interstage oil injection on efficiency

The variation of volumetric efficiency and adiabatic efficiency of first stage compressor under different rotating speed and interstage oil injection flow rate is shown in Figure 5(a) and (b), respectively. As is mentioned above, the variation of interstage oil injection flow rate does not influence the performance of first stage compressor directly, but it would change the oil injection temperature in the oil injection system at different rotating speed, which has a significant effect on the performance of compressors according to references. At a low rotating speed of 1200 rpm, with the decrease of interstage oil injection flow rate, the volumetric efficiency and adiabatic efficiency of first stage remain nearly unchanged due to a stable oil injection temperature. However, at higher rotating speeds of 2200 rpm and 3100 rpm, a colder oil in the first stage improves the compressor performance slightly, resulting from a better clearance-sealing and cooling effect. The maximum gain occurs at 3100 rpm with the volumetric efficiency enhanced by 1.04% and the adiabatic efficiency enhanced by 0.43%, when the interstage oil injection flow rate is reduced from 10 L/min to 2 L/min.

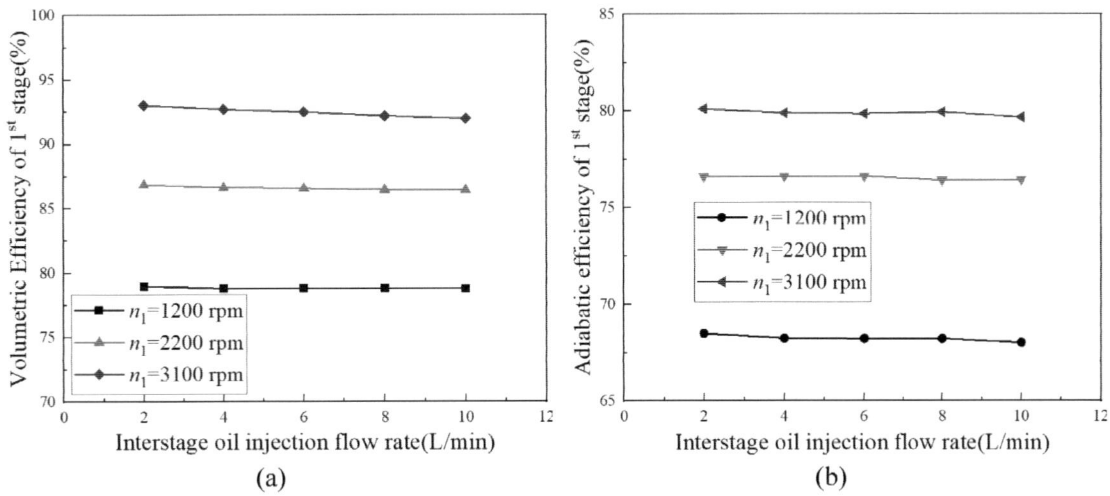

Figure 5. Variation of efficiencies of first stage compressor under different rotating speed and interstage oil injection flow rate: (a)volumetric efficiency; (b)adiabatic efficiency.

The variation of volumetric efficiency and adiabatic efficiency of second stage compressor under different rotating speed and interstage oil injection flow rate is shown in Figure 6(a) and (b), respectively. As it can be seen, the reduction of the interstage oil injection flow rate enhances the performance of the second stage compressor significantly, especially at a lower rotating speed. With the decrease of interstage oil injection flow rate, the volumetric efficiency of the second stage increases. It is improved by 1.52%, 0.79% and 0.48% at rotating speeds of 1200 rpm, 2200 rpm and 3100 rpm, respectively. Reducing the oil injected into the intermediate pipe decreases the oil entering the suction chamber of the second stage compressor, which occupies part of the chamber volume. Hence, the reduction of oil in the suction chamber supplies more volume for air, leading to the improvement of volumetric efficiency. But less oil in the working chamber of the screw compressor would weaken the clearance-sealing effect of oil. The proportion of oil in the suction chamber is defined as the following:

$$\tau = \frac{q_{\text{oil,1}} + q_{\text{oil,int}}}{q_{V2}} \tag{8}$$

where $q_{\text{oil,1}}$ and $q_{\text{oil,int}}$ refer to the oil injection flow rate in the first stage and interstage respectively, and q_{V2} refers to the actual volumetric flow rate of the second stage compressor.

With the decrease of interstage oil injection flow rate, the proportion of oil in the suction chamber decreases with a maximum change of 1.52%, 0.81% and 0.65% at rotating speeds of 1200 rpm, 2200 rpm and 3100 rpm respectively, which approaches the change amplitude of volumetric efficiency. Results demonstrate that the improvement of volumetric efficiency of the second stage is mainly caused by the reduction of the oil in the suction chamber in the variation range of oil injection flow rate. Therefore, the improvement of volumetric efficiency could not demonstrate the reduction of leakage in the working chamber of second stage compressor in this condition. But that might have an influence on the interstage pressure and rotating speed of second stage.

As shown in Figure 6(b), with the decrease of interstage oil injection flow rate, the adiabatic efficiency of the second stage increases. It is improved by 1.56%, 0.93% and 0.35% at rotating speeds of 1200 rpm, 2200 rpm and 3100 rpm, respectively. The gain in adiabatic efficiency of second stage compressor could be owing to the reduction of oil-induced power losses. An excess amount of oil from the first stage and interstage oil injection exists in the working chamber of second stage compressor, which causes considerable frictional losses and oil churning. Decreasing the interstage oil injection flow rate reduces the oil in the second stage significantly, and hence, reduces the oil-induced power losses and improves the adiabatic efficiency.

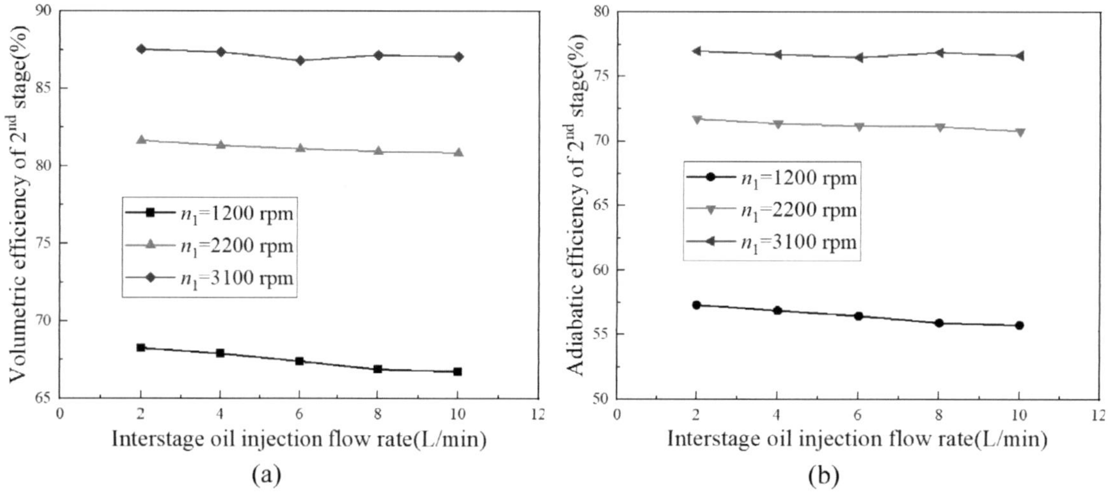

Figure 6. Variation of efficiencies of second stage compressor under different rotating speed and interstage oil injection flow rate: (a)volumetric efficiency; (b)adiabatic efficiency.

Figure 7 shows the variation of adiabatic efficiency of the entire two-stage compressor under different rotating speed and interstage oil injection flow rate. With the decrease of interstage oil injection flow rate, the adiabatic efficiency of the entire two-stage compressor increases. A maximum gain of 0.84%, 0.75% and 0.95% could be achieved at rotating speeds of 1200 rpm, 2200 rpm and 3100 rpm, respectively.

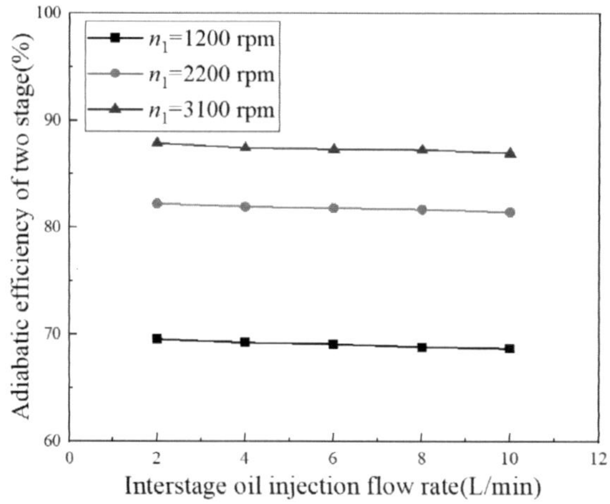

Figure 7. Variation of adiabatic efficiency of two-stage compressor under different rotating speed and interstage oil injection flow rate.

3.3. *Effect of interstage oil injection on specific power*

The variation of specific power under different rotating speed and interstage oil injection flow rate is shown in Figure 8. The specific power of the two-stage compressor unit increases with the decrease of interstage oil injection flow rate. A maximum gain of 1.25%, 0.66% and 0.62% could be achieved at rotating speeds of 1200 rpm, 2200 rpm and 3100 rpm, respectively. Results demonstrate that reducing the interstage oil injection flow rate moderately could decrease the power consumption of the air compressor unit, especially at a low rotating speed.

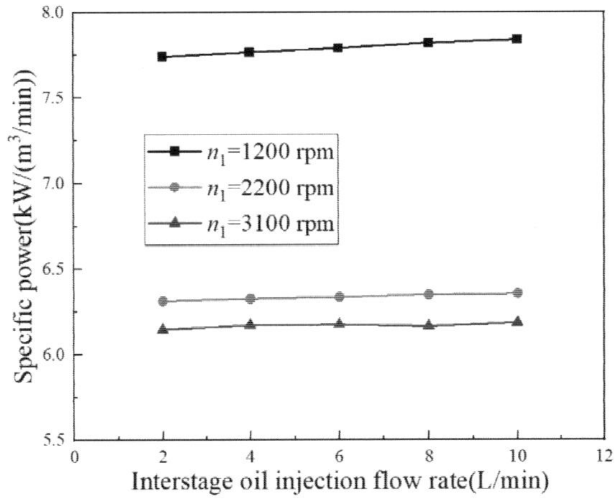

Figure 8. Variation of specific power under different rotating speed and interstage oil injection flow rate.

Based on the previous results, the decrease of specific power could be owing to the performance improvement of second stage compressor. It is found that an excess amount of oil exists in the second stage compressor and reducing the oil moderately could improve the compressor performance. But from the view of the functions of oil in the screw compressors, a lack of oil would result in a bad clearance-sealing and cooling effect, which aggravates the compressor performance. Consequently, there will be an optimal oil flow rate for the second stage compressor. However, according to the previous results, the turning point for the optimal oil flow rate did not appear in the variation range of interstage oil injection flow rate. In order to study further, an intermediate oil separator, which could remove the oil in the air before it enters the second stage compressor, was installed as it is shown in Figure 2. Investigation about this aspect will be done in the future work.

4. Conclusion

In this paper, an experimental test rig was established to study the effect of interstage oil injection on the performance of a dual-motor-driven two-stage screw air compressor. In order to acquire data close to reality, the compressor unit was tested under the actual service conditions and control modes. The compressor operated at rotating speeds of 1200 rpm, 2200 rpm and 3100 rpm, and the interstage oil injection flow rate varied from 2 L/min to 10 L/min. The following conclusions could be drawn from the experimental results:

- With the decrease of interstage oil injection flow rate, the interstage temperature drop decreases slightly, with a maximum change lower than 3 °C. It is found that the flow velocity has a comparatively higher effect on the heat transfer of interstage cooling than the oil/gas ratio.
- Reducing the interstage oil injection flow rate leads to an improvement in volumetric efficiency of the second stage compressor, mainly caused by the decrease of the proportion of oil in the suction chamber. But the improvement of volumetric efficiency could not demonstrate the reduction in leakage in the working chamber of the second stage compressor. Reducing the interstage oil injection flow rate also leads to an improvement in adiabatic efficiency of the second stage compressor, which could be owing to the reduction of oil-induced power losses. Additionally, the maximum gain occurs at a low rotating speed of 1200 rpm, with an improvement of up to 1.52% on volumetric efficiency and 1.56% on adiabatic efficiency.

- With the decrease of interstage oil injection flow rate, the specific power of the air compressor unit is also reduced. And a maximum gain of 1.25%, 0.66% and 0.62% in specific power could be achieved at rotating speeds of 1200 rpm, 2200 rpm and 3100 rpm, respectively.

The results in this paper might contribute to the oil injection control in the dual-motor-driven two-stage screw air compressor and achieve a performance improvement of the machine. Results also demonstrate that an excess amount of oil exists in the second stage compressor and reducing the oil moderately could improve the compressor performance. And there might be an optimal oil flow rate for the second stage compressor. Hence, a further investigation aiming at the oil in the second stage compressor will be done in the future work.

References

[1] Vittorini D and Cipollone R 2016 Energy saving potential in existing industrial compressors *Energy* **102** 502-15
[2] Deipenwisch R and Kauder K 1999 Oil as a design parameter in screw-type compressors: oil distribution and power losses caused by oil in the working chamber of a screw-type compressor *International Conference on Compressor and their Systems* pp 49-58
[3] Stosic N, Milutinovic L, Hanjalic K and Kovacevic A 1992 Investigation of the influence of oil injection upon the screw compressor working process *International Journal of Refrigeration* **15** 206-20
[4] Fujiwara M and Osada Y 1995 Performance analysis of an oil-injected screw compressor and its application *International journal of refrigeration* **18** 220-27
[5] Peng X, Xing Z, Cui T and Shu P 2000 Experimental study of oil injection and its effect on performance of twin screw compressors *International Compressor Engineering Conference at Purdue* p 1491
[6] Lee W, Ma R, Chen S, Wu W and Hsia H 2001 Numerical simulation and performance analysis of twin screw air compressors *International Journal of Rotating Machinery* **7** 65–78
[7] De Paepe M, Bogaert W and Mertens D 2005 Cooling of oil injected screw compressors by oil atomisation *Applied Thermal Engineering* **25** 2764-79
[8] Basha N, Kovacevic A, Stosic N and Smith I 2018 Effect of oil-injection on twin-screw compressor performance *IOP Conf. Ser.: Mater. Sci. Eng.* **425** 012009
[9] Basha N, Kovacevic A, and Rane S 2021 Numerical Investigation of Oil injection in Screw Compressors *Applied Thermal Engineering* **193** 116959
[10] He Z, Wang T, Wang X, Peng X and Xing Z 2018 Experimental investigation into the effect of oil injection on the performance of a variable speed twin-screw compressor *Energies* **11** 1342
[11] Wang X, Hwang Y and Radermacher R 2009 Two-stage heat pump system with vapor-injected scroll compressor using R410A as a refrigerant *International Journal of Refrigeration* **32** 1442-51
[12] Zlatanovic I and Rudonja N 2012 Experimental evaluation of desuperheating and oil cooling process through liquid injection in two-staged ammonia refrigeration systems with screw compressors *Applied Thermal Engineering* **40** 210-15
[13] Shen J, Feng G Xing Z and Wang X 2018 Theoretical study of two-stage water vapor compression systems *Applied Thermal Engineering* **147** 972-82
[14] Hirayama T, Hirano K, Shida S and Monasry J 2012 Development of a new mechanism for dual rotary compressor *Transactions of the Japan Society of Refrigerating and Air Conditioning Engineers* **30** 67-73
[15] Hauser J, Beinert M and Herlemann S 2013 Thermodynamic simulation of multi-stage screw compressors using chamber-based screw model *8th International Conference on Compressors and their Systems* pp 247-56
[16] Liu G, Yang Q, Zhao Y Wang L and Li L 2016 Analysis of performance of two-stage screw compressor under various operating conditions *Proceedings of the Institution of Mechanical Engineers, Part A: Journal of Power & Energy* **230** 660-68

[17] Andres R, Hesse J, Hetze F and Low D 2018 Cfd simulation of a two stage twin screw compressor including leakage flows and comparison with experimental data *IOP Conf. Ser.: Mater. Sci. Eng.* **425** 012018
[18] Cipollone R, Bianchi G, Battista D D, Contaldi G and Murgia S 2015 Model based design of an intercooled dual stage sliding vane rotary compressor system *IOP Conf. Ser.: Mater. Sci. Eng.* **90** 012035
[19] Murgia S, Valenti G, Costanzo I, Colletta D and Contaldi G 2017 An intracooling system for a novel two-stage sliding-vane air compressor *IOP Conf. Ser.: Mater. Sci. Eng.* **232** 012054
[20] Kline S, McClintock F 1953 Describing uncertainties in single-sample experiments *Mech. Engineering* **75** 3-8

AUTHOR INDEX

Institute of Physics
Dirac House, Temple Back
Bristol BS1 6BE UK

ISSN: 1757-8981
ISBN 978-1-7138-6426-4

CURRAN ASSOCIATES INC.
proceedings.com